Residential Wiring Concepts and Applications

BASED ON THE 2023 NATIONAL ELECTRICAL CODE®

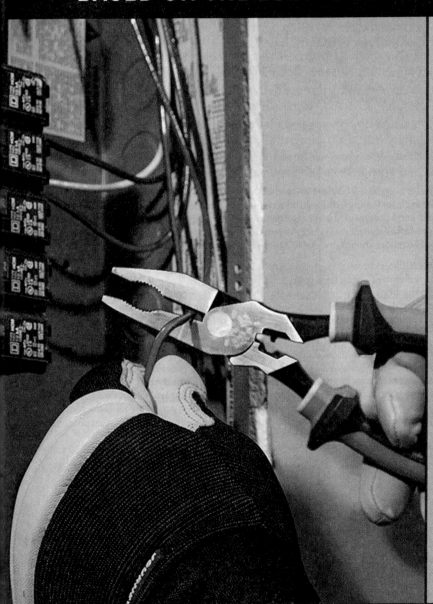

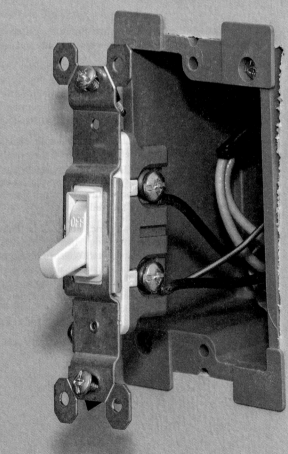

Quinton B. Phillips
Program Chair of Electrical
Systems Technology
Instructor at Athens Technical College
Athens, Georgia

Publisher
The Goodheart-Willcox Company, Inc.
Tinley Park, IL
www.g-w.com

Contributors

The author and publisher wish to thank the following teaching professionals for their valuable contributions to *Residential Wiring Concepts and Applications*:

Erik Bade, Electrical Instructor at Madison Area Technical College, for contribution of technical edits and second drafts in Sections 5, 6, 7, and 9.

Tom Kennedy, Electrical Instructor at Milwaukee Area Technical College, for authorship of Chapters 33–35 and technical edits and second drafts of Section 10.

Brian Khairullah, Training Specialist, for authorship of Chapter 32.

Matthew Wilkinson, Electrical Instructor at Madison Area Technical College, for authorship of Section 12 and technical edits and second drafts of Section 8.

Reviewers

The author and publisher wish to thank the following industry and teaching professionals for their valuable input into the development of *Residential Wiring Concepts and Applications*.

Timothy J. Brady
Lehigh Carbon Community College
Schnecksville, PA

Leonard J. Carey
Prairie State College
Chicago Heights, IL

Jesse Dahl
Minnesota North College–Hibbing
Hibbing, MN

David J. Dainelis
Grand Rapids Community College
Grand Rapids, MI

Randall D. Jacobs
Illinois State University
Normal, IL

Thomas Kennedy
Milwaukee Area Technical College
Milwaukee, IL

Brian Khairullah, MEd
Yucaipa, CA

Matthew R. Kiker
Harrisburg Area Community College
Harrisburg, PA

David Krzyston
SUNY College of Technology–Delhi
Delhi, NY

Robert Lang
College of DuPage
Glen Ellyn, IL

Harry McGuire
Valley Career and Technical Center
Fishersville, VA

Joseph Mollner
Dakota County Technical College
Rosemount, MN

Daniel Neff
Palm Beach State College
Lake Worth, FL

Eric Newcomer
Pennsylvania College of Technology
Williamsport, PA

David Newman
Polaris Career Center
Middleburg Heights, OH

Michael J. Olsen
William Rainey Harper College
Palatine, IL

Brian Rastede
Northeast Community College
Norfolk, NE

David Robinson
Los Angeles Trade Technical College
Los Angeles, CA

Nicholas A. Sinstack
SUNY College of Technology–Delhi
Delhi, NY

John G. Veitch
SUNY Adirondack
Queensbury, NY

Acknowledgments

The author and publisher would like to thank the following companies, organizations, and individuals for their contribution of resource material, images, or other support in the development of *Residential Wiring Concepts and Applications*.

Arlington Industries, Inc.

Carlon, Lamson & Sessions

Eaton

General Tools

Hubbell Inc.

Ideal Industries, Inc.

Klein Tools

Milwaukee Electric Tool Corporation

Stanley Black & Decker

The Garlinghouse Company

US Department of Energy

Werner Co.

The *National Electrical Code*®

The most informative and authoritative body of information concerning electrical wiring installation in the United States, and perhaps the world, is the *National Electrical Code*® (*NEC*). This code establishes a set of rules, regulations, and criteria for the installation of electrical equipment. Compliance with these methods will result in a safe installation.

The *NEC* is drafted by a team of experts assembled for this purpose by the National Fire Protection Association (NFPA). This team is formally called the *National Electrical Code* committee. They revise and update the *NEC* every three years. It is imperative that anyone installing electrical wiring obtains and studies the *NEC*. Articles and sections of the *NEC* are referred to throughout this text. Although certain portions, tables, and examples are directly quoted from its text, there is enough useful information in the *NEC* that not having it available would be a tremendous hindrance.

The latest edition of the *National Electrical Code* can be purchased from the National Fire Protection Association by visiting their website. Online access to the *National Electrical Code* and over 1,400 NFPA codes and standards is available with an NFPA LiNK® subscription. NFPA LiNK includes all current editions as well as a library of legacy codes and standards going back five editions. Subscribers also have early access to newly released editions before the printed book is available for purchase. To learn more about NFPA LiNK, visit the NFPA website at www.nfpa.org.

Features of the Textbook

The instructional design of this textbook includes student-focused learning tools to help you succeed. This visual guide highlights these features.

Chapter Opening Materials

Each chapter opener contains a chapter outline, a list of learning objectives, and a list of technical terms.

The **Chapter Outline** lists the topics that will be covered in the chapter.

Learning Objectives clearly identify the knowledge and skills to be gained when the chapter is completed.

Technical Terms list the key words to be learned in the chapter.

The **Introduction** provides an overview and preview of the chapter content.

Additional Features

Additional features are used throughout the body of each chapter to further learning and knowledge.

Pro Tips provide advice and guidance that is especially applicable for on-the-job situations.

Safety Notes alert you to potentially dangerous materials and practices.

Code Applications provide opportunities to apply understanding of the *National Electric Code* to different scenarios.

Procedures provide clear instructions for hands-on activities.

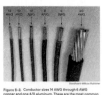

Illustrations and Photos

Illustrations have been designed to clearly and simply communicate the specific topic. Photographic images provided show the latest equipment. *NEC Tables* reflect changes made to the 2023 edition.

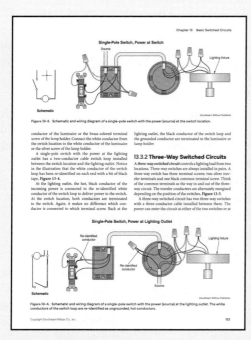

End-of-Chapter Content

End-of-chapter material provides an opportunity for review and application of concepts.

A concise **Summary** provides an additional review tool and reinforces key learning objectives. This helps you focus on important concepts presented in the text.

Know and Understand questions enable you to demonstrate knowledge, identification, and comprehension of chapter material.

Apply and Analyze questions extend learning and develop your abilities to use learned material in new situations and to break down material into its component parts.

Critical Thinking questions develop higher-order thinking and problem solving, personal, and workplace skills.

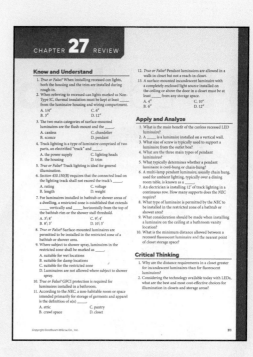

Student Tools

Student Text

Residential Wiring Concepts and Applications is a comprehensive text that focuses on the information and requirements, techniques, processes, and procedures used by residential electricians.

Lab Workbook

- Hands-on practice includes questions and activities.
- Projects offer students opportunities to work on challenges they may encounter on a jobsite.

G-W Digital Companion

For digital users, e-flash cards and vocabulary exercises allow interaction with content to create opportunities to increase achievement.

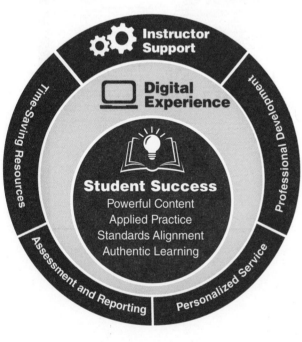

Instructor Tools

LMS Integration

Integrate Goodheart-Willcox content within your Learning Management System for a seamless user experience for both you and your students. EduHub® LMS–ready content in Common Cartridge® format facilitates single sign-on integration and gives you control of student enrollment and data. With a Common Cartridge integration, you can access the LMS features and tools you are accustomed to using and G-W course resources in one convenient location—your LMS.

G-W Common Cartridge provides a complete learning package for you and your students. The included digital resources help your students remain engaged and learn effectively:

- **Digital Textbook**
- Online **Lab Workbook content**
- **Videos**
- **Animations**
- **Drill and Practice** vocabulary activities

When you incorporate G-W content into your courses via Common Cartridge, you have the flexibility to customize and structure the content to meet the educational needs of your students. You may also choose to add your own content to the course.

For instructors, the Common Cartridge includes the Online Instructor Resources. QTI® question banks are available within the Online Instructor Resources for import into your LMS. These prebuilt assessments help you measure student knowledge and track results in your LMS gradebook. Questions and tests can be customized to meet your assessment needs.

Online Instructor Resources

- The **Instructor Resources** provide instructors with time-saving preparation tools such as answer keys, editable lesson plans, and other teaching aids.
- **Instructor's Presentations for PowerPoint®** are fully customizable, richly illustrated slides that help you teach and visually reinforce the key concepts from each chapter.
- Administer and manage assessments to meet your classroom needs using **Assessment Software with Question Banks**, which include hundreds of matching, completion, multiple choice, and short answer questions to assess student knowledge of the content in each chapter.

See www.g-w.com/residential-wiring-concepts-applications-2024 for a list of all available resources.

Professional Development

- Expert content specialists
- Research-based pedagogy and instructional practices
- Options for virtual and in-person Professional Development

Brief Contents

Contents

Introduction to Residential Wiring

kurhan/Shutterstock.com

A career in the electrical trades is a prudent decision in today's economy, as plenty of well-paying jobs are available throughout the industry. Skilled and knowledgeable residential electricians are always in demand. As new homes are constructed every day, older homes may need maintenance, upgrades, or additions that require electrical work. This textbook aims to introduce the student to all aspects of residential wiring.

Chapter 1, *Getting Started in the Industry*, details information regarding career opportunities in the electrical industry, the *National Electrical Code*, and electrical licenses and permits. The basics of power generation, transmission, and distribution as well as a summary of electrical circuits are included to build a solid electrical foundation.

Working with electricity is inherently hazardous, so a residential electrician must understand the potential dangers. Chapter 2, *Safety in Residential Wiring*, discusses general safety rules, safe practices when working with electricity, and proper personal protective equipment.

Getting Started in the Industry

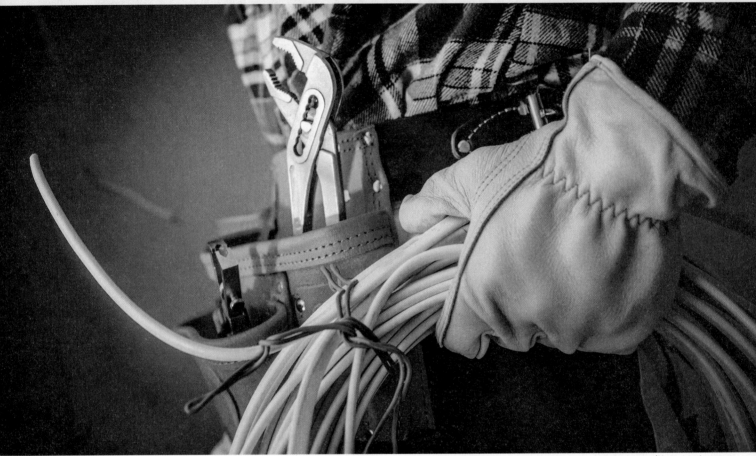

Virage Images/Shutterstock.com

CHAPTER OUTLINE

LEARNING OBJECTIVES

After completing this chapter, you will be able to:

- Discuss the career opportunities available in the electrical industry and why the residential wiring market remains stable.
- Understand the meaning of work ethics and how it applies to one's career pursuit.
- Understand the origin and organization of the *National Electrical Code*.
- Explain the electrical licensing and permitting process.
- Understand electrical power generation, transmission, and distribution basics.
- Describe how a simple electrical circuit operates.
- Define ungrounded conductor, grounded conductor, neutral conductor, and equipment grounding conductor.
- Define the terms volts, amps, resistance, and power.
- Describe how to use Ohm's law to determine unknown electrical values.
- Differentiate between Ohm's law and Watt's law.

TECHNICAL TERMS

alternating current (AC)
amp
apprentice
authority having jurisdiction (AHJ)
current
demand profile
direct current (DC)
distribution transformer
electrical construction permit
electrical fault
electrical license
equipment grounding conductor
explanatory material
ground fault
grounded conductor
load
mandatory rules
National Electrical Code (NEC)
neutral conductor
nominal
ohm
Ohm's law
overcurrent
peak hours
permissive rules
power
resistance
résumé
short circuit
substation
thermal generation
ungrounded conductor
volt
voltage
watt
Watt's law
work ethics
workplace skills

Introduction

The residential housing market comprises a large portion of the national economy. A career in the residential electrical sector is a wise choice for many reasons. New homes are being built every day, and existing homes need maintenance and upgrades. Everyone needs a home, and every home uses electricity. Additionally, electrical jobs cannot be outsourced to another country. There is a tremendous demand for competent and knowledgeable residential electricians.

This chapter will introduce students to the electrical industry and discuss where and how to begin seeking employment in the field. A brief overview of the *National Electrical Code* is intended to acquaint the reader with the basics before exploring it further in Section 5, The *National Electrical Code* and Construction Basics of this text. This chapter will also provide an overview of electrical fundamentals with the intent to build on this content in each chapter of the textbook. This will help students apply their knowledge and gain the necessary skills and knowledge to find success as a residential electrician.

1.1 Career Opportunities in the Electrical Industry

The residential electrical sector is a stable and reliable market that always needs qualified workers. As long as people live in homes, there is a need for electricians to install, maintain, and add to residential electrical systems. Many electricians find a niche market in which to specialize, such as swimming pool or hot tub wiring, solar power, high-end landscape lighting, basement buildouts, and emergency backup power systems, to name a few.

There are two broad categories of residential wiring: new construction and service, renovation, and repair. New construction can be further separated into custom-built homes and speculation-built, or spec-built, homes. Custom homes are built for a particular homeowner. Investors build spec homes with the intent to sell them at a profit when completed. Service, renovation, and repair refer to repairing, adding to, or extending electrical circuits or equipment. Such work may involve replacing a luminaire, or light fixture, adding a new circuit for additional lights and receptacles, or re-wiring an entire residence. Re-wiring a residence is sometimes necessary to upgrade an outdated electrical system.

The industry also relies on support positions, such as electrical utility technicians, electrical supply house salespeople, electrical estimators, electrical inspectors, and others. Many residential electricians move into these job areas as they progress through their careers. It is not unusual for electricians with several years of experience to obtain an electrical license from their state and start their own company.

1.1.1 Obtaining Employment

When seeking career options in the field, you must know how to find and apply for an electrical job. Job postings can be found through a variety of different media. One of the most reliable methods to find employment leads is through networking, i.e., talking to friends and family. These days, the internet may be one of the best sources to search for job openings. There are job-search engines and government agencies that can assist you in finding and applying for jobs.

Many electricians begin their career as an apprentice. An *apprentice* works alongside a skilled master tradesman to learn a trade firsthand. Most apprenticeships last a specified period. Four or five years is not uncommon for an electrical apprentice.

Once you find a job opening, you are generally asked to fill out a job application, which can be completed in person or online. It helps to already have a résumé prepared because many employers like to see this included in an application. A *résumé* is a document that outlines your professional qualifications and relevant work experience. See **Figure 1-1**. Employers often use résumés to screen applicants for the skills and education needed for the job. The appearance of the application form can give an employer the first impression about you. Fill out the form accurately, completely, and neatly. How well you accomplish this can determine whether you are called for an interview.

A job interview allows you to learn more about the company and convince the employer that you are the best person for the position. The employer wants to know if you have the skills needed for the job. Adequate preparation is essential for making a lasting, positive impression. It will also help you feel more confident and comfortable during the interview. A few helpful tips for performing well in the interview are as follows:

- Be polite, friendly, and cheerful during the process.
- Answer all questions carefully and as thoroughly as you can.
- Be honest about your abilities and shortcomings.
- Be prepared for a skills-related test. A prospective employer may ask you to take employee tests. Some employers administer tests to job candidates to measure their knowledge or skill level under stress.

Within 24 hours after the interview, be sure to send a letter to the employer, thanking them for the interview. If you get a job offer, respond to it quickly. If you do not receive an offer after several interviews, evaluate your interview techniques and seek ways to improve them.

1.1.2 Work Ethic Skills

Knowing and understanding electrical concepts is only one part of being a successful electrician. Work ethics and workplace skills play an equally important role in finding a job,

Michael J. Garcia

134 Lincoln Street
Wilton, CA 93232

(212) 555-1234
mjgarcia22@gmail.com

Career Objective
To obtain an entry-level position in the electrical wiring field.

Work Experience
Builder's Supply, Holloton, CA
Retail Sales Associate

- Provide assistance to customers in hardware, plumbing, and electrical departments.
- Review department inventory.
- Maintain neat and orderly displays.

Builder's Supply, Holloton, CA
Receiver/Stocker

- Unload, store, and stock merchandise.
- Operated fork trucks and hand trucks.

Tony's Grocery Store, Wilton, CA
Cashier

- Operated checkout scanner and bagged groceries.
- Answered customer questions.
- General stocking and cleaning throughout store.

Education
Associate Degree in Electrical Construction Technology
Oceanside Community College

- GPA: 3.22/4.0
- Coursework included residential wiring, commercial wiring, and solar photovoltaic systems.
- Participated in SkillsUSA Chapter.

References
Available upon request.

Goodheart-Willcox Publisher

Figure 1-1. An example of a sample résumé.

10 Work Ethics Traits

Appearance	Displays proper dress, grooming, hygiene, and manners
Attendance	Attends class arrives and leaves on time, tells instructor in advance of planned absences, and makes up assignments promptly
Attitude	Shows positive attitude, appears confident, and has true hopes of self
Character	Displays loyalty, honesty, dependability, reliability, initiative, and self-control
Communication	Displays proper verbal and non-verbal skills and listens
Cooperation	Displays leadership skills; properly handles criticism, conflicts, and stress; maintains proper relationships with peers and follows the chain of command
Organizational Skill	Shows skills in management, prioritizing, and dealing with change
Productivity	Follows safety practices, conserves resources, and follows instructions
Respect	Deals properly with diversity, shows understanding and tolerance
Teamwork	Respects rights of others, is a team worker, is helpful, is confident, displays a customer service attitude, and seeks continuous learning

Goodheart-Willcox Publisher

Figure 1-2. Work ethic skills are critical to career success and advancement.

keeping that job, and advancing your career. **Work ethics** are personal traits, as opposed to technical knowledge, that make you a more valuable employee. **Workplace skills**, or *employability skills*, are needed to succeed in a work environment. Employers want employees who are punctual, dependable, and responsible. They want their employees to be capable of taking the initiative and working independently. Other desirable employee qualities include organization, accuracy, and efficiency. See **Figure 1-2**.

1.2 Introduction to the National Electrical Code

The **National Electrical Code (NEC)**, published in 1896, is a standard for the safe installation of electrical wiring and equipment in the U.S. The first edition was created in response to the growing number of electrical fires. This occurred only 17 years after the invention of the incandescent light bulb. A group of insurance companies saw the need to standardize and consolidate the myriad of existing and often-conflicting standards of electrical installations. The *NEC* is now published by the National Fire Protection Association (NFPA) and is updated every three years to reflect industry changes. The current edition is 2023.

All residential electrical installations must be installed following the rules of the *National Electrical Code*. The stated purpose of the *NEC* is the "practical safeguarding of persons and property from hazards arising from the use of electricity." When the individual states, counties, and municipalities adopt the *NEC*, it becomes the legal standard that all electricians must follow. It is not intended to serve as an instructional manual for untrained persons.

Navigating the *NEC* can be challenging at times. Understanding its layout is the first step in mastering its use. See **Figure 1-3**. The *NEC* is divided into an introduction, nine chapters, and several annexes:

- Chapters 1 through 4 generally apply to all installations.
- Chapters 5, 6, and 7 apply to special occupancies, equipment, and conditions.
- Chapter 8 is dedicated to the installation of communication systems.
- Chapter 9 contains tables regarding conductor, cable, and conduit properties.
- The annexes relate to different referenced standards. They provide calculation examples, additional tables for proper implementation of various code articles, and a model adoption ordinance to guide municipalities in establishing an electrical inspections department.

Each chapter of the *NEC* is divided into articles. The articles are broken down into parts and parts into sections. The format starts from the most general topic to the most specific topic. For example, Article 250 addresses the entirety of "Grounding and Bonding." Part III of Article 250 covers just the "grounding electrode system" and the "grounding electrode conductor." Section 250.52(A), pronounced *250 dot 52, A*, specifies electrodes permitted for grounding. See **Figure 1-4**.

Information in the Code can fall into one of three categories: mandatory rules, permissive rules, and explanatory material. See **Figure 1-5**. **Mandatory rules** identify actions that are specifically required or prohibited. They are characterized by the phrases *shall* or *shall not*. **Permissive rules** describe actions that are allowed but not required. Permissive rules are indicated by the terms "shall be permitted" or

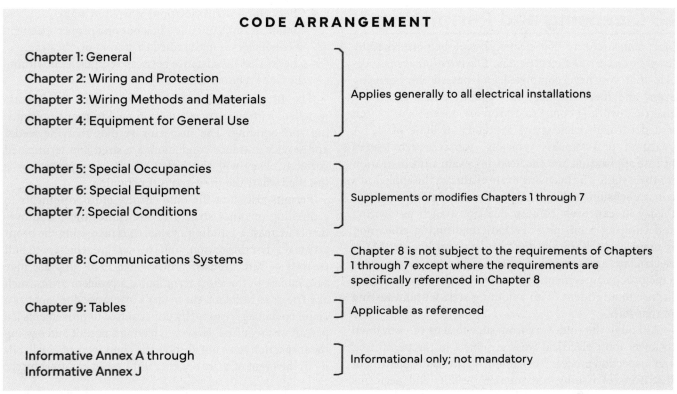

CODE ARRANGEMENT

Chapter 1: General
Chapter 2: Wiring and Protection
Chapter 3: Wiring Methods and Materials
Chapter 4: Equipment for General Use
} Applies generally to all electrical installations

Chapter 5: Special Occupancies
Chapter 6: Special Equipment
Chapter 7: Special Conditions
} Supplements or modifies Chapters 1 through 7

Chapter 8: Communications Systems
} Chapter 8 is not subject to the requirements of Chapters 1 through 7 except where the requirements are specifically referenced in Chapter 8

Chapter 9: Tables
} Applicable as referenced

Informative Annex A through Informative Annex J
} Informational only; not mandatory

Goodheart-Willcox Publisher

Figure 1-3. The *National Electrical Code* contains nine chapters. Understanding the basic structure of the *NEC* will make it easier to locate specific requirements.

NEC Outline Format
Chapter 2
  Article 250
    Part X
      Section 250.184
        (B) (Subdivision 1)
          (8) (Subdivision 2)
            a. (Subdivision 3)

Goodheart-Willcox Publisher

Figure 1-4. *NEC* chapters are divided into parts, and the parts contain sections.

"shall not be required." They are typically used to identify options or alternative methods. ***Explanatory material*** is included in the *NEC* in the form of "Informational Notes." These notes may reference additional codes or standards, other sections of the *NEC*, or further information that helps apply the particular Code section. Informational notes are, as their name states, informational only and are not enforceable requirements.

Using and understanding the *NEC* is emphasized throughout this text.

Categories of Information in the National Electrical Code

- Specifically required or prohibited
- Keywords: "shall" or "shall not"

Mandatory Rules

- Allowed but not required
- Identify options or alternative methods
- Keywords: "shall be permitted" or "shall not be required"

Permissive Rules

- Not enforceable requirements
- References, additional information
- Keywords: "informational note"

Explanatory Material

Goodheart-Willcox Publisher

Figure 1-5. Information in the *NEC* falls into one of three categories. Recognizing the different types of information will make it much easier for you to understand and interpret the Code.

1.3 **Licensing and Permitting**

Most states require that electrical work be performed by *licensed* or *certified* electricians. **Electrical licenses** indicate that you have completed and passed the licensing exam and meet the necessary qualifications to ensure electrical work is completed properly. Licenses are often issued through each state's Secretary of State office. A board of professionals typically oversees each trade's license application and the licensing exam administration for their state. The licensing exam evaluates the applicant's comprehension of the *NEC* and knowledge of electrical theory. It can even contain questions, such as owning and running a business. Periodic continuing education is required to maintain one's professional license. Many electricians are licensed in several states. Most states have a reciprocity agreement where they will grant an electrical license to a resident of an adjoining state without testing in that state.

Although the rules vary from state to state or even town to town, most electrical work is subject to the permitting and inspection process. Each municipality has a department that oversees issuing construction permits and performs the necessary inspections. The person who conducts these inspections is called the **authority having jurisdiction (AHJ)**. An electrical license is generally required to receive an electrical construction permit. An **electrical construction permit** is a document issued by the local municipality to a qualified electrician. The permit is, in essence, an agreement that the work will be done in a Code-compliant manner and will be subject to inspections by the AHJ. Some municipalities allow a homeowner to pull a permit for a house in which they own and live. The homeowner's work is still subject to inspection by the municipalities' AHJ.

Permits are generally required when the electrician extends a circuit or adds a new circuit. However, some electrical work does not require a permit, such as replacing light fixtures or devices like a damaged switch or receptacle.

If the electrical utility has disconnected the service for a panel change or service upgrade, a permit must be pulled, and the work must be inspected. The power cannot be restored until the AHJ approves and makes the reconnection request directly to the electrical utility. The inspector works in the interest of public safety and ensures that the work is done according to all applicable codes. Remember that the *NEC* is a code of *safety* standards.

Each new construction installation is subject to two inspections: a rough-in inspection and a final inspection. A rough-in inspection is for work that is rendered inaccessible by finished walls or direct burial. The electrical inspector completes the following tasks:

- Ensures that cabling is adequately secured and supported and that the correct size and type of wire is used for each circuit.

- Checks for box-fill violations where too many conductors are contained in a box and proper spacing of conductors is maintained to prevent overheating.
- Checks that the distance between receptacles is within the Code requirements.

The final inspection ensures the service is correctly installed, with particular attention paid to proper grounding and bonding. The inspector verifies that the feeder and service entrance conductors are sized and terminated correctly. They will check that shock-safety protection is installed where required and is functioning.

Permits also alert the municipality of improvements to a dwelling or other structure. Sometimes these improvements increase a building's value, thus increasing the property taxes. For this reason, some owners are reluctant to pull permits or hire licensed professionals. This practice may save money in the short term, but if a problem arises, such as a fire or an accident, the owner's insurance does not pay a claim on damages caused by unpermitted improvements if a permit was required. However, having a permit and passing the inspection does not absolve the contractor of any liability in the event of a fire or other unforeseen circumstance.

PRO TIP   **Building Inspectors**

The inspector is not an adversary. View the inspector as another set of eyes reviewing your work.

1.4 **Power Generation, Transmission, and Distribution Basics**

Most of the electricity in the United States is produced by thermal generation, **Figure 1-6**. In *thermal generation*, a form of fuel is used to produce steam, which drives a turbine that produces electricity. The most common fuel used in electrical power plants is natural gas, followed by coal and petroleum, **Figure 1-7**.

Nuclear power plants are a form of thermal generation. The heat produced by decaying nuclear material creates the steam that drives the turbines. Hydroelectric dams are a source of electrical generation. Utility-scale photovoltaics and wind turbines are becoming increasingly less expensive and play a large role in the future of electrical power generation. Once generation is complete, the next two stages of electrical power production are transmission and distribution.

Although the typical residence uses 120 volts and 240 volts for lights and appliances, the standard power plant generates electricity at 13,000 volts, or 13 kilovolts (kV). This voltage is too low to transmit over long distances without

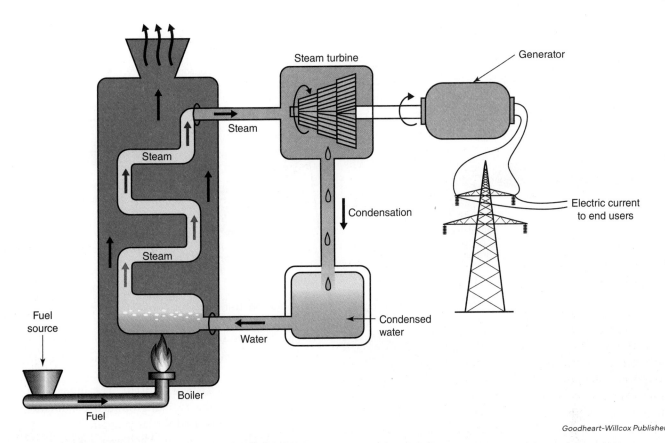

Figure 1-6. The process of thermal generation. Thermal generation uses heat to produce steam to drive turbines that produce electricity.

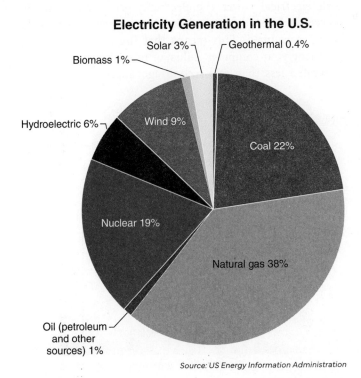

Electricity Generation in the U.S.

Solar 3%
Geothermal 0.4%
Biomass 1%
Hydroelectric 6%
Wind 9%
Coal 22%
Nuclear 19%
Natural gas 38%
Oil (petroleum and other sources) 1%

Source: US Energy Information Administration

Figure 1-7. Electrical generation in the United States by fuel type.

experiencing too much power loss. For this reason, transformers immediately convert the voltage to 138,000 volts, or 138 kV, for transmission. Transmitting electricity at such a high voltage allows smaller-sized conductors to carry the power great distances throughout the electrical grid, **Figure 1-8**, with minimal losses. The United States electrical grid comprises more than 360,000 miles of transmission lines connecting around 7000 power-generating stations to nearly every home and business in the country.

For residential and light commercial use, electricity is sent to a transmission *substation*, **Figure 1-9**, where transformers reduce the voltage to a lower distribution level (4160 to 12,500 volts is typical). This reduced voltage is distributed over power lines, either overhead or underground, that generally follows a street's path. *Distribution transformers*, **Figure 1-10**, located near the house, either on the power pole or pad-mounted on the ground, steps the voltage down to around 240/120 volts, nominal, for residential use. The word *nominal* means *in name only* and is a descriptive term rather than an actual measurement. The actual voltage might read 244/122 volts if you were to measure it with a voltmeter. (Just as 2×4 is a nominal term for a board that actually measures 1-1/2″ × 2-1/2″.)

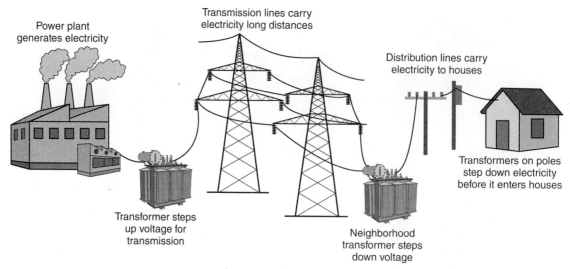

Goodheart-Willcox Publisher

Figure 1-8. Simplified electrical grid illustration showing electrical generation, transmission, and distribution.

KANITHAR AIUMLAOR/Shutterstock.com

Figure 1-9. A typical electrical substation.

The 240/120-volt, single-phase, 3-wire system is the standard electrical system used across the United States for powering homes. In some parts of the country, the actual voltage can be 230/115 or 220/110. The higher voltage is available between the two ungrounded conductors, Line 1 and Line 2, and is used for large appliances, such as the furnace or the clothes dryer. The lower voltage is available between the grounded, neutral conductor and either of the two ungrounded conductors and is used for lighting and receptacles. **Figure 1-11**.

Because there is no practical method of storing utility-scale electrical power, the electricity generated by all the power plants in the United States must exactly equal the power consumed. This means that the electricity that is powering your computer, lights, or television currently was generated less than one second ago. Electrical usage also

A

B

Roger Bruce/Shutterstock.com

Figure 1-10. Distribution transformers. A—Pole-mounted transformers are used for overhead services. B—Pad-mounted transformers are used for underground services.

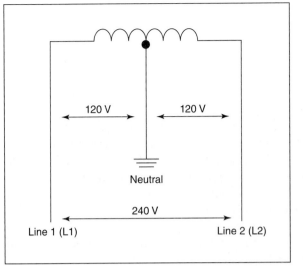

Goodheart-Willcox Publisher

Figure 1-11. The 240/120-volt schematic showing 120 volts from line to neutral and 240 volts from line to line.

varies throughout the day. For a residence, electrical usage is the lowest overnight while we sleep. Our demand for electricity increases in the mornings when we wake up and get ready for school or work. Residential electrical usage drops during the day when fewer people are at home but increases to its highest levels when we return from school and work

and begin dinner preparations. An electrical *demand profile*, **Figure 1-12**, demonstrates this pattern as it graphs the usage of electricity over time. An office or retail store shows a different demand profile where the major consumption of electricity would occur during business hours.

The portions of the day where the most electrical power is being consumed are referred to as *peak hours*. Different power generation types are used to match the amount of electricity generated to the amount of electricity consumed. Base load power plants operate during the non-peak hours to satisfy the minimum, or base, electrical demand. Coal-fired and nuclear power plants can take 12 or more hours to start and are used to meet base load demands. Hydroelectric plants and natural gas-fired power plants can begin generating power in under an hour and are brought online to satisfy peak demand hours.

1.5 How Electrical Circuits Work

Electricity is the flow of electrons, which make up an electrical *current*. Electrical current can be either alternating or direct. *Alternating current (AC)* reverses the direction at regular intervals, while *direct current (DC)* flows in only one direction. All electrical circuits have at least

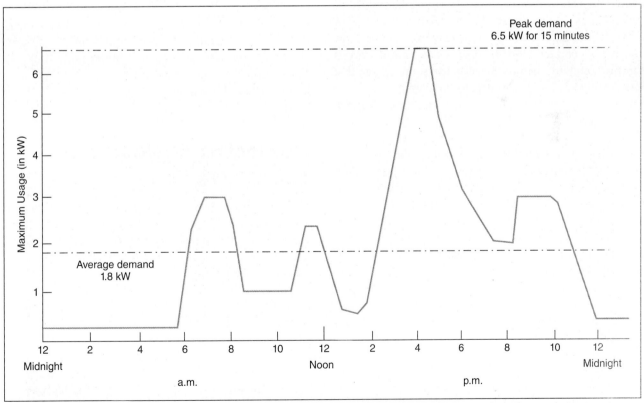

Goodheart-Willcox Publisher

Figure 1-12. A typical residential electrical demand profile showing the use of electricity throughout the day. Note the peak usage during dinner preparation.

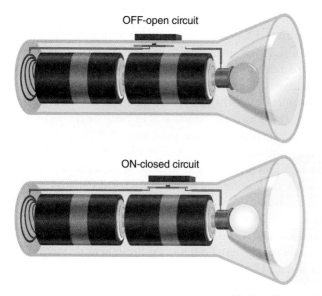

OFF-open circuit

ON-closed circuit

Colin Hayes/Shutterstock.com

Figure 1-13. An electrical circuit must have a power source, conductors, and a load. In a flashlight, batteries are the power source, wires are the conductors, and the bulb is the load. When the switch is open, the light is off. When the switch is closed, the light is on.

three components: a power source, conductors, and a load, **Figure 1-13**. Consider a flashlight. The battery is the power source, which makes the flashlight operate, the wires inside are the conductors, or the material that transmits electricity, and the bulb is the load that consumes the power.

In an electrical circuit, electrons leave the source, flow through the conductors to the load, where useful work is performed, and are then returned to the source. Note that electricity always tries to return to its source, not to the ground. This is important to reiterate: electricity does *not* flow to the ground.

The flow of electricity can be further manipulated by switches and overcurrent devices. A switch is used to open or close a portion or the entirety of an electrical circuit. When there is an open circuit, no current can flow. A closed circuit is one where the current can flow. This means a light turns on if the circuit is closed and off if the circuit is open. Excessive current flow can damage the conductors or equipment, which is a leading cause of electrical fires. Overcurrent devices, such as fuses and circuit breakers, act as a relief valve and automatically open the circuit if the current becomes too high. Fuses are generally one-time use and must be replaced when blown. Circuit breakers can be reset and reused when they trip.

An electrical circuit can display other conditions besides open or closed. *Electrical faults* occur when the electricity follows an unintended path. A *short circuit* is a fault condition where the load has been unintentionally bypassed, resulting in dangerously high current flow. It is called a short circuit because the electricity travels on a shorter route

than intended. A short circuit may result from an action, such as sticking a fork into an electric toaster, or from equipment failure, such as a damaged power cord.

A *ground fault* is defined in the *NEC* as "an unintentional, electrically conductive connection between an ungrounded conductor of an electrical circuit and the normally non-current-carrying conductors, metal enclosures, metal raceways, metal equipment, or earth." A ground fault could result from an energized conductor making contact with a clothes dryer's metal frame due to vibrations or other circumstances. Short circuits and ground faults result in *overcurrent*, which is any current over the equipment rating or the conductors' current-carrying capacity.

The terms ungrounded conductor, grounded conductor, neutral conductor, and equipment grounding conductor must be understood as they can confuse novice students. An *ungrounded conductor* is the hot conductor that carries electricity from the power source to the load. It usually has black or red insulation. A *grounded conductor* takes electricity from the load back to the source and has white or gray insulation. A *neutral conductor* is a grounded conductor that carries the current imbalance between two ungrounded, hot conductors. The neutral conductor is also white or gray. All neutral conductors are grounded, but not all grounded conductors are neutral. All these terms (ungrounded conductor, grounded conductor, and neutral conductor) are all current-carrying conductors under normal operation. An *equipment grounding conductor* helps the circuit breaker quickly clear ground faults and short circuits and does not carry current under normal operation. The equipment grounding conductor can be bare (no insulation) or have insulation that is green or green with one or more yellow stripes along its entire length.

1.6 Electrical Properties and Ohm's Law

Electricity has observable and measurable properties, and it acts in predictable ways. The three most common properties of electricity are voltage, current, and resistance. Comparing electricity to the flow of water in a pipe helps to understand these concepts.

Voltage is the pressure that pushes electrons through a circuit. It is like water pressure in a pipe. Voltage is measured by the *volt* and is abbreviated with the letter *E*, indicating electromotive force. *Current* is a quantity measurement and refers to the number of electrons passing a point over a period of time. It can be understood as the quantity of water passing, measured in gallons per minute. Unlike voltage that pushes, current flows. The unit of measure for current is the *amp* and is abbreviated with the letter *I* for intensity of current. *Resistance* is the opposition to current flow and acts similarly as a reduction in the pipe size that restricts the water flow. Resistance is measured in **ohms** and is denoted by the Greek letter omega Ω.

An electrical circuit has a load, which is the component being powered. The **load** converts electrical energy into some other form, such as light, heat, or motion. The type and size of the load determine the amount of current flow. Without a load in the circuit, there can be no current.

The relationship between current, voltage, and resistance can be understood by Ohm's law. **Ohm's law** states that current flow in a circuit is directly proportional to the applied voltage and inversely proportional to the resistance of a circuit. This law was founded in 1827 by German physicist Georg Ohm at a time when many scientists were independently conducting experiments with newly discovered electricity and its associated properties. This law was initially met with skepticism but became widely accepted in the 1840s.

If two of the values are known, Ohm's law can be used to determine the third value. The following

$$E = I \times R$$

where

E = voltage measured in volts (V)
I = current measured in amperes (A)
R = resistance measured in ohms (Ω)

Other equations can be derived to find any of the three values. The equations are as follows:

$$I = E/R$$
$$R = E/I$$

The best method to interpret Ohm's law is to represent it graphically as the Ohm's law wheel, shown in **Figure 1-14**. To use the wheel, simply cover the unknown value, and the

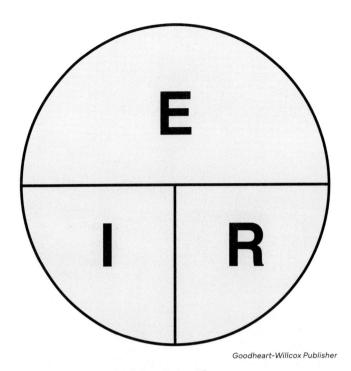

Goodheart-Willcox Publisher

Figure 1-14. The Ohm's law "wheel."

remaining two values can be multiplied or divided to determine the unknown value, **Figure 1-15**. For example, consider a circuit that has a current flow of 10 A and a resistance of 12 Ω. What is the applied voltage of this circuit?

To determine the answer, cover the unknown value (E for voltage), resulting in I and R side by side, meaning $E = I \times R$. Therefore, 10 A $\times$ 12 Ω = 120 V. In simpler terms, Ohm's law means that if the voltage is held constant, an increase in resistance must be accompanied by a decrease in current.

$$120\ V = 1\ \Omega \times 120\ A$$
$$120\ V = 120\ \Omega \times 1\ A$$

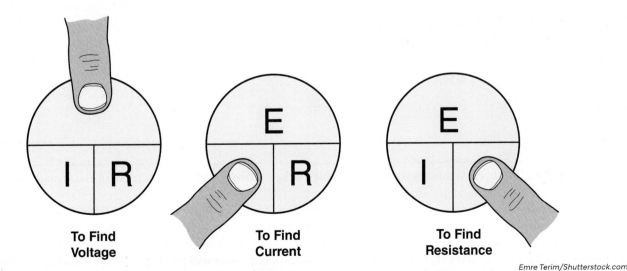

To Find Voltage   **To Find Current**   **To Find Resistance**

Emre Terim/Shutterstock.com

Figure 1-15. Using Ohm's law to determine an unknown value.

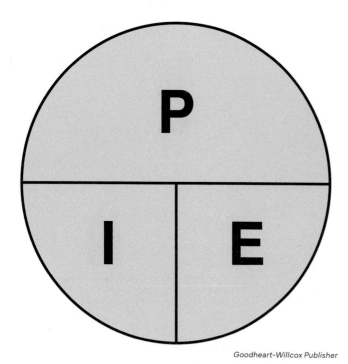

Goodheart-Willcox Publisher

Figure 1-16. The Watt's law wheel is used just like the Ohm's law wheel. Cover the unknown value and multiply or divide the remaining values to determine the unknown quantity.

Power is another electrical value and is defined as the rate at which work is performed. Its unit of measure is the *watt*. *Watt's law* is a variant of Ohm's law and is used to determine the values that define the amount of power in a circuit, **Figure 1-16.** Watt's law is as follows:

$$P = I \times E$$

where

P = power in watts (W)
I = current measured in amperes (A)
E = voltage measured in volts (V)

By the same logic of Ohm's law, other equations can be derived from Watt's law. The equations are as follows:

$$I = P/E$$
$$E = P/I$$

Electricians often use Watt's law to determine the amount of current when the watts and the amount of the source voltage are known. For example, consider a piece of machinery that is rated at 1500 watts and will be connected to 120 volts. How much current will the machine draw? Covering I (for amps) results in P over E, or 1500 watts ÷ 120 volts = 12.5 amps.

The Ohms law wheel and the Watt's law wheel can be combined to form the power wheel, **Figure 1-17.** Using the power wheel is a bit different from the Ohm's law or the Watt's law wheels and provides all equations for each of the four major electrical properties.

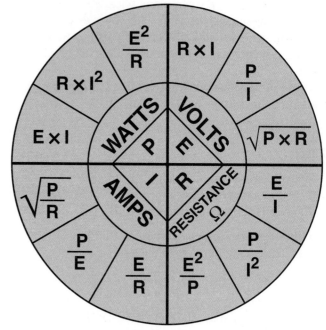

Goodheart-Willcox Publisher

Figure 1-17. Ohm's law and Watt's law can be combined into the power wheel. The power wheel gives all the formulas for each of the four major electrical quantities.

Summary

- The residential electrical sector is a stable and reliable market. There will always be a need for electricians to install, maintain, and add to residential electrical systems.
- Residential electrical jobs can be found through networking or by searching job posting websites on the internet.
- A job interview allows you to learn more about the company and to convince the employer that you are the best person for the position.
- Work ethics are personal traits, as opposed to technical knowledge, that will make you a more valuable employee.
- All residential electrical installations must be installed following the rules of the *National Electrical Code* or *NEC*.
- Most states require that electrical work be performed by licensed or certified electricians.

- Most of the electricity in the United States is produced by thermal generation, where some form of fuel is used to produce steam, which drives the turbine that produces electricity.
- Transmitting electricity at a high voltage allows for smaller-sized conductors to carry electrical power great distances with minimal losses.
- The 240/120-volt, single-phase, 3-wire system is the standard electrical system used across the United States for powering homes.
- The flow of electricity can be manipulated by switches and overcurrent devices such as fuses and circuit breakers.
- An open circuit is one where no current can flow. A closed circuit is one where the current can flow.
- Excessive current flow can damage conductors or equipment and is a leading cause of electrical fires.
- A short circuit is a fault condition where the load has been unintentionally bypassed, resulting in dangerously high current flow.
- A ground fault is an accidental contact between an energized conductor and a grounded conductor or grounded equipment frame.
- An ungrounded conductor is the hot conductor that carries electricity from the power source to the load.
- A grounded conductor takes electricity from the load back to the source.
- A neutral conductor is a grounded conductor that carries the current imbalance between two ungrounded, hot conductors.
- An equipment grounding conductor helps the circuit breaker quickly clear ground-faults and short-circuits and does not carry current under normal operation.
- Voltage is measured in volts and is abbreviated with the letter E for electromotive force.
- Current is measured in amps and is abbreviated with the letter I for intensity of current.
- Resistance is measured in ohms and is denoted by the Greek letter omega Ω.
- Ohm's law states that current flow in a circuit is directly proportional to the applied voltage and inversely proportional to the resistance of a circuit.
- Watt's law is a variant of Ohm's law and is used to determine the values that define the amount of power in a circuit.

CHAPTER 1 REVIEW

Know and Understand

1. Homes that are built for a particular customer are called _____ homes.
   - A. spec
   - B. modular
   - C. custom
   - D. mobile

2. A good method for finding a job in residential wiring is _____.
   - A. networking
   - B. internet job search engines
   - C. governmental agencies
   - D. All of the above.

3. The day after a job interview, you should _____.
   - A. call every hour to see if you got the job
   - B. send a letter thanking the interviewer for their time
   - C. assume you didn't get the job
   - D. assume you did get the job and show up for work the next day

4. The first edition of the *National Electrical Code* was published in _____. The current edition is dated _____.
   - A. 1896, 2023
   - B. 2023, 1896
   - C. 1776, 2023
   - D. 2020, 2023

5. *True or False?* Only some residential electrical installations must be installed following the rules of the *NEC*.

6. *True or False?* All residential electrical work is subject to the permitting and inspections process.

7. Most of the electrical power in the United States is produced by _____.
   - A. solar power
   - B. wind turbines
   - C. hydroelectric dams
   - D. thermal generation

8. The standard electrical system used in the United States is the _____.
   - A. 240/120-volt, three-phase, 3-wire system
   - B. 240/120-volt, single-phase, 3-wire system
   - C. 208/120-volt, single-phase, 3-wire system
   - D. 208/120-volt, three-phase, 4-wire system

9. The three components of an electrical system are _____, _____, and _____.
   - A. power source, grounding conductor, grounded conductor
   - B. light bulb, switch, fuse
   - C. conductors, fuse, circuit breakers
   - D. power source, conductors, load

10. *True or False?* Electrical current can flow in a closed circuit.

11. *True or False?* Circuit breakers must be replaced when they trip.

12. The _____ conductor carries the current from the source of power to the load.
    - A. equipment grounding
    - B. neutral
    - C. grounded
    - D. ungrounded

13. The _____ conductor carries the electrical current from the load back to the source.
    - A. equipment grounding
    - B. neutral
    - C. grounded
    - D. ungrounded

14. The _____ conductor connects the normally non-current-carrying metal parts of the electrical equipment together to help clear ground faults.
    - A. equipment grounding
    - B. neutral
    - C. grounded
    - D. ungrounded

15. The _____ conductor carries the imbalance of current between two ungrounded conductors.
    - A. equipment grounding
    - B. neutral
    - C. grounded
    - D. ungrounded

16. *True or False?* The grounded conductor carries current under normal operation.

17. *True or False?* Electrical current flows from the power source through the load and then to the ground.

18. The pressure that pushes the electrons through an electrical circuit is _____.
    - A. amps
    - B. volts
    - C. resistance
    - D. power

19. A quantity measurement of the number of electrons flowing past a point over a period of time is the _____.
    - A. current
    - B. voltage
    - C. resistance
    - D. watts

20. Opposition to the flow of current is known as _____.

    A. wattage

    B. amps

    C. voltage

    D. resistance

21. The _____ determines if current flows in an electrical circuit.

    A. resistance

    B. load

    C. voltage

    D. wattage

22. *True or False?* Ohm's law states that an increase in resistance must be accompanied by a decrease in current flow.

23. *True or False?* Watt's law can be used to determine the amount of current flow in an electrical circuit if the values of watts and voltage are known.

24. Voltage is abbreviated by the letter _____.

    A. E

    B. I

    C. R

    D. P

25. Current is abbreviated by the letter _____.

    A. E

    B. I

    C. R

    D. P

Apply and Analyze

1. What is meant by the term *nominal* voltage?

2. Why is the voltage generated at the power plant transformed to a higher level for transmission?

3. Why must the amount of electrical power that is generated precisely equal the amount of electricity consumed?

4. What is the stated purpose of the *National Electrical Code*?

5. A 240-volt electrical appliance produces a current flow of 50 amps. What is the resistance of the appliance?

6. How much power is used by a 120-volt electrical load that produces 15 amps of current flow?

7. An electrical circuit produces 5000 watts of power and has 11 ohms of resistance. What is the applied voltage of this circuit?

Critical Thinking

1. What are some methods the utilities use to match their power generation to the fluctuating usage throughout the day?

2. What function do fuses and circuit breakers perform in a residential electrical circuit?

3. How can an electrician use Ohm's law or Watt's law to properly size conductors for an electrical circuit?

CHAPTER 2 | Safety in Residential Wiring

rawf8/Shutterstock.com

SPARKING DISCUSSION

How does an understanding of electrical theory promote safe work practices?

LEARNING OBJECTIVES

After completing this chapter, you will be able to:

- List the safety rules to consider for residential electricians.
- Explain the mission of OHSA and its focus four work hazards.
- Describe how electric shocks occur and the effects on the human body.
- Define the three types of ground-fault protection devices.
- Discuss the process and purpose of lockout/tagout.
- Explain the basic principles of ladder safety.
- Give examples of the different classes of fires and the types of extinguishers used to fight them.
- Categorize the different types of personal protective equipment.
- Summarize basic first aid practices for common jobsite injuries.

TECHNICAL TERMS

arc blast
arc fault
arc-fault circuit interrupters (AFCI)
arc flash
electrocution
electrical arc
ground-fault circuit interrupters (GFCI)
lockout/tagout (LOTO)
Occupational and Health Administration (OSHA)
P.A.S.S.
parallel arc
personal protective equipment (PPE)
series arc
voltmeter

Introduction

Both general safety and electrical safety knowledge are the most important safeguard against accidents on the jobsite. Some safety rules and practices are common sense, such as "Do not work under the influence of drugs or alcohol." Others may require some form of education or training to be able to answer questions, such as "What is the highest usable step on a step ladder?" or "Is this the proper fire extinguisher to use on that fire?"

The *National Electrical Code* defines a "qualified person" as one who has the skills and knowledge related to the construction and operation of electrical equipment and installations and has received safety training to recognize and avoid the hazards involved. This chapter will introduce you to critical safety measures and practices to help you become a qualified person in the field.

2.1 General Safety Rules

- **Avoid drugs and alcohol at work.** The use of drugs and alcohol has no place on any jobsite, especially in electrical work. Mental and physical impairments from mood-altering substances can cause one's own injuries and lead to the injury of another person. This could occur, for example, by carelessly leaving energized wires exposed, or by not fully tightening a termination that could later start a fire. If a worker is injured on the job and it is found they were under the influence, they will likely lose their job and be responsible for their medical bills.

  Drugs and alcohol can impair one's judgment for days after the intoxicating effects have worn off. If there are detectable levels of a substance in your body, you are under the influence. Most electrical firms will conduct pre-hire drug screenings and random screenings throughout the year to ensure compliance.

- **Always be alert.** Alertness means being aware of your surroundings. This is especially true in a construction site where many trades are working alongside one another. Whether new construction or remodel work, it is the nature of a construction site that conditions are continually changing, and it is paramount that workers be aware of these changes. Many jobsite injuries are caused by inattention to one's surroundings.

- **Avoid working alone.** Although not always possible or practical, working with a partner is recommended for several reasons. Foremost is to have someone there to provide first aid or call for help if an injury occurs. This is especially true when working with electricity and applies when working at a height, using power tools, or engaging in other high-risk activities.
- **Work with one hand.** It is a good habit for electricians to become proficient in working with one hand only. It is often advised to keep one hand in your pocket to keep you from using both hands. An electrical shock from hand to hand is the most dangerous because it provides a path for current to travel directly through the heart, causing the heart to fibrillate or "flutter." A heart in fibrillation cannot pump blood and may lead to cardiac arrest, stroke, or death.
- **Practice general housekeeping.** A messy jobsite is a dangerous one. Accumulated debris such as conduit and wire scraps can be trip hazards. The best practice is to clean as you go or at the end of each day. A clean jobsite is a more pleasant workspace than one in disarray.

2.2 OSHA

The *Occupational and Health Administration (OSHA)* is an arm of the United States Department of Labor with a mission to ensure safe and healthful conditions in the workplace. OSHA sets and enforces these workplace standards and provides training, education, and assistance. Part of OSHA's mission is to assist employers in reducing or eliminating workplace hazards. OSHA has identified four hazards responsible for most construction site accidents: Fall Hazards, Caught-In or Between Hazards, Struck-By Hazards, and Electrical Hazards. See **Figure 2-1**.

Focus Four Hazards

Fall

Caught-in
or between

Struck-by

Electrical

Goodheart-Willcox Publisher

Figure 2-1. OSHA's focus four.

Falls, trips, and slips are the leading cause of workplace injuries. Over one-third, 34%, of all workplace fatalities are from falls, either falling from above or falling on the same level. Prevention methods include the use of guard rails, fall arrest systems, and ladder safety.

Caught-in or between injuries occur when a worker is crushed, squeezed, pulled into, rolled over, or involved in a cave-in. Using machines that are properly guarded and supported can reduce these types of injuries.

A struck-by accident occurs when a worker makes physical contact with an object. Seventy-five percent of the contacts are by heavy equipment, where falling objects account for the remaining 25% of these accidents. Keep clear of heavy equipment in use, such as earth-moving equipment, and avoid working directly under a suspended load.

A common type of electrical hazard, *electrocution*, is an exposure to a lethal amount of electricity. Always maintain a safe distance from energized power lines, inspect extension cords for damaged or missing grounding prongs on the plug, and use ground-fault circuit interrupters protection at all construction sites.

2.3 Electrical Safety

It is often said that in a typical workday, the most dangerous thing an experienced electrician will encounter is getting in their car and driving to work. This does *not* mean working with electricity is without danger. Instead, an experienced electrician is able to recognize hazards and take measures to mitigate any potential safety concerns.

To practice electrical safety, always assume an electrical circuit is energized until determined; otherwise, this can be checked by using a voltmeter. A *voltmeter* is an instrument that measures electric potential between two points. It will indicate if any voltage is present and at what levels. One of the most important safety rules is to never work on an energized circuit if the power can be safely disconnected. In some cases, work, such as troubleshooting, may require working on an energized circuit. However, working on an energized circuit that can be safely shut off is dangerous and irresponsible.

2.3.1 Electrical Hazards

There are three categories of electrical hazards: electrical shock, arc flash, and arc blast. An *electrical shock* occurs when a person becomes part of the electrical circuit through contact with an energized component. In many cases, the electrical shock is the precipitating event that leads to a more significant injury. Someone may get a shock and fall from a ladder, or a shock can cause someone to recoil their hand into a moving piece of machinery. An electrical shock of sufficient duration and intensity may result in an electrical burn that can damage skin, bones, and internal organs.

The amount of electrical current that hits the body determines the degree of severity. In the previous chapter, we discussed the concepts of electrical voltage, current, and resistance using a water analogy—voltage is electrical pressure, current is the rate of flow (similar to gallons per minute), and resistance is resistance to current flow. In an electrical circuit, high resistance results in less current flow, and low resistance results in a higher current flow. **Figure 2-2** shows the effects of current levels on the human body. A typical residential electrical circuit can draw as much as 15 or 20 amps. Human skin has an electrical resistance that varies with wetness. See **Figure 2-3**.

To illustrate the dangers of working on residential-level voltages, consider the following scenario. An electrician uses a faulty 120-volt corded drill to drill holes through the framing studs in an unfinished basement. He has a nail in his boot that makes intermittent contact with his foot. The faulty drill and the nail in the boot now provide an electrical path to ground that includes the electrician. With a dry skin resistance of 100,000 ohms, the current flow through the body would be 1.2 mA. 120 V ÷ 100,000 Ω = .0012 A or 1.2 mA. With this voltage amount, the electrician might feel a slight tingle.

On a hot and humid day where our electrician is profusely sweating, their skin resistance is lowered to 1000 Ω. Using the same drill, the current flow through the body has increased to .12 A, or 120 mA. 120 V ÷ 100,000 Ω = .12 A.

Effects of Electric Shock

Level of Current	Effects on Human Body
1 mA	Threshold of feeling. Slight tingling.
5 mA	Shock felt, but not painful yet. Involuntary muscle movements.
10–20 mA	Painful shock. Sustained muscle contraction. Inability to release grip.
100–300 mA	Paralysis of respiratory muscles. Can be fatal. Severe internal and external burns.
2 A	Cardiac arrest (heartbeat stops). Internal organ damage. Death is probable.

Goodheart-Willcox Publisher

Figure 2-2. Effects of electrical current on the human body.

Skin Condition	Resistance
Dry skin	100,00 to 500,000 Ω
Perspiring (sweaty hands)	1000 Ω
In water (completely wet)	150 Ω

Goodheart-Willcox Publisher

Figure 2-3. The resistance of the human body under wet and dry conditions.

Using **Figure 2-2** as a guide, this would lead to extreme pain, respiratory arrest, severe muscular contractions, and possible death.

If our electrician is wet with sweat on this same hot and humid day and is standing in a puddle of water while drilling holes with the faulty drill, their resistance is lowered to about 150 Ω. 120 V ÷ 100,000 Ω = .8 A or 800 mA, which will likely lead to heart failure and death. It bears repeating: *A 120-volt residential electric circuit can kill you.*

Also, note in **Figure 2-2** the range of 10–20 mA is the "freezing current" or "let go range." This level of current through the human body can cause muscular contractions or muscular lock, where you cannot let go of the energized component. Never touch a person who is "frozen to the line." If you do touch them, it automatically includes you in the circuit and causes you to experience muscular contraction, which then endangers both of you. If you cannot quickly disconnect the power, the proper way to free someone under muscular lock is to use a non-conductive material such as a wooden broom or a length of lumber to knock them free.

An *electrical arc* is when the electricity bridges a gap in the electrical circuit using the air as a conductor, **Figure 2-4**. Small arcs are common in electrical appliances with motors, such as a hair dryer or a vacuum cleaner, or when opening or closing a light switch. But larger arcing events can be extremely dangerous. An arc flash and arc blast both result from the same arc fault but are different events. An *arc fault*, or arcing fault, is an unwanted condition in the electrical circuit that causes arcing and sparking between two points. An *arc flash* is the light and heat from an arc that is large enough to ignite the surrounding air and cause fires and injuries. The temperature of an arc flash can reach 35,000°F. The *arc blast* is the explosion of the electrical equipment as a result of the arc flash. Together with the flash's high temperatures, the flying debris can kill or maim

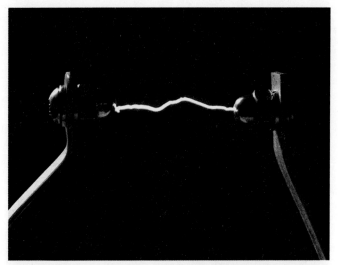

Vladimir Breytberg/Shutterstock.com

Figure 2-4. An electric arc between two points.

a worker that is not adequately protected. **Figure 2-5** shows the standard protective equipment when working on equipment that could cause an arc flash. Although the arc flash and arc blast are dangerous occurrences, they are unlikely to occur at residential wiring voltage levels.

2.3.2 Circuit Protection

In residential wiring, shock is prevented by using ground-fault circuit interrupters, or GFCIs, and tamper-resistant receptacles. Arc-fault circuit interrupters, or AFCIs, protect against an arcing fault condition.

A GFCI receptacle, **Figure 2-6**, is identified by its rectangular shape and the TEST and RESET buttons on the receptacle's face. A *ground-fault circuit interrupter (GFCI)* works by monitoring the current flow in both the grounded and ungrounded conductors attached to it. If the currents are the same, as they should be, the GFCI is not activated. If, however, the device senses a current imbalance of 4–6 mA, it opens the circuit, shutting the power off. Remember, a ground fault occurs when a path to ground is other than the intended path. A particular type of circuit breaker, **Figure 2-7**, can provide ground-fault protection to the entire circuit. GFCI-protected extension cords, **Figure 2-8**, are also available. More on GFCI receptacles and breakers will be included in future chapters.

Nutthapat Matphongtavorn/Shutterstock.com

Figure 2-5. An electrician is wearing arc blast protective clothing. Note the full face and neck coverage, gloves, and boots.

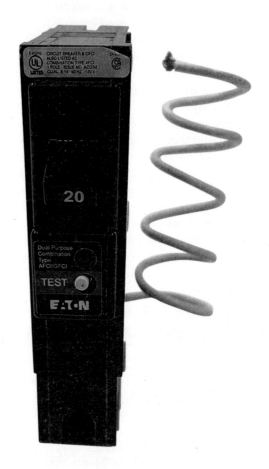

Eaton

Figure 2-7. A typical GFCI breaker.

The Toidi/Shutterstock.com

Figure 2-6. A GFCI receptacle being reset.

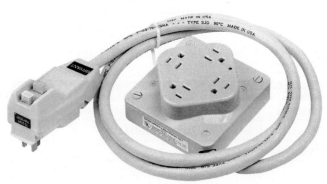

Hubbell Inc.

Figure 2-8. GFCI extension cord.

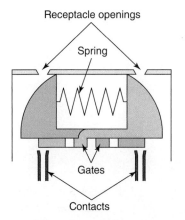
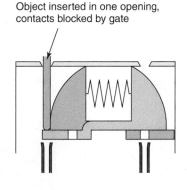
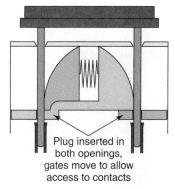

Receptacle openings

Spring

Gates

Contacts

Object inserted in one opening, contacts blocked by gate

Plug inserted in both openings, gates move to allow access to contacts

Goodheart-Willcox Publisher

Figure 2-9. A cutaway view of the tamper-resistant receptacle shutter mechanism.

Tamper-resistant receptacles, **Figure 2-9**, were developed to protect small children from electrical shock resulting from inserting keys, bobby pins, paper clips, and the like into receptacle openings. These receptacles contain an ingenious shutter mechanism that will accept a plug inserted into both slots simultaneously but will reject a foreign object from being inserted into a single slot. Tamper-resistant receptacles are required in all newly constructed dwellings.

The *arc-fault circuit interrupter (AFCI)* is another safety device that opens the electrical circuit in the event of an arcing fault. Arc-fault protection is usually provided by an arc-fault circuit breaker, **Figure 2-10**. Although the AFCI breaker looks similar to a GFCI breaker, AFCI protection is different than GFCI protection. The arc-fault circuit interrupter monitors the circuit for the distinctive features of an arcing fault. An arcing fault occurs when the conductor has been damaged in some way.

There are two types of arc faults: a series arc and a parallel arc, **Figure 2-11**. A *series arc* is an arc that occurs across a break in the same conductor. A series arc may also result from a loose connection at the terminal screws of a switch or receptacle and/or a loose splice. A *parallel arc* is an arc that occurs across two or more conductors of different potential, such as a grounded and ungrounded conductor or ungrounded conductors from different phases. Arcing faults can lead to injuries or fires.

All AFCIs are now of the combination-type AFCI, or CAFI, which detects both series and parallel arcs. Do not confuse the combination-type AFCI, which provides only arc-fault protection, with the dual-purpose GFCI/AFCI, which provides both ground-fault and arc-fault protection.

An AFCI can distinguish between the normal arcing of a hairdryer or vacuum cleaner motor and the arc caused by a damaged conductor. A common cause of conductor damage results from an extension cord being run between a rug and

the floor. The cord's conductors can become compromised by walking on the rug, or it might be pinched by a bed frame or other furniture, resulting in an arcing condition that may ignite the rug or other surrounding combustible items, such as bed sheets or curtains.

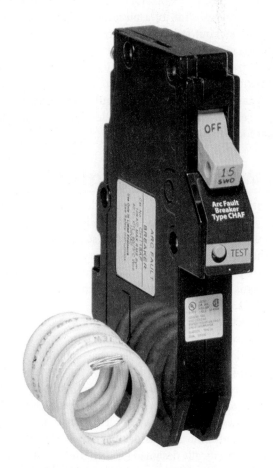

Eaton

Figure 2-10. An AFCI circuit breaker. Note the similarity in appearance with the GFCI breaker.

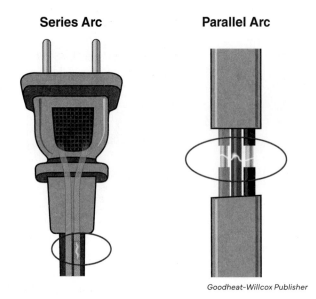

Series Arc   Parallel Arc

Figure 2-11. A series arc across a gap in the same conductor and a parallel arc across multiple conductors.

2.3.3 Lockout/Tagout

Lockout/tagout (LOTO) is a safety practice used extensively in commercial and industrial electrical work to prevent the accidental energization of a piece of equipment being serviced. The lockout step in this procedure locks the equipment's disconnecting means in the open position, and the tagout step sets up a tag that identifies who is working on the equipment and the estimated time of completion. Although lockout/tagout is not performed regularly in residential wiring, it is recommended that the residential electrician have a breaker lock, **Figure 2-12**.

2.4 Working at Heights

It is critical to understand proper ladder construction and safety in the construction trades. Electricians are often called upon to work from ladders or scaffolding, and working at heights is one of the most dangerous tasks electricians regularly perform. The two types of ladders most used by electricians are the stepladder and the extension ladder, **Figure 2-13**. Electricians should only use ladders made of nonconductive fiberglass. Metal or wooden ladders should never be used when working with electricity.

Stepladders are folding ladders that commonly come in heights of 4′, 6′, 8′, 10′, 12′, 14′, 16′, 18′, or 20′. They should only be used on stable, level ground with the leg spreaders fully engaged. The highest usable step on the stepladder is the second-highest step from the top cap. Never stand on the top cap, and do not climb a stepladder that is leaned against the wall unless it is specifically designed for such use. You should never keep or leave tools and materials on top of a ladder as they can fall and cause damage or injuries. Always reposition the ladder when you need to reach your work, and never try to over-reach.

Figure 2-12. A circuit breaker lock is installed to prevent the energization of the circuit that is being worked on.

When using a straight ladder or an extension ladder, the correct position for the ladder's base is one-fourth of the ladder's height where it meets the structure, **Figure 2-14**. This involves moving the base of the ladder one foot from the structure for every four feet of height. For example, if the wall the ladder is propped on is 20 feet, the ladder must be 5 feet from the wall. The ladder should also extend at least three feet above the highest point of support, which allows for a much safer entry and exit point when using the ladder. When climbing, always face the ladder and keep three points of contact, either two hands and one foot, or two feet and one hand.

2.5 Fire Safety

There are three ingredients to any fire: fuel, heat, and oxygen. Fuel is any material that can burn. Heat is necessary to raise the fuel to its ignition temperature. Oxygen is required to sustain the combustion of the fuel. All three components are required to start a fire; removing any one ingredient will extinguish the fire. In fact, when you douse a fire with

Stepladder

Extension ladder

Werner Co.

Figure 2-13. A step ladder and an extension ladder.

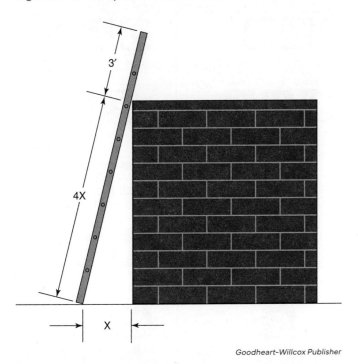

Goodheart-Willcox Publisher

Figure 2-14. The proper positioning of a straight ladder. If this wall is sixteen feet high, the ladder's base should be four feet from the wall's bottom.

water, you are lowering the temperature of the fuel below its combustion point.

Fires come in different classes, and the type of fuel determines the class of fire and the type of extinguisher used to fight it, **Figure 2-15**. There are several classes of fire:

Class A fires are of ordinary combustibles such as wood, paper, or fabrics.

Class B fires are of flammable liquids and gases.

Class C fires are of burning electrical equipment.

Class D fires are those of burning metal.

Class K fires are those of burning cooking oils and fats in a commercial kitchen and are technically a sub-class of class B.

Many fire extinguishers can be used on different classes of fires. For example, an extinguisher marked as class ABC can be used on any of those classes of fire. Never use an extinguisher on a fire for which it is not rated. Using pressurized water on a grease fire, for example, can spread the fire rather than put it out as the burning grease will float on the surface of the water. Using water on an energized electrical fire could result in water flowing back to the operator, resulting in a shock.

Fires can start suddenly and spread quickly. When fighting a fire, it is essential to stay focused and do not panic. The acronym *P.A.S.S.* can be used as a guide to proper fire extinguisher use:

- **P**—Pull the pin on the extinguisher so you can squeeze the handle.
- **A**—Aim the nozzle of the fire extinguisher to the base of the fire.
- **S**—Squeeze the handle to deploy the extinguisher. Stay six to eight feet from the fire if possible.
- **S**—Sweep the nozzle back and forth at the base of the fires.

2.6 **Personal Protective Equipment**

One of the easiest ways to prevent injuries on the job is by wearing the appropriate *personal protective equipment (PPE)*. OSHA defines PPE as "equipment worn to minimize exposure to hazards that cause serious workplace injuries and illnesses." Typical PPE includes eye protection, hearing protection, gloves, proper footwear, and head protection.

Safety glasses should always be worn while working, **Figure 2-16**. They are available with prescription lenses for those needing sight correction. Safety glasses protect the eyes from flying hazards that can pierce, bruise, or otherwise damage the eyes. The lenses and frames should be marked as Z87+, indicating that the eyewear has been tested and meets high-velocity impact requirements. These tests include dropping a pointed 500-gram weight onto the lens from a height of about 5 feet. The lenses must also withstand

Fire Classifications

Class	Description	Requires	New Symbol	Old Symbol
A	Ordinary Combustibles (Materials such as wood, paper, and textiles)	Cooling/quenching		A
B	Flammable Liquids (Liquids such as grease, gasoline, oils, and paints)	Blanketing or smothering		B
C	Electrical Equipment (Wiring, computers, switches and any other energized electrical equipment)	A nonconducting agent		C
D	Combustible Metals (Flammable metals such as magnesium and lithium)	Blanketing or smothering		D
K	Kitchen Fires (Grease, fat, and oil fires in commercial kitchens)	Blanketing or smothering		K

Fire Extinguishers

Type	Description		Typically Approved for Use On	Not for Use On
Pressurized Water	Water under pressure		A	B C D K
Carbon Dioxide (CO_2)	Carbon dioxide (CO_2) gas under pressure		B C	A D K
Foam	Aqueous film-forming foam (AFFF) or film-forming fluoroprotein (FFFP)		A B	C D K
Dry Chemical, Multipurpose Type	Typically contains ammonium phosphate		A B C	D K
Dry Chemical, BC Type	May contain sodium bicarbonate or potassium bicarbonate		B C	A D K
Dry Powder	May contain sodium chloride, sodium carbonate, copper, or graphite		D	A B C K
Wet Chemical	May contain potassium acetate, potassium citrate, or potassium carbonate		K	B C D

Goodheart-Willcox Publisher

Figure 2-15. Different types of fires require different types of fire extinguishers.

Purple Clouds/Shutterstock.com

Figure 2-16. Typical safety glasses. Note the wrap-around lens.

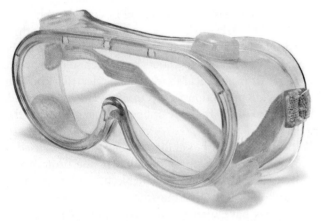

Gavran333/Shutterstock.com

Figure 2-17. Safety goggles. Note how the glasses will form-fit to the wearer's face.

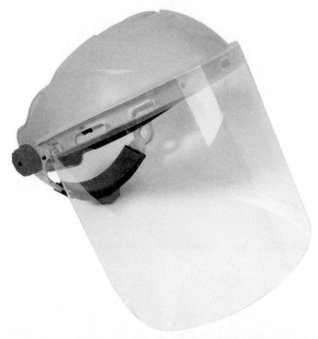

WhiteJack/Shutterstock.com

Figure 2-18. Typical protective face shield. Note that face shields alone do not protect the eyes. Separate eye protection is required.

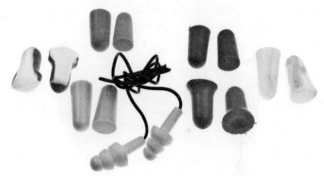

ribeiroantonio/Shutterstock.com

Figure 2-19. Assorted disposable earplugs.

a 1/4" steel ball being shot at the lens from distances ranging from 150 to 300 feet.

Other types of eye protection include safety goggles and face shields. See **Figures 2-17** and **Figure 2-18**. Safety goggles are different from safety glasses in that goggles fit snugly on the face while glasses have gaps between the frames and on one's face. Many safety goggles have a gasket around the frames that provide a snugger fit around the eyes and across the brow. Safety goggles offer additional protection over safety glasses. They prevent liquids from being splashed in the eyes and offer protection against fine dust and other particulate matter from entering the eyes. A face shield protects more of the face from splashes and flying debris, but it is not considered a substitute for safety glasses or goggles. Preferably, the face shield is used with safety glasses or goggles to provide greater protection for the wearer.

Hearing protection is required when working where there is excessive or prolonged noise. Loud noise can damage hearing and lead to fatigue. Hearing protection PPE is categorized by its noise reduction rating (NRR), a unit of measurement that determines the device's ability to limit sound exposure. The higher the NRR, the better the protection. OSHA requires hearing protection for any worker exposed to 85 decibels or higher over 8 hours, or 100 decibels or greater over 2 hours. Eighty-five decibels are the equivalent of loud traffic or a noisy restaurant, whereas 100 decibels is equal to a lawnmower or chainsaw. Earplugs and earmuffs are the most common forms of hearing protection. See **Figure 2-19** and **Figure 2-20**.

Wearing the proper work gloves can protect hands and fingers from abrasions, cuts, burns, and electrical shocks, among other injuries, **Figure 2-21**. No single glove type can protect against all hazards. It might make sense to always use gloves that provide the most protection, but big, bulky leather gloves can severely restrict movement. Thinner gloves may improve your grip and provide greater dexterity, but they may not protect against splinters, nails, and sharp

Shutter Baby photo/Shutterstock.com

Figure 2-20. Typical earmuffs for hearing protection.

A

objectsforall/Shutterstock.com

B

dezign56/Shutterstock.com

Figure 2-21. A—Leather work gloves. B—Cotton work gloves.

edges. Electrical utility linemen wear two pairs of gloves–a voltage-rated rubber inner glove and a pair of leather gloves on the outside, **Figure 2-22**.

PPE can also apply to everyday work wear, such as shirts, pants, and shoes or boots. Shirts and pants should be made from natural fibers such as 100% cotton or wool. In the event of a fire, synthetic fibers such as polyester, poly-cotton blends, or nylon can ignite and melt to the skin and cause more severe burns than what would occur from the initial fire. Clothing made of fire-resistant material is available but is typically not necessary for a residential electrician. The possible electrical arc conditions a residential electrician would be exposed to are too low to ignite clothing.

Proper footwear is another PPE consideration. See **Figure 2-23**. OSHA requires anyone working with or around live electrical wires to wear footwear with an EH rating. EH-rated shoes and boots have thick soles to protect the wearer from electrical shock or static electricity buildup. They provide protection for up to 600 volts in dry conditions. A label inside the shoe or boot indicates whether they are EH rated.

Where there is a possibility of foot injuries caused by falling heavy objects or rolling objects, toe protection is required. Toe protection is available in most work boots and shoes and even in some sneaker styles, **Figure 2-24**. Safety shoes have internal toe tips traditionally made of steel, but non-metallic options made from plastic, carbon fiber, Kevlar, or other strong material are also available. OSHA does not generally consider wearing steel-toed safety shoes to be a hazard as long as the shoe's conductive portion does not make direct contact with the wearer's foot and is not exposed to the outside of the shoe. However, common sense says that electrical workers should avoid steel-toed safety shoes.

Narin Eungsuwat via Getty Images

Figure 2-22. Electrical linesmen's gloves are voltage-rated inner gloves protected by leather gloves.

VladaKela/Shutterstock.com

Figure 2-23. Assorted safety shoes and boots. Note the thick soles.

Fahroni/Shutterstock.com

Figure 2-24. Sneaker-style safety shoes.

NAR studios/Shutterstock.com

Figure 2-25. Items in the basic first aid kit.

2.7 Workplace Injuries

Practicing proper workplace safety and wearing PPE are the first lines of defense against workplace injuries; however, accidents can and do occur. Although there is comprehensive first aid instruction that extends beyond the scope of this text, there are several basic actions you can take to minimize the severity of an accident or injury. Contact your local Red Cross, YMCA, or the Emergency Care and Safety Institute for proper first aid training and certification. At a minimum, every work truck should be supplied with a basic first aid kit, **Figure 2-25**.

Items in the basic first aid kit include assorted sized self-adhesive bandages, assorted sized sterile gauze bandages, medical adhesive tape, scissors, anti-bacterial ointments,

burn ointments, cream or spray for insect bites, alcohol wipes, tweezers, painkillers such as aspirin or ibuprofen, antihistamines, distilled water and syringes for cleaning wounds, eyewash, and a digital thermometer.

If you are involved in or come upon an accident with injuries, take a few seconds to assess the scene, and scan the immediate area for danger to the victim or any rescuer. If you see an unsafe condition, make it safe, if possible. For example, if there was an electrical mishap, can the power be disconnected? If this is not possible, do not enter the area. If there are two or more victims, check on those who are not moving or talking first. These are the individuals who may need the most help.

Call 911 for help *before* attempting to assist anyone who is injured. Tell the dispatcher your name and phone number, the exact location or address of the accident, what happened, the number of people who need assistance, their condition, and what is being done to help them. Do not hang up until the dispatcher instructs you to do so. They may be able to tell you how to care for the victim until the ambulance arrives.

Some basic first aid for common worksite injuries include the following:

- **Electrical shock**—Turn off the power source if possible. Do not approach the victim until the power has been turned off. *Never* make contact with someone who is "frozen" to the electrical line. Do not move the victim unless there is an immediate danger of further injury. The dispatcher may instruct you on how to perform CPR until help arrives.
- **Bleeding**—If the wound is deep, do not try to clean it. Instead, apply pressure to stop the bleeding. Use a clean cloth to absorb the blood. If the cloth becomes soaked, do not remove it. Instead, keep adding layers as this helps with clotting. For a shallow wound, wash it with soapy water and then flush the wound with clean water. Apply an antibiotic ointment, if available, and cover the wound with a clean cloth or bandage.

- **Broken bones**—Do not try to straighten or realign the bones or push in a bone that has pierced the skin. Only move the victim if it is necessary to avoid further injury. Keep the victim still until help arrives.
- **Heat exhaustion**—The symptoms of heat exhaustion include painful muscle spasms, heavy sweating, severe thirst, headache, nausea, and vomiting. Have the victim stop their activity and rest in a cool area. Remove or loosen any tight or excess clothing, such as coats or safety vests. If available, apply wet, cool cloths to the skin, particularly the underarms and the inside thigh where the leg meets the pelvis. If the victim is responsive, provide them with small sips of water or a sports drink, such as Gatorade® or Powerade®.
- **Choking**—When an adult appears to be choking, ask them, "Are you choking?" If they cannot speak, cough, breathe, or are coughing weakly or making high-pitched noises, begin abdominal thrusts, known as the Heimlich Maneuver. See **Figure 2-26**. Place your arms around the victim's torso. Make a fist with one hand and place the thumb side of the fist just above the navel. Grasp the fist with your other hand and forcefully apply quick, upward thrusts. Repeat until the object is dislodged.

Heimlich Maneuver

 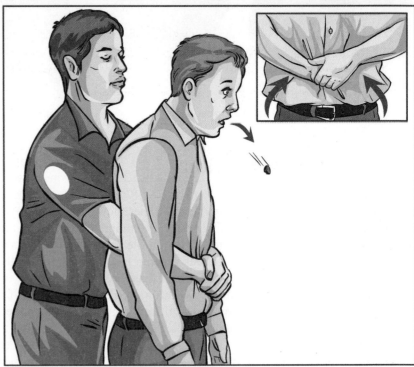

Drp8/Shutterstock.com

Figure 2-26. The Heimlich Maneuver illustrated.

- **CPR training** —CPR, cardiopulmonary resuscitation, can help save lives during a cardiac or breathing emergency such as a heart attack or electrical shock. CPR training is available from the American Red Cross or other civic organizations. In addition, the Mayo Clinic gives the following advice for untrained individuals when performing CPR: "If you are not trained in CPR or worried about giving rescue breaths, then provide hands-only CPR. That means uninterrupted chest compressions of 100 to 120 a minute until paramedics arrive. You do not need to try rescue breathing."

Summary

- Many jobsite injuries are caused by inattention to one's surroundings.
- An electrical shock from hand to hand is the most dangerous as it provides a path for current to travel directly to the heart.
- OSHA is an arm of the United States Department of Labor with a mission to ensure safe and healthful working conditions.
- Falls, trips, and slips are the leading cause of workplace injuries.
- Electrocution is exposure to a lethal amount of electricity.
- An electrical shock occurs when a person becomes a part of the electrical circuit through contact with an energized component.
- In residential wiring, shock prevention is accomplished using ground-fault circuit interrupters (GFCIs) and tamper-resistant receptacles.
- The GFCI works by monitoring the current flow in both the grounded and ungrounded conductors attached to it.
- Tamper-resistant receptacles have a shutter mechanism that will accept a plug inserted into both slots simultaneously but reject a foreign object from being inserted into a single slot.
- The arc-fault circuit interrupter (AFCI) is a safety device that opens the electrical circuit in the event of an arcing fault.
- A series arc occurs when the arc is across a break in the same conductor. A parallel arc is an arc across two or more conductors. Arcing faults can lead to injuries or fires.
- Lockout/tagout is a safety practice used extensively in commercial and industrial electrical work to prevent the accidental energization of a piece of equipment being serviced.
- Electricians should only use ladders made of non-conductive fiberglass. Metal or wooden ladders should never be used when working with electricity.
- When using a straight ladder or an extension ladder, the correct position for the ladder's base is one-fourth of the ladder's height where it meets the structure.
- Fires come in different classes, and it is the type of fuel that determines the class of fire and the type of extinguisher used to fight it.
- OSHA defines personal protective equipment, or PPE, as "equipment worn to minimize exposure to hazards that cause serious workplace injuries and illnesses."
- Typical PPE includes eye protection, hearing protection, gloves, proper footwear, and head protection.
- Call 911 to summon help *before* attempting to assist anyone who is injured on the jobsite. Do not hang up until the dispatcher tells you to.

Know and Understand

1. *True or False?* Drugs and alcohol can impair one's judgment for days after the effects have worn off.
2. *True or False?* The physical conditions of a construction worksite can change rapidly.
3. *True or False?* When working on electrical circuits, it is always best to work with two hands.
4. OSHA's Focus Four hazards include slip, trip, and fall accidents, struck-by accidents, caught-in or caught-between accidents, and _____.

   A. splinters
   B. cuts and bruises
   C. electrocution
   D. amputations
5. A(n) _____ occurs when a person becomes a part of an energized electrical circuit.

   A. electrical shock
   B. arc fault
   C. arc blast
   D. arc flash
6. The current threshold for sustained muscular contractions that result in a painful shock and the inability to let go of an energized electrical line is _____.

   A. 1 mA
   B. 4–6 mA
   C. 10–20 mA
   D. 100–300 mA
7. *True or False?* The electrical resistance of skin increases with increasing moisture, such as perspiration.
8. *True or False?* A typical residential 120-volt circuit is generally considered to be not lethal.
9. The GFCI device will open the electrical circuit if it senses a current imbalance between _____.

   A. 1–3 mA
   B. 4–6 mA
   C. 10–12 mA
   D. 15 amps
10. *True or False?* A series arc is an electrical arc across two or more conductors.
11. Electricians should only use ladders made of _____.

    A. wood
    B. metal
    C. plastic
    D. fiberglass
12. If the upper point of support of an extension ladder is 20 feet high, the bottom of the ladder should be _____ feet from the base of the structure.

    A. 4
    B. 5
    C. 6
    D. 7
13. A fire of ordinary combustibles such as wood or paper is a Class _____ fire.

    A. A
    B. B
    C. C
    D. D
14. *True or False?* Any type of fire extinguisher can be used on burning electrical equipment.
15. *True or False?* Safety glasses are designed specifically to protect the wearer from liquids being splashed into the eyes.
16. *True or False?* For proper protection, a face shield should always be worn in conjunction with safety glasses or safety goggles.
17. OSHA requires hearing protection for workers exposed to 85 decibels or higher for _____ hours or more.

    A. 2
    B. 4
    C. 6
    D. 8
18. *True or False?* It is always safest to work with the thickest gloves possible.
19. *True or False?* Electrical workers should always wear clothing made from 100% natural fibers such as cotton or wool.
20. *True or False?* Always call 911 to summon help *before* attempting to assist anyone who has been injured in a worksite accident.

Apply and Analyze

1. What methods do companies employ to maintain a drug-free workplace?
2. Explain the mission of OSHA.
3. What is the leading cause of workplace injuries?
4. Relate how an electrical shock can be the triggering event to a more significant injury.
5. Describe the operation of a GFCI device.
6. Describe the operation of an AFCI device.
7. What are the three ingredients needed for a fire, and how do they interact to maintain the fire?
8. List seven items that should be included in a basic first aid kit.

Critical Thinking

1. Evaluate the statement that the most dangerous activity an experienced electrician will undertake in the typical workday is to drive their car to work.
2. Correlate the relationship between electrical resistance and electrical current as it applies to the effects of an electric shock on the human body.

Tools Used in House Wiring

Lyudmila Zavyalova/Shutterstock.com

CHAPTER 3
Hand Tools

CHAPTER 4
Power Tools

CHAPTER 5
Electrical Test Equipment

Every residential electrician uses tools on the job. Some tools are general-purpose tools, while others are designed and used for specific tasks. In addition, there are hand tools and power tools. This section introduces the student to the myriad of tools used in residential wiring.

Chapter 3, *Hand Tools*, begins with an overview of hand tool safety and tool organization and storage. Then, tools that are commonly used and unique to the electrical trades are discussed.

Chapter 4, *Power Tools*, discusses those tools commonly used by electricians. Most power tools used by electricians are classified as cutting tools, such as circular saws or reciprocating saws, or drilling tools, which are the most common power tools used in residential wiring.

Chapter 5, *Electrical Test Equipment*, introduces the student to the electrical meters and measuring devices commonly used in residential wiring.

CHAPTER **3** | Hand Tools

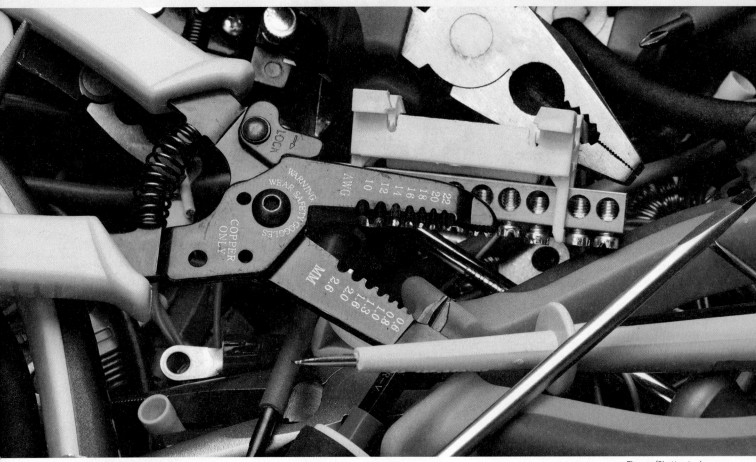

Flegere/Shutterstock.com

SPARKING DISCUSSION

How are hand tools used in residential wiring?

LEARNING OBJECTIVES

After completing this chapter, you will be able to:

- List basic rules and practices of hand tool safety.
- Discuss the different types of screwdrivers and their purposes.
- Explain the different types of pliers used by electricians and their purposes.
- Describe when insulated tools are used.
- Explain the uses of a jab saw and hacksaw.
- Define aggressiveness as it applies to saw blades.
- Identify tools used to fish conductors through the voids in walls and ceilings.
- Identify different tools to check for plumb and level.
- Identify tools an electrician may use in demolition.
- Explain why an electrician may need patching and cleaning materials.

TECHNICAL TERMS

adjustable wrench
aggressiveness
box wrench
cable cutter
cable ripper
cat's paw
combination wrench
crimp connector
crimping pliers
diagonal cutting pliers
electrician's hammer
electrician's tool belt
end-cutting pliers
fish stick
fish tape
hacksaw
hand bender
hex wrench
insulated tool
jab saw
laser plumb
level
lineman (side cutter) pliers
long-nose (needle-nose) pliers
masonry chisel
metal file
multipurpose tool
nut driver
open-end wrench
pendant chain tool
plumb
plumb bob
precision screwdriver
prybar
pump pliers
quick-rotating driver
rasp
ratchet set
scratch awl
screwdriver
screw-holding driver
sledgehammer
stubby screwdriver
tape measure
tapping tool
teeth per inch (tpi)
terminal block screwdriver
tin snip
torpedo level
torque screwdriver
torque wrench
utility knife
wire stripper

Introduction

The function of any tool is to allow you to do something you could not do with your bare hands, or if you tried, it would be extremely difficult or cause you harm. With time, the tools become an extension of your body, and using them comes as naturally as snapping your fingers. An understanding of a particular tool's usage and limitations is critical for any tradesman. Electricians are called upon to perform a multitude of tasks and must be proficient with the wide variety of tools necessary for this wide array of work.

Tools are generally divided into two categories: hand tools and power tools. Hand tools will be discussed in this chapter, and power tools will be discussed in the next chapter.

SAFETY NOTE   **NIOSH**

The National Institute for Occupational Safety and Health (NIOSH) is the research agency of OSHA and is part of the Centers for Disease Control in the U.S. Department of Health and Human Services. According to NIOSH, about 2,000 U.S. workers a day have job-related eye injuries that require medical treatment. Most of these eye injuries are a result of dust, wood chips, metal slivers, or other small objects striking or scraping the surface of the eye. The type of work performed and the environmental conditions should dictate the proper type of eye protection required. Although no single form of eye protection can protect against all injuries, it is critical that workers in the construction trades wear eye protection.

3.1 Hand Tool Safety

Hand tools are not often considered dangerous. Still, misuse of simple tools can lead to many types of injuries, including eye injuries, cuts, pinches and puncture wounds, and repetitive motion disorders. Most accidents caused by hand tools are a result of misuse or improper maintenance. Misuse includes using the wrong tool for the job or using the proper tool in an incorrect manner. Some basic rules of hand tool safety include:

- Always wear your eye protection and other proper PPE.
- Use the correct tool for its intended purpose and understand the tool's uses and limitations.
- Do not use broken or damaged tools, such as dull cutting tools or screwdrivers with worn tips.
- Do not use a "cheater bar" to gain increased leverage when using a wrench.
- Always cut in a motion away from your body when using a utility knife or other blade.

- When handing a tool to another person, keep sharp points and cutting edges away from yourself and others.
- Secure and keep track of tools when working at a height; a falling tool can cause great harm.
- Avoid using tools with wet or greasy hands.
- Return tools to their proper place after use.
- Clean and oil your tools as needed.
- Inspect tools regularly for wear or damage.
- Replace damaged or broken tools.
- Avoid carrying tools, particularly sharp pointed tools, in your pants pockets.

3.2 Tool Belts and Toolboxes

Electricians keep their hand tools organized in an ***electrician's tool belt***, **Figure 3-1**, which supports a pouch designed with pockets to hold the unique tools used by electricians. If you watch an experienced electrician at work, you may notice they can easily select and replace tools from their belt without even looking.

Most electricians will have a large assortment of tools, and to lighten the load, they will only carry what they need for the task. You'll likely need a toolbox, tool bag, backpack, or bucket sleeve, **Figure 3-2**, for tool storage in addition to

Delmas Lehman/Shutterstock.com

Figure 3-1. The electrician's tool belt is made specifically for the tools used by electricians.

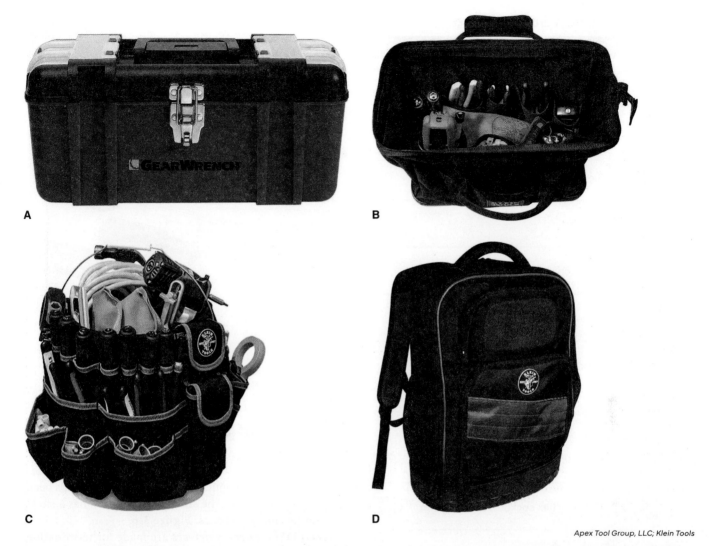

A

B

C

D

Apex Tool Group, LLC; Klein Tools

Figure 3-2. A toolbox (A), tool bag (B), bucket sleeve (C), and tool backpack (D) can be used to transport tools to the worksite and help to keep tools organized.

the tool belt. A side pouch holds screws, wire nuts, staples, and other incidentals and is used in conjunction with the tool pouch, **Figure 3-3**.

3.3 Screwdrivers

Screwdrivers, often called *drives* for short, are tools used for the installation and removal of screws. They come in different tip styles and sizes, **Figure 3-4**. The tip of the screwdriver is called the head, and there are many types of screwdriver heads. The two most common types of screwdrivers are flathead and Phillips. The flathead, or slotted, screwdrivers fit with the slotted screw and are sized by the tip's actual width, usually fractional inches. Phillips head and square drives have four standard tip sizes from #0 to #4, with #0 being the smallest, **Figure 3-5**. A #2 is the standard size for most everyday screws, and the #1 is a tad smaller.

Klein Tools

Figure 3-3. A side pouch on a tool belt can hold fasteners and connectors.

Common Screw Heads

Head	Type
☐	Square
⬡	Hex
⊖	Slotted
⊕	Phillips
✚	Frearson
▣	Robertson/square recess
✶	Torx®
✶	Security Torx®
⬢	Hex socket
⬡	Security hex socket
▲	Tri-wing
✳	Pozidriv
⊙	Spanner
⊹	Torq-set
✺	Spline
◗	One-way
⧖	Clutch

Goodheart-Willcox Publisher

Figure 3-4. Common screwdriver tip includes the flathead, Phillips head, torx, and hex head.

Klein Tools

Figure 3-5. Phillips head and slotted screwdrivers are available in several standard sizes.

3.3.1 Types of Screwdrivers

There are various screwdrivers available in different lengths, from stubby to long, and have different uses. At a minimum, an electrician should have a #2 and a #1 Phillips head and a #2 square drive. The ***stubby screwdriver*** has a short shaft for use in tight spaces, **Figure 3-6.**

Terminal block screwdrivers are made for terminating conductors in terminals found in many electrical and electronic components, **Figure 3-7.** ***Precision screwdrivers***,

Klein Tools

Figure 3-6. Stubby screwdrivers are used in tight spaces where a traditional screwdriver would not fit.

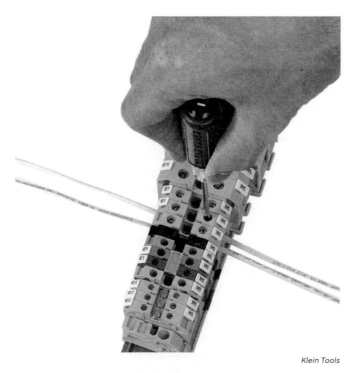

Klein Tools

Figure 3-7. Terminal block screwdrivers are used on smaller screws that are found in many electronic devices.

General Tools

Figure 3-8. Electrician's precision screwdrivers can be a handy addition to any toolset.

also called *jeweler's screwdrivers*, are screwdrivers with tiny tips, **Figure 3-8**. They are handy for a variety of uses and should be in every electrician's toolset. Some kits include picks and hooks that can solve many problems.

3.3.2 Screw-holding and Quick-rotating Drivers

Screw-holding drivers and quick-rotating drivers are two convenient specialty drivers. The *screw-holding drivers* hold and release the screw using some mechanical form, **Figure 3-9**. This feature is beneficial in tight situations where it is hard to get the screw started or a task where you are working over your head, such as installing a ceiling fan or a ground screw into the back of a junction box. It should be noted that some screw-holding divers are to get the screw started and can be damaged if used to torque the screw fully.

Quick-rotating drivers have an offset shaft that is free to rotate in the handle, **Figure 3-10**. With practice, this feature allows the user to install and remove screws quickly using only one hand.

Klein Tools

Figure 3-9. The screw-holding driver grasps the screw so it can be inserted into a tight space.

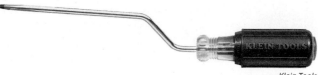

Klein Tools

Figure 3-10. A quick-rotating driver allows the user to install or remove screws more quickly than a traditional screwdriver.

PRO TIP   **Flathead Screwdrivers**

Do not use your good flathead screwdrivers as a chisel, as this will damage the tool. However, it is a good idea to have an old screwdriver for this purpose.

3.4 **Nut Drivers**

Nut drivers are another standard tool used by electricians, **Figure 3-11.** They are similar to screwdrivers but have a hex socket at their tip. Nut drivers are used to install or remove nuts, bolts, and hex-head screws found in electrical equipment and materials. They can have either magnetic tips, hollow shafts, or both. Most nut driver sets are color-coded by size, **Figure 3-12,** and are often purchased in sets containing several sizes. Two of the driver sizes are used more than any other: the yellow 5/16″ and red 1/4″.

Klein Tools

Figure 3-11. Nut driver tightens and loosens nuts and hex-head fasteners.

Klein Tools

Figure 3-12. Nut drivers are often color-coded as to their different sizes. This makes selecting the proper size quicker and easier.

3.5 **Pliers**

Pliers are another type of hand tool and there are many types. Different types can perform specific functions such as turning, cutting, bending, clamping, crimping, and tightening and loosening bolts and nuts. *Lineman (side cutter) pliers* are heavy pliers used to cut cables, conductors, and small screws, and pull and hold conductors, **Figure 3-13.** They may also have additional helpful features like crimping dies, insulation strippers, or a fish tape grip. A feature of the linesman's pliers is that the jaws' tips do not meet when closed. There is a gap of approximately 1/16″, which prevents excessive damage to conductors in the jaws' grip. It also allows the cutting edges to meet firmly to ensure a clean cut.

Long-nose (needle-nose) pliers have long, thin noses. They are used to form small conductors, cut, hold, and pull conductors, and are ideal for gripping or retrieving objects in tight spaces. Mini-needle nose pliers and bent-nose pliers are also available. See **Figure 3-14.**

Diagonal cutting pliers are used for cutting cables and conductors in tight spaces, **Figure 3-15.** They are also an excellent choice for removing staples in wood framing members. An angled head style is also available. The angled head provides extra leverage when pulling staples and nails and makes it easier to work in tight spaces.

Pump pliers are often called *channel-lock* or *tongue-and-groove pliers,* **Figure 3-16.** They have a sliding jaw that locks in place at specific intervals to size to different nuts, bolts, or other items. They are used to hold and tighten raceway couplings and connectors and hold and turn conduit and

Klein Tools

Figure 3-13. Lineman pliers have many uses in the field. They are used to grasp, pull, and cut conductors and materials.

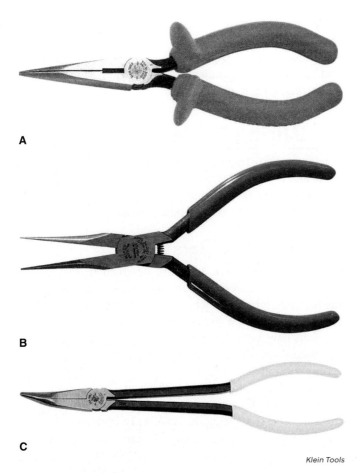

A

B

C

Klein Tools

Figure 3-14. Various types of long-nose pliers. A—Long nose pliers. B—Mini needle nose pliers. C—Bent nose pliers.

Klein Tools

Figure 3-16. Standard pump pliers. Standard pump pliers can be cumbersome to use and often require two hands to adjust. Self-adjusting pump pliers automatically adjust to the proper size using one hand.

Klein Tools

Figure 3-15. Diagonal-cutting pliers easily cut through electric cables and conductors.

tubing. Of course, they are also used to tighten and loosen nuts and bolts. They are a bit cumbersome to use and sometimes require two hands to adjust. A self-adjusting style automatically fits the required size and is ideal for frequent change over to different-sized workpieces or one-handed use. Pump pliers are often used in pairs with wrenches.

3.5.1 Specialty Pliers

Specialty pliers are used in addition to the standard types described above and generally have one specific use that is unique to the tool. **End-cutting pliers**, also called *nippers*, are an excellent alternative to linesman's and diagonal

cutting pliers. See **Figure 3-17**. They are used to trim and pull nails and other items close to the surface without damage. In older electrical installations, boxes installed overhead for a luminaires' support were often nailed to the framing member. The nail head in the box is inaccessible to the claw of a hammer, but end-cutting pliers can be used to pull the nail straight out. This is very common when replacing a luminaire box with a ceiling fan-rated box in an existing home.

Klein Tools

Figure 3-17. End-cutting pliers can be used to pull nails where a hammer claw cannot reach.

The *pendant chain tool* has only one purpose—to open and close the chain used to hang pendant luminaires without scratching or otherwise damaging or deforming the chain, **Figure 3-18**. Chain-hung chandeliers are very common in residential dining rooms.

Tin snips are used to cut sheet metal and other thin, rigid materials such as metal straps and the like, **Figure 3-19**. They are not an everyday tool for the electrician and are not technically considered a plier, but they can be very handy in the right circumstances. The color of the handle indicates the direction it can cut. For example, a red handle means the tool cuts to the left, a green handled tool cuts to the right, and a yellow handle indicates it is a combination type that cuts straight, right, or left.

3.6 **Wiring Tools**

Wiring tools are those tools that are unique to the electrical trade and are specifically designed to strip insulation from conductors, cut various sizes of cables and conductors, and crimp assorted fittings on conductors, among other uses.

3.6.1 **Wire Strippers**

Wire strippers, **Figure 3-20**, are designed to remove the insulation from several different wire sizes without damaging the underlying conductor. With practice, you will be able to strip any size wire without even looking at the tool. Most strippers have holes for quickly forming a terminal loop in the conductor. Many have shears for cutting different size screws without damaging the threads. Some stripper models will also remove the outer sheathing of NM (non-metallic) cable without damaging the conductor insulation. Non-metallic cable is a wiring method mostly used in residential wiring and will be discussed further in Chapter 8, *Electrical Conductors and Connectors.*

Goodheart-Willcox Publisher

Figure 3-18. The pendant chain tool is used solely to open and close the chain of a pendant luminaire without marring the finish of the chain.

Klein Tools

Figure 3-19. Tin snips are used to cut thin sheet metal.

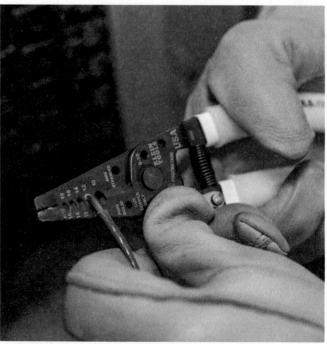

Klein Tools

Figure 3-20. Wire strippers are used to remove the insulation from electrical conductors. Most strip multiple wire sizes with a single tool.

Klein Tools

Figure 3-22. Crimping pliers crimp termination and butt-splice fittings onto conductors.

3.6.2 Cable and Crimping Tools

Cable cutters are used to cut cables or large conductors without distorting the cable or conductor, **Figure 3-21**. They range in size from small, one-handed size for branch circuit cables and conductors to the larger, two-handed model used for service-entrance conductors and cables. The ratcheting style can easily cut nearly any size conductor or cable using one hand.

Crimping pliers crimp bare or insulated terminal devices, **Figure 3-22**. They are also used with crimp sleeves for connecting multiple grounding conductors in an electrical box. *Crimp connectors* are a solderless method of terminating electrical conductors. Most crimping pliers also have conductor cutting capabilities, and some can strip conductor insulation.

Cable rippers quickly slice the outer sheathing of NM cable using a barb, **Figure 3-23**. The user squeezes the tool and pulls the ripper. The sheathing is then cut off with a knife or snippers.

3.6.3 Multipurpose Tools

The electricians' *multipurpose tool*, **Figure 3-24**, combines several tools' functions into a single tool. This tool is used to cut and strip different sizes of solid and stranded conductors. Different sized, color-coded crimping dies are used for insulated and non-insulated crimp terminations. They may also cut small screw sizes without damaging the threads. They work well on cutting bolts to install light fixtures or fans. Carrying a multipurpose tool frees up space in your toolbelt, lightening your load or making room for other tools.

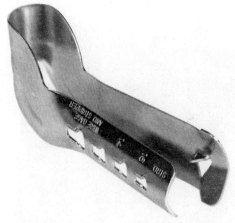

Klein Tools

Figure 3-23. A cable ripper opens the outer sheathing of non-metallic sheathed cable without damaging the insulated conductors inside.

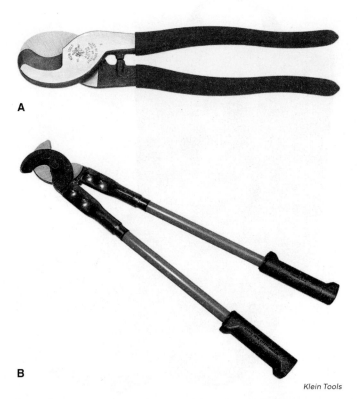

A

B

Klein Tools

Figure 3-21. Cable cutters cut larger sizes of cables and conductors cleanly. A—Hand-held cable cutters. B—Cable cutters.

Klein Tools

Figure 3-24. The electrician's multipurpose tool combines the functions of many tools, such as cutting, stripping, and crimping conductors, in a single tool.

3.6.4 Insulated Tools

Although rare for the residential electrician, at times, one must work on an energized circuit. When this is the case, OSHA requires the use of insulated tools, **Figure 3-25**. *Insulated tools* are covered in a nonconductive coating, which allows the electrician to work on energized circuits more safely. Although the majority of residential wiring is performed on unenergized circuits, at a minimum, a residential electrician should have an insulated screwdriver for working in energized panels.

3.7 Wrenches

Wrenches are tools used for tightening and loosening nuts and bolts. There are many types of wrenches used in electrical work. *Adjustable wrenches*, are a cousin of the pump plier, **Figure 3-26**. They have a moveable jaw that is adjusted using a thumbscrew. They are used to tighten conduit couplings and connectors, tighten pressure-type wire connectors, and remove and hold nuts and bolts. When using adjustable wrenches, hold the handle as far away as possible from the jaws and pull toward you, **Figure 3-27**.

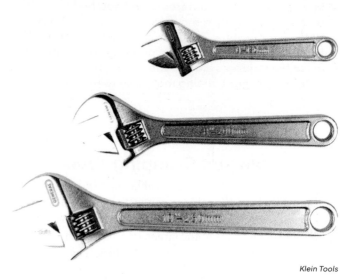

Klein Tools

Figure 3-26. Adjustable wrenches are used to tighten and loosen bolts. They are often used in combination with standard wrenches.

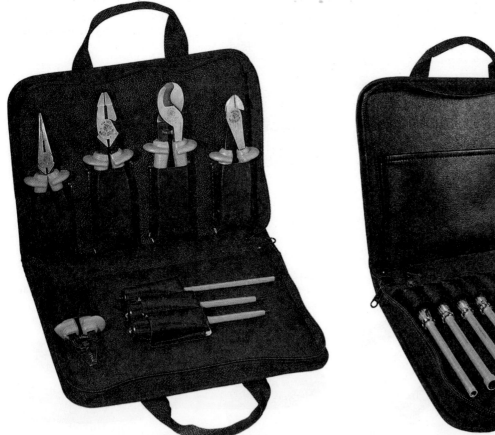

Klein Tools

Figure 3-25. Insulated tools are required when working on energized circuits. The comfort grips on standard tools are not designed to protect against electrical shock.

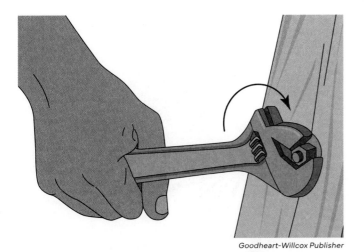

Goodheart-Willcox Publisher

Figure 3-27. Always position the tool properly and pull a wrench toward your body to avoid tool damage and personal injury.

3.7.1 Box and Open-end Wrenches

The box wrench, **Figure 3-28**, and the open-end wrench, **Figure 3-29**, have fixed heads designed to fit a particular sized fastener. A **box wrench** completely encircles, or boxes, the nut or the bolt's head, which allows for a better grip than an open head. An **open-end wrench** has an open end that slips onto the nut or bolt from the side. Most wrenches have a head at both ends of the tool. If both ends are of the box type, they will be different sizes. If one end is a box wrench and the other is an open-end wrench, known as a **combination wrench**, they will likely be the same size.

Box wrenches and open-end wrenches are usually bought in sets and are available in standard and metric sizes. The standard size, referred to as the SAE (Society of Automotive Engineers), is measured in fractions of an inch and inches, while metric wrenches are sized in millimeters. SAE and metric sizes are not interchangeable. Injury or damage can occur from using a metric wrench on an SAE nut or vice versa. The electrician should have a set of wrenches in the most common SAE and metric sizes.

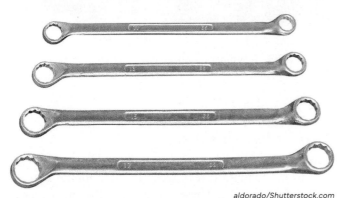

aldorado/Shutterstock.com

Figure 3-28. Box wrenches have ends that completely encircle the fastener. They are available in standard and metric sizes.

Klein Tools

Figure 3-29. Open-end wrenches are quick to fit the fastener and are easy to use.

3.7.2 Ratchet Set

The **ratchet set**, **Figure 3-30**, also called a *socket set* or *socket wrench*, is an alternative tool used for tightening and loosening nuts and bolts. It is a two-piece tool with

Klein Tools

Figure 3-30. Ratchet sets will have an assortment of head sizes to fit a variety of fasteners. Most ratchet sets will come with extenders to reach where a standard wrench cannot.

a ratcheting handle and interchangeable, various-sized socket heads. The ratcheting handle allows you to torque the tool in one direction to engage the fastener while it freely swings in the opposite direction when tightening a bolt or nut. A switch on the handle reverses the action for loosening. This ratcheting action allows you to use the tool more quickly because you do not have to remove and reposition the tool when there is not enough room to turn it in a full circle. The socket heads come in standard and metric sizes. Ratcheting box wrenches, **Figure 3-31**, are also available.

3.7.3 Hex Wrenches

Hex wrenches, **Figure 3-32**, are also called *Allen wrenches* or *hex keys*. They have a hexagonally shaped cross-section that fits the recessed holes in hex fasteners. In electrical work, hex fasteners are commonly found in lugs used to terminate conductors in-service equipment and other electrical components. Hex wrenches are commonly sold in three different forms: sets of assorted sized, individual L-shaped tools; sets of assorted sizes combined into a folding tool; or an individually sized T-handle style, **Figure 3-33**.

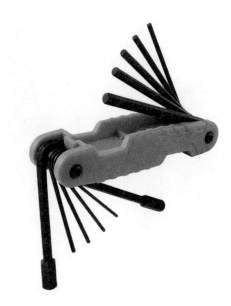

Klein Tools

Figure 3-32. Hex wrenches are used in electrical work to tighten or loosen lugs in meter cabinets and distribution panels.

Klein Tools

Figure 3-33. A typical T-handle hex wrench provides a great deal of torque.

Klein Tools

Figure 3-31. Ratcheting box wrenches can be used in tight spaces to tighten or loosen fasteners.

3.7.4 Torque Wrenches and Screwdrivers

Torque wrenches tighten hex heads fasteners and bolt-type lugs to the manufacturer's recommended torque requirements, **Figure 3-34**. These are used in the main service panel where the conductors attach to the main breaker and in the service and metering equipment. There are two basic designs of torque wrench: the beam-type and the

Klein Tools

Figure 3-34. A torque wrench can be used to apply an exact amount of torque to a fastener. The *NEC* requires all terminations be made to the manufacturer's required torque, if given.

ratcheting-type. The beam-type is the most basic form of torque wrench and is made of two beams. One beam is a lever used to apply the torque to the fastener being tightened and serves as the tool's handle. Attached to the wrench head, the second beam is as free as the indicator beam. The indicator beam's free end travels over a calibrated scale attached to the handle to read the amount of torque being applied.

PRO TIP   **Torque**

Torque is the amount of force or tightness exerted on a fastener and is measured in foot-pounds and inch-pounds. Sometimes the terminology is reversed to pound-feet and pound-inches. There are 12 inch-pounds in one foot-pound. The metric unit of torque is the newton-meter.

The ratcheting-style torque wrench has an internal mechanism that allows the wrench to slip when the desired torque is reached. Torque using the wrench is set by turning a knob or dial on the tool. *Torque screwdrivers*, **Figure 3-35**, are similar to the ratchet-style torque wrench in that they will slip at the preset torque but are used on smaller-sized screws. The *NEC* requires all terminations to be torqued to the manufacturer's specifications.

3.8 Utility Knives

Utility knives, **Figure 3-36**, are a type of cutting hand tool with a folding or retractable blade. Many knives have a magazine that holds extra blades. Some have a feature for changing blades without resorting to other tools to open the knife, eliminating the need for disassembly to change blades. Electricians primarily use utility knives to strip cable sheathing and conductor insulation.

3.9 Tapping Tool

A *tapping tool*, **Figure 3-37**, taps threaded holes for securing equipment to metal, enlarges existing holes, and re-taps damaged threads. The blades on a tapping tool are brittle

Klein Tools

Figure 3-35. A torque screwdriver for tightening screws to the proper torque.

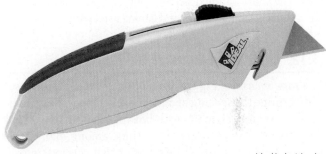

Ideal Industries, Inc.

Figure 3-36. An electrician's utility knife is used to remove the sheathing from non-metallic sheathed cable, among many other uses.

Klein Tools

Figure 3-37. A tapping tool set is used to make standard-sized threads for machine screws. Top—Tapping tool. Bottom—Tapping tool replacement blade.

and likely to break if dropped or otherwise mishandled. There are replacement blades available as needed.

3.10 Scratch Awl

The *scratch awl*, **Figure 3-38**, is a pointed tool used to start screw holes, make pilot holes for drilling, and mark metal, wood, and other materials. The awl is a versatile tool and a good addition to any toolset.

3.11 Hammers

Hammers are available in different sizes and types for various work purposes. An *electrician's hammer*, **Figure 3-39**, is commonly used in electrical work. It has a longer nose than a traditional framer's hammer and has a smooth striking face. They are used to drive and pull nails and staples, pry electrical boxes from framing members, and strike awls and chisels. A proper electrician's hammer should have long, straight claws that easily remove electrical equipment.

The *sledgehammer*, **Figure 3-40**, has a large, flat, heavy head attached to a long handle. The weight of the head and the length of the handle combine to create a tremendous striking force. Electricians use the sledgehammer to pound 8-foot ground rods into the earth. They are also used in demolition work to break up concrete and other tough materials. The short-handled sledge, often called a Thor hammer, has a slightly smaller head, and can be used with one hand.

3.12 Tape Measure

The *tape measure*, **Figure 3-41**, often called a *tape*, is used to determine the spacing of outlet boxes along walls, to measure the proper height of outlet box installation, or to check measurements on construction prints, among many other uses. A tape has a rolled metal ruler that retracts into its case. Most are 25-35 feet in length and have a standout of up

Klein Tools

Figure 3-39. The electrician's hammer has a long, straight claw that aids in removing electrical boxes.

A

B

Apex Tool Group, LLC

Figure 3-40. Sledgehammers are used by electricians to drive the 8′ ground rods into the earth or for light demolition of concrete or masonry block. A—A sledgehammer. B—A short handled sledgehammer.

Klein Tools

Figure 3-38. The scratch awl can be used to start screw holes in wood or sheetrock and to make markings on different materials.

Klein Tools

Figure 3-41. The tape measure is used during the rough-in stage to determine the proper placement of device boxes and to check measurements on plans. Most are 25-35 feet in length.

to 10 feet or more. They measure feet, inches, and fractions of an inch in 1/2″, 1/4″, 1/8″, and 1/16″ increments. A tape measure with a magnetic tip is made for use on metal studs but can also be used to retrieve a dropped screw from the top of a ladder. An open reel field tape, **Figure 3-42,** can measure distances of over 300′.

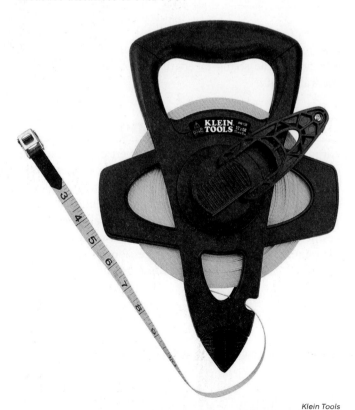

Klein Tools

Figure 3-42. Open-reel field tapes can measure distances up to 300 feet or more.

3.13 **Saws**

A *jab saw*, **Figure 3-43,** also known as the *sheetrock saw* or *keyhole saw*, is used to cut holes for device boxes in sheetrock and thinner wood paneling and can also be used to make holes for toggle bolts. It has a sharp, pointed tip and aggressive teeth. A saw blade is rated on its level of aggressiveness. *Aggressiveness* indicates how aggressive, or intense, a saw cuts based on the number of teeth it has. A saw blade with few *teeth per inch (tpi)* is considered more aggressive than one with more teeth per inch. A saw blade with more teeth per inch results in a finer cut, but it takes more strokes of the saw than a blade with fewer teeth per inch. A saw blade with fewer teeth per inch results in a rougher cut.

A *hacksaw*, **Figure 3-44,** is used to cut metal or PVC conduit, strut channel, and larger conductors and cables. It has a fine-toothed, replaceable blade. Some blades have more teeth at the front of the blade to help get the cut started. **Figure 3-45** shows the blade starts with 32 teeth per inch, then 24 tpi, and ends with 18 tpi. A quality hacksaw has a thumbscrew or crank for tensioning the blade. When replacing a hacksaw blade, make sure the blade is facing the proper direction; the cut stroke is the forward stroke.

3.14 **Wire Fishing Tools**

When adding circuits, switches, or receptacles to an existing home, electricians are often called upon to "fish" cables through the voids in walls and ceilings or pull conductors

Klein Tools

Figure 3-43. Jab saws are used to cut holes in sheetrock and thin paneling in remodel work.

Klein Tools

Figure 3-44. Hacksaws are used to cut metallic and nonmetallic conduit and large cables and conductors.

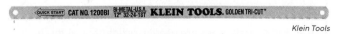

Klein Tools

Figure 3-45. On some hacksaw blades, the number of teeth per inch varies along the blade's length.

through a conduit. There are several tools designed specifically for this purpose.

In residential wiring, the *fish tape*, **Figure 3-46**, is used in pushing or pulling electric wire through thermal installation and "fishing" conductors through the voids in wall cavities and conduit. Some electricians call the fish tape a *snake* or *snake wire*. It is made of either steel or nonconductive fiberglass in lengths from 25′ to over 200′ and is very rigid and flexible. The tape is stored and deployed from a tough plastic reel. The tip of the tape has a loop or hook for attaching the conductor. Using the fish tape in walls and ceilings can be challenging and is a skill that takes practice and patience to master.

Another wire fishing tool is fiberglass rods, **Figure 3-47**, commonly referred to as fish sticks. **Fish sticks** have threaded ends that allow the individual pieces to be coupled into a single, longer length. They usually come in a case containing several lengths of 4- or 5-feet rods. Some are made of glow-in-the-dark material to make them easy to locate in dark attics or crawl spaces. An assortment of tips can be installed depending on the application. Blunt tips with an attachment hole, hooks, and even lights are common. These are a must-have tool for remodeling work.

When installing a switch or receptacle in a finished wall, fishing the conductors can be accomplished using a weighted string or small chain that is dropped through a hole in the wall frame's top plate. The free end is retrieved

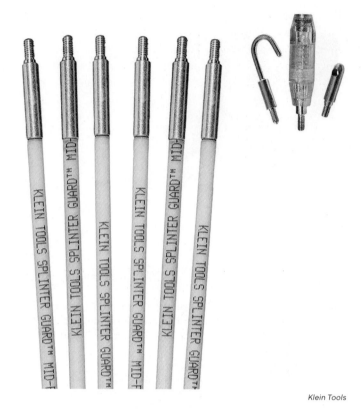

Klein Tools

Figure 3-47. Fish sticks are an alternative to fish tape. Fish sticks come in 3′ to 5′ lengths that can be coupled together. In addition, several types of tips are available such as hooks and rings.

Klein Tools

Figure 3-46. Fish tapes are used in residential wiring to "fish" conductors through wall and ceiling cavities and through conduit.

through the hole cut for the electrical box. The cable is attached to the string or chain and pulled back through the wall cavity.

3.15 Conduit Bending Tools

Although used primarily for commercial or industrial electrical installations, it is not uncommon for residential electricians to install conduit in certain instances. Electrical metallic tubing (EMT) is used in residential wiring to protect exposed cables from damage. It can also be used for aesthetic purposes. Thus, learning proper conduit bending techniques is necessary for a professional-looking installation. EMT is bent using a **hand bender**, **Figure 3-48**.

Klein Tools

Figure 3-48. The EMT hand bender makes bends in EMT conduit to the proper *NEC* radius.

Conduit comes in several different sizes, and each size conduit has its corresponding bender. PVC conduit may be used in underground electrical installations. PVC is heated with a heat gun or heating blanket, **Figure 3-49**, until pliable and bent to the desired angle.

3.16 Plumb Bobs, Lasers, and Levels

Plumb bobs have been used for thousands of years to ensure buildings and structures are vertically true or plumb to the earth. See **Figure 3-50**. The *plumb bob* is a weight with a pointed bottom supported at the top by a length of string. An electrician may use the plumb bob to position a luminaire in the ceiling over a particular point below. In some instances, it may be easier to lay out the entire lighting plan on the floor and then transfer the points to the ceiling using a plumb bob. These materials are simple to use and inexpensive to purchase.

The plumb bob is being replaced by a *laser plumb*, which shoots a laser to transfer points from floor to ceiling. See **Figure 3-51**. Some models have multiple lasers to align objects to several points. Others use a rotating laser to "paint" a level line on a wall.

A *torpedo level*, **Figure 3-52**, is a small leveling tool used to check electrical conduit, equipment, appliances, switch cover plates, and receptacles for level and plumb. When an item is *level*, it is parallel with the floor or ground. When an item is *plumb*, it is perpendicular, or at a right angle, to the floor or ground. Most torpedo levels have one magnetic edge and a grooved edge to sit on the conduit. Ensure the magnetic edge is clear of metal debris before using, or you

BaLL LunLa/Shutterstock.com

Figure 3-49. PVC conduit can be bent using a heat gun or heat blanket. This figure shows a worker using a heat gun to bend PVC conduit.

NP27/Shutterstock.com

Figure 3-50. Plumb bobs can be used to transfer a point on the ceiling to its corresponding point on the floor or vice versa.

Klein Tools

Figure 3-51. The laser plumb is the modern version of the plumb bob and uses lasers to transfer points.

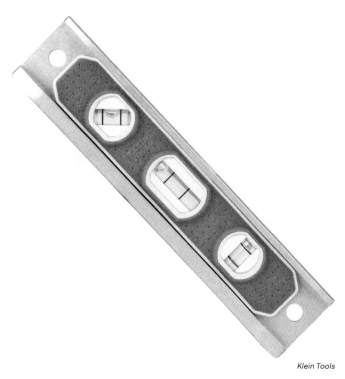

Figure 3-52. Torpedo levels are used in residential wiring to ensure equipment is installed level and true.

will get an inaccurate reading. The *NEC* requires all installations to be completed in a neat and workmanlike manner.

Torpedo levels are a type of "spirit level" in which alcohol partially fills a vial, leaving a bubble to float around freely. The vial has two markings that are about the size of the bubble. When the bubble rests between the lines, the level has "bubbled out," and the object being measured is level or plumb. If the bubble is to the left of the center, then the object slopes downward left-to-right ("downhill"). A bubble on the right indicates a slant upward slant from left to right ("uphill"). The level will generally have three vials, one for level (horizontal trueness), one for plumb (vertical trueness), and one at a 45° angle. There are also specialty levels designed explicitly for bending conduit with vials at 0°, 90°, 45°, and 30°. See **Figure 3-53**.

Figure 3-53. The electrician's level is another type of level used primarily for conduit bending.

3.17 Metal Files and Wood Rasps

Metal files and rasps are simple tools used to shape certain materials. *Metal files*, **Figure 3-54**, have closely spaced, hardened-steel grooves that shape, trim, and smooth any material made of metal, wood, or plastic. Metal files can also sharpen tools and some drill bits, deburr conduit, and cut and shape metal. *Rasps*, **Figure 3-55**, are files made specifically for shaping wood. Their coarse teeth are punched up from the steel surface. Wood rasps can enlarge holes cut or drilled in wood.

3.18 Demolition Tools

Prybars and crowbars, **Figure 3-56**, are used to pull or pry two objects apart. Electricians often use prybars and crowbars in a residential attic to remove floorboards from the joists to access the ceiling below. Prybars are generally made of wide, flat metal and are smaller and shorter than the octagonal or round crowbar. Prybars are generally 18 inches or less, whereas the crowbar is three feet or more in length. The bends in the bar provide extra leverage, and these tools can exert tremendous pressure.

A *cat's paw* is a type of pry bar used to remove embedded nails from wood. It has a split, chiseled tip that is driven with a hammer under the nail head. The cat's paw will damage the wood around the nail and should only be used where this damage is acceptable.

Figure 3-54. Metal files are used to remove burrs from freshly cut metals and sharpen some drill bits, such as auger bits.

Figure 3-55. Wood rasps are mostly used in remodel work to enlarge holes for device boxes that are cut into wooden baseboards.

Apex Tool Group, LLC

Figure 3-56. The pry bar and crowbar are used to remove floorboards and baseboards to access the interior of the wall or ceiling. This photo is showing a pry bar.

The **masonry chisel** is a demolition tool used to break up concrete, brick, or tiles. The chisel is generally held with one hand and struck with a hammer. Never strike a chisel with a deformed head, as the chisel may shatter, causing injury to the user.

3.19 **Patching and Cleaning**

Electricians are often called to patch holes made in walls and floors. Thus, always have patching tools and materials kit with you or on every service vehicle, which includes putty knives, wood filler, sheetrock spackle, and an assortment of wood stain touch-up pens.

It does not matter how good a job you have done if you do not clean up after yourself. Cleaning up after oneself shows you respect the homeowner and take pride in your work. Hand smudges on walls should be cleaned with an all-purpose cleaner and paper towels. A broom and a dustpan or a filtered shop-vac are used to clean up sheetrock dust and sawdust. Whether service work or new construction, the condition you leave your jobsite has a lasting impression on the customer and is likely to be the one thing they remember.

Summary

- Most accidents caused by hand tools are a result of misuse or improper maintenance.
- Electricians keep their hand tools organized in an electrician's tool belt.
- Screwdrivers are tools used for the installation and removal of screws and come in different tip sizes and styles.
- Lineman pliers, also called side-cutting pliers, are heavy pliers used to cut cables, conductors, and small screws, and pull and hold conductors.
- Diagonal cutting pliers are used for cutting cables and conductors in tight spaces.
- Pump pliers have a sliding jaw that locks in place at intervals to accommodate different sizes of nuts, bolts, or other items.
- Wire strippers are designed to remove the insulation from several different wire sizes without damaging the underlying conductor.
- Cable cutters are used to cut cables or large conductors without distorting the cable or conductor.
- Insulated tools are covered in a nonconductive coating, which allows the electrician to work on energized circuits more safely.
- The box wrench and the open-end wrench have fixed heads designed to fit a particular sized fastener.
- An electrician's hammer has a longer nose than a traditional framer's hammer and will not have a waffle face.
- The tape measure is used to determine the spacing of outlet boxes along walls, measure the proper height of outlet box installation, and check measurements on construction prints.
- The jab saw is used to cut holes for device boxes in sheetrock and thinner wood paneling.
- Hacksaws are used to cut metal or PVC conduit, strut channel, and larger conductors and cables.
- A saw blade is rated on its level of "aggressiveness" based on the blade's number of teeth per inch.
- The fish tape, made of steel or fiberglass, is used to pull cables through the voids in walls or ceilings or to pull conductors through a conduit.
- A hand bender is a tool designed to bend electrical metallic conduit.
- PVC conduit is heated by a heat gun or heating blanket until pliable and is then bent to the desired angle.
- When an item is level, it is parallel with the floor or ground. When an item is plumb, it is perpendicular, or at a right angle, to the floor or ground.

- The torpedo level is a small leveling tool used to check electrical conduit, equipment, appliances, and cover plates for switches and receptacles for level and plumb.
- Metal files have closely spaced, hardened steel grooves that shape, trim, and smooth anything made of metal, wood, or plastic. Rasps are files with coarse teeth made specifically for shaping wood.
- Demolition tools used by residential electricians include crowbars and pry bars, the cat's paw for removing embedded nails, and the masonry chisel for breaking concrete, brick, and tile.
- Residential electricians should keep a patching kit to repair holes that were intentionally or unintentionally made in walls and floors.

Know and Understand

1. *True or False?* Most accidents caused by hand tools are a result of misuse or improper maintenance.
2. The standard size Phillips head screwdriver is the _____.
   - A. #1
   - B. #2
   - C. #3
   - D. #4
3. Electricians can use _____ to keep their tools organized.
   - A. tool belts
   - B. toolboxes
   - C. backpacks
   - D. All of the above.
4. Small screwdrivers with tiny tips are called _____.
   - A. screw-holding drivers
   - B. quick-rotating drivers
   - C. stubby screwdrivers
   - D. precision screwdrivers
5. *True or False?* A flathead screwdriver can be used as a chisel with no damage to the tool.
6. *True or False?* Most nut driver sets are color-coded by size.
7. Heavy pliers used to cut cables, conductors, and small screws are called _____.
   - A. linesman's pliers
   - B. long-nosed pliers
   - C. pump pliers
   - D. diagonal cutting pliers
8. The _____ is used to grip or retrieve items in tight spaces.
   - A. linesman's pliers
   - B. long-nosed pliers
   - C. pump pliers
   - D. diagonal cutting pliers
9. *True or False?* Pump pliers and adjustable wrenches are two terms for the same tool.
10. The tool designed to strip insulation from several different wire sizes without damaging the underlying conductor is the _____.
    - A. cable ripper
    - B. wire stripper
    - C. crimping plier
    - D. None of the above.
11. *True or False?* If both ends of a wrench are of the box type, they will be different sizes.
12. *True or False?* There is no need for an electrician to have a set of both SAE and metric wrenches because they are interchangeable.
13. A two-piece tool that uses a ratcheting handle and interchangeable, different-sized socket heads is the _____.
    - A. box wrench
    - B. open-end wrench
    - C. ratchet-box wrench
    - D. ratchet set

14. *True or False?* The electrician's hammer should have long, curved claws to simplify the removal of electrical equipment.
15. The smallest increment on a tape measure is _____.
    - A. 1'
    - B. 1"
    - C. 1/2"
    - D. 1/16"
16. *True or False?* An aggressive saw blade will have many teeth per inch.
17. PVC conduit is _____ until it is pliable enough to be bent.
    - A. hilled
    - B. heated
    - C. notched
    - D. rolled
18. A weight with a pointed bottom and supported by a length of a string is called a _____.
    - A. cat's paw
    - B. fish tape
    - C. plumb bob
    - D. spirit level
19. *True or False?* A rasp is a type of file designed specifically for shaping metal.
20. A type of pry bar used to remove embedded nails from wood is called a _____.
    - A. cat's paw
    - B. masonry chisel
    - C. crowbar
    - D. plumb bob

Apply and Analyze

1. How does carrying a multitool make more space in your toolbelt?
2. What is the function of the torque wrench and torque screwdriver?
3. Describe the features of an electrician's hammer.
4. Explain the concept of "fishing" wires as it applies to residential wiring.
5. Demonstrate an understanding of the concepts of level and plumb.
6. Why is it essential for a residential electrician to carry cleaning supplies?

Critical Thinking

1. Explain what is meant by the term "aggressiveness" as it pertains to saw blades.

CHAPTER **4** | # Power Tools

Chepko Danil Vitalevich/Shutterstock.com

LEARNING OBJECTIVES

After completing this chapter, you will be able to:

- List some rules of power tool safety.
- Describe the advantages and disadvantages of corded and cordless power tools.
- Compare the different types of power cutting tools.
- Explain the different types of drills and their uses.
- List the parts of a rod-type drill bit.
- Distinguish between the various types of drill bits and their applications.
- Discuss basic drilling techniques.

TECHNICAL TERMS

auger bit
bit extender
body
brad-tip bit
chuck
circular saw
clutch
cordless screwdriver
driver bit
flex bit
general-use drill
hammer drill
high-speed steel (HSS)
hole saw
impact driver
jigsaw
low-speed, high-torque drill
masonry bit
oscillating tool
paddle bit
point
reciprocating saw
right angle drill
rod-type bit
rotary hammer drill
saw bit
self-feed bit
shank
step bit
stripped screw extractor
tear-out and blow-out
tile bit
twist bit

Introduction

The previous chapter introduces the types of hand tools used in residential electrical work. Unlike a hand tool, a power tool uses a form of energy to operate other than the user's physical strength. Most power tools use an electric motor, although some are powered by compressed air or other means. Power tools are either portable or stationary. Portable power tools are hand-held tools such as drills or jigsaws. Stationary power tools include items such as table saws and lathes. This chapter focuses on the typical portable power tools used by residential electricians. These tools are used on nearly every residential wiring job, and it is unimaginable to today's electricians that power tools were not always available.

4.1 Power Tool Safety

Power tools are very dangerous, which means extra care and caution must be used when operating them. These tools generally operate at extremely high speeds, and catastrophic accidents can happen in an instant, resulting in severe injuries. It is critical the user understands how to operate the tools properly and safely. Follow these rules to ensure proper tool safety:

- Always wear the appropriate PPE, such as eye protection and gloves, when using power tools.
- Read and understand the operator's manual that comes with every power tool.
- Do not use power tools in wet conditions.
- Plug corded power tools into properly installed grounded receptacle outlets.
- All 120-volt corded power tools should be plugged into GFCI receptacles at a residential construction site.

- Disconnect tools when changing accessories such as saw blades and drill bits with drills that use a chuck key.
- Always switch off the power tool before plugging it in.
- Keep hands and other body parts away from all cutting edges and moving parts.
- Do not force a power tool; let the tool do the work.

4.2 Corded and Cordless Power Tools

Most portable power tools are available in corded and cordless versions. There are benefits and drawbacks for each type, with the tradeoff being one of power compared to convenience. A corded power tool delivers more power and is more lightweight than an equivalent cordless tool. The corded power tool must be plugged into a power source, whereas the cordless tool has a rechargeable battery for its power. A corded tool can be used nearly continuously with no downtime, while the cordless tool's batteries need periodic charging throughout the day. Many cordless tools are supplied with multiple batteries, so one can be charging while the other is in use.

Cordless power tools are available in many battery voltages and capacities. The higher the battery's voltage, the more power the tool can deliver. Capacity is like the battery's gas tank and is measured in amp-hours, abbreviated Ah. Amp-hours refer to a battery's capacity or how long it will stay charged while in use. When using a corded cutting tool such as a jigsaw or a circular saw, care must be taken not to cut the cord. The length of the power cord limits the corded tool's use. Using an extension cord that is too long or has too small a gauge can cause an under-voltage and permanently damage the tool.

4.3 Power Cutting Tools

There are many types of power saws used in electrical work. *Reciprocating saws* are saws where the blade moves back and forth to make a cut, **Figure 4-1**. They are most often used during demolition—it is not the go-to saw for precision work. Sawzall® is a product name from Milwaukee® tools and is now universally applied to all reciprocating saw brands. An important distinction among power saws is their cutting speed. The saw's cutting speed is determined by the amount of force applied to the trigger. The faster the blade moves, the faster the cutting action.

Blades are interchangeable among brands, and most saws today have toolless, quick-changing blades, **Figure 4-2**. There are other attachments for reciprocating saws that extend the tool's capability beyond cutting, including scrapers, sanders, and scouring brushes, **Figure 4-3**.

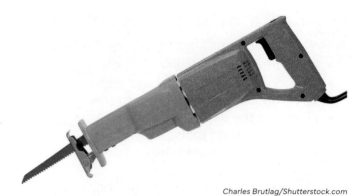

Charles Brutlag/Shutterstock.com

Figure 4-1. Reciprocating saws are used to cut wood, metal, and other materials.

Courtesy of Milwaukee Electric Tool Corporation

Figure 4-2. Reciprocating saw blades are generally interchangeable between brands. Different types of blades are used on different materials.

Jigsaws are a smaller type of reciprocating saw that make cuts with more precision than the larger reciprocating saw, **Figure 4-4**. The saw's base, often called the foot, is adjustable to allow for cutting angles. Similar to the reciprocating saw, the blade's speed can be controlled by the amount of force applied to the trigger. However, jigsaws have a wider range of speeds and can be set by turning a dial, usually located on the trigger. In addition to the reciprocal action, the blade has an optional range of orbital motion that allows the blade to move forward on the upstroke, creating a greater cutting force. The orbital motion is generally selected with a switch on the tool. T-Shank blades are the most common, but there is variation among manufacturers, **Figure 4-5**.

Circular saws use a disk-shaped blade with a spinning, rotary motion to cut different materials, such as wood, plastic, metal, concrete, and other masonry items, **Figure 4-6**.

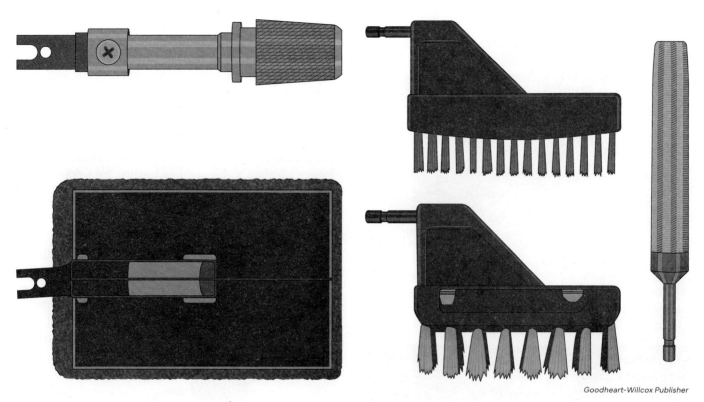

Goodheart-Willcox Publisher

Figure 4-3. Reciprocating saw attachments extend the ability of the reciprocal saw beyond cutting to include scraping, scrubbing, and filing.

przymat/Shutterstock.com

Figure 4-4. Jigsaws are used where more precision is needed or for lighter-duty cutting than the reciprocal saw.

mihalec/Shutterstock.com

Figure 4-5. Jigsaw blade types. Different manufacturers use different types of blade shanks. Universal blades combine the T-shank with the U-shank.

The standard saw blade is 7 1/4″, although other size models are available. The base of the saw can be adjusted for blade depth and beveling. The blade depth should be set so that it is only slightly beyond the thickness of the workpiece. The spring-loaded blade guard should *never* be removed, wedged open, defeated, or bypassed in any manner.

The *oscillating tool* uses a slight, high-speed side-to-side motion to cut holes for switch and receptacle boxes in a variety of wall surfaces, **Figure 4-7.** The stroke is about 3° at about 20,000 strokes per minute and feels more like a vibration. The oscillating tool is so versatile and easy to use that it is replacing the jigsaw for many uses.

phoMAKER/Shutterstock.com

Figure 4-6. Circular saws are used to cut lumber and plywood.

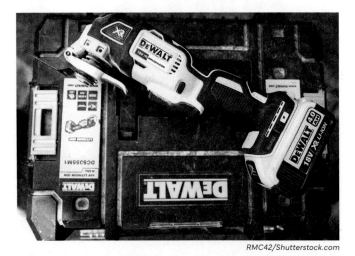

RMC42/Shutterstock.com

Figure 4-7. The oscillating tool is replacing the jigsaw for many applications, including cutting sheetrock and paneling to add device boxes in remodel work.

4.4 Drills

Drills are the most common power tool used in electrical work. There are many types of drills and drill bits, and each has its specific purpose and use. Some models are designed to perform multiple functions. It is not uncommon for an electrician to have five or more different types of drills on the truck. A quality, heavy-duty, professional drill will be more expensive to purchase but will be better built and have a longer life than a cheaper, low-quality, light-duty home-owner's drill. Do not skimp on your tools!

4.4.1 General-Use Drills

Most electricians will have both corded and cordless drills on the truck for general use. *General-use drills* are your every-day go-to drill for most tasks, **Figure 4-8**. Most general-use drills have a pistol grip, although D-handle and combination grips are also available, **Figure 4-9**. The general-purpose drill can be used to drill holes in various materials

Dinga/Shutterstock.com

Figure 4-8. A cordless, general-use drill runs on rechargeable batteries.

A

B

Courtesy of Milwaukee Electric Tool Corporation

Figure 4-9. D-grip handles (A) and pistol grip (B) on a drill are a matter of personal preference.

and to drive and remove screws. Electric drills are generally classified as high-speed and can deliver speeds that range from 2000–4500 rpm. Most have reversible, variable-speed motors in which the drill's speed is related to the force put on the trigger.

Where and how you use the drill will determine whether a cordless or corded drill is used. A cordless drill offers greater mobility but requires periodic charging. Most will come with an extra battery, but it still takes longer to charge a battery than drain it, so downtime is a real possibility. A corded drill can be used continuously with no downtime but requires a receptacle to plug it in. The corded drill is most used when drilling multiple holes in the framing members during rough-in, while a cordless drill might be more convenient to use in the attic or crawlspace during remodel work.

4.4.2 Special-Use Drills

Special-use drills have features that set them apart from general-use drills. They may be operational or design features that allow the drill to perform functions that a general-use drill cannot accomplish. As with general-use drills, many special-use drills also come in corded and cordless models.

Hammer Drills

Hammer drills, **Figure 4-10**, use a "percussive" motion, which better penetrates brick or concrete. They can make holes of up to 1/2″ or more in diameter. They are used when installing materials on brick or concrete with the use of anchors. Some hammer drills will include a depth-stop rod, **Figure 4-11**, that can be adjusted to the desired depth. Some drills are exclusively hammer drills, although many general-purpose drills will have a hammer-drill setting.

Goodheart-Willcox Publisher

Figure 4-11. The depth-stop rod is used to drill holes to a certain depth.

Low-Speed, High-Torque Drills

Low-speed, high-torque drills, **Figure 4-12**, drill larger holes in hardwood types. Other trades use them for mixing paint, plaster, and mortar. These drills are physically larger than other drills because they use gearing to slow the rotation, which provides a considerable turning force or torque. Most have two speeds: 0–300 RPM and 0–1200 rpm. An extended handle is included on the drill for leverage against higher torque and provides stability and control for the user.

Right Angle Drills

Right angle drills, **Figure 4-13**, drill holes between framing studs or in other tight places. The angled head gives them a slimmer profile so they can fit where other drills will not. Most right-angle drills are standard, general-purpose drills

Courtesy of Milwaukee Electric Tool Corporation

Figure 4-12. Low-speed, high-torque drills use internal gearing to slow the bit rotation, which allows for more torque. These drills can drill large diameter holes in thick lumber without bogging down.

Courtesy of Milwaukee Electric Tool Corporation

Figure 4-10. The hammer drill is used to drill holes in hard surfaces such as brick, masonry, and stone.

Courtesy of Milwaukee Electric Tool Corporation

Figure 4-13. Right-angle drills can reach into tight spaces such as between wall studs.

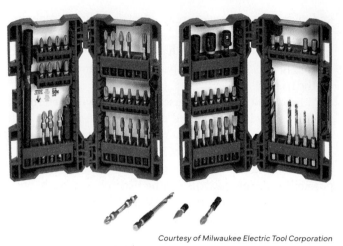

Courtesy of Milwaukee Electric Tool Corporation

Figure 4-15. Impact drivers use a percussive action to help drive screws and fasteners into wood. Shown is a driver set.

with a right-angle function adapter. They are used extensively in the rough-in stage of new construction wiring.

Rotary Hammer Drills

Rotary hammer drills, **Figure 4-14**, are used for heavy-duty masonry work and are much larger and more powerful than the standard hammer drill. They are more like a small, handheld jackhammer. They can drill, core, and chisel concrete, stone, and similar materials. Most have the capability to switch between a hammer-only function for chiseling, a hammer-drill setting for drilling holes up to 1 1/2″ diameter or more, and drill-only function for coring larger diameter holes. Some models can drive ground rods with a special bit. Examples of rotary hammer drill bits include various chisel types like a spade. A ground rod driver, core drilling up to 6″, and drill bits up to 2″ diameter and 22″ in length are also available.

Impact Driver and Cordless Screwdrivers

The *impact driver*, **Figure 4-15**, drives or removes screws. They have a typical high-speed drill's rotary function until more torque is needed to drive in a screw. At this point, a

hammering impact action is automatically engaged. The impact action increases the amount of torque applied and allows it to drive large, long lag screws.

Cordless screwdrivers, **Figure 4-16**, are more portable and can be used to drill small-diameter holes, such as pilot holes for screws, without needing a cord to plug them in. Nearly all these tools come with a 1/4 ″ quick-change chuck.

4.4.3 Drill Chucks

A *chuck* is the part of the drill that grasps a drill bit. The standard drill will have either a keyless chuck or a keyed chuck. Keyed chucks are generally made of solid metal and are very durable. The use of the key allows for a tighter grip on the bit. Turning the key opens or closes the jaws of the chuck. Keyless chucks are operated by hand and allow for a quicker change of bits. The user grabs the chuck with one hand while running the drill forward or reverse to open or close the chuck.

Drill chucks are sized by the maximum diameter drill bit. They accept either 3/8″ or 1/2″. Electricians should have a drill with a 1/2″ chuck. The quick-change chuck has a spring-loaded collar that is slid back, allowing the user to swap bits easily. They only accept quick-change bits, which will be discussed later in this chapter.

4.4.4 Other Drill Features

Some drills allow for additional features to be included. Some combine the features of several drills into a single unit. Other features make the drill easier to use; these features are mostly on cordless drills but may be included on some corded drills. The function selection is to choose different drill settings.

A *clutch* only engages in screw-driving mode and uses internal gearing to control the amount of torque delivered,

Courtesy of Milwaukee Electric Tool Corporation

Figure 4-14. The rotary hammer drill is a heavy-duty tool that can drill, core, and chisel concrete, masonry, and stone.

A

Figure 4-17. Drill clutch adjusts the amount of torque delivered in the screw-driving mode. The clutch will slip at the preset torque to prevent driving the screw too far into the material.

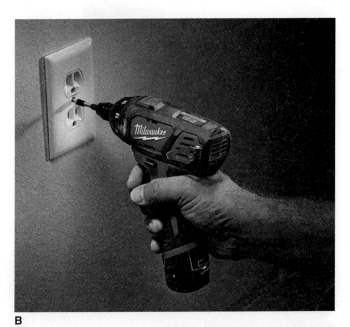

B

Figure 4-16. A—A hex cordless screwdriver. B—A cordless screwdriver used to secure a lighting outlet.

Figure 4-17. The drill slips at the preset torque to prevent driving the screw too deep into different materials. Sheetrock workers use drivers with a clutch to set the sheetrock screws at the proper depth. The speed selector sets the maximum speed of rotation. Higher drilling speeds produce less torque and drain the batteries of cordless drills more quickly.

Nearly every drill available today has a reversing switch to reverse the direction of rotation and can be used to remove driven screws or remove a hole-making bit easily. Many will also have a trigger lock button that keeps the drill running when you release the trigger. The drill is stopped by fully depressing and releasing the trigger. Extreme caution should be exercised when using the trigger lock function, as severe accidents can occur in the blink of an eye. Always make certain the trigger lock is off before plugging in the drill.

4.5 Drill Bits and Their Uses

Drill bits are used with the drill to make holes in different materials. Although there are many different types of drill bits, they all operate in a similar manner to cut the hole and dispel the cutting debris. A drill bit's size refers to the diameter of the hole it can cut. There is a seemingly endless selection of drill bits for innumerable uses. Understanding the types of bits and their uses is essential to the residential electrician. Using the incorrect bit for the job can ruin the bit, damage the toll, and waste time and money.

4.6 Drill Bit Types

Most drill bits are made from **high-speed steel (HSS)**. High-speed steel is a type of steel. It is *not* the speed at which the bit is used. In fact, when drilling hardened metals like stainless steel, a slow rotation is required, or the bit is not cut. A drill bit must be a harder material than the material it is drilling into, or the bit wears down rather than cutting into the material. Stainless steel is a challenging material to drill and requires a cobalt bit. A bit made from a harder metal also breaks more easily, especially under pressure. A good rule for any drilling is to let the drill do the work. Do not push too hard!

The *rod-type bit*, **Figure 4-18**, is made up of three basic parts:

1. **Shank**—The part grabbed by the jaws of the drill chuck. The shank is usually round but may have a hex configuration that allows the chuck's jaws to grab them more firmly and prevent slippage.
2. **Body**—The part between the shank and the point. It has the spirals, called "flutes," that remove the dust and chips.
3. **Point**—The part of the drill bit that does the actual cutting.

Kutcher Serhii/Shutterstock.com

Figure 4-18. A rod-type bit comprises a shank, body, and point.

A ***twist bit*** is the most common rod-type drill bit and is used for drilling into soft materials like wood, soft iron, and aluminum. Cobalt-tipped bits can drill harder materials, such as stainless steel. Twist bits are often sold and organized according to size in a case called *an index*. Twist bits can be sharpened with a special tool.

The ***brad-tip bit*** is for drilling wood. The center point keeps the bit from "wandering," and the pointed flutes produce a cleaner exit point. They have wider flutes to expel large chips. Typically, they are made from high-speed steel. The gold-colored bits, **Figure 4-19**, have a titanium coating to reduce friction and prolong sharpness.

At first glance, a ***masonry bit*** looks like an ordinary twist bit, but the tips are different. Masonry bits have a flared, tungsten-carbide cutting tip, **Figure 4-20**. Masonry bits generally are used with a hammer drill to drill into hard surfaces such as concrete, brick, or stone.

Auger bits, also called *ship auger bits*, are exclusively for drilling wood, **Figure 4-21**. These bits have a screw tip that pulls the bit through the wood, making a very clean hole.

Photo Win1/Shutterstock.com

Figure 4-20. Masonry bit with tungsten-carbide tip. The flared tip distinguishes the masonry bit from the standard twist bit.

Courtesy of Milwaukee Electric Tool Corporation

Figure 4-21. The auger drill bit has wide flutes to expel the large chips created when drilling.

They make larger chips, so the flutes are larger. They can be as long as 18″ or more. With a powerful drill, you could make a hole through wood blocks that are just a bit smaller than the length of the bit. The larger flutes also allow for deeper boring because they can remove more chips. When making deep bores, it is good practice to periodically pull the drill out of the hole to expel more chips.

Paddle bits, also called *spade bits* or *butterfly bits*, have a wide, flat profile for boring wood, **Figure 4-22**. They generally splinter the wood as the bit exits, but some brands have spurs on the cutting tip edges to minimize splintering. Spade bits have pointed tips to start the hole. Some have screw tips to help pull the bit through the wood. Paddle bits are easily sharpened with a regular metal file.

Another wood-only bit, the ***self-feed bit***, **Figure 4-23**, is used to bore through framing lumber such as 2×4s or larger. They come in many sizes, but they are mainly used

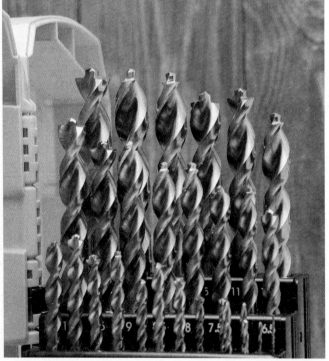

PromKaz/Shutterstock.com

Figure 4-19. Titanium-coated brad-tip drill bits. The brad tip keeps the bit from "wandering" when starting a hole.

Charoen Krung Photography/Shutterstock.com

Figure 4-22. Example of a paddle bit.

Rvector/Shutterstock.com

Figure 4-23. A self-feed bit is used to bore large diameter holes in lumber and is typically used with a low-speed, high-torque drill. The threaded tip helps pull the bit through the wood.

Courtesy of Milwaukee Electric Tool Corporation

Figure 4-24. Hole saws are used to cut different diameter holes in thinner materials such as sheetrock, paneling, and plywood.

by electricians to make large-diameter holes of 2″ or more. These bits have a screw tip to pull them through the wood. They also have cutting teeth on the edges that cut the wood and one or more chisel faces to scoop out the chips. They have puck-shaped cutting heads but no flutes, so the user must pull the bit back periodically to remove chips and dust. Failure to do so may cause the bit to become jammed in the hole from the waste material build-up.

Hole saws are a cylindrical saw blade used in power drills, **Figure 4-24.** The blade is mounted on an arbor with an attached pilot bit that acts as the center of the hole and stabilizes the saw. Electricians use the hole saw to cut through thinner material such as plywood siding and roof decking; however, they are cumbersome to use in framing material. Hole saws can also cut through the soft steel sheet metal used in many electrical cabinets. They also tend to "kick back," so a firm grip on the drill is critical. Once the material is penetrated, the hole saw contains a slug that can be removed with a screwdriver through the slots in the side.

Special hole saws are available that have an abrasive edge instead of saw teeth. These are made for cutting into plaster walls and ceilings that would wear down the teeth of a standard hole saw.

4.7 Miscellaneous Drill Bits

Step bits, also called *unibits*, have a stepped, conical shape that drills a wide range of hole diameters, generally in sheet metal, **Figure 4-25.** Many commercial electrical cabinets do not come with preformed knockouts, and the installer must create them. The smaller diameter step bits come to a point and are self-starting, while the larger bits have more blunt tips and are generally used to enlarge existing holes. They can also be used to deburr existing holes drilled with regular bits.

Driver bits, **Figure 4-26**, are available in any tip style that screwdrivers come in. It is faster and easier to drive screws using a drill or a cordless screwdriver than by hand. There is a standard color-coding for the different sizes of hex head driver bits. Many electricians refer to the driver size by color rather than its inch size.

Flex bits, **Figure 4-27**, are flexible bits up to five-feet long that allow you to drill holes through several stud-bays of residential framing or through blocks in walls or ceilings that are unreachable with standard bits. This bit is used extensively in residential remodel work where conductors must be fished through finished walls and ceilings. They have holes in the tip and shaft for the attachment of wires or pull strings. Care must be taken when using these as it is very easy to lose track of where the bit is drilling, and as a result, it can damage walls, ceilings, and roofs.

n_defender/Shutterstock.com

Figure 4-25. Step bits cut different sizes of holes with a single bit. Used exclusively in sheet metal.

Nikita Rublev/Shutterstock.com

Figure 4-26. Driver bits are used with drills to insert or remove screws in a variety of surfaces.

Klein Tools

Figure 4-27. Flex bits are used extensively in remodel work.

Bit extenders, **Figure 4-28**, are a drill bit accessory that allows you to increase the reach of a bit or drill deeper holes as long as the bit's cutting head is larger than the receiving end of the extender bit. It is advisable to have an assortment of different lengths of extenders.

A *saw bit*, **Figure 4-29**, has the tip of a regular twist bit, but the middle of the shank has a knurled pattern that can cut plywood or paneling when a lateral force is applied. They can be challenging to control and are best suited for cutting rough openings, not precision work. *Tile bits* are used for drilling ceramic tiles and glass, **Figure 4-30**.

Stripped screw extractors are used with power drills to remove screws when the heads of the screws are stripped

Goodheart-Willcox Publisher

Figure 4-28. Bit extenders can increase the reach of the drill.

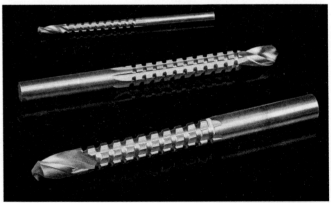

KPixMining/Shutterstock.com

Figure 4-29. Saw bits can be used as a jigsaw to cut large holes in thin material such as plywood, sheetrock, and wood paneling.

yarm_sasha/Shutterstock.com

Figure 4-30. Tile bits have a flared, triangular tip and are used exclusively for drilling in ceramic tile.

and do not respond to the screwdriver's tip. They have two ends—one end drills out the worn slots of the screw head, and the other end is the extractor. The extractor is used with the drill in reverse. As the extractor turns, it digs into the screw head and pulls out the damaged screw.

4.8 Drilling Basics

When drilling metal or other hard surfaces, drill bits tend to wander when starting a hole. Creating a dimple in the

workpiece using a scratch awl or nail punch, **Figure 4-31,** will give the drill bit a place to ride as you begin to drill.

SAFETY NOTE	Proper PPE

Wearing proper eye protection when drilling is critical. Always wear safety glasses when drilling any material. One tiny metal fragment or wood splinter can cause a severe eye injury. Metal drill shavings can be extremely hot and can embed in the eyeball very quickly. Safety glasses with side shields that wrap around the sides of your face provide the best protection.

Lubrication is critical when drilling steel that is 1/8″ or thicker. Lubricating the bit with cutting oil reduces friction and heat build-up making drilling easier and extending the bit's life. Any lubrication is better than none. A squirt of penetrating oil such as WD-40® works in a pinch. Remember to drill metal at a *slow speed*: 350–1000 rpm is recommended for drilling most metals.

To drill larger diameter holes in metal, start with a smaller hole and drill successively larger holes until you reach the size hole required. This results in a cleaner hole with less effort.

Tear-out and blow-out is damage that occurs when the bit exits the other side of the piece, **Figure 4-32.** If possible, clamp a sacrificial board to the back of the workpiece to minimize tear-out on the workpiece. When drilling masonry blow-out occurs when the bit exits the block or brick. Ensure any tear-out or blow-out is not on the show side of what you are drilling.

Finally, sharp drill bits save time and aggravation. Sharpen or replace dull bits when necessary. If you can smell burning wood, your bit is dull!

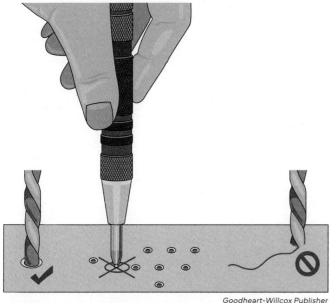
Goodheart-Willcox Publisher

Figure 4-31. Before drilling, create a dimple in the workpiece using a scratch awl or nail punch.

Goodheart-Willcox Publisher

Figure 4-32. The back surface of a workpiece being drilled may be damaged by tear-out or blow-out as the drill bit exits.

Summary

- Power tools operate at extremely high speeds, and accidents can happen in an instant and may result in catastrophic injuries. It is critical the user understands how to operate the tool properly and safely.
- A corded power tool will deliver more power and will be lighter in weight than a cordless tool. It can be used nearly continuously with no downtime, while the cordless tool's batteries need periodic charging.
- On reciprocating saws, the blade moves back and forth to make a cut.
- The jigsaw is a type of reciprocating saw.
- The circular saw uses a flat, round spinning blade to cut different materials.
- The oscillating tool uses a slight, high-speed side-to-side motion to make its cut.
- The electric drill is the power tool most used by residential electricians.
- The general-use drill is used for most tasks, such as drilling holes or driving and removing fasteners.
- Hammer drills use a rotary, percussive motion to drill holes in brick or concrete.
- Low-speed, high-torque drills are used to drill larger diameter holes in wood.
- Rotary hammer drills are much larger and more powerful than the standard hammer drill and can drill, core, and chisel concrete, stone, and similar materials.
- The chuck is the part of the drill that holds the drill bit.
- The "rod-type" bit comprises three basic parts, the shank, the body, and the point.
- High-speed steel is a type of steel, *not* the speed at which the drill bit is used.
- The twist bit is the most common drill bit.
- There are other types of bits used in electrical work, including the brad-tip, masonry, auger, paddle, and self-feed bits.
- Hole saws are a cylindrical saw blade used in power drills. The blade is mounted on an arbor with an attached pilot bit that acts as the center of the hole and stabilizes the saw.
- Step bits have a stepped, conical shape that drills a wide range of hole diameters, generally in sheet metal.
- Flex bits are long, flexible drill bits that allow you to drill holes through several stud-bays or through blocks in walls or ceilings that are unreachable with standard bits.
- Create a dimple in the workpiece to prevent the drill bit from wandering when starting a hole.
- To drill clean, large diameter holes in metal, start with a smaller hole and drill successively larger holes until you reach the size hole required.
- Tear-out and blow-out damage occurs when the bit exits the other side of the piece.

Know and Understand

1. All corded tools used in a residential construction site should be plugged into a _____ receptacle.
   - A. duplex
   - B. tamper-resistant
   - C. GFCI protected
   - D. indoor

2. A good choice of saw for demolition work is the _____.
   - A. reciprocating saw
   - B. jigsaw
   - C. circular saw
   - D. oscillating tool

3. The drill best suited for drilling into concrete is the _____.
   - A. general-use drill
   - B. hammer drill
   - C. right-angle drill
   - D. low-speed, high-torque drill

4. The drill type most used in residential rough-in wiring is the _____.
   - A. general-use drill
   - B. hammer drill
   - C. right-angle drill
   - D. low-speed, high-torque drill

5. The part of the drill that tightens and holds the bit is called the _____.
   - A. clutch
   - B. chuck
   - C. vise
   - D. grip

6. *True or False?* Higher drilling speeds produce less torque and drain batteries of cordless drills more quickly.

7. The part of the drill bit that does the cutting is the _____.
   - A. shank
   - B. body
   - C. point
   - D. flute

8. *True or False?* High-speed steel is a type of steel alloy and does *not* refer to the speed of the drill.

9. The most common style drill bit is the _____.
   - A. twist bit
   - B. masonry bit
   - C. auger bit
   - D. paddle bit

10. *True or False?* Twist bits cannot be sharpened when they become dull.

11. *True or False?* An auger bit can be used on wood or metal.

12. A drill bit accessory that allows you to increase the reach of a bit or drill deeper holes is the _____.
   - A. flex bit
   - B. bit extender
   - C. driver bit
   - D. drill index

13. The damage that occurs when the bit exits the other side of the piece is called _____.
   - A. blow-back
   - B. blow-out
   - C. tear-out
   - D. Both B and C.

Apply and Analyze

1. Explain the orbital action available on most jigsaws.
2. Compare and contrast the hammer drill and the rotary hammer drill.
3. Compare and contrast the benefits and drawbacks of corded tools and cordless tools.

Critical Thinking

1. Summarize why a residential electrician needs several different types of drills.

CHAPTER 5 | Electrical Test Equipment

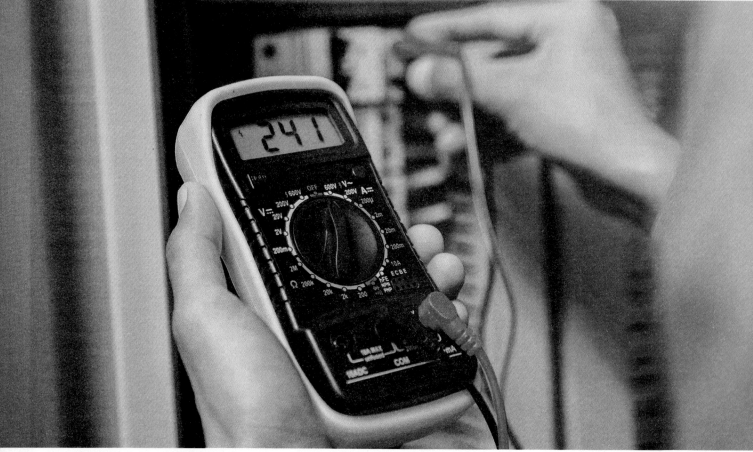

A_stockphoto/Shutterstock.com

SPARKING DISCUSSION

How does a foundation in electrical theory aid in understanding electrical test equipment readings?

Introduction

It is nearly inconceivable to think that the early experimenters in electricity had no measuring instruments at their disposal and had to invent the meters needed to measure the new phenomenon they were exploring. On the other hand, electricians today have a wide array of testing devices available. These devices can measure the electrical values of volts, amps, ohms, continuity, and frequency and test electrical components such as capacitors and inductors. Individual meters are available to test these values, but most electricians use a meter that combines most, if not all, of these functions. An understanding of electrical meters and test equipment use is essential for anyone working in the electrical industry.

5.1 Voltage Testers and Meters

The terms *testers* and *meters* are often used interchangeably, but they are not the same. *Voltage testers* are used to indicate the presence or absence of voltage. The neon tester, **Figure 5-1**, glows when connected to an energized circuit, but it offers no indication of the voltage value. A *solenoid voltage tester*, **Figure 5-2**, uses an electric plunger called a solenoid to indicate several common nominal voltage values. Both testers need to be connected to exposed energized components to get a reading. A *noncontact voltage tester*, also called a *pen*, *ticker*, or *screamer*, can detect voltage without making physical contact with the circuit. See **Figure 5-3**. They operate by detecting the magnetic field that develops around a current-carrying conductor. This tester glows at the tip and emits a beeping noise when voltage is detected. A limitation of the noncontact

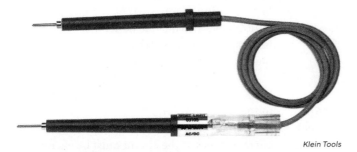

Klein Tools

Figure 5-1. This neon circuit tester will illuminate to indicate the presence of voltage from 80 to 600 volts AC or DC.

Klein Tools

Figure 5-2. The solenoid tester has a plunger-type readout and will vibrate when voltage is detected.

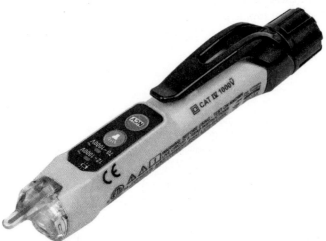

Klein Tools

Figure 5-3. The noncontact voltage tester can sense the presence of voltage through the conductor's insulation. They are often used in troubleshooting to determine where a circuit loses its power.

voltage tester is that it cannot read through certain types of electrical cables, such as metal-clad and Type SEU. At times, the meter may illuminate but not make an audible sound. In this case, further testing must be done with a different type of meter.

The **voltmeter** is a voltage tester that senses voltage and also provides a readout of the amount of voltage present.

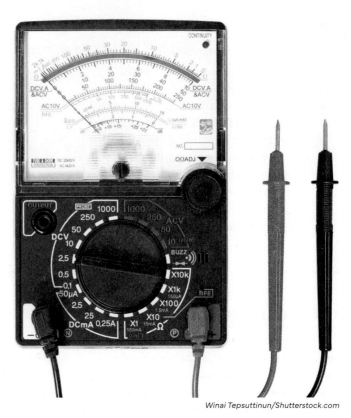

Winai Tepsuttinun/Shutterstock.com

Figure 5-4. The analog meter uses a pointer and scale rather than a digital readout. This particular meter measures DC voltage and current, AC voltage, and resistance.

Although most multimeters today have digital readouts, some electricians still prefer using the analog meter, which uses a pointer and a scale to indicate the measured values, **Figure 5-4.** Standalone voltmeters are somewhat rare as most voltmeters are combined with other meters in a single unit.

5.2 **Multimeters**

A **multimeter** is an all-in-one device that measures volts, values of resistance in ohms, and current in amps. This device is often called a *VOM (volt-ohm-milliammeter)* or a *DMM (digital multimeter)*, **Figure 5-5.** There are many functions of this device beyond measuring the basic electrical values. For example, the resistance function can determine **continuity**, or whether the circuit is complete. Some multimeters also include noncontact voltage detection capabilities. A selector switch on the unit's face is used to set the meter to the desired measurement.

All measurements on a multimeter are made using test leads. Test leads are conductors that have plugs to connect to the meter and probes on the other end to connect to the test points. Care must be used when measuring voltage, and current values as the circuit must be energized to get a reading. Conversely, resistance/continuity measurements should never be taken on an energized circuit.

Klein Tools

Figure 5-5. The digital multimeter combines several features into one easy-to-read meter. The function of the meter is selected by the dial.

5.3 **Clamp Meters**

A residential electrician uses a ***clamp meter*** to measure current, **Figure 5-6**. A clamp meter is similar to a digital multimeter but also includes clamps or jaws that measure current. It is known that when current flows in a conductor, a magnetic field is established around that conductor, and the strength of the magnetic field is in direct proportion to the amount of current. The clamp meter is a non-contact ammeter that calculates the current by measuring the magnetic field's strength. It is a much safer alternative to the ***in-line ammeter***. The conductor being measured is clamped between the meter's spring-loaded jaws, and a current reading is displayed.

SAFETY NOTE   **In-line Ammeter Readings**

Measuring current using a multimeter's test leads instead of the clamp can be very dangerous for the residential electrician and should be avoided. When measuring current using the test leads, the meter must be inserted in series with, or "in line" with, the load, which puts the user in danger of shock or electrocution.

Current can be tested independently in either the ungrounded conductor or the grounded conductor. However, no reading is displayed if both conductors are clamped, as the magnetic fields in the two conductors cancel one another out. Wrapping the conductor twice around the

Klein Tools

Figure 5-6. The clamp meter allows the user to measure current without contacting the exposed conductors.

clamp will result in a reading that is twice the actual current flow. Some brands of clamp meters use forks instead of spring-loaded clamps, **Figure 5-7**.

Konstantin Batrakov/Shutterstock.com

Figure 5-7. Some clamp meters use forks instead of spring-loaded jaws. The space between the forks limits the size conductor that can be measured.

Klein Tools

Figure 5-8. A three-prong receptacle tester can determine if a receptacle is properly wired. The illumination of the lights indicates different conditions such as an open ground, open neutral, open hot, hot/ground reversed, hot neutral reversed, and correct wiring.

5.4 Receptacle Testers

A three-prong *receptacle tester* is a simple, compact device for checking a receptacle for proper wiring, **Figure 5-8**. It is plugged into an energized grounded receptacle, and the lamps on the unit light up to indicate the receptacle's status. Newer models have an LCD (liquid crystal display) that displays the voltage and wiring status. Some models have a button for testing the operation of a GFCI receptacle or GFCI circuit breaker. However, if the GFCI receptacle is connected to an ungrounded, two-wire circuit with no equipment grounding conductor, the push-button GFCI test will not function. The GFCI receptacle's proper operation can be tested using the test/reset buttons on the receptacle's face.

5.5 Circuit Breaker Finder

A *circuit breaker finder* is a two-piece diagnostic tool that uses a plug-in transmitter and a hand-held receiver to determine which circuit breaker protects a given circuit, **Figure 5-9**. The transmitter is plugged into an energized receptacle and draws rapid pulses of current from the circuit to produce the signal. The battery-powered receiver is used at the breaker panel to scan for the breaker receiving the transmitter's signal. When the proper breaker is detected, the receiver beeps. A screw-in adapter, **Figure 5-10**, allows the transmitter to be plugged into a luminaire socket. Some models have clips that can attach directly to the conductors. When troubleshooting a circuit or making a repair or addition, the electrician should never work on an energized circuit. Using the circuit breaker finder allows the electrician to zero in on the proper circuit breaker, so they do not have to resort to turning them off one by one. This can save time and prevent frustration.

Klein Tools

Figure 5-9. The two-piece circuit breaker finder uses a transmitter and receiver to trace the wiring back to the circuit breaker that is protecting the circuit.

Klein Tools

Figure 5-10. An accessory kit for a circuit breaker finder includes adapters to be used with luminaires and two-prong receptacles and alligator clips to be used on bared conductors.

5.6 Continuity Testing

Continuity testing is one of the most common tests an electrician conducts. Continuity tests the circuit for completeness using the ohmmeter function of a multimeter. Using this feature of a meter can help in tracing conductors, testing the functionality of switches, and determining the condition of a fuse, among many other uses. It is a very valuable tool for troubleshooting a circuit or piece of equipment to determine open circuit and short circuit conditions.

The concepts of *zero resistance* and *infinite resistance* must be mastered to understand continuity. Resistance is the opposition to the flow of current in an electrical circuit. Think of a switch that controls the luminaire in any room. If the switch is closed and the light is glowing, the resistance across the switch is very close to zero ohms. When the switch opened and the light is off, the resistance across the switch is infinite. No amount of voltage can push the current through an open switch.

Continuity testing is a test of electrical resistance and, as such, should only be conducted on a de-energized circuit. When testing for resistance or continuity, the tester's battery emits a known voltage into the circuit. If the circuit is complete, the meter will read the battery's voltage as it completes the circuit. The meter calculates the amount of resistance based on the difference between the known voltage imposed on the circuit and the resulting voltage returned from the circuit. Most ohmmeters emit a beep to indicate continuity. All ohmmeters will give an ohmic value of resistance if it is within the meter's range of measurement. For an open circuit, the meter will display some form of non-numeric reading, such as a dash or *INF* for infinity or *OL*.

Summary

- Voltage testers are used to indicate the presence or absence of voltage.
- The voltmeter senses voltage and provides a readout of the amount of voltage present.
- The multimeter is an all-in-one device that measures volts, values of resistance in ohms, and current in amps.
- The in-line ammeter uses the test leads of a multimeter to measure electrical current.
- The clamp meter is a noncontact ammeter that measures electrical current by sensing an energized circuit's magnetic field.
- When a switch is closed and the load is energized, the resistance across the switch is very close to zero ohms.
- When the switch opened and the load is de-energized, the resistance across the switch is infinite.
- Continuity is a test of the circuit for completeness. The ohmmeter function of a multimeter is used for testing for continuity.

Know and Understand

1. Types of voltage testers include _____.
   A. solenoid voltage tester
   B. noncontact voltage tester
   C. voltmeter
   D. All of the above.

2. *True or False?* The noncontact voltage tester can detect the presence of voltage in all types of electrical cables.

3. The "M" in VOM stands for_____
   A. meter
   B. millivolt
   C. micro
   D. milliammeter

4. The type of meter display that uses a pointer and a scale to indicate the measured values is called a(n) _____ meter.
   A. analog
   B. digital
   C. solenoid
   D. DMM

5. *True or False?* To read the electrical current, a clamp meter must be connected in series with the load.

6. A clamp meter reads six amps of current in a conductor. If the conductor is wrapped twice around the clamp, the meter will read _____ amps.
   A. 6 amps
   B. 12 amps
   C. 3 amps
   D. 0 amps

7. *True or False?* The plug-in receptacle tester contains a battery and should never be used on an energized receptacle.

8. The best tool for determining which breaker protects a given circuit is the _____.
   A. voltmeter
   B. ammeter
   C. circuit breaker finder
   D. noncontact voltage tester

9. Testing for continuity is testing the circuit to see if it is _____.
   A. energized
   B. complete
   C. hazardous
   D. overloaded

10. *True or False?* Continuity testing should only be performed on a live circuit.

Apply and Analyze

1. What is the resistance of an open switch?
2. What is the resistance of a closed switch?
3. Continuity testing is actually a test of _____.
4. An ammeter reads five amps when clamped around the ungrounded conductor and five amps when clamped around the grounded conductor. What will the ammeter read if clamped around both conductors?

Critical Thinking

1. Explain how one could remove a switch from a circuit and test it for proper operation using a multimeter.

SECTION **3** Materials Used in House Wiring

Dmitry Melnikov/Shutterstock.com

CHAPTER 6
Fasteners and Anchors

CHAPTER 7
Electrical Boxes

CHAPTER 8
Electrical Conductors and Connectors

CHAPTER 9
Electrical Conduit

CHAPTER 10
Electrical Devices

CHAPTER 11
Panelboards, Overcurrent Protective Devices, and Disconnects

CHAPTER 12
Lamps and Lighting

It takes a great deal of materials to complete a residential electrical installation. This section introduces the student to these materials:

- Fasteners and anchors, which include screws and nails.
- Electrical boxes, which secure switches and receptacles.
- Conductors and connectors to be properly selected for their intended use.
- Conduit and fittings, which require various installation considerations.
- Electrical devices, which include assorted switches and receptacles.
- Overcurrent protection housed in panelboards.
- Lamps and lighting included for an entire residence.

CHAPTER **6** | # Fasteners and Anchors

Atlantist Studio/Shutterstock.com

SPARKING DISCUSSION ⚡

How is the best fastener or anchor chosen for the task at hand?

CHAPTER OUTLINE

Introduction

Fasteners are any mechanical components used to join, or fasten, parts together. They will be encountered regularly in your daily work. Understanding the types of fasteners available and their uses is critical for anyone working in the construction trades. Most projects will require some form of nail, screw, bolt, or anchor, which are used for various fastening purposes. The number and types of fasteners can be overwhelming, but remember, they all perform the same function. In electrical work, you will be working with primarily threaded fasteners. Threaded fasteners have a threaded design, or surface, on their component. The most common is a screw, and another is the bolt. In general, a bolt is used in conjunction with a nut, whereas a screw is used without a nut. In short, if you turn a head, it is a screw; if you turn a nut, it is a bolt. It is perfectly acceptable, however, to use the terms interchangeably.

6.1 Machine and Coarse-Thread Screws

Electricians generally make the distinction between machine screws and coarse threaded screws. The *threads* of a screw are the raised spirals along its length that provide the holding power. *Machine screws* have threads that are designed to fit a matching nut or another threaded component that is generally made of metal. *Coarse-thread screws* do not use a matching nut; instead, they make their threading as they are driven into a material, generally wood. Some machine screws, called *self-tapping screws*, tap their threads into the material as they are installed.

6.1.1 Machine Screws

A machine screw is the most common threaded fastener in electrical work. These screws are commonly referred to by their size and length, a sequence that is known as *callout*. Consider two common callouts, 6-32×1 or 10-24×4. The first number of the callout is the screw gauge or diameter. The higher the gauge number, the larger the diameter of the screw. Screw gauges larger than 12 are listed in fractional inches, such as 3/16, 1/4, 5/16, 1-1/4, etc. The second number of the callout is the number of threads per inch. For instance, a 10-24 screw has 24 threads per inch of screw length. The last digit of the callout is the length of the screw, usually measured in inches from the bottom of the head. Therefore, a 6-32×2 is a 6-gauge screw with 32 threads per inch, two inches long.

6.1.2 Machine Screws for Electrical Work

The electrician uses machine screws to attach devices such as switches and receptacles to electrical boxes, secure cover plates to devices and boxes, and install luminaires and ceiling fans. Machine screws come in various gauges and thread counts, although only four or five sizes are used in electrical work: 6-32, 8-32, 10-24, 10-32, and occasionally the 1/4-20. Each size screw has a specific use. The skilled electrician should be able to identify a screw size by sight.

The different screw types are used for distinct functions. 6-32 screws are used to attach devices to the box and attach cover plates to devices. Nearly all switches and receptacles are shipped with captive 6-32 screws held in place using thin plastic washers. Always align cover plate screws for a neat appearance. An exception to *Section 314.27(A)(1)* in the *NEC* allows two or more 6-32 screws to support a wall-mounted luminaire that weighs six pounds or less.

Size 8-32 screws are slightly thicker in diameter than 6-32 and can support more weight. Therefore, they are used to support most ceiling-mounted luminaires up to 50 pounds. Metal octagonal boxes and 4×4 boxes are shipped with 8-32 screws installed in the mounting ears. Non-metallic boxes do not come with the screws installed, but most luminaires will include them with the mounting hardware.

Size 10-24 screws have a greater diameter than 8-32 and fewer threads per inch. They are required for ceiling fan mounting because they can support more weight and are less affected by vibrations. Most boxes rated for ceiling fan support come with the required screws. Ceiling fan boxes and 10-24 screws are also recommended for the support of extra-heavy chandeliers.

All metal devices and junction boxes must be bonded to the equipment grounding conductor with a 10-32 green grounding screw or other approved means. All metal devices and junction boxes have one or more threaded holes for this purpose. Do *not* use this hole for mounting the box. **Figure 6-1** summarizes the common machine screws used by electricians.

Screw Size	Use
6-32	For the installation of switches and receptacles and their cover plates
8-32	Used to secure luminaires to the outlet box
10-32	Thread size of green grounding screw used in metallic boxes for bonding purposes
10-24	Used to secure ceiling fans up to 70 pounds to the outlet box

Goodheart-Willcox Publisher

Figure 6-1. A summarization of the common types of machine screws used by electricians.

6.1.3 Coarse-Thread Screws

Coarse-thread screws have very sharp tips and more space between each thread than the machine thread. These screws take less time to insert or remove, as the screw travels further in one rotation due to the greater space between threads. Coarse-thread screws are used in softer materials such as wood, plastic, or plasterboard because their coarse threads are able to grip the soft fibers of the material and hold the screw in place. A fine-thread screw is more likely to slip out of soft material.

Coarse-thread screws come in various head shapes, including the bugle head, hex head, pan head, and flat head, **Figure 6-2**. The different drive styles for these screws include the Phillips head drive, flathead drive, star-drive head, and square drive.

6.1.4 Sheetrock Screws

The most common coarse thread screw used is the sheetrock screw, **Figure 6-3**, also called a *drywall screw*. These screws were initially created to attach sheetrock to the

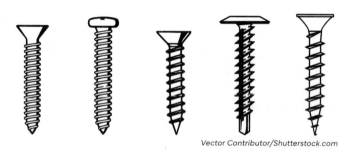

Vector Contributor/Shutterstock.com

Figure 6-2. Assorted head shapes for screws.

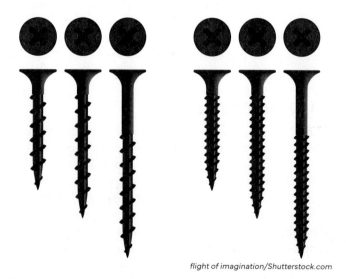

flight of imagination/Shutterstock.com

Figure 6-3. Assorted sheetrock screws. Note that they come in different degrees of thread coarseness.

framing studs, but workers in other trades quickly realized their versatility. Because they are sharp, they can be driven into wood easily by hand. Electricians use them to install metal and plastic device boxes and junction boxes. They *cannot* be used to support the device in the box. Most sheetrock screws have fully threaded shanks. Longer screws may be fully or partially threaded. They nearly always have Phillips head drives, but some are square drive or star-drive heads.

6.1.5 Sheet Metal Screws

A sheet metal screw is used to connect sheet metal to another sheet metal like a metal wireway to its cover or sheet metal components to a wood backing board. These screws have fully threaded shanks, and the tip and threads are sharp, enabling them to cut through a hard metal surface easily. Generally, there are two kinds of metal sheet screws: the self-tapping screw and the self-drilling screw. They are determined by their type of tip. The self-tapping screw has a standard, pointed tip, and the self-drilling screw has a flanged tip. See **Figure 6-4**. Both penetrate sheet metal easily without a pilot hole. If used in thicker metals, the self-tapping screw will need a pilot hole, while the self-drilling screw will make its own hole.

6.2 Anchoring Hardware

Electricians are called to install electrical equipment on a variety of surfaces. The equipment must be "rigidly and securely fastened" to whatever surface they are installing

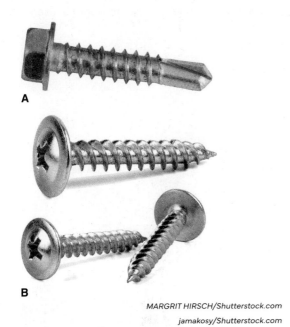

A

B

MARGRIT HIRSCH/Shutterstock.com

jamakosy/Shutterstock.com

Figure 6-4. The two types of sheet metal screws. A—The self-drilling screw. B—The self-threading screw. Note the flanged tip of the self-drilling screw.

on, as required by the *NEC* in *Section 314.23(A)*. The term *securely fastened* is not defined in the *NEC*, but most installers and authorities having jurisdiction (AHJs) interpret this to mean at least two fasteners adequately tightened to prevent movement.

At times electrical equipment can be quite heavy or subjected to additional force as the user may be required to open or close covers and activate external switches and levers. Other times, the material a screw is driven into may be of inadequate strength to hold the screw. *Anchors* are used to provide additional support than just the screw to withstand these additional forces or weight. There is no single anchor that works on every surface, so installers need to be familiar with the many different anchors used with a wide range of surfaces. This section covers some of the more commonly used anchors.

Plastic anchors come in conical, ribbed, and toggle, **Figure 6-5**. The conical and ribbed plastic anchors work similarly and are intended to be used with solid surfaces, such as brick or concrete, and can also be used in hollow walls, such as drywall. The toggle-style plastic anchors are used in hollow walls only. Heavy items should not be mounted on heavy walls using only anchors. If you cannot embed at least one screw in a framing stud, you may need to find a way to add some form of blocking for additional support. The *NEC* specifically prohibits the use of wooden plugs for use as an anchor in *Section 110.13(A)*.

Screw-in anchors come in metal or plastic and are intended for use in drywall, **Figure 6-6**. They are designed to be directly screwed into the wall, but a small pilot hole may help. They can also be used to install surface-mounted boxes and equipment. They may provide more support than plastic inserts. Some screw-in anchors are also expanding to provide even greater support.

Toggle bolts, **Figure 6-7**, are the traditional anchor method for mounting heavier items on hollow walls and are made of two parts: the folding toggle and the machine screw. The size of the toggle bolt is determined by the diameter and length of its screw. The folding toggle increases in size as the diameter of the screw increases. A longer toggle bolt can be installed in walls with a thicker surface material.

Toggle bolts require a pilot hole that can accommodate the folded toggle. The wings unfold and lock into place behind the drywall and provide a strong holding power by distributing the weight over a greater area than other types of anchors. Care must be taken that the toggles are fully open before tightening, otherwise they will come back through the hole. Generally, toggle bolts are one-time use, as the toggle falls into the wall when the screw is extracted. Toggle bolts are ideal for installing surface-mounted items on a ceiling.

Molly bolts are another option for hollow walls, **Figure 6-8**. Molly bolts are easy to install, and the largest mollies can hold up to 50 lb. A molly bolt adds permanent

Vinokurov Alexandr/Shutterstock.com; Atlantist Studio/Shutterstock.com; ANDY RELY/Shutterstock.com

Figure 6-5. Plastic anchors can be used to secure equipment to solid surfaces such as brick or masonry or used in hollow walls such as sheetrock or plaster.

Cico/Shutterstock.com

Figure 6-6. The screw-in anchor is used in sheetrock walls and can support more weight than the plastic anchor.

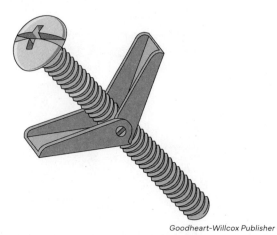

Goodheart-Willcox Publisher

Figure 6-7. Toggle bolts are used exclusively in hollow walls and ceilings.

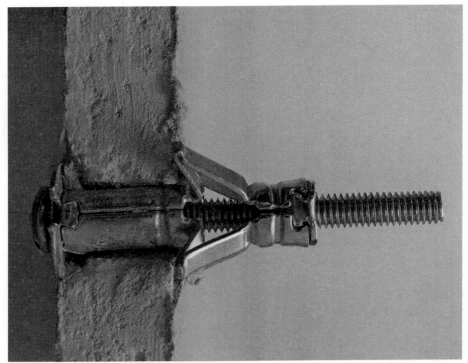

ANDY RELY/Shutterstock.com

Figure 6-8. A molly bolt installed in sheetrock. The screw can be removed and replaced with no loss in strength to the anchor.

screw threads to any material to which it is attached. Thus, anything installed with a molly can be installed and taken down repeatedly with no loss of strength. You should pre-drill the hole for all mollies.

Sleeve anchors are expansion anchors used in various masonry materials, including brick, concrete, and stone, **Figure 6-9**. They are mostly used to secure materials to solid walls or flooring. Sleeve anchors feature a split expansion sleeve over a machine stud bolt with a flared end. The entire unit is driven into a pre-drilled hole, and when the nut is tightened, the threaded bolt is pulled upwards, and the flared end of the bolt expands the sleeve, providing a firm grip. A stainless-steel style is available for outdoor applications but can also be used indoors. The zinc-plated steel anchors are for indoor, dry environments only and are great for installing boxes, supports, and braces onto concrete walls.

The lag shield, **Figure 6-10**, is an anchoring device designed for use with coarse-threaded lag bolts rather than machine screws. Similar to plastic anchors, they are inserted into the pilot hole. When the lag screw is inserted, the hinged lag shield expands, producing a great holding force. It is suitable for use in concrete, brick, and the mortar joints of block or brick walls. Lag shield anchors are made from zinc/aluminum alloy. In harder masonry materials, short-style lag shields are used and can reduce drilling time. The long-style version is used in soft or weak masonry to develop better holding strength. It is essential to properly insert the lag shield so the expanding end is deepest in the hole and the hinged end is closest to the surface.

Narubaet panyasing/Shutterstock.com

Figure 6-9. Sleeve anchor. Note the flared end opposite the nut.

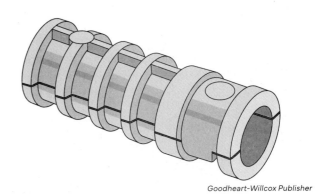

Goodheart-Willcox Publisher

Figure 6-10. Lag shields are similar to sleeve anchors but use a coarse-threaded lag bolt.

Drive pin anchors, **Figure 6-11**, also called hammer pins or hammer drives, are generally nonremovable. Hammer drive pins support lighter loads and attach fixtures to solid concrete, concrete blocks, brick, or other masonry surfaces. They are installed in a pre-drilled hole and set with hammer strikes that drive the pin into the anchor expanding the anchor to form a tight fit. A setting tool, **Figure 6-12**, is available to help set the pin, especially when installing where your hammer can't reach the pin.

Concrete screws, **Figure 6-13**, typically called Tapcons®, require a pilot hole and are used in solid concrete, brick, or block without an accompanying anchor. The screw has double threads to cut into the material and provide high holding power. When installing the screw in concrete, it is essential to install the screw by hand or to use a low-torque drill or a drill with a clutch to prevent the screw head from snapping off.

Although not an anchoring system, the pop rivet kit, **Figure 6-14**, is a welcome addition to your toolset. Although

Matt Benoit/Shutterstock.com

Figure 6-13. Concrete screws. Note the double threads. This type of fastener is inserted into a pre-drilled hole in the concrete.

Shahril KHMD/Shutterstock.com

Figure 6-14. A pop rivet tool and rivets.

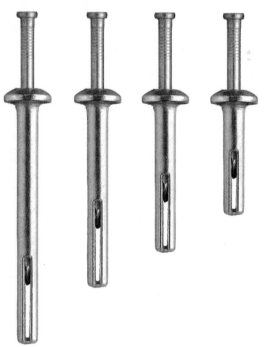

Ktuec/Shutterstock.com

Figure 6-11. Drive pin anchors are a type of sleeve anchor used in solid surfaces such as concrete and masonry.

Goodheart-Willcox Publisher

Figure 6-12. Drive pin anchors can be installed using a special setting tool.

not often called upon, they can be handy when installing or repairing a variety of products. Pop rivets allow you to fasten objects together even when there is no access to the rear side of the pieces being joined. They are inexpensive, compact, and easy to use, and leave a clean, professional-looking finish.

6.3 Nails

Nails are generally used to attach two pieces of lumber together. Electricians use nails to install wood blocking in walls and ceilings to position outlet boxes where they are needed. Blocking is also used as a reinforcement when hanging ceiling fans or heavy chandeliers. In the United States, nails are sized by length according to the *penny system*, **Figure 6-15**, which, confusingly, is abbreviated with the lower-case letter *d*. The sixteen-penny nail is written as 16d. Two nail types often used in electrical work are the framing nail, also called a common nail, and the finish nail.

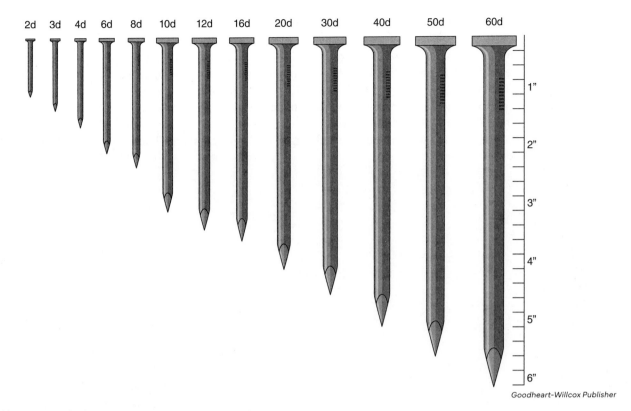

Figure 6-15. The penny system is a method for sizing nails.

Goodheart-Willcox Publisher

Framing nails, also called *common nails*, are pointed on one end and have a smooth, flattened head on the other end. They are mostly used in framing 2×-dimensional lumber. The 16d common nail is about 3 1/2 inches long and is the standard size used in residential framing, and should be stocked by the residential electrician. A smaller 8d nail can be used to secure 1×-dimensional used as "running boards" to protect electrical cables from physical damage in attics and crawl spaces. *Finish nails* do not have the flattened head of a framing nail. Instead, they have a rounded, dimpled head that is meant to be counter-sunk below the surface using a nail setting tool. Finish nails generally have a thinner shank than the same length framing nail. The thinner shank is less likely to split the wood when installing the nail.

Summary

- Machine screws have threads designed to fit a matching nut or other threaded components.
- Coarse-thread screws make their threading as they are driven into a material, generally wood.
- The callout of a screw describes its diameter, length, and thread count.
- The electrician uses machine screws to attach devices such as switches and receptacles to electrical boxes, secure cover plates to devices and boxes, and install luminaires and ceiling fans.
- 6-32 screws are used to attach devices to the box and attach cover plates to devices.
- 8-32 screws are used to support ceiling-mounted luminaires up to 50 pounds.
- 10-24 screws are required for ceiling fan mounting because they can support more weight and are less affected by vibrations.
- 10-32 screws are used to bond metallic device boxes and electrical equipment to the equipment grounding conductor.
- Coarse-threaded screws have more space between each thread than the machine thread.
- Two commonly used coarse-threaded screws are sheetrock screws and sheet metal screws.
- Electricians use sheetrock screws to install metal and plastic device boxes and junction boxes.
- The sheet metal screw has a fully threaded shank, and the tip and threads are sharp, enabling them to easily cut through the hard metal surface.
- There are two kinds of metal sheet screws, the tapping screw and the self-drilling screw.

- The plastic anchor is intended for solid surfaces such as brick or concrete but can also be used in hollow walls such as drywall.
- Screw-in anchors are intended for use in drywall.
- Toggle bolts are a method for mounting heavier items on hollow walls and are made of two parts: the folding toggle and the machine screw.
- Sleeve anchors are expansion anchors used in various materials, including brick, concrete, and stone. They are mostly used to secure materials to solid walls or flooring.
- In the United States, nails are sized by length according to the "penny system." The penny is abbreviated with the lower-case letter *d*.

Know and Understand

1. The screw used to secure switches and receptacles to the outlet box is _____.
   - A. 6-32
   - B. 8-32
   - C. 10-32
   - D. 10-24

2. _____ sized screws are used to secure ceiling fans to their outlet box.
   - A. 6-32
   - B. 8-32
   - C. 10-32
   - D. 10-24

3. The screw most often used for the installation of luminaires is the _____.
   - A. 6-32
   - B. 8-32
   - C. 10-32
   - D. 10-24

4. The holes in metallic boxes for the green grounding screw are tapped for _____ size screws.
   - A. 6-32
   - B. 8-32
   - C. 10-32
   - D. 10-24

5. *True or False?* An 8-gauge screw has the same diameter as an 8-gauge conductor.

6. The most commonly used coarse thread screw used is the _____.
   - A. 8-32
   - B. sheet metal screw
   - C. sheetrock screw
   - D. green grounding screw

7. A good choice of anchor for a hollow wall is the _____.
   - A. toggle bolt
   - B. sleeve anchor
   - C. lag shield
   - D. drive pin

8. *True or False?* Toggle bolts can be removed and reused many times.

9. A fastener that can be used to fasten objects together even when there is no access to the rear side of the pieces being joined is the _____.
   - A. machine screw
   - B. molly bolt
   - C. sleeve anchor
   - D. pop rivet

10. Nails lengths in the United States are sized according to the _____ system.
    - A. penny
    - B. nickel
    - C. dime
    - D. quarter

Apply and Analyze

1. A screw with the callout 8-32×3 has a length of _____, _____ threads per inch, and a(n) _____ gauge shaft.
2. Explain the difference between a self-tapping screw and a self-drilling screw.
3. Detail the proper method of installing a lag shield anchor.

Critical Thinking

1. What section of the *NEC* allows no fewer than two 6-32 screws to support a wall-mounted luminaire that weighs 6 lb or less?

CHAPTER **7** | # Electrical Boxes

ungvar/Shutterstock.com

SPARKING DISCUSSION

Why is it essential to be able to select the appropriate type of electrical box for the task at hand?

LEARNING OBJECTIVES

After completing this chapter, you will be able to:

- Identify types of electrical boxes.
- Summarize the features of a metallic and nonmetallic box.
- Define the term ganging in relation to a metallic box.
- Explain the requirements for boxes used with suspended ceiling fans.
- Differentiate between new work and old work electrical boxes.
- Explain the requirements for boxes and covers used in wet locations.
- Describe the purpose and use of covers and plaster rings.

TECHNICAL TERMS

damp location
electrical box
F-clip
ganging
handy box
junction box
luminaire
metallic box
new work box
nonmetallic box
old work box
plaster ring
raised cover
surface mounting
weatherproof box
wet location

Introduction

With the exception of wires, electrical boxes are the most common component installed in residential wiring. Therefore, it is critical for the electrician to understand the different types of electrical boxes, their uses, and safety concerns in order to complete a safe and code-compliant installation. This chapter introduces you to some of the most used electrical boxes in residential wiring. Box-fill calculations and other *NEC* rules concerning the installation and use of electrical boxes will be discussed in Chapter 22, *Outlet Box Installation*.

7.1 Electrical Boxes

An *electrical box* is an enclosure that holds various electrical connections. They are either metallic or nonmetallic in construction and come in a staggering number of types and sizes. Boxes are used at specific locations to install receptacles, switches, and *luminaires*, or light fixtures. They can also be used for a *junction box*, known as a *J-box*, where two or more wires are joined with no device attached. An electrical device box is used to house electrical devices, such as switches and receptacles, and is described by how many devices it holds. For example, an electrical device box holding only one device is a single-gang box, one with two devices is a two-gang or double-gang box, and up through three- and four-gang. It is important to identify the correct box needed for each specific project.

7.2 Nonmetallic Boxes

In most residential wiring today, *nonmetallic boxes* are used far more than metallic boxes to install switches and receptacles. This is because they are inexpensive, lightweight, and durable. They are also shipped with captive nails, making installation so quick and easy that they are often referred to as *nail-on boxes* by many electricians.

Most nonmetallic boxes are made of PVC plastic, as depicted in **Figure 7-1**, but fiberglass boxes are also available and used in firewalls found between occupancies

Goodheart-Willcox Publisher

Figure 7-1. PVC electrical boxes.

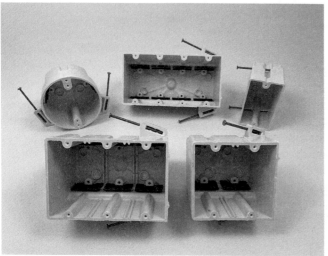

Goodheart-Willcox Publisher

Figure 7-2. Fiberglass electrical boxes. Note the thicker walls than on the PVC boxes.

in duplexes, apartments, and other multi-family dwellings. See **Figure 7-2.** Many electricians prefer fiberglass over PVC plastic because the fiberglass box is more rigid, holding its shape even if installers are not careful.

Nonmetallic boxes come in single-gang through four-gang or more in various depths. A round nonmetallic box is used for luminaires. Nonmetallic boxes also come in the 4″ × 4″ square size but are scarcely used in residential wiring.

7.3 **Metallic Boxes**

Metallic boxes, usually made of steel, are stronger, fireproof, and more secure than plastic boxes. A standard **metallic box** is used to install switches and receptacles and can have many features, such as mounting straps, knockouts and

pryouts where the cables enter the box, internal clamps to secure the cables, threaded holes for mounting the device, and a different threaded hole for the equipment grounding conductor. See **Figure 7-3.** A metallic box can be installed by driving nails through it and into the stud. It might also come with convenient mounting straps.

A metallic box has a 3″ × 2″ opening with a depth that ranges from 1 1/2″ to 3 1/2″. The sides of most metallic boxes can be removed to join several boxes together to make a larger box that can accommodate more than one device. This process is called **ganging.** Many electricians refer to these boxes as GEM (gangable metallic) boxes. See **Figure 7-4.**

Metal boxes are rarely used in new construction, but a residential electrician will encounter them in service and retrofit work.

7.3.1 **Handy or Utility Boxes**

A **handy box** or *utility box* is a common type of metallic box. Handy boxes are primarily used for **surface mounting** where the installation is exposed and on the wall's surface. They are characterized by rounded corners and

Goodheart-Willcox Publisher

Figure 7-3. A single-gang metallic device box. Note the installed equipment grounding pigtail.

Goodheart-Willcox Publisher

Figure 7-4. Two metallic device boxes ganged together (left). A single-gang device box with one wall removed (right).

Goodheart-Willcox Publisher

Figure 7-5. A handy or utility box and assorted covers.

Goodheart-Willcox Publisher

Figure 7-6. Assorted metallic boxes: an octagonal (OCT) box (top), 4″ × 4″ metal box (middle), and a shallow, pancake box (bottom).

multiple knockouts on all the walls. They can accommodate a device, such as a receptacle or a switch. For example, the handy box might be used in a residence to install a switch or receptacle on the concrete wall of an unfinished basement. They have their own style of cover-plate for receptacles, switches, or blank. See **Figure 7-5**. Various configurations are available, including handy boxes that can hold more than one device.

7.3.2 Other Metallic Boxes

Other types of metallic boxes, depicted in **Figure 7-6**, include the 4″ × 4″ square box, also known as the four-by-four, four-square, or 1900 box; the 4″ octagonal-shaped (OCT) box; and the shallow pan or "pancake" box used exclusively for the installation of luminaires. The 4″ × 4″ square box is used as a junction box or may contain devices if used with the proper cover. OCT boxes are used for installing luminaires or serving as junction boxes. When used as a junction box, blank covers are available.

7.4 Ceiling Fan-Rated Boxes

Ceiling fan-rated boxes are sturdier than a traditional round nail-on or a standard OCT box. These boxes use a larger diameter screw to support and withstand the additional

Goodheart-Willcox Publisher

Figure 7-7. Two examples of metal ceiling fan electrical boxes. This particular brand is shipped with four mounting screws to attach the box to the house framing.

weight and vibrations of the ceiling fan. These features also make them ideal for hanging heavy chandeliers. Ceiling fan-rated boxes must be marked by the manufacturer to indicate they are suitable for such use. See **Figure 7-7.**

7.5 New Work and Old Work Boxes

In residential, the terms *new work* and *old work* have important distinctions. **New work boxes** are for new construction when open studs are available, and drywall or plaster has not been added yet. While this is convenient for placing boxes in an organized layout, new work boxes can warp or twist if installed incorrectly. Nail-on boxes, for example, are all new work boxes.

Old work boxes, also called *cut-in* or *pop-in* boxes, are meant for installations in existing dwellings where the drywall has already been added. Old work boxes have "ears" that allow the box to be positioned in the wall without falling through. There is also some form of flap or clip that can be tightened and pinches the wallboard to hold the box in place. See

Figure 7-8. Old work boxes come in single- and multi-gang boxes, and round luminaire boxes. There is even an old work bracket that is used with a fan-rated box to install ceiling fans.

Metallic boxes can be used as old work boxes by using F-clips. **F-clips** are made from easily-bendable sheet metal and used to secure metallic boxes to the wall's surface. See **Figure 7-9.**

Goodheart-Willcox Publisher

Figure 7-8. Old work electrical boxes. The round box is used for luminaires, and the rectangular boxes are used to hold devices.

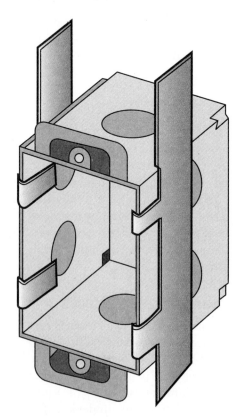

Old work electrical box retrofit wall clips

Goodheart-Willcox Publisher

Figure 7-9. A special support clip for securing boxes to a wall.

7.6 **Weatherproof and Outdoor Boxes**

Weatherproof boxes, generally called *bell boxes*, are sealed enclosures used for surface mounting devices and fixtures to exterior walls, decks, or other outdoor structures or in wet or damp locations. A **wet location** is defined as any area subject to saturation or is directly impacted by the weather. A **damp location** is subject to moderate degrees of moisture, such as a roofed open porch. Boxes used in wet or damp locations must be rated for such use.

Weatherproof boxes come in single-gang, double-gang, and round configurations and are made of cast aluminum. See **Figure 7-10**. The boxes are readily available in gray, white, and bronze colors. They come in many combinations of threaded entry numbers and sizes and will come with threaded plugs to use on the unused openings. There will usually be one fewer plug than there are openings, as at least one of the openings will be used to bring the conductors into the box. All bell boxes will have an equipment grounding conductor screw installed or have a raised, threaded hole for one.

Round weatherproof boxes have integral mounting tabs, and the boxes come with optional mounting straps that may be installed with the supplied screws. Using these external mounting tabs maintains the weatherproof integrity of the box.

Weatherproof boxes have dimples on the back that can be easily removed to provide mounting holes in the back of the box. Although these boxes are listed as weatherproof, the screw penetrations in the back of the box, if used, will need caulking. Also, the threaded plugs will need caulking to prevent rainwater seepage.

Houses that are to be finished with vinyl siding will have special mounting blocks for receptacle and luminaire installation. These mounting blocks generally have a self-contained box for the device or luminaire and are made to seamlessly integrate with the vinyl siding.

7.7 **Electrical Box Covers**

Section 314.25 requires each box in a completed installation to have a cover, faceplate, lampholder, or luminaire installed. To accommodate the installation of devices or luminaires, raised covers and plaster rings are used with 4″ × 4″ square boxes. **Raised covers** are used for surface-mounted boxes, while **plaster rings** are used for flush applications. They come in various depths and are available in single-gang and double-gang configurations. Square to round plaster rings are used for luminaires. Plaster rings can be metallic or nonmetallic. Metallic plaster rings are depicted in **Figure 7-11**.

There are many cover options for many different applications. When a standard electrical box is used as a junction box, blank covers are available. Blank covers are also used when a device or fixture has been removed, but it is impractical to remove the box. Covers with threaded openings can be used for flood fixtures, light sensors, or motion sensors.

In weatherproof boxes, the cover or fixture must be appropriately rated for its intended location. To prevent water intrusion a rubber or foam gasket is installed between the cover or accessory and the box. Extension rings, combination covers, and covers with a vertical or horizontal orientation are also available.

Equipment grounding screws

Goodheart-Willcox Publisher

Figure 7-10. Assorted weatherproof boxes and two plugs. Note the installed green equipment grounding screws.

Goodheart-Willcox Publisher

Figure 7-11. 4″ × 4″ metal boxes with plaster rings.

Weatherproof covers can be rated as "weatherproof" or "weatherproof while in use." Weatherproof covers have a spring-loaded flap that protects the receptacle from the weather when not in use. These types of covers can be used in damp locations but not wet locations. Wet locations require the use of a cover that is weatherproof while in use. These types of covers either expand to accommodate the plug or have a hood that is large enough to accommodate the plug. For types with a hood, the hood shall be listed and identified as "extra-duty" per *Section 406.9(B)(1)*.

Summary

- Electrical boxes are used in every residential electrical installation. They come in a variety of shapes and sizes and will be either metallic or nonmetallic in construction.
- Metallic boxes are usually made of steel, while nonmetallic boxes are made of PVC plastic or fiberglass. Metallic boxes are strong, fireproof, and secure, but nonmetallic boxes are cheaper and more lightweight.
- Most electrical boxes used in residential wiring are the nonmetallic, nail-on type.
- Each type of box has a purpose. Boxes are used to house switches and receptacles; handy boxes are metallic boxes used when surface mounting; ceiling fan-rated boxes must be marked by the manufacturer to indicate suitable use for ceiling fans; junction boxes are used to contain the joining or splicing of two or more wires; the "pancake" box used exclusively for installing luminaires; the standard 4″ × 4″ square box or OCT box; and weatherproof boxes must be rated for use in wet or damp locations.
- Most metallic boxes are gangable, meaning they can be combined to make a larger box to accommodate more than one device. One-gang, two-gang, three-gang, and so forth describe how many devices a box will contain.
- The term *new work* refers to installations and materials used in new construction, whereas *old work* refers to installations and materials used in existing dwellings.
- Plaster rings and raised covers are used with 4″ × 4″ square boxes to attach devices or luminaires.

Know and Understand

1. What type of box is used to join two or more wires and has no device installed?
   - A. A pan box
   - B. A joiner box
   - C. A new work box
   - D. A junction box

2. Where four switches are required, you should use a _____ box.
   - A. single-gang
   - B. double-gang
   - C. three-gang
   - D. four-gang

3. *True or False?* Nonmetallic boxes are more expensive than metallic boxes.

4. The most common electrical boxes installed today are constructed of _____.
   - A. steel
   - B. fiberglass
   - C. rubber
   - D. cast aluminum

5. *True or False?* Handy boxes are mostly surface-mounted rather than flush-mounted.

6. What shape is the nonmetallic box used for the installation of luminaires?
   - A. Square
   - B. Octagonal
   - C. Round
   - D. Oval

7. *True or False?* Switches and receptacles may be installed in a shallow pan box.

8. *True or False?* Any box that is suitable for a luminaire can be used to support a ceiling fan.

9. What type of box should be used for installing a very heavy chandelier?
   - A. A ceiling fan-rated box
   - B. A round old work box
   - C. A metal box
   - D. A round nail-on box

10. Cut-in and pop-in boxes are other names for _____ boxes.
    - A. old work
    - B. new work
    - C. weatherproof
    - D. device

11. Which of the following may be required to ensure a weatherproof box will not allow for water intrusion?
    - A. Plastic wrap
    - B. A rubber cover
    - C. Caulking
    - D. A small roof

Apply and Analyze

1. Describe the process of ganging metallic boxes.
2. What is the difference between a wet location and a damp location?
3. Identify two components that can be added to a 4″ × 4″ square box for the installation of devices.

Critical Thinking

1. Why might an electrician need to know about metallic boxes even though they are not often used in new construction these days?
2. You are working as an electrician, and the customer requests you remove a luminaire and install a ceiling fan. Explain how you would accomplish this in an *NEC*-compliant manner. Provide as much detail as possible.

CHAPTER **8**

Electrical Conductors and Connectors

P A/Shutterstock.com

SPARKING DISCUSSION

What considerations must be made when determining a conductor size for a particular branch circuit?

CHAPTER OUTLINE

8.1   Conductor Sizing

8.2   Cables and Conductors

    8.2.1   Solid and Stranded Conductors

8.3   Ampacity

    8.3.1   *NEC Table 310.16*

8.4   Conductor Material

8.5   Conductor Connectors and Fittings

Introduction

Electrical conductors are the wires that tie all the other electrical components together to form the residence's electrical system. There are many different sizes and types of electrical conductors, and the electrician must understand the functions and limitations of each used in electrical installations. It may seem daunting, but selecting the correct size and type of conductor is one of the most important jobs of an electrician. This chapter will focus on the basics of electrical conductors and cables. We will revisit conductor calculations in Chapter 19, *Conductor Sizing and Ampacity Adjustments*.

8.1 Conductor Sizing

Electricians use different sizes of conductors for different applications. Smaller-sized conductors are used to power smaller loads, while larger-sized conductors are used to power larger loads. For example, the conductor size used for general lights and receptacles in a dwelling would be insufficient to power the cooking range in the kitchen. While using too large of a conductor for a particular load would merely be a waste of money, an undersized conductor could very well result in an electrical fire.

Electrical conductors in the United States are sized according to the **American Wire Gauge (AWG)**. In the manufacturing process, different sizes of conductors are made by drawing material, copper or aluminum, through successively smaller dies, **Figure 8-1**. A die is a device that forms the conductor to a certain shape and size. A conductor is given a numerical designation based on the number of dies it takes to create it. If the conductor is drawn through 14 dies, it is called a 14 AWG, whereas a conductor drawn through six dies is a 6 AWG. Therefore, the larger the

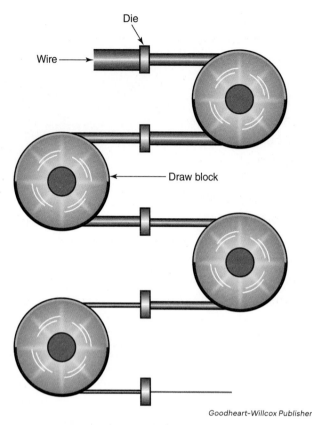

Figure 8-1. AWG wire sizes are based on the number of dies the conductor is drawn through.

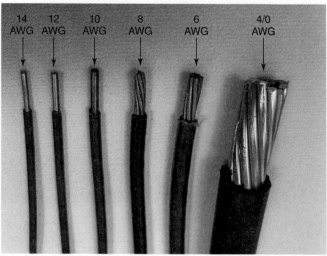

Goodheart-Willcox Publisher

Figure 8-2. Conductor sizes 14 AWG through 6 AWG copper and one 4/0 aluminum. These are the most common conductor sizes used in residential wiring. AWG sizes greater than 6 AWG and smaller than 4/0 are not shown.

AWG number, the smaller the wire size. Residential wiring employs conductors sized 14 AWG through size 4/0 (pronounced "four-aught"), **Figure 8-2**.

Conductors are sized according to the American Wire Gauge, but their diameter is measured in *circular mils* *(cmils)*. One mil equals 1/1000 of an inch, which is about the same diameter as a human hair. This means a 14 AWG conductor measures 4110 cmils, and 4/0, the largest AWG size, is 211,600 cmils. For conductors larger than 4/0, they are sized in thousands of circular mils, abbreviated kcmil, where the letter *k* is the metric designator for 1000 units. For instance, the next sized conductor after 4/0 is 250 kcmil, which means it has a diameter of 250,000 circular mils. The *NEC Chapter 9, Table 8*, shown in **Figure 8-3**, contains information on conductor properties. The first three columns show the conductor AWG size and its area in square millimeters and cmils.

CODE APPLICATION   Finding the Largest Solid Conductor

Scenario: An electrician needs to find the largest solid conductor listed in *NEC Chapter 9, Table 8*.
Chapter 9, Table 8: This table outlines conductor properties such as AWG and kcmil sizes, overall area, diameter, and stranding.
Solution: The *NEC* requires conductors 6 AWG and larger to be stranded. Referring to *NEC Chapter 9, Table 8*, columns 4 through 6 show the conductor stranding properties. Notice under the Quantity column, conductor sizes 18 AWG through 8 AWG show two numbers: 1 and 7. The number 1 indicates there is

only one strand, meaning the conductor is actually a solid conductor, and the number 7 indicates there are seven strands to the conductor. This means that an 8 AWG conductor is the largest solid conductor show. Columns 5 and 6 show the diameter of the individual strands in millimeters and inches. Also, note that columns 7 and 8 show the conductor's overall diameter. A stranded conductor has a slightly larger overall diameter than the same AWG size solid conductor but is still considered to have the same circular mils.

Table 8 Conductor Properties

Size (AWG or kcmil)	Area mm²	Area Circular mils	Stranding Quantity	Stranding Diameter mm	Stranding Diameter in.	Overall Diameter mm	Overall Diameter in.	Overall Area mm²	Overall Area in.²	Copper Uncoated ohm/km	Copper Uncoated ohm/kFT	Copper Coated ohm/km	Copper Coated ohm/kFT	Aluminum ohm/km	Aluminum ohm/kFT
18	0.823	1620	1	—	—	1.02	0.040	0.823	0.001	25.5	7.77	26.5	8.08	42.0	12.8
18	0.823	1620	7	0.39	0.015	1.16	0.046	1.06	0.002	26.1	7.95	27.7	8.45	42.8	13.1
16	1.31	2580	1	—	—	1.29	0.051	1.31	0.002	16.0	4.89	16.7	5.08	26.4	8.05
16	1.31	2580	7	0.49	0.019	1.46	0.058	1.68	0.003	16.4	4.99	17.3	5.29	26.9	8.21
14	2.08	4110	1	—	—	1.63	0.064	2.08	0.003	10.1	3.07	10.4	3.19	16.6	5.06
14	2.08	4110	7	0.62	0.024	1.85	0.073	2.68	0.004	10.3	3.14	10.7	3.26	16.9	5.17
12	3.31	6530	1	—	—	2.05	0.081	3.31	0.005	6.34	1.93	6.57	2.01	10.45	3.18
12	3.31	6530	7	0.78	0.030	2.32	0.092	4.25	0.006	6.50	1.98	6.73	2.05	10.69	3.25
10	5.261	10380	1	—	—	2.588	0.102	5.26	0.008	3.984	1.21	4.148	1.26	6.561	2.00
10	5.261	10380	7	0.98	0.038	2.95	0.116	6.76	0.011	4.070	1.24	4.226	1.29	6.679	2.04
8	8.367	16510	1	—	—	3.264	0.128	8.37	0.013	2.506	0.764	2.579	0.786	4.125	1.26
8	8.367	16510	7	1.23	0.049	3.71	0.146	10.76	0.017	2.551	0.778	2.653	0.809	4.204	1.28
6	13.30	26240	7	1.56	0.061	4.67	0.184	17.09	0.027	1.608	0.491	1.671	0.510	2.652	0.808
4	21.15	41740	7	1.96	0.077	5.89	0.232	27.19	0.042	1.010	0.308	1.053	0.321	1.666	0.508
3	26.67	52620	7	2.20	0.087	6.60	0.260	34.28	0.053	0.802	0.245	0.833	0.254	1.320	0.403
2	33.62	66360	7	2.47	0.097	7.42	0.292	43.23	0.067	0.634	0.194	0.661	0.201	1.045	0.319
1	42.41	83690	19	1.69	0.066	8.43	0.332	55.80	0.087	0.505	0.154	0.524	0.160	0.829	0.253
1/0	53.49	105600	19	1.89	0.074	9.45	0.372	70.41	0.109	0.399	0.122	0.415	0.127	0.660	0.201
2/0	67.43	133100	19	2.13	0.084	10.62	0.418	88.74	0.137	0.3170	0.0967	0.329	0.101	0.523	0.159
3/0	85.01	167800	19	2.39	0.094	11.94	0.470	111.9	0.173	0.2512	0.0766	0.2610	0.0797	0.413	0.126
4/0	107.2	211600	19	2.68	0.106	13.41	0.528	141.1	0.219	0.1996	0.0608	0.2050	0.0626	0.328	0.100
250	127	—	37	2.09	0.082	14.61	0.575	168	0.260	0.1687	0.0515	0.1753	0.0535	0.2778	0.0847
300	152	—	37	2.29	0.090	16.00	0.630	201	0.312	0.1409	0.0429	0.1463	0.0446	0.2318	0.0707
350	177	—	37	2.47	0.097	17.30	0.681	235	0.364	0.1205	0.0367	0.1252	0.0382	0.1984	0.0605
400	203	—	37	2.64	0.104	18.49	0.728	268	0.416	0.1053	0.0321	0.1084	0.0331	0.1737	0.0529
500	253	—	37	2.95	0.116	20.65	0.813	336	0.519	0.0845	0.0258	0.0869	0.0265	0.1391	0.0424
600	304	—	61	2.52	0.099	22.68	0.893	404	0.626	0.0704	0.0214	0.0732	0.0223	0.1159	0.0353
700	355	—	61	2.72	0.107	24.49	0.964	471	0.730	0.0603	0.0184	0.0622	0.0189	0.0994	0.0303
750	380	—	61	2.82	0.111	25.35	0.998	505	0.782	0.0563	0.0171	0.0579	0.0176	0.0927	0.0282
800	405	—	61	2.91	0.114	26.16	1.030	538	0.834	0.0528	0.0161	0.0544	0.0166	0.0868	0.0265
900	456	—	61	3.09	0.122	27.79	1.094	606	0.940	0.0470	0.0143	0.0481	0.0147	0.0770	0.0235
1000	507	—	61	3.25	0.128	29.26	1.152	673	1.042	0.0423	0.0129	0.0434	0.0132	0.0695	0.0212
1250	633	—	91	2.98	0.117	32.74	1.289	842	1.305	0.0338	0.0103	0.0347	0.0106	0.0554	0.0169
1500	760	—	91	3.26	0.128	35.86	1.412	1011	1.566	0.02814	0.00858	0.02814	0.00883	0.0464	0.0141
1750	887	—	127	2.98	0.117	38.76	1.526	1180	1.829	0.02410	0.00735	0.02410	0.00756	0.0397	0.0121
2000	1013	—	127	3.19	0.126	41.45	1.632	1349	2.092	0.02109	0.00643	0.02109	0.00662	0.0348	0.0106

Notes:

1. These resistance values are valid **only** for the parameters as given. Using conductors having coated strands, different stranding type, and, especially, other temperatures changes the resistance.
2. Equation for temperature change: $R_2 = R_1[1 + a(T_2 - 75)]$, where $\alpha_{cu} = 0.00323$, $\alpha_{Al} = 0.00330$ at 75°C.
3. Conductors with compact and compressed stranding have about 9 percent and 3 percent, respectively, smaller bare conductor diameters than those shown. See Table 5A for actual compact cable dimensions.
4. The IACS conductivities used: bare copper = 100%, aluminum = 61%.
5. Class B stranding is listed as well as solid for some sizes. Its overall diameter and area are those of its circumscribing circle.

Informational Note: NEMA WC/70-2009, *Power Cables Rated 2000 Volts or Less for the Distribution of Electrical Energy*, or ANSI/UL 1581-2017, *Reference Standard for Electrical Wires, Cables, and Flexible Cords*, is the source for the construction information. *National Bureau of Standards Handbook 100*, dated 1966, and *Handbook 109*, dated 1972, is the reference where the resistance is calculated.

Figure 8-3. *NEC Chapter 9, Table 8* contains information on conductor properties such as the circular mils and resistance.

8.2 Cables and Conductors

In residential wiring, branch circuit conductors from the distribution panel to the outlets are usually installed in a cable system. A **cable system**, or *cable*, is a factory-made assembly of two or more conductors that are contained in an outer sheathing. The outer sheathing is typically made of metallic or nonmetallic material. Residential wiring cables have two or more color-coded insulated conductors and a bare, uninsulated equipment grounding conductor, all contained inside a nonmetallic sheathing. The nonmetallic-sheathed cable is called **Type NM cable** and is also known as *Romex* to electricians. The outer sheathing of NM cable is color-coded based on its size—white for 14 AWG, yellow for 12 AWG, orange for 10 AWG, and black for 8 AWG and 6 AWG.

Type NM cable is sized according to the AWG size and the number of insulated conductors. A 14-2 NM cable has one white and one black insulated 14 AWG conductor and one bare equipment grounding conductor. A 14-3 NM cable includes a red insulated conductor to the black, white, and bare conductors. The bare equipment grounding conductor is the same size as the insulated conductors for 14 AWG, 12 AWG, and 10 AWG cables. Any 8 AWG and 6 AWG cables have a 10 AWG bare equipment grounding conductor.

The *NEC* requires that insulated conductors be color-coded to indicate their function in the circuit, **Figure 8-4**. White and gray are used for the grounded, neutral conductor. The equipment grounding conductor can be bare or have insulation that is green or green with one or more yellow stripes. The ungrounded, hot conductors can be any color other than white, gray, green, or green with yellow stripes.

Conductor Color Coding in NM Cable

Color	Function	Purpose
Black	Ungrounded, hot conductor	Used as an ungrounded or "hot" conductor. Supplies the current to the load in 120-volt branch circuits
Red	Ungrounded, hot conductor	Used as an ungrounded or "hot" conductor. When used in conjunction with the black conductor, it carries current to the load in 120/240-volt multi-wire branch circuits, as in an electric cooking range circuit.
White or gray	Grounded conductor	The return path for current from the load back to the source. The grounded conductor is a current-carrying conductor. Often called the "neutral" conductor, but only truly a neutral when used with black and red wire in a multiwire circuit.
Bare, green, green with yellow stripes	Equipment grounding conductor	Bonds all normally noncurrent carrying metal parts of a circuit together; does not carry current except in a ground fault situation.

Goodheart-Willcox Publisher

Figure 8-4. Conductor insulation color chart. The *NEC* requires conductors to be color-coded as to their use.

CODE APPLICATION   Means of Identifying the Grounded Circuit Conductor

Scenario: What are the *Code*-compliant methods of identifying the grounded circuit conductor?

Section 200.6: This section outlines acceptable methods of identifying the grounded conductor in an electrical circuit.

Solution: The *NEC* requires grounded conductors sized 6 AWG and smaller, commonly used in residential wiring, to be identified by one of the following means:

1. A continuous white outer finish
2. A continuous gray outer finish
3. Three continuous white or gray stripes along the conductor's entire length on other than green insulation

The *NEC* requires conductors 4 AWG and larger to be identified by one of the following means:

1. A continuous white outer finish
2. A continuous gray outer finish
3. Three continuous white or gray stripes along the conductor's entire length on other than green insulation
4. At the time of installation, by a distinctive white or gray marking at its terminations that encircles the conductor or insulation

Type NM cable is permitted to be installed either exposed or concealed in normally dry locations. If exposed, it must be protected so that it is not subject to physical damage. It is not rated for damp or wet locations. *Type UF cable* is similar to NM, but it is rated for exposure to direct sunlight and is suitable for direct burial. The type UF cable has a nonmetallic sheathing that is thicker than traditional NM cable and is molded around the individual conductors to protect them from water intrusion. There is no need for the cable to be encased in a conduit, but the *NEC* does specify burial depths for underground conductors. Burial depths will be discussed in Section 10, *Outdoor Circuits*.

Service conductors and feeder cables are not exactly a type of cable but a designation given the cable based on its use. The same cable type can be used for both service entrances and feeders. *Service conductors* are large diameter cables, usually 1/0 to 4/0 AWG, installed to carry the entire load of a residence. The service conductors run from the utility service point to the line side of the service disconnect. The *utility service point* is where the utility's jurisdiction ends, and the premise's wiring begins. *Feeder cables* are also usually large diameter cables and may carry the entire load or only a portion of the entire load of a residence. They feed the panel from the main service disconnect and are connected to the service disconnect's load side on one end and to the main distribution panel on the other end. Feeders can also supply power to a subpanel's branch circuits. A *subpanel* is a distribution panel that is additional to the main panel and usually fed from a breaker in the main panel.

Two common types of service and feeder cables are Type SEU and Type SER, **Figure 8-5**. *Type SEU cable* has two sunlight-resistant, insulated conductors and a bare concentrically wound grounded conductor. A gray sunlight-resistant PVC outer jacket covers the conductors. SE stands for service entrance cable, and although the cable has a U-shaped profile, the U in SEU stands for unarmored, meaning there is no metallic sheathing. *Type SER cable* has three insulated sunlight-resistant conductors and one uninsulated equipment grounding conductor. All the conductors are wrapped in a gray sunlight-resistant PVC jacket. The cable has a round profile, but the R in SER stands for "Reinforced tape" for the fiberglass wrapping tape surrounding the conductors. Type SEU and Type SER cables are not suitable for underground use with or without a raceway.

Another type of service entrance cable, *Type USE cable*, **Figure 8-6**, is constructed of multiple insulated conductors twisted together and has no outer sheathing. USE stands for underground service entrance. Type USE cable is rated for direct burial and is sunlight resistant. Three-conductor Type USE cable is called triplex, and four-conductor is called

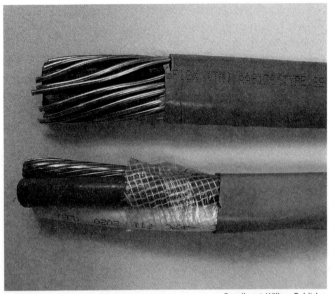

Goodheart-Willcox Publisher

Figure 8-5. Type SEU and Type SER cables. Note that Type SEU has a *U*-shaped cross-section, where Type SER has a round cross-section. The concentrically wound bare conductor in Type SEU cable.

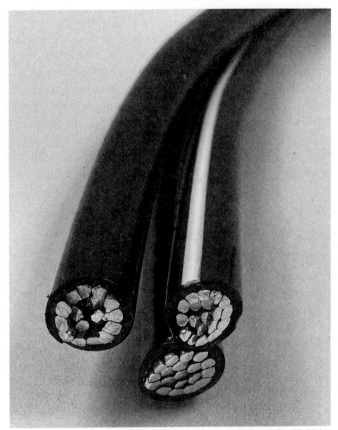

Goodheart-Willcox Publisher

Figure 8-6. Three-conductor Type USE cable is also called triplex. Although there is no outer sheathing, Type USE is still referred to as a cable.

quadruplex. A residential electrician may use quadruplex as a feeder to run underground power to a subpanel in an outdoor workshop.

8.2.1 Solid and Stranded Conductors

Individual electrical conductors are either solid or stranded. Solid wire is heavy, weather-resistant, and anti-corrosive that is used mainly outdoors or for heavy-duty applications with higher currents. Stranded wire is flexible and does not split. It is used in smaller applications such as circuit boards and speaker wires.

In cable systems, sizes 14 AWG through 10 AWG contain all solid conductors. Sizes 8 AWG and 6 AWG have stranded insulated conductors and a solid, bare equipment grounding conductor. Stranding the larger conductors makes them easier to handle and install because stranded conductors are more pliable.

8.3 Ampacity

In residential wiring, it is important to use the correct size conductors for the current (load) requirement of a circuit to prevent the conductors from overheating, which could possibly result in a fire. This means ampacity must be considered. *Ampacity* is the current-carrying capacity of a conductor and is defined in the *NEC* as "the maximum current, in amps, that a conductor can carry continuously under the conditions of use without exceeding its temperature rating." In this definition, the phrase "conditions of use" refers specifically to the *ambient temperature*, or the *air temperature*, surrounding the conductor and the number of conductors contained in a cable or conduit. Under equal temperature ratings and conditions of use, a larger conductor has a higher ampacity than a smaller conductor.

The *NEC* has correction factors to adjust the ampacity of conductors depending on the conditions of use. These adjustments will be discussed later in Chapter 19, *Conductor Sizing and Ampacity Adjustments*.

8.3.1 *NEC Table 310.16*

In the *NEC, Chapter 9, Table 8* is one of many tables that covers conductors and conductor properties. However, the most referenced table in the *NEC* is *Table 310.16*, **Figure 8-7**, which lists the allowable ampacities of electrical conductors. It can be used in many circumstances to find various information. It is also one of the most misapplied, especially when the user does not understand that the ampacities listed are valid only under the conditions of use specified in the table's heading. These conditions specified include not having more than three current-carrying conductors in a raceway, cable, or earth (direct buried) and are based on an ambient temperature of 30°C (86°F). If these conditions are not met, the ampacity of the conductor must be adjusted or corrected. These adjustments and corrections will be discussed at length in Chapter 19, *Conductor Sizing and Ampacity Adjustments*.

The information in *Table 310.16*'s temperature rating of the conductors is an additional area that requires clarification. There are three temperature ratings listed in the table: 60°C (140°F), 75°C (167°F), and 90°C (194°F). The temperature ratings are based on the conductor's insulation type and are the temperatures at which the insulation begins to break down or degrade. The heat implied by the temperature ratings results from excess current or high ambient temperatures.

The table is divided straight down the middle—with copper conductors on the left and aluminum or copper-clad aluminum on the right. If the conductor size and insulation

---

CODE APPLICATION   Determining Conductor Size for Electrical Application

Scenario: An electrician needs to determine what size copper conductor with Type UF insulation would be required to serve a 30-amp load for an electric clothes dryer.

Table 310.16: This table in the *NEC* is primarily used to select the appropriately sized conductor to serve a particular current level.

Solution: Scan down the Type UF copper column until you find 30 amps, then check the AWG size conductor needed, which, in this case, is a 10 AWG. If the required ampacity does not correspond with a current listed in the table, you will choose the conductor with an

ampacity of at least the load to be served. For example, if you needed to determine what size copper with a Type USE insulation is required to serve a 125-amp load, you must find Type USE copper under the 75°C column. Scan down the column until you find the AWG size with an ampacity of at least 125 amps, which is a 1 AWG with an ampacity of 130 amps.

It also should be noted for future reference that although the individually insulated conductors in Type NM cable have an insulation rating of 90°C, we **must** choose our ampacities from the 60° column as per *Section 334.80*.

Table 310.16 Ampacities of Insulated Conductors with Not More Than Three Current-Carrying Conductors in Raceway, Cable, or Earth (Directly Buried)

Size AWG or kcmil	Temperature Rating of Conductor [See Table 310.4(A)]						Size AWG or kcmil
	60°C (140°F)	75°C (167°F)	90°C (194°F)	60°C (140°F)	75°C (167°F)	90°C (194°F)	
	Types TW, UF	Types RHW, THHW, THW, THWN, XHHW, XHWN, USE, ZW	Types TBS, SA, SIS, FEP, FEPB, MI, PFA, RHH, RHW-2, THHN, THHW, THW-2, THWN-2, USE-2, XHH, XHHW, XHHW-2, XHWN, XHWN-2, XHHN, Z, ZW-2	Types TW, UF	Types RHW, THHW, THW, THWN, XHHW, XHWN,USE	Types TBS, SA, SIS, THHN, THHW, THW-2, THWN-2, RHH, RHW-2, USE-2, XHH, XHHW, XHHW-2, XHWN, XHWN-2, XHHN	
	COPPER			ALUMINUM OR COPPER-CLAD ALUMINUM			
18*	—	—	14	—	—	—	—
16*	—	—	18	—	—	—	—
14*	15	20	25	—	—	—	—
12*	20	25	30	15	20	25	12*
10*	30	35	40	25	30	35	10*
8	40	50	55	35	40	45	8
6	55	65	75	40	50	55	6
4	70	85	95	55	65	75	4
3	85	100	115	65	75	85	3
2	95	115	130	75	90	100	2
1	110	130	145	85	100	115	1
1/0	125	150	170	100	120	135	1/0
2/0	145	175	195	115	135	150	2/0
3/0	165	200	225	130	155	175	3/0
4/0	195	230	260	150	180	205	4/0
250	215	255	290	170	205	230	250
300	240	285	320	195	230	260	300
350	260	310	350	210	250	280	350
400	280	335	380	225	270	305	400
500	320	380	430	260	310	350	500
600	350	420	475	285	340	385	600
700	385	460	520	315	375	425	700
750	400	475	535	320	385	435	750
800	410	490	555	330	395	445	800
900	435	520	585	355	425	480	900
1000	455	545	615	375	445	500	1000
1250	495	590	665	405	485	545	1250
1500	525	625	705	435	520	585	1500
1750	545	650	735	455	545	615	1750
2000	555	665	750	470	560	630	2000

Notes:
1. Section 310.15(B) shall be referenced for ampacity correction factors where the ambient temperature is other than 30°C (86°F).
2. Section 310.15(C)(1) shall be referenced for more than three current-carrying conductors.
3. Section 310.16 shall be referenced for conditions of use.
*Section 240.4(D) shall be referenced for conductor overcurrent protection limitations, except as modified elsewhere in the *Code.*

Reproduced with permission of NFPA from NFPA 70, National Electrical Code, 2023 edition. Copyright © 2022, National Fire Protection Association. For a full copy of the NFPA 70, please go to www.nfpa.org

Figure 8-7. *NEC Table 310.16* is known as the Ampacities Table.

type are known, the conductor's allowable ampacity will be the number where the column and row meet. For example, a 6 AWG copper conductor with Type TW insulation has an ampacity of 55 amps. An aluminum 500 kcmil conductor with type THHN insulation has an ampacity of 350 amps.

8.4 Conductor Material

In theory, electrical conductors can be made from any type of metal. Pure silver is the best conductor of electricity, followed by pure copper and pure gold. For reasons of cost and abundance, only copper and aluminum are used. In residential wiring, copper is used for branch circuits, generally in sizes ranging from 14 AWG to 6 AWG. Aluminum is used for service and feeder conductors, usually sized from 2/0 to 4/0. Aluminum and copper do not perform equally, though. A copper conductor has a higher ampacity than an aluminum conductor of the same size.

From the late 1960s to the early 1970s, aluminum wiring was used in residential branch circuit wiring. Although there is nothing inherently dangerous in using aluminum branch circuit wiring, problems have occurred at the device terminations. All metals expand when heated and contract when cooled, and we know that current flow in a conductor creates heat in a conductor. Aluminum expands and contracts to a greater degree than copper, which is not a problem with larger service and feeder conductors but can present problems on smaller conductors at the switch and receptacle terminations of branch circuits. This expansion and contraction at the device terminals caused the conductors to loosen themselves from the terminal screws, resulting in a poor connection. A poor electrical connection may cause excessive heating, and this excessive heat may lead to a fire.

The *NEC* explicitly prohibits the direct connection of conductors made of dissimilar metals, such as copper to aluminum. Dissimilar metals can have a deteriorating effect on one another when exposed to moisture, which becomes problematic because there is always some moisture content in the air. For instance, most times, when copper and aluminum are connected, the aluminum begins eating away at the copper. Special connectors are made specifically for copper to aluminum wiring connections, and devices that are rated for aluminum conductors are available.

8.5 Conductor Connectors and Fittings

In residential wiring, the most common type of wire connector used to splice two or more branch circuit conductors is the **twist-on connector**, commonly called a *wire nut*, **Figure 8-8**. These connectors come in different sizes and are used to connect conductors 18 AWG through 6 AWG. Wire nuts have a square-wire spring in the tip that grips the wire as you twist in, making a firm connection. Wire nuts can be used on solid or stranded conductors or a combination of solid and stranded.

PRO TIP	Pretwisting Wires

Pretwisting the wires is not required by some manufacturers, but it is a best practice, especially when joining three or more wires.

Figure 8-8. Assorted twist-on wire connectors are known in the trade as *wire nuts*.

All wire nuts have a limit to the number and size of wires they can accommodate. Never put more than five conductors under a wire nut. If more than five conductors need to be spliced, use a short "pigtail" with four of the conductors and make a second splice.

Push-in connectors, **Figure 8-9**, are increasing in popularity. Like wire nuts, push-in connectors can be used on solid or stranded conductors or a combination of solid and stranded. They are available with two to eight ports for the joining of multiple conductors. Proponents of the push-in connectors claim they are faster and easier to use than wire nuts and take up less volume in the device box.

Larger diameter conductors are spliced using insulated lugs or by using split bolts. Current-carrying conductors spliced with split bolts are insulated with rubber mastic tape and then wrapped in standard vinyl electrical tape.

In *Section 300.15*, the *NEC* requires wire splices to be in a box or a conduit body. Two exceptions are the use of rated splice kits and direct-burial rated connectors. The rated splice kit is designed for splicing two or three conductor Type NM cables that can then be rendered inaccessible, as behind a finished wall, for example. Twist-on connectors rated for direct burial are available. Direct-burial twist-on connectors are filled with a silicone sealant to prevent moisture intrusion. Although further moisture-proofing is not required, it is recommended that the splice be wrapped with rubber mastic and nylon tape in the same manner as the split bolt.

Cable connectors secure cables where they enter electrical boxes and equipment. One type of metallic connector uses screws to "pinch" the cable in the connector. Care must be taken to not overtighten the connector since damage to the cable and a possible short circuit can result.

Nonmetallic push-in connectors are available and do not pose the danger of causing a short-circuit. Nonmetallic push-in connectors are less expensive and have a faster installation time than the metallic screw-down type of connector, **Figure 8-10**.

Grigvovan/Shutterstock.com

Figure 8-9. Push-in wire connectors are becoming more and more popular. They are easy to use and take up less room in the box.

Goodheart-Willcox Publisher

Figure 8-10. Metallic and nonmetallic cable connectors in sizes 1/2″ and 3/4″.

Summary

- Electrical conductors in the United States are sized according to the American Wire Gauge (AWG). The larger the AWG number, the smaller the diameter of the conductor.
- The outer sheathing of NM cable is color-coded as to its size: white for 14 AWG, yellow for 12 AWG, orange for 10 AWG, and black for 8 AWG and 6 AWG.
- The *NEC* requires that insulated conductors be color-coded to indicate their function in the circuit. White and gray are used for the grounded, neutral conductor. The equipment grounding conductor can be bare or have insulation that is green or green with one or more yellow stripes. The ungrounded, hot conductors can be any color other than white, gray, green, or green with yellow stripes.
- The same cable types can be used for service conductors and feeders and include Type SEU cable, Type SER cable, and Type USE cable.
- Ampacity is the current-carrying capacity of a conductor. Ambient temperature is the temperature that surrounds an object on all sides. Under equal temperature ratings and conditions of use, a larger conductor will have a higher ampacity than a smaller conductor.
- Individual electrical conductors are either solid or stranded. In cable systems, sizes 14 AWG through 10 AWG contain all solid conductors, while sizes 8 AWG and 6 AWG will have stranded insulated conductors in addition to the solid, bare equipment grounding conductor.
- The purpose of *NEC Table 30.16* is to list the allowable ampacities of electrical conductors. The temperature ratings are based on the conductor's insulation type and are the temperatures at which the insulation begins to break down or degrade.
- *NEC Table 310.16* is also used to select the appropriately sized conductor to serve a particular current level. If the conductor size and insulation type are known, the conductor's allowable ampacity will be the number where the column and row meet.
- In residential wiring, the most common type of wire connector used to splice two or more branch circuit conductors is the twist-on connector, commonly called a *wire nut*. Larger diameter conductors are spliced using insulated lugs or by using split bolts.
- Cable connectors secure cables where they enter electrical boxes and equipment. Do not overtighten the connector since damage to the cable and a possible short circuit can result.

Know and Understand

1. *True or False?* The larger the AWG number, the larger the wire size.
2. What color is the outer sheathing of a 12 AWG Type NM cable?
   A. White
   B. Yellow
   C. Orange
   D. Black
3. A 14-3 Type NM Cable would have a total of _____ conductors.
   A. three
   B. four
   C. fourteen
   D. seventeen
4. *True or False?* The white grounded conductor supplies current to the load in a 120-volt branch circuit.
5. What is the conductor that bonds all normally noncurrent-carrying metal parts of an electrical installation?
   A. The equipment grounding conductor
   B. The grounded conductor
   C. The ungrounded conductor
   D. The neutral conductor
6. What cable type is rated for direct burial?
   A. Type NM cable
   B. Type UF cable
   C. Type SEU cable
   D. Type SER cable
7. *True or False?* Stranded conductors have a smaller diameter than the same AWG-sized solid conductor.
8. *True or False?* The larger the conductor, the greater the conductor's ampacity.
9. A conductor's ampacity can be determined by using _____.
   A. *NEC Chapter 9, Table 8*
   B. *NEC Table 320.26*
   C. *NEC Table 310.16*
   D. an ammeter

10. The temperature rating of a conductor is based on which of the following?
    A. Ambient temperature
    B. Its size in AWG
    C. Its ampacity
    D. Its insulation type
11. *True or False?* Twist-on wire connectors cannot be used with stranded conductors.
12. *True or False?* Any type of twist-on connector can be used for direct burial in the earth.

Apply and Analyze

1. What is the area in cmil of a 1 AWG conductor?
2. A conductor that measures 300 kcmil has how many strands?
3. What color is the outer sheathing of a 12 AWG cable?
4. What type of service entrance cable is rated for direct burial in the earth?
5. According to *NEC Table 310.16*, what is the ampacity of an aluminum 1/0 conductor with Type THHN insulation?
6. What AWG size copper conductor with Type THW insulation would be sufficient to power a 125-amp load?
7. What are two instances where a conductor splice does not have to be accessible for servicing?
8. What are three methods of securing and supporting Type NM cable?
9. Type NM cable must be supported at intervals not exceeding _____.

Critical Thinking

1. Explain why conductors with a large AWG size are smaller in size than conductors with a smaller AWG size.
2. Compare service conductors to feeder conductors.

CHAPTER 9

Electrical Conduit

ballykdy/Shutterstock.com

Introduction

The primary purpose of conduit in an electrical system is the physical protection of conductors. Although few municipalities require the use of conduit as the primary wiring method for residential installations, certain conduit types will be used in some capacity in most dwellings. Therefore, the residential electrician must understand the types of electrical conduit and raceways, their associated components, and how they contribute to the electrical system. This chapter will introduce the student to the most common types of conduit used in residential wiring.

9.1 Electrical Conduit

An *electrical conduit* is a tubular raceway or enclosed channel designed to hold and protect electrical conductors or cables. They can be classified based on wall thickness, mechanical stiffness, and material used. They are made of rigid or flexible metallic or nonmetallic material but are not defined in Article 100 of the *NEC*. The conditions of use will generally dictate which type of conduit is to be used for a specific application. For example, some metallic conduit is unsuitable for direct burial in the earth, and certain types of conduit must be used in wet locations. Electrical conduit is usually limited to a few specific applications in residential wiring.

9.2 Electrical Conduit Sizing

Electrical conduit is sized based on the inner diameter (ID) of the conduit, not its outer diameter (OD). The inner diameter determines the number of conductors a conduit can safely accommodate. The outer diameter of the conduit changes based on the thickness of the conduit wall.

For many years, electrical conduit sizing was measured in inches, called the *electrical trade size*. Typical trade sizes were 1/2″, 3/4″, 1″, 1 1/2″, etc. Metric designators were later added to the *Code*, and conduit was referred to by its electrical trade size listed first, followed by the metric designator in parentheses. For example, 2″ (metric designator 53). In the 2002 edition of the *NEC*, the convention changed to list the metric designation first, followed by the electrical trade size in parentheses.

Now, a typical conduit is referred to as a metric designator (trade size) without specifying units of measurement, for example, metric designator 53 (trade size 2). No unit of measurement, such as millimeters or inches, is specified because the measurements are nominal measurements. Nominal, meaning "in name only," is an approximation of the conduit's actual inner diameter.

Conduits of the same trade size but composed of different materials have different inner diameters. For example, a trade size 1 Type EMT conduit has an actual ID of 1.049″, whereas a trade size 1 Type RMC has an actual ID of 1.063″. In *Table 300.1(C)*, the *NEC* reinforces that the metric designators and trade sizes are not exact dimensions. **Figure 9-1** lists some of the typical metric designators and their equivalent in electrical trade size.

9.3 Conduit Used in Residential Wiring

There are various types of conduits used in residential wiring. As mentioned earlier, the conditions of use will determine the type of conduit used. This section will review metallic and nonmetallic conduit, including rigid and intermediate metal conduit (RMC and IMC), electrical metal tubing (EMT), rigid PVC, and flexible conduit (LFNC) for underground or wet locations. Note that the *NEC* refers to types of conduit by these abbreviations or prefaced with "Type."

Metric Designators and Electrical Trade Size Equivalences

Metric Designator	Trade Size
12	3/8
16	1/2
21	3/4
27	1
35	1–1/4
41	1–1/2
53	2
63	2–1/2
78	3
91	3–1/2
103	4
129	5
155	6

Goodheart-Willcox Publisher

Figure 9-1. Comparison of conduit sizes expressed as metric designators and trade sizes.

9.3.1 Rigid Metal Conduit (RMC) and Intermediate Metal Conduit (IMC)

Rigid metal conduit (RMC) and *intermediate metal conduit (IMC)*, **Figure 9-2**, are steel conduits that provide the most protection for electrical conductors. These two types of conduits are nearly identical and have the same definition in the *NEC*. However, Type RMC has the thickest wall and is the heaviest steel conduit. The thinner wall in IMC does not affect its strength, and IMC is considered just as strong as RMC.

Type RMC is available in sizes from metric designator 16 (trade size 1/2) to metric designator 155 (trade size 6), and Type IMC is available in sizes from metric designator 16 (trade size 1/2) to metric designator 103 (trade size 4). RMC and IMC are manufactured in ten-foot lengths, use the same fittings and couplings to join pieces of conduit together, and come factory-threaded on each end.

In residential wiring, the most common use of RMC and IMC is the service mast to protect the service entrance conductors in an overhead service. When a through-the-roof service is required, it is the only conduit that has the strength to support the weight of the utility's conductors without guy wires or other additional support. In certain municipalities, RMC or IMC is required to be used in underground installations, such as from a home to an external garage feed.

RMC and IMC are difficult or impossible to bend with hand benders, so mechanical benders are used. Factory-made bends are available and can be coupled onto the conduit. The conduit can be cut to length with a hacksaw or other means and rethreaded in the field with the proper equipment. The conduit must be securely fastened within 3′ of any conduit termination, such as an outlet or junction box, or another piece of electrical equipment. It should be supported at intervals not exceeding 10′.

9.3.2 Electrical Metallic Tubing (EMT)

In residential wiring, *electrical metallic tubing (EMT)* is used to protect cables and conductors from physical damage. It may also be used for aesthetic purposes as an exposed

Goodheart-Willcox Publisher

Figure 9-2. Intermediate metal conduit (IMC) and rigid metal conduit (RMC) can use the same fittings. Shown is a length of IMC conduit, threaded on both ends, a one-hole strap, a threaded coupling, and a bushing.

run of conduit has a neater, more professional appearance than a run of exposed NM cable. EMT is much thinner and lighter than RMC and IMC and is available in ten-foot lengths in sizes from metric designator 16 (trade size 1/2) to metric designator 103 (trade size 4).

EMT conduits of metric designator size 16 (trade size 1/2) through size 35 (trade size 1 1/4) are easily bent by hand using a conduit bending tool. Larger diameters require a mechanical bender or the use of factory-made bends. All sizes of EMT can be easily cut with a hacksaw.

Like other metal conduits, EMT can also be used for a service mast that is attached to the side of a house, but EMT does not penetrate the roof. In this case, the conduit must be securely fastened within 3′ of any conduit termination, such as an outlet or junction box, or another piece of electrical equipment. EMT should be supported at intervals not exceeding 10′.

9.3.3 Rigid Polyvinyl Chloride Conduit (PVC)

Rigid polyvinyl chloride conduit (PVC) is a nonmetallic conduit often used in underground installations where more protection is needed for conductors than that provided by direct-burial cable, such as underground feeder (UF) cable. It is available in sizes from metric designator 16 (trade size 1/2) to metric designator 155 (trade size 6). See **Figure 9-3**. PVC is manufactured in 10′ lengths and comes with a prefabricated, integral coupling on one end, called the *bell end*.

PVC is sometimes used as the overhead mast for residential services that do not penetrate the roof. Sizes up to metric designator 41 (trade size 1-1/2) can be cut with ratcheting PVC cutters. Larger sizes are cut with a hacksaw or PVC saw.

PVC must be securely fastened within 3′ of any conduit termination, such as an outlet or junction box, or another piece of electrical equipment.

9.3.4 Liquidtight Flexible Nonmetallic Conduit (LFNC)

Liquidtight flexible nonmetallic conduit (LFNC) is a weather-tight, flexible conduit with a round cross-sectional area. It is used in short lengths of 6′ or less to connect to vibrating equipment. It is often found in exposed, wet locations, is weather-proof when used with the proper fittings, and is sunlight resistant. It is also rated for direct burial in the earth. See **Figure 9-4**. LFNC is available in sizes from metric designator 12 (trade size 3/8) through metric designator 103 (trade size 4) and sold in coils of various lengths depending upon its diameter. This type of conduit is easily cut to length using a utility knife or PVC cutters.

The most common use for LFNC in residential wiring is for a whip from the disconnecting means to an outdoor air conditioning condenser unit, **Figure 9-5**. A *whip* is a short length, three to six feet, of flexible conduit between a junction box or disconnecting means and a piece of electrical equipment. These whips can be made in the field or bought pre-assembled.

Goodheart-Willcox Publisher

Figure 9-3. PVC conduit and fittings. The connector used in PVC is called a *terminal adapter,* or *TA.* Also shown are a coupling and a cap.

Goodheart-Willcox Publisher

Figure 9-4. Liquidtight flexible nonmetallic conduit (LFNC) and fittings. A 90° connector and straight connector are also shown. This type of conduit is also called *seal-tight.*

CODE APPLICATION   Direct-Buried Conduit

Scenario: An electrician needs to know what type of cable to install in a conduit that is to be directly buried underground.
Section 300.5(B)
    Section 300.5 pertains to the underground installation of conductors.

Solution: *Section 300.5(B)* classifies the inside of a direct-buried conduit to be a wet location, so all conductors or cables installed must be listed as suitable for use in wet locations. Type UF cable is listed for use in wet locations. Type NM cable is not listed as suitable for wet locations.

ARENA Creative/Shutterstock.com

Figure 9-5. A whip from the disconnecting means to the air conditioning compressor unit.

When installed in lengths greater than 6′, the conduit should be secured within 12″ of any conduit termination, such as an outlet or junction box or another piece of electrical equipment, and securely fastened at intervals not exceeding 3′.

9.4 Fittings and Supports for Conduit

Fittings are parts of the electrical system that perform a mechanical function rather than an electrical function, such as couplings, connectors, and straps. There are many types of fittings available for each type of conduit, and it is important to understand when and how they should be used.

Bushings are used on the conduit ends to protect the conductors from abrasion where the conductors exit the conduit at a piece of electrical equipment. A **coupling** is a conduit fitting used to join two pieces of conduit together. A **connector** is used to join a conduit to a junction box, outlet box, or piece of equipment. Threaded conduit, such as RMC and

IMC, typically use a pair of locknuts as a connector. RMC and IMC also use a plastic cap to protect the threads on one end of the conduit, while a coupling protects the threads on the other end. When exposed to the elements, such as when used as a service mast, the conduit is threaded into a weather-proof hub to prevent water intrusion into the equipment.

For unthreaded conduit such as EMT, couplings and connectors are slipped on the conduit ends and secured with set screws or compression. Set screw fittings are used in dry locations, while compression fittings are used in wet locations. See **Figure 9-6**. Fittings used in wet locations must be marked for use in wet locations.

PVC connectors are called **terminal adapters (TAs)**. A **female adapter (FA)** is a PVC fitting threaded on the inside. All PVC fittings are glued to the conduit using PVC cement that melts the two pieces together. Connectors for LFNC are either straight or have a 90-degree bend for tight spaces.

Straps are support fittings used to fasten conduit in place. RMC and IMC can be supported using any number of approved straps marked for heavy wall (HW) conduit. Straps used to secure EMT are marked as thin wall (TW) or EMT. LFNC and PVC use straps identified for heavy wall conduit.

Rigid PVC can be supported using heavy wall conduit straps or two-hole PVC straps. Because PVC sags between supporting straps to a greater degree than metallic conduit, the maximum spacing for supports for horizontal runs of PVC varies depending on the conduit size.

Additionally, PVC tends to expand and contract its length due to changes in temperature. This movement can be severe enough to pull fittings loose or otherwise damage the integrity of a conduit run. Therefore, when the change in length is expected to be 1/4″ or greater, a PVC expansion fitting should be used to compensate for the thermal expansion.

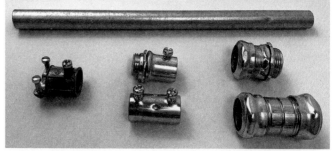

Goodheart-Willcox Publisher

Figure 9-6. Electrical metallic tubing (EMT) and fittings. The set-screw fittings on the left are used for dry locations, whereas the compression fittings on the right are used for damp and wet locations.

9.4.1 Conduit Bodies

A *conduit body* is defined as a portion of the conduit system
that provides access through a removable cover to its interior. Conduit bodies can provide additional pull points for
installing conductors in a conduit system. Pull points provide an opening in the conduit system where the conductors
can be accessed and pulled through the conduit. Regardless
of the conduit system used, there can be no more than 360°
of bends between pull points, **Figure 9-7**. Over 360° of bends
can hinder the pulling of the conductors through the conduit and may damage the conductor.

Conduit bodies are available for metallic and PVC conduits. Some conduit bodies are used to provide a tight, 90°

change in the direction of a conduit run. They are classified
by where the removable cover is located. The following are
common conduit bodies found in residential wiring:

- **LB**. An L-shaped conduit body that has a cover on the
  back when held as shown in **Figure 9-8**.
- **LR**. An L-shaped conduit body with the cover on the
  right.
- **LL**. An L-shaped conduit body with the cover on the
  left.
- **T**. A T-shaped conduit body.
- **C**. A straight conduit body used exclusively as a pull
  point or junction point on long runs of conduit.

Note that each cover must be accessible for removal and
inspection after installation.

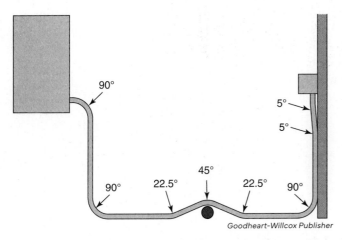

Goodheart-Willcox Publisher

Figure 9-7. The total degree of bends is not to exceed 360°.
In this example, the total number of bends is 340°.

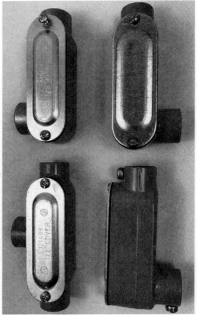

Goodheart-Willcox Publisher

Figure 9-8. Four common conduit bodies are the LL, LR, T,
and LB.

Summary

- The primary purpose of electrical conduit in the residence's electrical system is the physical protection of conductors.
- Electrical conduit is classified according to wall thickness, mechanical stiffness, and material used. Conduit size is based on the inner diameter of the conduit.
- The types of conduit used in residential wiring include rigid and intermediate metal conduit (RMC and IMC), electrical metal tubing (EMC), rigid polyvinyl chloride conduit (PVC), and liquidtight flexible nonmetallic conduit (LFNC).
- RMC and IMC are heavy-wall conduits with threaded ends. RMC has a thicker wall than IMC. RMC and IMC are difficult or impossible to bend with hand benders, so mechanical benders are used when bending is required. The most common use of RMC and IMC is the service mast to protect the service entrance conductors in an overhead service.
- EMT is a thin-wall conduit that is not threaded. Depending on the size, this conduit can be bent by hand using the proper tool, a mechanical bender, or factory-made bends. Like other metal conduits, EMT can also be used for a service mast that is attached to the side of a house, but EMT does not penetrate the roof.
- PVC is a lightweight, nonmetallic conduit often used in direct burial applications. PVC is sometimes used as the overhead mast for residential services that do not penetrate the roof.
- LFNC is often used in exposed, wet locations. It is weather-proof when used with the proper fittings, is sunlight resistant, and is rated for direct burial. LFNC is mostly used in short lengths to connect to vibrating equipment, like outdoor air conditioning condenser units.
- When installing RMC, IMC, and EMT, the conduit must be securely fastened within 3′ of any conduit termination, such as an outlet or junction box, or another piece of electrical equipment. It should be supported at intervals not exceeding 10′. PVC conduit must be securely fastened within 3′ of any conduit termination. When LFNC is installed in lengths greater than 6′, the conduit should be secured within 12″ of any conduit termination and securely fastened at intervals not exceeding 3′.
- Fittings are parts of the electrical system that performs a mechanical rather than an electrical function. A coupling is a conduit fitting that joins two pieces of conduit together; a connector is a conduit fitting used to join a conduit to a junction box, outlet box, or piece of equipment; bushings are fittings used on the conduit ends to protect the conductors from abrasion; and straps are support fittings used to secure the conduit in place.
- A conduit body is defined as a portion of the conduit system that provides access through a removable cover to its interior. Common types of conduit bodies found in residential wiring are LB, LR, LL, T, and C.

Know and Understand

1. The size of a conduit is based on the _____ of the conduit.
   A. inner diameter
   B. outer diameter
   C. inner radius
   D. outer radius

2. What is the best type of conduit to use for a through-the-roof service mast?
   A. Rigid metal conduit
   B. Electrical metallic tubing
   C. Rigid PVC conduit
   D. Liquidtight flexible nonmetallic conduit

3. *True or False?* Type EMT conduit can be easily bent by hand when using the appropriate tool.

4. What type of conduit is often used in exposed locations because it has a neater appearance than exposed cable?
   A. Rigid metal conduit
   B. Electrical metallic tubing
   C. Rigid PVC conduit
   D. Liquidtight flexible nonmetallic conduit

5. What is the best type of conduit to use for an underground installation?
   A. Rigid metal conduit
   B. Electrical metallic tubing
   C. Rigid PVC conduit
   D. Intermediate metal conduit

6. What type of conduit is mostly used to connect the outdoor air conditioning condenser unit to its disconnecting means?
   A. Rigid metal conduit
   B. Electrical metallic tubing
   C. Rigid PVC conduit
   D. Liquidtight flexible nonmetallic conduit

7. *True or False?* A connector is a conduit fitting used to connect two pieces of conduit together.

8. *True or False?* A coupling is a conduit fitting used to couple the conduit to an outlet box, junction box, or piece of equipment.

9. *True or False?* The fittings used for IMC and RMC are interchangeable.

10. What part of a conduit system provides access to the interior through a removable cover?
    A. A coupling
    B. A connector
    C. A terminal adapter
    D. A conduit body

Apply and Analyze

1. What is the trade size equivalent of a metric designator 21 conduit?

2. How does the *NEC* classify the interior of a direct-buried conduit?

3. Which conduit fitting is designed to protect the conductors from abrasion at the conduit's end?

4. What type of fitting is used with PVC conduit to counteract the changes in the length of the conduit resulting from temperature variations?

5. Which types of conduit types discussed in this chapter use heavy wall straps, and which use thin wall straps?

Critical Thinking

1. What are three uses for conduit bodies?

sockagphoto/Shutterstock.com

CHAPTER OUTLINE

Introduction

The *National Electrical Code* defines a device as "a unit of an electrical system, other than a conductor, that carries or controls electric energy as its principal function." Typical devices used in residential wiring are receptacles and switches. Electrical equipment, such as disconnects, motion sensors, and photosensors, are used for automatic lighting control and also considered devices but are more commonly referred to as switches. This chapter introduces the most common devices and switches used in residential wiring. Connections and installation methods will be covered in Chapter 26, *Device and Appliance Trim Out.*

10.1 Receptacles

Receptacles account for the majority of the devices installed in a residence. A *receptacle* is a device installed for the cord and plug connection of a piece of electrical equipment. The *NEC* has strict rules defining the number of receptacle outlets required based on the available wall space. Do not confuse the terms receptacle and outlet. Although all receptacles are outlets, not all outlets are receptacles. An *outlet* is defined as a point in the electrical system where electrical energy is taken to supply utilization equipment. For example, luminaires, smoke detectors, and ventilation fans are outlets. There are many types of receptacles that will be explored in the following sections. *NEC* receptacle requirements are covered in Chapter 18, *Branch Circuit/Receptacle Requirements.*

10.1.1 Duplex Receptacles

Duplex receptacles are the most common devices installed in a residence and are designed with two electrical receptacle outlets connected with a mounting strap, or yoke. See **Figure 10-1**. The *yoke* contains two captive 6-32 machine screws that mount the device to its box. A duplex receptacle has two slot shapes of different sizes to prevent plugs from improper insertion. The short slot is connected to the ungrounded,

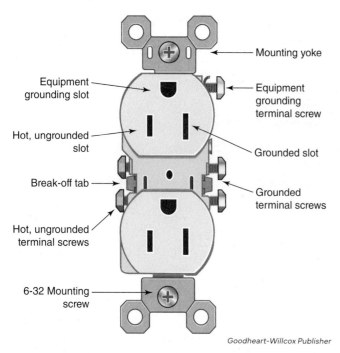

Mounting yoke

Equipment grounding slot

Equipment grounding terminal screw

Hot, ungrounded slot

Grounded slot

Break-off tab

Grounded terminal screws

Hot, ungrounded terminal screws

6-32 Mounting screw

Goodheart-Willcox Publisher

Figure 10-1. The front of a 15-amp duplex receptacle.

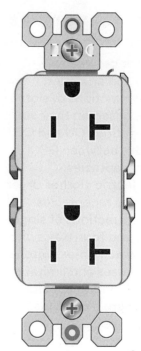

Goodheart-Willcox Publisher

Figure 10-2. A 20-amp duplex receptacle. Note the T-shaped slot for the grounded prong.

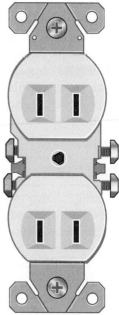

Goodheart-Willcox Publisher

Figure 10-3. A 15-amp two-prong, ungrounded duplex receptacle. Note the absence of the horseshoe-shaped slot for the equipment grounding prong.

hot conductor. The long slot is connected to the grounded conductor, which is sometimes misidentified as the neutral conductor.

Duplex receptacles of 120 volts are manufactured with 15-amp and 20-amp ratings. A receptacle rated 20-amps has a T-shaped grounded slot, **Figure 10-2**. The horseshoe shaped slot is connected to the equipment grounding conductor. Older residences built before the 1962 *Code* may have outdated two-prong receptacles installed, **Figure 10-3**. These receptacles are not aligned with the *NEC* requirement for separate equipment grounding conductor.

Duplex receptacles have five terminal screws on the device—two brass, two silver, and one green. The brass-colored screws are connected to an ungrounded, hot conductor with black or red insulation. The silver screws are used for white, grounded conductors. The green hexagonal screw is connected to a bare or green equipment grounding conductor. The screw is located on the receptacle's yoke that functions as a component in the equipment grounding system. A cover plate also screws directly into the yoke. A removable tab is located between the two silver-colored terminal screws and the two brass-colored terminal screws. Removing the tab allows the receptacle to split so that one of the outlets can be switch-controlled while the other remains energized at all times.

The back of a duplex receptacle is shown in **Figure 10-4**. Note the push-in wiring terminals, also called *back-stabs*, are rated only for 14 AWG conductors. The proper amount of insulation must be removed before using push-in terminals. A strip gauge indicates the amount of insulation that needs to be stripped to use push-in connectors. Most

manufacturers allow for either push-in terminals or terminal screws for conductor terminations, but not both. The back of the receptacle is also stamped to indicate the type of conductor permitted to be attached. For example, *CU* indicates the receptacle is rated for copper conductors only, and *CO/ALR*

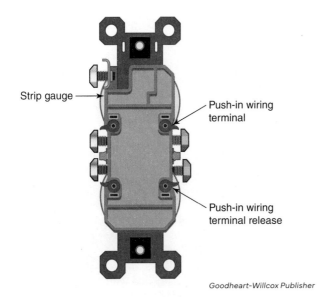

Figure 10-4. Back of a duplex receptacle.

Strip gauge

Push-in wiring terminal

Push-in wiring terminal release

indicates the receptacle is rated for copper, aluminum, or copper-clad aluminum conductors.

Note that since most duplex receptacles are installed closer to the ground, many are required to be tamper-resistant for safety.

10.1.2 Tamper-Resistant Receptacles

Tamper-resistant receptacles were developed to protect children from electrical shock caused by inserting a metal object, such as a key or a hairpin, into the slots of a receptacle. These receptacles employ a spring-loaded shutter device that rejects a foreign object's insertion into either slot, only allowing the insertion of a two- or three-pronged plug, **Figure 10-5**.

All new residences are required by the *NEC* to have these receptacles installed, even in homes without children, waiting areas of certain facilities, and anywhere children may not be closely supervised. More specifically, *Section 406.12* requires all 15-amp and 20-amp, 125-volt and 250-volt nonlocking-type receptacle outlets in a residence to be tamper-resistant.

There are exceptions to this general rule. Tamper-resistant receptacles are not required if they are located higher than 5 1/2′ from the floor, if they are part of a luminaire or appliance, or if the receptacle is behind a piece of equipment that is not easily moved, such as a refrigerator. Non-grounding receptacles can be used for replacements as permitted in *Section 406.4(D)(2)(a)*.

10.1.3 Ground-Fault Circuit Interrupter (GFCI) Receptacles

Another receptacle with unique safety features is a ground-fault circuit interrupter (GFCI) receptacle. Recall that a ground fault is an unintended, electrically conductive connection between an ungrounded, hot conductor and normally non-current-carrying conductors or metallic components of an electrical system. *Ground-fault circuit interrupter (GFCI) receptacles*, **Figure 10-6**, help prevent electrical shocks by opening the circuit when a ground fault condition is detected. They are recognized by their rectangular shape and the TEST and RESET buttons on the front. They function by employing an internal electronic circuit to monitor the amount of current flowing in and out of the device. In normal operation, the currents should be the same. However, if the device senses a current imbalance of 6 milliamps or more, it will act quickly to open the circuit and shut off the power within 25 milliseconds.

Where receptacles require GFCI protection, a GFCI device can provide protection to other standard receptacles by using a feed-through feature known as a *line-load connection*, which will be discussed in Chapter 26, *Device and Appliance Trim Out. Section 210.8(A)* requires GFCI protection for receptacles that serve any kitchen countertop and dishwasher circuits, bathrooms, and basements (finished or unfinished, outdoors, or in crawl spaces) within six feet of a sink, and laundry areas—basically any location where standing water may be present. GFCI receptacles can also replace a two-prong, ungrounded receptacle where there is no equipment grounding conductor present.

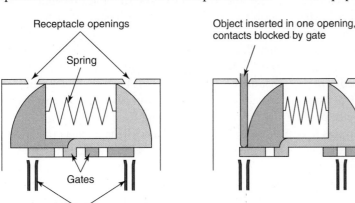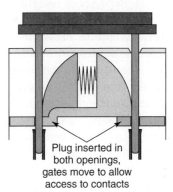

Figure 10-5. Cutaway view of a tamper-resistant duplex receptacle.

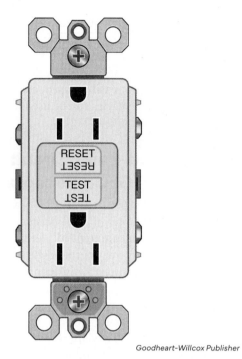

Figure 10-6. A 15-amp GFCI receptacle. Note the TEST and RESET buttons on the face of the receptacle.

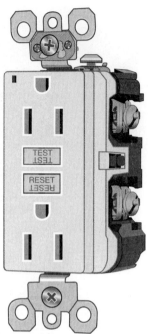

Figure 10-7. A 15-AFCI Receptacle. Note the similarity to the GFCI receptacle.

It should be noted that the GFCI will not protect against short circuits between two hot conductors. Nor does it protect against shock when a user comes into contact with both the grounded and the ungrounded circuit conductors simultaneously, as there would be no current imbalance under this condition. Furthermore, the GFCI does not limit the magnitude of the ground fault but rather limits the amount of time the fault current is allowed to flow.

The wiring connections for GFCI receptacles will be discussed in Chapter 28, *Distribution Panel Trim Out*.

10.1.4 Arc-Fault Circuit Interrupter (AFCI) Devices

Recall that an arc fault is an electrical discharge created when current flows through an unplanned path, and an *arc-fault circuit interrupter (AFCI) receptacle* is a device intended to protect against electrical fires caused by arc faults. Visually, an AFCI receptacle, **Figure 10-7**, is very similar to the GFCI receptacle, but the two should not be confused. They differ in important functional ways. While GFCI devices protect people from electrical shocks, an AFCI device prevents electrical accidents within the walls. They do this by recognizing the characteristics unique to arcing and then de-energizing the circuit when an arc fault is detected.

Typically ARCI protection is provided by an AFCI circuit breaker in the distribution panel but can also be provided by an AFCI receptacle installed at the first outlet in the circuit if several other conditions are met. In dwelling units, arc-fault circuit protection is required in all 15-amp

and 20-amp, 125-volt circuits supplying outlets and devices installed in kitchens, family rooms, dining rooms, living rooms, parlors, libraries, dens, bedrooms, sunrooms, recreational rooms, closets, hallways, laundry areas, or similar rooms or areas. Refer to *Section 210.12(A)* for more information on AFCI receptacle installation requirements.

10.1.5 Electric Dryer and Electric Range Receptacles

Because they are used to produce a great deal of heat, electric clothes dryers and kitchen ranges have higher current requirements than other residential plug-connected loads. Therefore, receptacles that serve these appliances are physically larger than the standard duplex receptacle, requiring a double-gang box or cover for installation. Surface-mounted range and dryer receptacles do not need a box for installation. See **Figure 10-8**. Receptacles for electric clothes dryers and kitchen ranges are dual rated at 125/250 volts, which means the appliance needs both the higher and lower voltages to operate. Ranges and dryers use the higher voltage for the heating elements and the lower voltages for the lights, clock, and timer. The dryer receptacle is rated for 30 amps, and the range receptacle is rated at 50 amps.

Two different wiring methods are used for these appliances: the three-wire circuit and the four-wire circuit. Prior to the 1996 *Code*, three-wire branch circuits were acceptable using the bare conductor as both the grounded, neutral conductor and the equipment grounding conductor under certain conditions. As a result, there are millions of existing

A

B   *Goodheart-Willcox Publisher*

Figure 10-8. A—Four-wire range receptacles. B—Dryer receptacles. Note the L-shaped neutral slot on the dryer receptacles. When purchasing a new range or dryer, the cord is sold separately.

homes with three-wire circuits for the clothes dryer and kitchen range. Since the 1996 *Code*, all range and dryer circuits are required to be of the four-wire type with two hot conductors, an insulated grounded neutral conductor, and a bare or insulated equipment grounding conductor.

10.2 **Switches**

Switches are devices used to open or close a part of an electrical circuit. In residential wiring, switches are mainly used to control light fixtures, but they can also be used with appliances and equipment, such as a garbage disposal or a whole-hose house ventilation fan.

Switches all function similarly—to control the flow of current to a lighting load—but come in different designs.

The most common design is a standard toggle switch, where the fixture is turned on or off by moving the switch up or down. A rocker switch, also called a *decorator-style switch*, is popular alternative that functions identically to the toggle switch but is more appealing for incorporating into interior design. A rocker switch might also be ideal for people with limited mobility as they require minimal movement to operate. Most device manufacturers also offer a line of high-end programmable switches that connect to Wi-Fi. For historical renovations, push-button switches are popular with homeowners restoring a residence built in the 1920s and 30s, when these switches were the standard. Push-button switches are still common in a wide range of various applications, including portable equipment and power tools, electronics, and other household appliances, but are not widely utilized for light fixtures in residential wiring.

Section 210.70(A) requires at least one wall switch-controlled lighting outlet to be installed in every habitable room, kitchen, and bathroom. In rooms other than the kitchen or bathroom, a wall switch that controls one or more receptacles can take the place of the lighting outlet, so long as the wall switch controls one or more cord and plug-connected lamps. Almost all switches are shipped with captive 6-32 screws used to mount the device in the box, but some higher-end devices require special mounting brackets and do not include mounting screws. In residential wiring, electricians will install different types of switches depending on locations or functionality, including single-pole, three-way, and four-way switches, as well as dimmers and combination devices.

10.2.1 **Single-Pole Switches**

A *single-pole switch*, **Figure 10-9**, is used to control a load from a single wall location. It has two brass-colored terminal screws and a green equipment grounding conductor terminal screw. It is a toggle switch with designations on the toggle for ON in the up position and OFF in the down position. On rocker switches, these ON/OFF markings may not appear.

10.2.2 **Three-Way Switches**

A *three-way switch*, **Figure 10-10**, is always installed in pairs and used to control loads from two wall locations, such as a long hallway or at the top and bottom of a staircase. It has three terminal screws and a green equipment grounding conductor screw. Two of the terminal screws are brass colored, and the third terminal screw is black. The brass screws are for the "traveler" conductors, and the black screw is for the "common" conductor. Three-way switches have no ON and OFF designation on the toggle because either switch position, up or down, can turn the fixture on or off depending on the condition of the other switch in the circuit. Chapter 13, *Basic Switched Circuits,* will cover installing these switches more in detail.

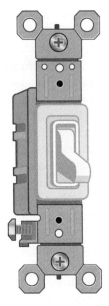

Goodheart-Willcox Publisher

Figure 10-9. The single-pole switch.

Goodheart-Willcox Publisher

Figure 10-10. The three-way switch.

10.2.3 Four-Way Switches

A *four-way switch*, **Figure 10-11**, controls a lighting load from three or more locations. A four-way switch is always installed between a pair of three-way switches, and there is no limit as to how many can be used. The four-way switch has four terminal screws and the green equipment grounding screw. Typically, two of the screws will be brass-colored, and two are black-colored and often labeled input and output. As with the three-way switch, there is no ON/OFF designation on the four-way switch because either position may turn the fixture on or off based on the conditions of the other switches in the circuit.

Goodheart-Willcox Publisher

Figure 10-11. The four-way switch.

10.2.4 Dimmers

A *dimmer switch*, **Figure 10-12**, can turn a lighting load on or off and reduce the output level of light emitted. In the past, dimmers used resistors or diodes to reduce the amount of voltage delivered to the light bulb but not the power consumption—it merely diverted some energy to be lost as heat. Modern dimmers use electronic circuitry to manipulate the voltage sent to a luminaire, reducing the light output and conserving energy. Dimmers are frequently installed to control the ambiance in dining rooms.

Dimmers are available in a wide variety of styles. The most common is the "push-push" rotary dimmer, in which the light is turned off and on by pushing the rotary knob. The dimming function is accomplished by rotating the knob. Other styles include the sliding knob, the toggle/slide, the rocker/slide, and the toggle-only dimmer.

Goodheart-Willcox Publisher

Figure 10-12. Assorted dimmer styles. From left to right; Rotary dimmer, toggle switch with slide dimmer, rocker switch with slide dimmer.

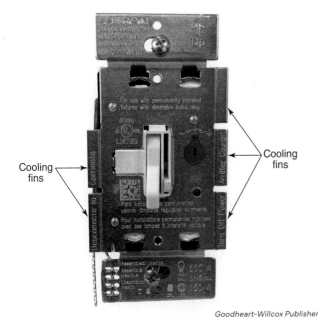

Goodheart-Willcox Publisher

Figure 10-13. Dimmer with removable cooling fins.

Dimmers have a wattage rating that should not be exceeded. A typical dimmer is rated at 600 watts, but larger ratings are available. The wattage rating is based on the sum of the watts of all the bulbs being dimmed. For example, a large chandelier may have twelve sixty-watt bulbs installed, resulting in a total of 720 watts. Thus, this fixture would need a dimmer rated for at least 720 watts.

It is common for dimmers to become warm during normal operation. Many dimmers use cooling fins to help dissipate this heat, **Figure 10-13**. However, these fins may need to be removed to fit multiple devices in a box, which as a result, reduces the dimmer's rating. This concept will be covered in the manufacturer's installation instructions.

Most dimmers are connected to the circuit conductors using *pigtails,* which are short lengths of conductor, rather than terminal screws. Three-way dimmers are available, but only one three-way dimmer is used in conjunction with a standard three-way switch in a three-way dimming circuit. So, for example, if two three-way dimmers were used, the brightest lighting level would be whatever the lowest level is on the other switch.

10.2.5 Combination Devices

Combination devices have two devices mounted on the same yoke. There can be two single-pole switches, a single-pole switch and a three-way switch, a switch and a receptacle, or a single-pole switch with a pilot light. Receptacles with integrated USB ports are available in many configurations, **Figure 10-15**. Combination devices are often used where a location requires multiple devices, but there is not enough room between the studs for a multi-gang box. Using stacked switches, for example, can allow for two switches in a single-gang box or four switches in a double-gang box.

Dimmers are *not* to be used for ceiling fan speed control. If used for this purpose, the fan motor will be damaged. Special speed-control switches are available to control fan speed without pulling the speed-control chain on the fan. The model shown in **Figure 10-14** also has a light switch/dimmer incorporated.

Goodheart-Willcox Publisher

Figure 10-14. Fan speed controller with light dimming capability.

Goodheart-Willcox Publisher

Figure 10-15. Assorted combination devices. From left to right—stacked switch with two single-pole switches on a common yoke, switch with a pilot light to indicate the load is on, stacked switch and single receptacle on the same yoke.

Summary

- Although all receptacles are outlets, not all outlets are receptacles. A receptacle is a device installed for the cord and plug connection of a piece of electrical equipment. An outlet is defined as a point in the electrical system where electrical energy is taken to supply utilization equipment.
- In a receptacle outlet, the short slot is connected to the ungrounded, hot conductor, and the long slot is connected to the grounded conductor. A receptacle rated 120 volt/20-amps will have a T-shaped grounded slot. The horseshoe-shaped slot is connected to the equipment grounding conductor.
- The back of duplex receptacles are stamped to indicate the type of conductor permitted to be attached. CU indicates the receptacle is rated for copper conductors only, and CO/ALR indicates the receptacle is rated for copper, aluminum, or copper-clad aluminum conductors.
- Visually, the AFCI receptacle is very similar to the GFCI receptacle, but they differ in important functional ways. Mainly, an AFCI device protects against electrical fires within the walls while GFCI receptacles protect people from electrical shocks.
- The receptacles for electric clothes dryers and kitchen ranges are dual rated at 125/250 volts; the dryer receptacle will be rated for 30 amps, and the range receptacle is rated at 50 amps. The dual-voltage rating means the appliance needs both the higher and the lower voltages to operate.
- A single-pole switch is used to control a lighting load from a single location; three-way switches are used to control lighting loads from two locations, such as a long hallway or staircase, and they are always installed in pairs; and four-way switches control a lighting load from three or more locations, and they are always installed between a pair of three-way switches.
- Dimmer switches use electronic circuitry to manipulate the voltage sent to a luminaire, reducing the light output and conserving energy. Dimmer styles include a rotary knob, sliding knob, the toggle/slide, the rocker/slide, and the toggle-only dimmer.
- Three-way dimmers are available, but only one three-way dimmer is used in conjunction with a standard three-way switch in a three-way dimming circuit.
- Combination devices have two or more devices mounted on the same strap. They are often used where a location requires multiple devices, but there is not enough room between the studs for a multi-gang box.

Know and Understand

1. A point in the electrical system where electrical energy is taken to supply utilization equipment is a(n) _____.
   A. receptacle
   B. outlet
   C. luminaire
   D. branch circuit

2. *True or False?* All outlets are receptacles.

3. On the face of a duplex receptacle, the long slot is the _____.
   A. grounded slot
   B. ungrounded slot
   C. equipment grounding slot
   D. hot conductor slot

4. On the face of a duplex receptacle, the short slot is the _____.
   A. grounded slot
   B. ungrounded slot
   C. equipment grounding slot
   D. tamper-resistant slot

5. On the face of a duplex receptacle, the horseshoe-shaped slot is the _____.
   A. grounded slot
   B. ungrounded slot
   C. equipment grounding slot
   D. hot conductor slot

6. When wiring a duplex receptacle, the white conductor is connected to the _____ terminal screw.
   A. silver
   B. brass
   C. green
   D. black

7. When wiring a duplex receptacle, the black or red conductor is connected to the _____ terminal screw.
   A. silver
   B. brass
   C. green
   D. black

8. When wiring a duplex receptacle, the bare or green conductor is connected to the _____ terminal screw.
   A. silver
   B. brass
   C. green
   D. black

9. *True or False?* A single GFCI receptacle can provide ground-fault protection to multiple standard receptacles.

10. *True or False?* A GFCI receptacle is designed to protect against shock when someone simultaneously contacts both the grounded and ungrounded conductors.

11. What type of switch is used to control a lighting load from one location?
    A. A single-pole switch
    B. A double pole switch
    C. A three-way switch
    D. A four-way switch

12. What type of switch is used to control a lighting load from two locations?
    A. A single-pole switch
    B. A double-pole switch
    C. A three-way switch
    D. A four-way switch

13. What type of switch is used to control a lighting load from three or more locations?
    A. A single-pole switch
    B. A double-pole switch
    C. A three-way switch
    D. A four-way switch

14. What switches are always installed in pairs?
    A. Single-pole switches
    B. Double-pole switches
    C. Three-way switches
    D. Four-way switches

15. *True or False?* Dimmer switches can be used to control the speed of a ceiling fan.

Apply and Analyze

1. What is the yoke of a device?
2. What are tamper-resistant receptacles and why are they required in all new residences?
3. What does the rating CO/ALR mean on a receptacle?
4. Where is GFCI protection required in a residence?
5. What is the difference between GFCI and AFCI receptacles?
6. What does a dual-voltage rating mean in relation to electric clothes dryers and electric ranges?
7. When installing a dimmer on a three-way lighting circuit, why is only one dimmer installed?

Critical Thinking

1. You are an electrician roughing in a lighting circuit in a new home. The plans call for four switches to be installed by the front door, but there is only enough space between the studs for a two-gang device box. What can be done to fit four switches in the space available?

CHAPTER **11**

Panelboards, Overcurrent Protective Devices, and Disconnects

The Toidi/Shutterstock.com

SPARKING DISCUSSION

How does an overcurrent protective device determine the rating of the branch circuit?

After completing this chapter, you will be able to:

- Identify the components in a distribution panel.
- Summarize the main breaker and main lug classifications of distribution panels.
- Explain the purpose of overcurrent protective devices in an electrical circuit.
- Describe how a circuit breaker operates under thermal overload, ground fault, and short circuit conditions.
- Describe how an inverse-time breaker operates.
- Explain the AIC rating of a circuit breaker.
- Identify when to utilize single-pole, double-pole, and twin circuit breakers.
- Describe how a dual-element time-delay fuse operates.
- Discuss the benefits of using the Type S fuse over the Edison base fuse.
- Identify three disconnecting means and appliances or equipment that may utilize them.
- Explain a NEMA 3R rating on enclosures.

AIC rating
arc-fault circuit interrupting (AFCI) breaker
bimetallic strip
circuit breaker
disconnect
double-pole circuit breaker
dual-element time-delay fuse
dual-purpose GFCI/AFCI breaker
Edison base fuse
fuse
ground-fault circuit interrupting (GFCI) breaker
inverse-time breaker
main circuit breaker
NEMA 3R rating
overcurrent protective device (OCPD)
panelboard
pull-out disconnect
service disconnecting means
single-pole circuit breaker
twin circuit breaker
Type S fuse

Introduction

In an electrical distribution system, a ***panelboard*** is an enclosure that serves as the origin for the branch- and feeder-circuits wiring that feed various loads throughout a house. It also houses fuses and circuit breakers, or ***overcurrent protective devices (OCPD)***, that provide protection for these branch circuits from overloads and short circuits. These devices function to open the circuit under a fault condition. Another device, called a ***disconnect***, removes power from a piece of equipment for servicing and is required on all large appliances.

This chapter covers foundational information on distribution panels, overcurrent protective devices, and disconnects. The installation and wiring of this equipment will be covered in later chapters.

11.1 Panelboards

Panelboards, also called *distribution panels*, house fuses or circuit breakers that protect the branch circuit conductors in a residence. Panelboards are available with or without a main circuit breaker. The ***main circuit breaker*** is crucial in the electrical system because it offers a way to disconnect power from the entire panelboard. A main breaker feeds and protects the entire panel from an overcurrent condition. Individual circuits will likely trip before the main breaker.

A panel with the main circuit breaker is called a main breaker panelboard. A panel with no main breaker is called an MLO, or main lugs only. These two types of panels are functionally identical, **Figure 11-1**, but have different installation requirements. These requirements will be discussed in Chapter 24, *Service Equipment*.

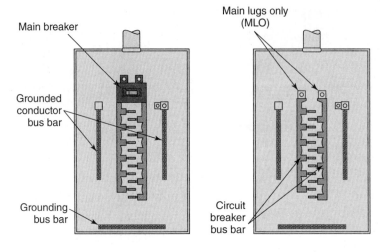

Figure 11-1. Comparison of a main breaker panel and an MLO panel.

Recall that electrical systems in the United States are 3-wire, 120/240-volt, meaning there are two ungrounded, hot conductors and one grounded, neutral wire conductor. There is a nominal 240 volts between the two ungrounded conductors and 120 volts between either ungrounded conductor and the grounded conductor, **Figure 11-2**. Although this system is called a 3-wire system, there is a fourth wire—the equipment grounding conductor.

The two hot, ungrounded conductors, Line 1 (L1) and Line 2 (L2), are connected to the main circuit breaker's lugs or the main lugs of the MLO panel. The lugs feed the main buses where circuit breakers attach. There are two separate buses, and they each supply alternating spaces. This arrangement allows a double-pole breaker to receive voltage from each ungrounded conductor to provide 240 volts. A single-pole circuit breaker attaches to just one bus to provide 120 volts. Breaker spaces are numbered from top to bottom with odd numbers on the left and even numbers on the right.

In an MLO panelboard, the grounded conductor bus is isolated from the metallic enclosure using a nonmetallic standoff, **Figure 11-3**. In a main breaker panel, the grounded conductor bus can be bonded to the metallic enclosure, or it might be isolated, depending on the location of the service

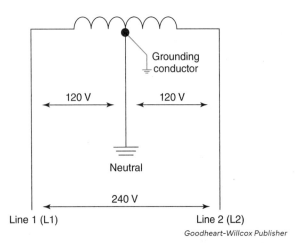

Goodheart-Willcox Publisher

Figure 11-2. Schematic of a 120/240-volt electrical system. Note the neutral and the equipment grounding conductor are connected.

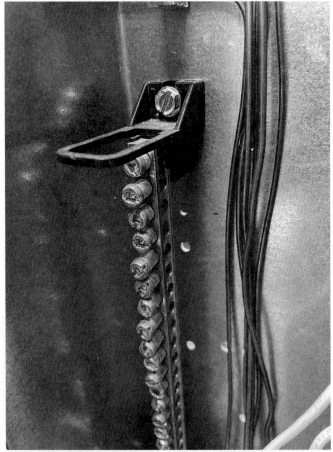

Goodheart-Willcox Publisher

Figure 11-3. The neutral bus of the MLO panel is isolated from the panel enclosure using a nonmetallic standoff.

disconnecting means. The **service disconnecting means** disconnects the entire electrical service from the utility. It may be located outside the house at the service equipment, or the service disconnecting means could be the main breaker in the panelboard. After the service disconnecting means, the white grounded conductors and the bare equipment grounding conductors are forever isolated from one another. This topic will be discussed in detail in Chapter 25, *Service Installation*.

The equipment grounding conductor bus is attached directly to and bonded to the enclosure. This arrangement allows overcurrent protective devices to operate under a ground fault and is an example of bonding normally noncurrent-carrying metallic electrical equipment.

11.2 Overcurrent Protective Devices

Electricians size conductors based on the expected current the conductors can carry. Recall from Chapter 8, *Electrical Conductors and Connectors*, conductors have a specific ampacity that cannot be exceeded. Overcurrent protective devices, such as circuit breakers and fuses, protect the conductors by acting as a safety valve that opens the circuit when the current flow exceeds its allowable limit. This can result from overloading the circuit with too many cord-and-plug loads, for example. Faulty equipment can also cause currents to rise to unexpected levels and trip a circuit breaker or blow a fuse.

An overcurrent protective device establishes the rating of the branch circuit, not the size of the conductor or the size of the load. The ampacity of the conductor must be greater or equal to the overcurrent device protecting it. Otherwise, the conductor insulation can be degraded and deteriorate by the excess current before the breaker can respond.

Circuit breakers and fuses are the most common types of overcurrent protection in the field. In modern residential wiring, circuit breakers are the primary type of overcurrent protection, while older homes use fuses. The following sections will outline these two devices and discuss their function and considerations for use.

11.3 Circuit Breakers

Circuit breakers are the most common type of overcurrent protection and are required in all new construction. A **circuit breaker** is a device that employs a trip mechanism to open the circuit and shut off the supply of power when the breaker detects an overload, short circuit, or ground fault condition, **Figure 11-4.**

Circuit breakers use thermal and magnetic properties to operate. A circuit breaker has a **bimetallic strip** that

Goodheart-Willcox Publisher

Figure 11-4. Assorted circuit breakers used in residential wiring.

monitors the thermal properties of the device, **Figure 11-5**. Recall that any current flow in a conductor produces heat, and the amount of heat is directly proportional to the amount of current. A bimetallic strip is made by fusing two dissimilar metals together. When the strip is heated by excessive current flow, it deflects or bends, and this movement is used to open a set of spring-loaded contacts inside the circuit breaker, shutting off the power to the circuit.

In cases of short circuits or ground faults, the trip mechanism uses the breaker's magnetic properties to open the circuit. When current flows in a conductor, a magnetic field is established around the conductor, and the magnetic field is directly proportional to the amount of current. The extremely high currents of a ground fault or short circuit pull the magnetic armature away from the latching lever, which opens the spring-loaded contacts and shuts off the power to the circuit. Spring-loading the contacts facilitates the speed at which they open and close and reduces the amount of arcing as they do.

To avoid unnecessary tripping, all circuit breakers are designed to have a slight time delay. Some loads, notably motor loads, exhibit an inrush of current upon start-up that is greater than their normal operating current. The time delay allows the momentary high current to flow and subside without tripping the circuit breaker. Circuit breakers

Line terminal

Bimetallic strip

Load terminal

Magnetic pole piece

Contacts

Goodheart-Willcox Publisher

Figure 11-5. Cutaway view of a typical residential circuit breaker. This breaker is in the OFF position and the contacts are open.

used in residential wiring are classified as ***inverse-time breakers***, which means the time delay decreases as the magnitude of the current increases. In other words, the higher the fault current, the faster the device opens.

When circuit breakers trip, they are designed to be reset rather than replaced. Circuit breakers are reliable devices and should last for the life of the electrical system.

11.3.1 Circuit Breaker Ratings

All circuit breakers have specific ratings for ampere and voltage. The amperage rating is the continuous current-carrying ability of a circuit breaker. As mentioned, the breaker's amperage rating should never be greater than the ampacity of the wire it is protecting (an exception is the conductors that serve the outdoor air conditioning compressor unit, which will be discussed in a future chapter). For example, a 60-amp breaker could not protect an overloaded conductor rated for 20 amps. The voltage rating of a circuit breaker must be equal to or greater than the circuit voltage. This is not an issue for residential wiring, as all residential breakers are rated for 120/240 volts.

PRO TIP   **Standard Fuse and Circuit Breaker Sizes**

Standard sizes for fuses and fixed trip circuit breakers, per *Section 240.6*, are 15, 20, 25, 30, 35, 40, 45, 50, 60, 70, 80, 90, 100, 110, 125, 150, 175, 200, 225, 250, 300, 350, and 400.

Circuit breakers are also rated according to the level of fault current they can safely interrupt, which is called the ***AIC rating***, or ampere interrupting capacity, **Figure 11-6**. Fault currents can reach several thousand amps. Most residential circuit breakers have an AIC rating of 10,000 A, meaning the breaker can withstand an inrush current of up to 10,000 A before vaporizing in an arc blast explosion.

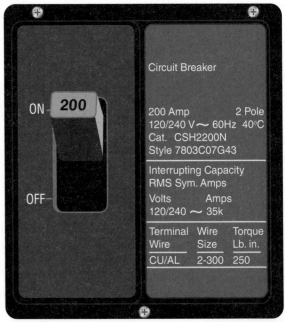

ON 200

OFF

Circuit Breaker

200 Amp   2 Pole
120/240 V ~ 60Hz 40°C
Cat. CSH2200N
Style 7803C07G43

Interrupting Capacity
RMS Sym. Amps
Volts   Amps
120/240 ~ 35k

Terminal Wire	Wire Size	Torque Lb. in.
CU/AL	2-300	250

Goodheart-Willcox Publisher

Figure 11-6. Circuit breaker AIC rating.

11.3.2 Types of Circuit Breakers

There are many types of circuit breakers available for different uses and purposes in residential wiring. This section will cover single-pole and double-pole breakers, twin breakers, GFCI and AFCI breakers, and dual-purpose GFCI/AFCI breakers.

Single-pole circuit breakers are used on 120-volt circuits, **Figure 11-7**. They take one space in the distribution panel. In a residence, single-pole breakers are used exclusively for 15- and 20-amp circuits. Most branch circuits in a house are fed from single-pole breakers. *Double-pole circuit breakers* are used for 240-volt and 120/240-volt circuits, **Figure 11-8**. A double-pole breaker takes two spaces in the distribution panel. They generally feed a specific appliance such as the electric range or the clothes dryer. *Twin circuit breakers* are two single-pole breakers that take only one space in the panel. Some manufacturers refer to these as *tandem breakers*. See **Figure 11-9**. Each breaker delivers 120 volts but using both will not provide 240 volts. However, twin breakers are available for 240-volt circuits, **Figure 11-10**. These double-pole twin breakers can have two different ampere ratings. Twin breakers are primarily used when adding a circuit to a panelboard with no available spaces and where the twin replaces an existing breaker. Most panelboards limit the number of twin breakers allowed, **Figure 11-11**.

Goodheart-Willcox Publisher

Figure 11-7. A typical 20-amp single-pole breaker.

Goodheart-Willcox Publisher

Figure 11-8. A typical 20-amp double-pole breaker.

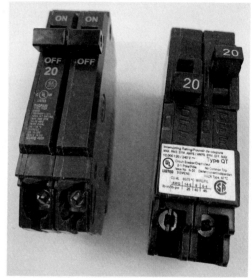

Goodheart-Willcox Publisher

Figure 11-9. 120-volt twin breakers. These breakers take up one space in the distribution panel.

Goodheart-Willcox Publisher

Figure 11-10. A 240-volt twin 30-amp/30-amp breaker. Note the outer two handles are tied together as are the inner two handles. This breaker takes two spaces in the distribution panel.

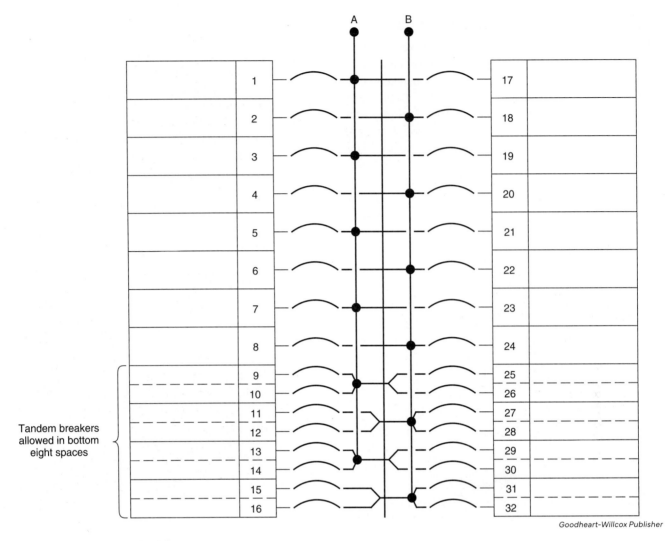

Tandem breakers
allowed in bottom
eight spaces

Goodheart-Willcox Publisher

Figure 11-11. Panel door schedule.

Ground-fault circuit interrupting (GFCI) breakers are devices designed to provide ground-fault protection to the entire branch circuit. Ground faults, GFCI receptacles, and ground-fault protection requirements have been discussed in Chapter 2, *Safety in Residential Wiring*, and Chapter 10, *Electrical Devices*. Similar to the GFCI receptacle, the GFCI circuit breaker functions by monitoring the current flow in the grounded and the ungrounded conductor and automatically opens the circuit if it senses an imbalance of 4-6 milliamps. The grounded and the ungrounded branch circuit conductors are both connected to the GFCI breaker. The grounded conductor's connection to the electrical service's grounding system is accomplished with a pigtail from the breaker to the neutral bus bar in the distribution panel. Some GFCI breakers have a mechanism that clips on the neutral bus bar and eliminates the pigtail. See **Figure 11-12**.

Arc-fault circuit interrupting (AFCI) breakers, **Figure 11-13**, are devices designed to provide arc-fault

protection to the entire circuit. Arcing faults, AFCI receptacles, and arc-fault protection requirements have been discussed in Chapter 2, *Safety in Residential Wiring*, and Chapter 10, *Electrical Devices*. AFCI circuit breakers are the most common method of AFCI protection. AFCI breakers are installed in a similar manner to GFCI breakers. Both branch circuit conductors are brought to the AFCI breaker, and the connection is made to the electrical service's grounding system with a pigtail or a plug-on neutral feature.

> **PRO TIP**   **AFCI Circuit Breakers**
>
> ACFI circuit breakers are sometimes referred to as *CAFIs* or *combination arc-fault circuit interrupters*. Recall from Chapter 2 that this device protects against both series and parallel arcing faults. However, it does *not* mean that the breaker provides both AFCI and GFCI protection.

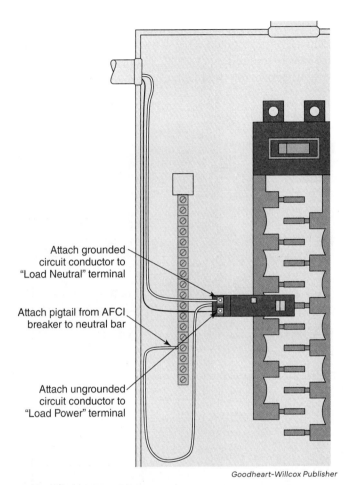

Attach grounded circuit conductor to "Load Neutral" terminal

Attach pigtail from AFCI breaker to neutral bar

Attach ungrounded circuit conductor to "Load Power" terminal

Goodheart-Willcox Publisher

Figure 11-12. The grounded conductor's connection to the electrical service's grounding system is accomplished with a pigtail from the breaker to the neutral bus bar in the distribution panel.

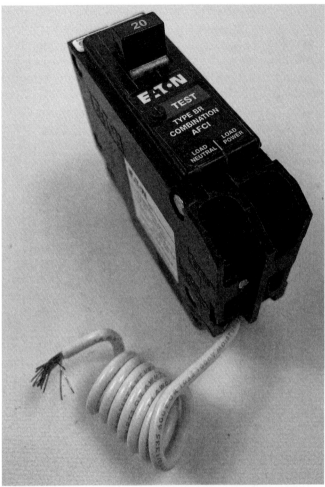

Goodheart-Willcox Publisher

Figure 11-13. An AFCI circuit breaker with a pigtail.

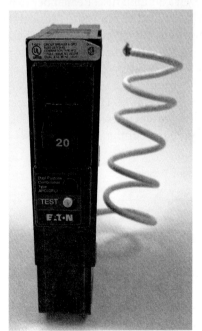

Goodheart-Willcox Publisher

Figure 11-14. A dual-purpose combination AFCI/GFCI circuit breaker.

Dual-purpose GFCI/AFCI breakers, **Figure 11-14**, are available and protect against both AFCI and GFCI hazards. The NEC requires some circuits, notably in residential kitchens and laundry areas, to have both GFCI and AFCI protection. This can be accomplished using a GFCI receptacle in combination with an AFCI circuit breaker or using a dual-purpose GFCI/AFCI circuit breaker.

11.4 Fuses

A *fuse* is a form of circuit protection found in many older homes. Fuses open the circuit by means of a fusible link that melts under high current conditions and opens the circuit. The fusible link is the smallest wire in an electrical circuit designed to open quickly under a short circuit or ground fault condition. Fuses are one-use only and need to be replaced when blown.

Time-delay fuses can provide overload protection, as well as short circuit and ground-fault protection, in circuits where surges and temporary overloads might be more common, such as in motor loads. These fuses allow for momentarily high startup currents before tripping the device. The ***dual-element time-delay fuse*** uses both a fusible link and a solder link to operate, **Figure 11-15**. A high current will melt the fusible link, but the solder link has a low melting point. Under a sustained thermal overload condition, the current eventually becomes hot enough to melt the solder link and release a spring that opens the circuit.

11.4.1 Types of Fuses

There are many types of fuses, but the two major types used in residential wiring are the screw-in plug fuse and the cartridge fuse. The screw-in plug fuse is the most common type used on 120-volt branch circuits and, as the name suggests, screws into the panel like a lightbulb in a lighting fixture. The cartridge fuse is used in pairs to protect 240-volt branch circuits and as the service disconnect. A pair of cartridge fuses are usually installed using a "pull-out" fuse holder, **Figure 11-16**. Typically, the cartridge fuses would be rated 30 amps to 60 amps.

Plug fuses are available in two styles: the Edison base fuse and the Type S fuse. An ***Edison base fuse*** is the original style of fuse but is undesirable because it can be used interchangeably with ratings. Any size fuse can be screwed into the fuse holder. This can cause a significant danger of over-fusing the circuit if, for example, a 15-amp fuse was replaced by a 20-amp or even a 30-amp fuse. Remember that the overcurrent device is installed to protect the conductor. A 30-amp fuse can never fully protect a conductor rated only to carry 15 amps, and an electrical fire could certainly result. Therefore, the Edison base fuse is generally discouraged from use in residential wiring.

A ***Type S fuse***, **Figure 11-17**, uses different base sizes intentionally to prevent interchangeability. Fuse holders still use the Edison base, but the Type S fuse uses adapters that

accept only the properly sized fuses. The adapters cannot be removed once installed. They have a barb that allows them to be easily screwed into the base but prevents their removal without destroying the adapter.

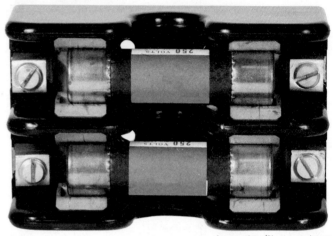

dcwcreations/Shutterstock.com

Figure 11-16. Cartridge fuses and a fuse holder.

Goodheart-Willcox Publisher

Figure 11-17. Type S fuse and adapter. Note the adapter has a barb that prevents its removal once installed.

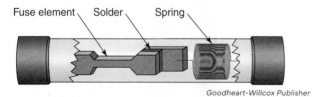

Fuse element   Solder   Spring

Goodheart-Willcox Publisher

Figure 11-15. Cutaway view of a time-delay fuse.

11.5 **Disconnects**

All electrical equipment is required to have a disconnecting means to remove power from equipment that is faulty or so that a technician can safely service it. For many electric appliances, the disconnecting means is simply a plug. In some instances, the circuit breaker can be used as the disconnect if it is within view of the equipment being serviced.

In residential wiring, the most common type of disconnect is the pull-out disconnect. The *pull-out disconnect* uses a removable block to disconnect an appliance from its source of power and can come with fuses or be nonfused. The nonfused pull-out disconnect is mainly used for large appliances in a residence. Household appliances that typically require a disconnecting means are the water heater, the furnace, and the air conditioning condenser unit.

If a disconnect is installed outside and exposed to the weather, it must have a National Electrical Manufacturers Association (NEMA) 3R rating, **Figure 11-18**. An enclosure with a *NEMA 3R rating* is designed to be installed where exposed to the elements. A disconnect must have an amperage rating equal to or greater than its associated equipment. For example, a 60-amp disconnect can be safely used on a water 30-amp rated water heater. The 60-amp disconnect is the standard size for residential use.

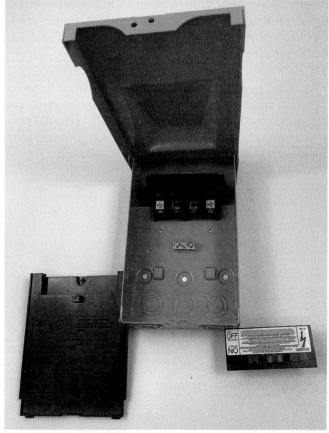

Figure 11-18. NEMA 3R disconnect. This disconnect could be used indoors or outdoors.

Summary

- Panelboards house fuses or circuit breakers that protect the branch circuit conductors in a residence. A panel with the main circuit breaker installed is called a main breaker panelboard. A panel with no main breaker is called an MLO, or main lugs only.
- Overcurrent protective devices, such as circuit breakers and fuses, protect the conductors by acting as a safety valve that opens the circuit when the current flow becomes excessive.
- A circuit breaker has a bimetallic strip that monitors the thermal properties of the device. When the strip is heated by excessive current flow, it deflects or bends, and this movement is used to open a set of spring-loaded contacts inside the circuit breaker, shutting off the power to the circuit. In cases of short circuits or ground faults, the trip mechanism uses the breaker's magnetic properties to open the circuit.
- Circuit breakers used in residential wiring are classified as inverse-time breakers, which means the time delay decreases as the magnitude of the current increases. The higher the fault current, the faster the device opens.
- The ampere interrupting capacity (AIC) rating identifies the level of fault current a circuit breaker can safely interrupt. Most residential circuit breakers have an AIC rating of 10,000 A.
- Single-pole circuits are used on 120-volt, 15-amp and 20-amp circuits. Double-pole breakers are used for 240-volt and 120/240-volt circuits, generally an electric range or clothes dryer. Twin circuit breakers are used when adding a circuit to a panelboard with no available spaces and where the twin replaces an existing breaker.

- Fuses open the circuit by means of a fusible link that melts under high current conditions and opens the circuit. A dual-element time-delay fuse uses both a fusible link and solder link to operate. A high current will melt the fusible link but not the solder link. Under a sustained thermal overload condition, the current eventually becomes hot enough to melt the solder link and open the circuit.
- An Edison base fuse can be used interchangeably with ratings, which can cause a significant danger of over-fusing the circuit. A Type S fuse uses different base sizes to prevent interchangeability, so only the properly sized fuse can be used.
- All electrical equipment is required to have a disconnecting means, either by plugs, circuit breakers, or pull-out disconnects. Household appliances that typically require a disconnecting means are the water heater, the furnace, and the air conditioning condenser unit.
- An enclosure with a NEMA 3R rating is designed to be installed where exposed to the elements.

Know and Understand

1. *True or False?* All distribution panels will contain a main breaker.
2. A circuit breaker functions to open the circuit under which of the following conditions?
   - A. Short circuit
   - B. Ground fault
   - C. Thermal overload
   - D. All of the above.
3. *True or False?* Circuit breakers wear out with regular use and frequently need replacing.
4. *True or False?* The circuit breaker uses a bimetallic strip to detect ground faults and short circuits.
5. Which of the following is a standard size for a fuse of a circuit breaker?
   - A. 75-amp
   - B. 55-amp
   - C. 110-amp
   - D. 225-amp
6. What type of circuit breaker is used for 120-volt circuits?
   - A. Single-pole breaker
   - B. Double-pole breaker
   - C. Three-pole breaker
   - D. Quad-pole breaker
7. What type of circuit breaker is used for 240-volt circuits?
   - A. Single-pole breaker
   - B. Double-pole breaker
   - C. Three-pole breaker
   - D. Quad-pole breaker
8. *True or False?* The combination AFCI, or CAFI, circuit breaker provides both GFCI and AFCI protection for the entire branch circuit.
9. *True or False?* Fuses are nonresettable and must be replaced when blown.
10. A dual-element time-delay fuse uses which of the following to open under a sustained overload condition?
    - A. Fusible link
    - B. Solder link
    - C. Bimetallic strip
    - D. Both fusible and solder links
11. What fuses are used in pairs for the service disconnect and to protect 240-volt circuits?
    - A. Edison base fuses
    - B. Type S fuses
    - C. Cartridge fuses
    - D. Screw-in fuses
12. What type of fuse uses a nonremovable adapter base?
    - A. Edison base fuse
    - B. Type S fuse
    - C. Cartridge fuse
    - D. Screw-in fuse
13. Which of the following can be used as the disconnecting means for an appliance?
    - A. Circuit breaker
    - B. Cord and plug
    - C. Pull-out disconnect
    - D. All of the above.

Apply and Analyze

1. Explain the operating principle of the bimetallic strip.
2. Explain the term "inverse-time" as it applies to a circuit breaker's operation.
3. Why are spring-loaded contacts used in circuit breakers?

Critical Thinking

1. What two physical phenomena occur because of current flow through a conductor? What role does this play in the operation of a circuit breaker?

CHAPTER 12

Lamps and Lighting

Valkantina/Shutterstock.com

CHAPTER OUTLINE

LEARNING OBJECTIVES

After completing this chapter, you will be able to:

- Describe the basic role of a lamp in an electrical circuit.
- Describe the concepts of color temperature and color rendering.
- Define wattage and lumens.
- Explain what determines a lamp's efficacy.
- Explain how incandescent, fluorescent, and LED lamps operate.
- Explain how bulbs of incandescent lamps are classified according to their physical size and shape.
- Describe how a bulb's finish affects its light output in incandescent lamps.
- Identify standard bases for incandescent lamps.
- Describe the operating voltage of an incandescent lamp.
- Explain the purpose of the ballast in fluorescent lamps.
- Explain the purpose of the LED driver.
- Identify the three classifications of lighting.

TECHNICAL TERMS

accent lighting
ballast
Class P ballast
color temperature
color-rendering index (CRI)
compact fluorescent lamp (CFL)
driver
fluorescent lamp
foot-candle
general lighting
incandescent lamp
lamp
LED lamp
lumen
luminous efficacy
task lighting
wattage
white light

Introduction

In the dawn of the electrical age, the primary use for electricity was lighting homes and workplaces. Today, this remains true as lighting accounts for up to 15% of the average residential electric bill. In an electrical circuit, the *lamp* will light up as the current passes through it and usually functions for lighting or indicating purposes. Residential electricians must have a conceptual understanding of visible light, different lighting sources, and proper field terminology. Among certified electricians, the standard convention is to use the term *lamp* for a receptacle that produces light from electric power. Lamps are commonly called *light bulbs*, but for this textbook, and to ensure technical accuracy, a lamp is any artificial light source. They can have metallic or non-metallic bases that are secured in a light fixture using a threaded base, pins, or caps.

12.1 Basics of Visible Light

The lights in your home, traffic lights and advertisements, and even the stars and some insects emit visible light. Visible light is only a small segment of the electromagnetic spectrum that can be seen by the human eye, **Figure 12-1**. The human eye detects different colors as light is reflected from an object and into our eyes. Any object will absorb and reflect the different colors of light that strike it. For example, a fire engine appears red because it absorbs all the blue and green light and reflects the red back to our eyes. For an object to appear as its true color, it must be viewed under *white light*, where all three primary colors are present in the proper proportions and appears purely white. If the red fire engine were viewed in light that contains only green and blue, it would appear black since a red object can only reflect red light.

Visible Spectrum

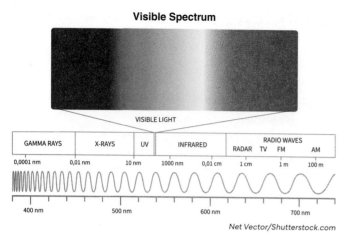

Net Vector/Shutterstock.com

Figure 12-1. Visible light is only a tiny portion of the electromagnetic spectrum.

There are two characteristics of the color of any light source: color temperature and color rendering. *Color temperature* is a measure of a light's coolness or warmth. Cool light is more blue in color, and warm light is more yellow. Color temperature is measured in degrees Kelvin, abbreviated with the letter *K*. The following are examples of color temperature for everyday light sources (**Figure 12-2**):

- A clear blue sky has a 10,000 K.
- A cool white fluorescent lamp has an approximate color temperature of 4200 K.
- A general-purpose incandescent lamp is 2500–3000 K.
- A candle flame is about 1800 K.

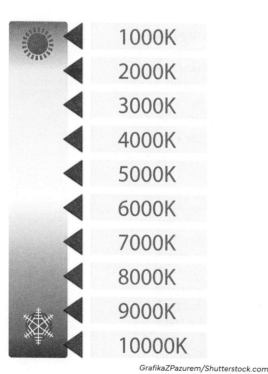

GrafikaZPazurem/Shutterstock.com

Figure 12-2. Color temperature scale in Degrees Kelvin.

On a temperature scale, note that the higher the temperature in degrees Kelvin, the cooler the light appears.

The *color-rendering index (CRI)*, also called *color rendering scale*, is a measure of how well a light source reproduces an object's actual color compared to how it appears in natural daylight. In other words, it is a measure of how a light source's color temperature affects how our eyes see an object's color. The CRI is a scale from 0 to 100, with 100 being the equivalent of direct sunlight. Direct sunlight is the optimum light source and faithfully renders an object's true color. Generally, incandescent lamps will have a higher CRI than fluorescent lamps, with the incandescent lamp at close to 100 on the CRI and some fluorescent lamps as low as 50.

12.2 **Wattage and Lumens**

In an electrical circuit, a lamp contains a filament that heats up and produces light when the current runs through it. The *wattage* of a lamp is a measure of its energy consumption. In other words, wattage is the input of power needed to produce light. To identify a lamp's brightness level, or output, check the lumens. *Lumens* are the measure of the amount of light a lamp produces. For lumens, the higher the number, the brighter the lamp.

A lamp's quality, or how well it produces visible light, is determined by its efficacy. *Luminous efficacy* is the ratio of the quantity of light output, lumens, to the amount of electric power consumed in creating the light, watts. The efficacy of a lamp is measured in lumens per watt (LPW). To calculate the efficacy of a lamp, divide lumens by the wattage. The higher the lumens per watt, the higher the lamp's efficacy. **Figure 12-3** compares the efficacy of three lamps of similar output in lumens.

Lamp type	Output in lumens	Efficacy
100-watt incandescent	1750	17.5 lumens/watt
26-watt fluorescent	1700	65 lumens/watt
18-watt LED	1620	90 lumens/watt

Goodheart-Willcox Publisher

Figure 12-3. A comparison of the efficacy of three lamps of similar lumens. The lamp with the highest lumens per watt is the most energy efficient and converts most of its consumed energy into light.

PRO TIP   **Efficacy and Efficiency**

Do not confuse the term *efficacy* with *efficiency*. While both terms measure performance as a ratio of input to output, efficacy measures how much gets done, and efficiency measures how well it is done. Efficacy is expressed in LPW, and efficiency is expressed as a percentage. The lamp with the highest efficacy is also the most efficient at converting electric power into light.

12.3 Types of Lamps

The three most common light sources used in a dwelling are incandescent lamps, fluorescent and compact fluorescent lamps, and LED lamps. Incandescent lamps have been around for over a century and account for the majority of lamps in use, but they are the least efficient. Compact fluorescent lamps (CFLs) are more energy-efficient than the incandescent lamp, and they were popular until the LED lamp surpassed them in light quality, price, and efficiency.

12.3.1 Incandescent Lamps

Incandescent lamps are the least efficient of the lamps. An *incandescent lamp* uses a filament suspended in the vacuum of a glass bulb. When current flows through the filament, it glows and produces visible light. However, less than 5% of the energy consumed is converted into light, and over 95% is lost as heat.

Incandescent lamps can be classified by the shape of the bulb or their type of base, **Figure 12-4**. The glass bulb of an incandescent lamp is given a letter designation to describe its shape, followed by a number that denotes the bulb's diameter in 1/8″ at its widest point, **Figure 12-5**. The standard incandescent lamp is known as an A19. This designation means it has an *A* style shape and is 19/8″, or 2-3/8″, in diameter. The G40 lamp has a globular, or round, shape and is 40/8″, or 5″, in diameter. Other bulb shape designations are *F* for flambeau or flame-shaped and PAR for a parabolic reflector lamp popular for outdoor floodlights.

The glass bulb of an incandescent lamp can be finished in several different ways. The two most common are clear and frosted. A clear glass bulb has no finish, and the filament is exposed to view. A lamp with a clear bulb emits an intense, bright light and casts crisp, clean shadows. On the other hand, the frosted bulb emits a more diffused light and casts muted shadows. The daylight bulb is another lamp finish that uses a blue-green glass that simulates the noonday sun's color temperature. Bulbs with metallic reflector coatings can concentrate and direct the light output of a lamp.

The base of an incandescent lamp is the metallic portion that inserts into the luminaire and supplies power to the filament. In residential lighting, the most common lamp bases used are the E12 candelabra base and the E26 medium base, **Figure 12-6**. The candelabra base is smaller in size and is primarily used in multi-bulb chandeliers. The medium base is for standard, everyday use found in most luminaires.

The incandescent lamp has a useful life of between 1000 and 2500 hours. Operating an incandescent lamp at less than its rated voltage can prolong its life but reduces its light output. Many manufacturers produce lamps with a rating of 130 volts but are intended to operate at 120 volts. A 60-watt, 130-volt lamp consumes just over 50 watts of power when connected to a 120-volt system with only a slight reduction in light output. Lowering the voltage by 8%, from 130 volts to 120 volts, can lead to a 300% increase in lamp life and a nearly 15% decrease in energy usage with a loss of less than 25% in light output.

12.3.2 Fluorescent Lamps

Fluorescent lighting has limited use in residential lighting and is employed mainly in garages, workshops, and kitchens, where bright light with limited shadowing is desired. A *fluorescent lamp*, also called a *fluorescent tube*, is gas-discharge lamp that produces visible light through a reaction with the lamp's phosphor coating. They come in several sizes and shapes but are commonly a tube. There are also U-bent-shaped fluorescent lamps, circular fluorescent lamps, and compact fluorescent lamps.

A *compact fluorescent lamp (CFL)* is designed to replace an incandescent lamp in many floor and table lamp fixtures and luminaires. Compared to incandescent lighting, fluorescent lighting is more efficient. Generally, 85% to 90% of the energy applied to the lamp is converted into usable light. The rest is dissipated as heat, mainly from the fixture's ballast, **Figure 12-7**.

All fluorescent lights require a ballast to operate. A *ballast* provides high voltage to start the lamp and regulates the current flow through the lamp while in operation. The ballast is rated according to its operating voltage and the number and size of tubes it powers.

The original type of ballast was magnetic and is characterized by its weight and emission of a hum when operating. An electronic ballast was developed in the 1990s and,

Incandescent Light Bulb

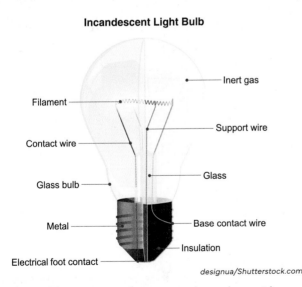

Inert gas

Filament

Support wire

Contact wire

Glass

Glass bulb

Metal

Base contact wire

Insulation

Electrical foot contact

designua/Shutterstock.com

Figure 12-4. The anatomy of a common incandescent lamp. One end of the filament contact wire is attached to the foot contact, and the other end is connected to the lamp's screw shell.

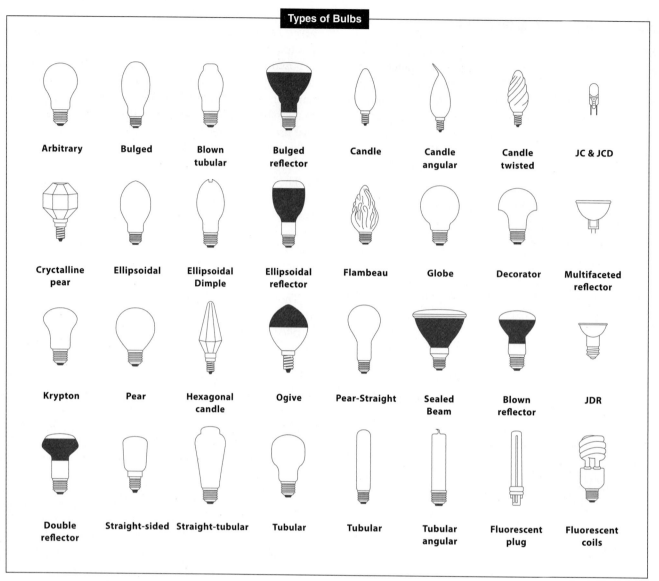

Figure 12-5. Bulb shape designations for incandescent lamps.

Lazuin/Shutterstock.com

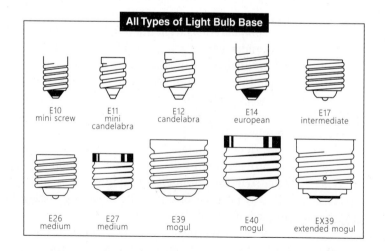

Figure 12-6. Base shapes and designations for incandescent lamps.

Lazuin/Shutterstock.com; VitaminCo/Shutterstock.com

available with a CRI of 95% or higher. A CFL with a color temperature of 2700 K will emit light that is indistinguishable from that of an incandescent lamp.

Although millions of fluorescent light fixtures are installed worldwide, their use is declining as the LED lamp increases in popularity. CFLs are becoming difficult to find in the lighting aisle of your local hardware store or home center.

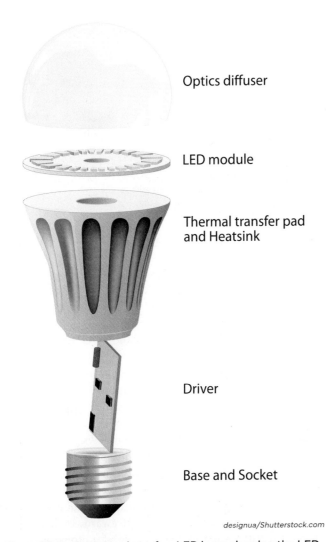

Optics diffuser

LED module

Thermal transfer pad and Heatsink

Driver

Base and Socket

designua/Shutterstock.com

Figure 12-7. Cutaway view of an LED lamp showing the LED chips and the electronic driver.

> **SAFETY NOTE**
>
> ## Disposing of Fluorescent Lamps
>
> Because fluorescent lamps contain mercury, they are classified as hazardous waste and have special requirements for disposal.

12.3.3 LED Lamps

An *LED lamp* produces light using light-emitting diodes (LEDs). LED lighting is quickly becoming the standard lighting method in homes and businesses, replacing inefficient incandescent lamps and less efficient fluorescent fixtures. It is the most efficient lighting method on the market and has nearly 85% of its energy converted into light. LEDs can operate for 100,000 hours or more, making them ideal for hard-to-reach fixtures where relamping can be difficult or dangerous.

The LED lamp requires a driver to operate. LEDs are operated using DC power, but the electrical utility supplies homes with AC power. A *driver* is an electronic device that converts the incoming AC power into the DC power needed for LED operation. The LED bulb has its driver installed in its base, **Figure 12-8**. Undercabinet LED lighting, which is also common, has a remote driver.

compared to the magnetic ballast, is lighter in weight, more efficient, has a longer life, and does not produce an audible hum under operation. The *NEC* requires fluorescent ballasts to be of the *Class P* type. A **Class P ballast** means that the ballast has thermal protection that disconnects it from the power if it overheats. A compact fluorescent lamp has a built-in ballast at its base.

When manufactured, a fluorescent tube or a CFL will have its air evacuated and replaced with an inert gas such as argon or xenon and mercury vapor. This gas must be ionized or given a high-voltage charge from the ballast before the lamp can illuminate. The inside surface of a fluorescent tube or CFL is coated with a phosphor powder that reacts to the energized mercury-vapor gas to create light. Fluorescent lighting has a reputation for poor color rendering.

While lamps at the cooler end of the color temperature scale do have poor color rendering, fluorescent lamps are

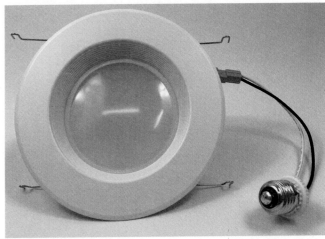

Goodheart-Willcox Publisher

Figure 12-8. Recessed light LED replacement lamp. The driver and can trim are integrated into the lamp.

LED lighting can be used throughout the home in many colors. An increasingly popular residential use for LEDs is in recessed lighting. Replacement LED lamps are available for use in existing recessed cans, **Figure 12-9**. Canless and ultra-thin "wafer" recessed LEDs are used in new construction and remodel work, **Figure 12-10**. LED tubes are made to replace the fluorescent tubes in a fluorescent fixture, **Figure 12-11**. Some are "drop-in" replacements and require no rewiring of the fixture, while others need a bit of rewiring to remove the ballast.

12.4 Classifications of Lighting

In residential lighting design, there are three classifications of lighting: general lighting, accent lighting, and task lighting. Each classification of lighting has different considerations to achieve the desired illumination. Illumination, or

Marko Poplasen/Shutterstock.com

Figure 12-11. An example of ambient, or general lighting, which illuminates an entire space.

Goodheart-Willcox Publisher

Figure 12-9. Canless ultra-thin "wafer" recessed light with remote driver.

light intensity, is measured in foot-candles. One *foot-candle* is equal to one lumen per square foot.

General lighting, also called *ambient lighting*, is the overall lighting of a room or space, see **Figure 12-12**. Overhead luminaires such as recessed lights, surface-mounted fixtures, and chandeliers provide general lighting, as well as floor and table lamp fixtures and windows. The primary consideration for general lighting is lumens per square foot of living space. A kitchen or bathroom needs more lumens per square foot than a living room or a bedroom.

Accent lighting is used to illuminate a point of interest, such as a piece of artwork, an award, or an architectural feature, such as the texture of a stone wall, see **Figure 12-13**. Accent lighting must be about three times brighter than the ambient lighting to be effective. Accent lighting is accomplished using directional recessed lights or track lighting where the individual track heads can be

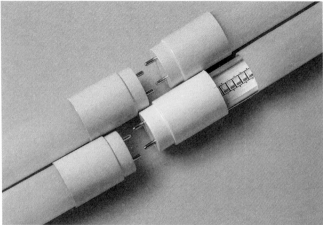

preedee anantuntikul/Shutterstock.com

Figure 12-10. LED tube lights for replacing a fluorescent tube.

Toms Svilans/Shutterstock.com

Figure 12-12. Accent lighting is used to highlight a particular item or surface texture.

Photographee.eu/Shutterstock.com

Figure 12-13. Task lighting is used to illuminate a workspace such as a kitchen countertop surface.

aimed at the point of focus. The lamp's color rendering capabilities should be considered when using accent lighting on a work of art.

Task lighting illuminates workspace surfaces such as a kitchen countertop, desk, or reading chair, see **Figure 12-14**. Task lighting can be accomplished using desktop or floor standing luminaires or under-cabinet lighting. If the task lighting source is overhead, like a recessed light or a track light, the fixture should be installed to minimize casting shadows on the workspace.

Summary

- The role of a lamp in an electrical circuit is to produce light when the current passes through it.
- There are two characteristics of the color of any light source: color temperature and color rendering. Color temperature is a measure of a light's coolness or warmth in Kelvins. The color rendering scale is a measure of how well a light source reproduces an object's actual color compared to how it appears in natural daylight.
- Wattage is a measure of a lamp's energy consumption or input. Lumen is the measure of the amount of light a lamp produces or outputs.
- A lamp's quality, or how well it produces visible light, is determined by its efficacy. Luminous efficacy is the ratio of the quantity of light output, lumens, to the amount of electric power consumed in creating the light, watts. It is expressed in lumens per watt (LPW).
- The three most common light sources used in a dwelling are incandescent lamps, fluorescent lamps, and LED lamps. An incandescent lamp uses a filament suspended in the vacuum of a glass bulb and is the least efficient. When current flows through the filament, it glows and produces visible light, but most energy is lost as heat. A fluorescent lamp is a gas-discharge lamp that produces visible light through a reaction with the lamp's phosphor coating. An LED lamp produces light using light-emitting diodes.
- The bulb of an incandescent lamp can be classified by the shape of the bulb or its type of base. The glass bulb of an incandescent lamp is given a letter designation to describe its shape, such as A, G, F, or PAR, followed by a number that denotes the bulb's diameter.
- The two most common ways an incandescent bulb can be finished are clear or frosted. A clear bulb has no finish and exposes the filament, which emits an intense, bright light and casts crisp, clean shadows. The frosted bulb emits a more diffused light and casts muted shadows. The daylight bulb is another lamp finish that uses a blue-green glass that simulates the noonday sun's color temperature.
- The base of an incandescent lamp is the metallic portion that inserts into the luminaire and supplies power to the filament. The most common lamp bases used in residential lighting are the E12 candelabra base and the E26 medium base.
- Operating an incandescent lamp at less than its rated voltage will prolong its life but reduce its output.
- All fluorescent lights require a ballast to operate. The ballast provides the high voltage needed to start the lamp and regulates the current flow through the lamp while in operation.
- The LED lamp requires a driver to operate. A driver is an electronic device that converts the incoming AC power into the DC power needed for LED operation.
- General lighting, also called ambient lighting, is the overall lighting of a room or space. Accent lighting is used to illuminate a point of interest. Task lighting is used to illuminate workspace surfaces.

Know and Understand

1. A lamp's _____ is a measure of its energy consumption.
   A. wattage
   B. lumens
   C. foot-candle
   D. efficacy

2. The measure of the amount of visible light a source produces is measured in _____.
   A. watts
   B. lumens
   C. foot-candles
   D. efficacy

3. A lamp's _____ is the ratio of the amount of light a lamp produces to the amount of power consumed.
   A. wattage
   B. lumens
   C. foot-candles
   D. efficacy

4. *True or False?* A lamp's efficacy is expressed as a percentage.

5. What is the least efficient type of lamp?
   A. Incandescent lamp
   B. Fluorescent lamp
   C. LED lamp
   D. All of these are equally inefficient.

6. *True or False?* An incandescent lamp with a clear bulb finish will emit a diffused light and cast muted shadows.

7. *True or False?* The higher the color temperature on the Kelvin scale, the warmer the light.

8. *True or False?* Operating an incandescent lamp below its rated voltage will prolong the life of the lamp.

9. All fluorescent lamps require which of the following to operate?
   A. Driver
   B. Base
   C. Emitter
   D. Ballast

10. What is the most efficient type of lamp?
    A. Incandescent lamp
    B. Fluorescent lamp
    C. LED lamp
    D. All of these are equally efficient.

11. LED lighting requires which of the following to operate?
    A. Driver
    B. Base
    C. Emitter
    D. Ballast

12. The overall lighting of a room or a space is called which of the following?
    A. General lighting
    B. Accent lighting
    C. Task lighting
    D. Spot lighting

13. What type of lighting is used to illuminate a point of interest?
    A. General lighting
    B. Accent lighting
    C. Task lighting
    D. Ambient lighting

14. What type of lighting is used to illuminate a workspace such as a kitchen countertop?
    A. General lighting
    B. Accent lighting
    C. Task lighting
    D. Ambient lighting

15. *True or False?* Accent lighting should be brighter than the ambient lighting if it is to be effective.

Apply and Analyze

1. For an object to appear as its true color, it must be viewed under what type of light?
2. A lamp's efficacy is measured in _____.
3. Light intensity is measured in _____.
4. An incandescent lamp may convert only 5% of the energy it consumes into visible light. What happens to the other 95% of the energy?
5. When referring to the lamp shape, what is meant by the term "PAR"?
6. What are the two most common lamp bases used in residential lighting?
7. What two functions does the fluorescent ballast perform?
8. Explain the purpose of the Class P ballast.
9. What is the purpose of the driver in LED lighting?

Critical Thinking

1. Why would a red object appear back to our eyes under a light containing only blue and green light?

Basic Circuit Wiring

grandbrothers/Shutterstock.com

All residential electrical installations include switched circuits. For example, the *National Electrical Code* requires a switched lighting outlet in every habitable room in a dwelling. Switched circuits range from basic, simple circuits to more complicated combination circuits.

Chapter 13, *Basic Switched Circuits*, covers basic switched circuits, including single-pole switching, three-way switching, and four-way switching. Single-pole switches control a lighting load from one location, three-way switching controls a lighting load from two locations, and four-way switching controls a lighting load from more than two locations.

Chapter 14, *Advanced Switched Circuits*, covers combination circuits including switch-controlled receptacles, circuits with constant-powered receptacles downstream from a switched lighting load, and controlling two or more different lighting loads from a single feed. A mastery of switched circuits is essential to anyone working in residential wiring.

CHAPTER **13** | Basic Switched Circuits

SPARKING DISCUSSION

How does a foundation in electrical theory help the electrician wire a basic switched circuit?

Introduction

An understanding of basic and advanced switched circuits is a valuable asset for the residential electrician. This chapter will focus on basic switched circuits, and advanced switched circuits will be covered in Chapter 14, *Advanced Switched Circuits*. A **switched circuit** uses a variety of switches to open and close an electrical circuit, usually to control lighting loads. Basic switched circuits are the single-pole, the three-way switch, and the four-way switch. Each of these will be discussed in turn in this chapter.

Every room in every home has at least one switch-controlled lighting circuit, and many rooms have more than one. Lighting circuits can be as simple as controlling a single luminaire from one location or a more complicated installation where multiple lighting loads are controlled from several locations. This chapter will provide an overview of the most common basic switched circuits found in residential wiring.

13.1 "Be the Electron"

Think of the wiring in your home as a highway on which electrons travel. Switched circuits act as a detour that directs the electrons to a lighting load before returning on their journey home. A complete circuit must exist for electrical current, or electrons, to flow and ultimately light your home. An excellent method to determine if the circuit is complete is to "be the electron" and trace the current path through the circuit. Recall that electrical current leaves its source, travels through the circuit, energizes a load, and returns to its source, **Figure 13-1**. You will be reviewing how current flows through each type of switched circuit in this chapter.

13.2 *NEC* Requirements for Switched Circuits

It is critical to understand the various requirements for switched circuits outlined by the *NEC*. When using nonmetallic cable as the wiring method, current is supplied to the load with the black or red ungrounded conductor, and the white, grounded conductor returns the current to the source. *Section 404.2(B)* does not permit switches to disconnect the grounded conductor. All switches are to be connected to the ungrounded, hot conductor.

Fundamentally, a switched circuit can receive its power at the switch location or the lighting outlet. Therefore, there are at least two types of wiring setups for the circuit. When the switched circuit's power is at the lighting outlet, a switch loop is installed to the switch location. A **switch loop** is wiring that creates a path for the ungrounded

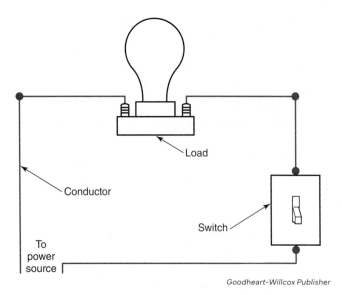

Figure 13-1. A circuit is simply a path from a source of power, through energy-using devices, back to the power source.

conductor from the lighting outlet to the switch location, and then back to the lighting outlet.

As a general rule, *Section 404.2(C)* requires a grounded conductor to be installed at a switch location. If a grounded conductor is required at the switch location, a three-conductor cable is used as the switch loop. In situations that do not require the grounded conductor to be installed at the switch location, a two-conductor can be used as the switch loop.

A two-conductor cable supplies the switch with constant power on one conductor and returns switched power to the lighting outlet on the other conductor. In such an arrangement, the switch loop has no grounded conductor.

When using two-conductor NM cable as the wiring method for a switch loop, the white conductor must be re-identified, or recolored, as an ungrounded conductor using tape, paint, or other effective means at each termination and wherever it is visible and accessible. The re-identification

may be any color except white, gray, or green. Furthermore, the re-identified white conductor is used to feed the switch, not as a return to the lighting outlet, per *Section 200.7(C)(1)*. Electricians use the phrase "down on white, back on black" to remember this wiring scheme.

13.3 **Basic Switched Circuits**

Recall that power for a switched circuit originates at either the lighting outlet or the switch location. The best practice is to install the power at the switch location rather than at the lighting outlet. This brings the grounded circuit conductor to each switch location, satisfying *Section 404.2(C)*. However, supplying power at the lighting outlet and installing a switch loop is also common, especially in existing installations. This chapter will illustrate both methods.

13.3.1 **Single-Pole Switched Circuits**

A *single-pole switched circuit* controls a lighting load from a single location. The single-pole switch operates by opening and closing the circuit in one location. Inside the switch is a single set of contacts that open and close, **Figure 13-2**. Be the electron and follow the path of the current through the circuit. Power enters the circuit at the switch location. When the switch is in the down or OFF position, the internal contacts are open, and no current can flow. When the switch is in the up or ON position, the internal contacts close, and current flows to the lighting outlet.

A single-pole switch with the power at the switch location has a two-conductor cable installed from the switch location to the lighting outlet, **Figure 13-3**. In the switch box, both white grounded conductors are spliced together, and the hot black conductors are terminated to the switch. It does not matter which conductor is under which terminal screw.

The luminaire is wired next. Most luminaires have pig-tail connections, while lamp holders have terminal screws. Connect the black conductor from the switch to the black

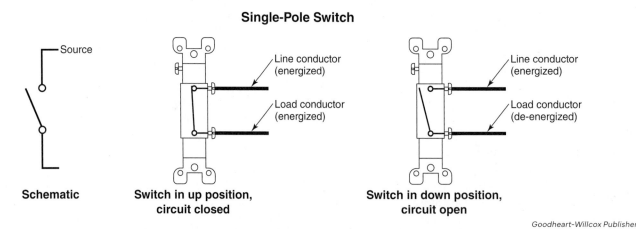

Figure 13-2. The internal operation of a single-pole switch.

Single-Pole Switch, Power at Switch

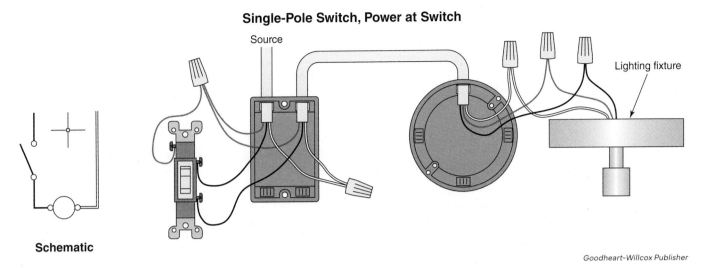

Schematic

Goodheart-Willcox Publisher

Figure 13-3. Schematic and wiring diagram of a single-pole switch with the power (source) at the switch location.

conductor of the luminaire or the brass-colored terminal screw of the lamp holder. Connect the white conductor from the switch location to the white conductor of the luminaire or the silver screw of the lamp holder.

A single-pole switch with the power at the lighting outlet has a two-conductor cable switch loop installed between the switch location and the lighting outlet. Notice in the illustration that the white conductor of the switch loop has been re-identified on each end with a bit of black tape, **Figure 13-4**.

At the lighting outlet, the hot, black conductor of the incoming power is connected to the re-identified white conductor of the switch loop to deliver power to the switch. At the switch location, both conductors are terminated to the switch. Again, it makes no difference which conductor is connected to which terminal screw. Back at the

lighting outlet, the black conductor of the switch loop and the grounded conductor are terminated to the luminaire or lamp holder.

13.3.2 Three-Way Switched Circuits

A *three-way switched circuit* controls a lighting load from two locations. Three-way switches are always installed in pairs. A three-way switch has three terminal screws: two silver traveler terminals and one black common terminal screw. Think of the common terminals as the way in and out of the three-way circuit. The traveler conductors are alternately energized depending on the position of the switches, **Figure 13-5**.

A three-way switched circuit has two three-way switches with a three-conductor cable installed between them. The power can enter the circuit at either of the two switches or at

Single-Pole Switch, Power at Lighting Outlet

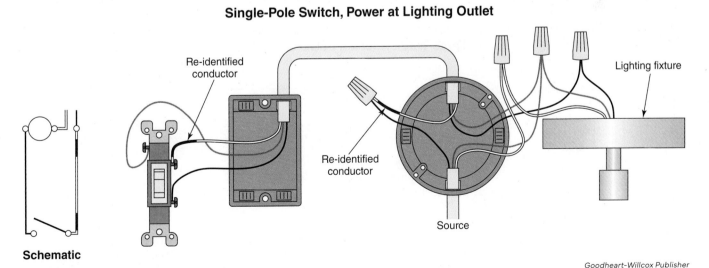

Schematic

Goodheart-Willcox Publisher

Figure 13-4. Schematic and wiring diagram of a single-pole switch with the power (source) at the lighting outlet. The white conductors of the switch loop are re-identified as ungrounded, hot conductors.

Three-Way Switch

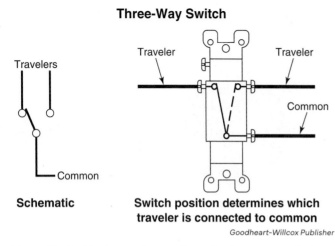

Schematic

Switch position determines which traveler is connected to common

Goodheart-Willcox Publisher

Figure 13-5. The internal operation of a three-way switch. Note the common is always connected, and the switching action alternates between the two travelers.

PRO TIP

Re-identifying the Traveler Wires

Although not a *Code* requirement, it is a best practice to re-identify the black traveler with colored tape or some other method. This better produces the switch connection during the trim-out phase when the devices are installed, which can be weeks or months after the rough-in stage when the outlet conductors are prepped.

---

the lighting outlet location. The most straightforward wiring method is to bring the power into the switched circuit at the first switch and connect the switched power to the lighting outlet at the second switch, **Figure 13-6**.

Be the electron and follow the path of the current through the circuit. Power enters the circuit on the common terminal of the first switch and travels to the second switch on one of the two traveler conductors based on the position of the first switch. The position of the second switch determines whether power is halted, indicating the load is off, or leaves the switched circuit on the common terminal of the second switch and flows on the black conductor to energize the luminaire. Power then returns to its source on the white, grounded conductor. Note that with this wiring scheme, a grounded conductor is present in each switch location, satisfying the requirements of *Section 404.2(C)*.

A second wiring method for a three-way switched circuit involves the power to the circuit at the lighting outlet, **Figure 13-7**. This is known as the *dead-end* three-way switch because the second switch location has only three conductors: the common and two travelers. A two-conductor cable supplies the power to the circuit at the lighting outlet location, and a two-conductor switch loop is run from the lighting outlet to the first switch location. A three-conductor cable is run from the first switch location to the second switch location. Since there is a switch loop, there is no grounded conductor in the switch boxes, and the white conductor must be re-identified as an ungrounded conductor. Again, it is also recommended to identify the black traveler conductor so as not to confuse it with the common conductor upon trim-out.

Be the electron and follow the path of current through the circuit. Power enters the switched circuit at the lighting outlet. The black, ungrounded power conductor is connected to the re-identified white conductor of the switch loop. The re-identified white conductor passes through the first switch box and becomes the common conductor of the second, dead-end, switch. The position of this switch determines whether power travels on either the black or red

Three-Way Switches, Power at Switch

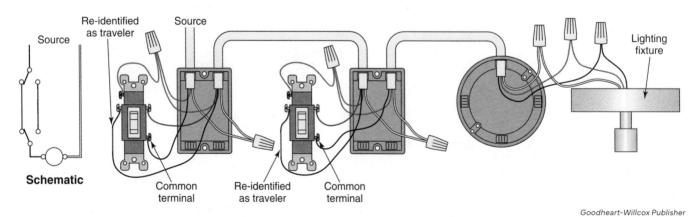

Goodheart-Willcox Publisher

Figure 13-6. Schematic and wiring diagram of a three-way switched circuit with the power at the first switch and switched power to the lighting load at the second switch. Note the re-identified black traveler conductor.

Three-Way Switches, Power at Lighting Outlet

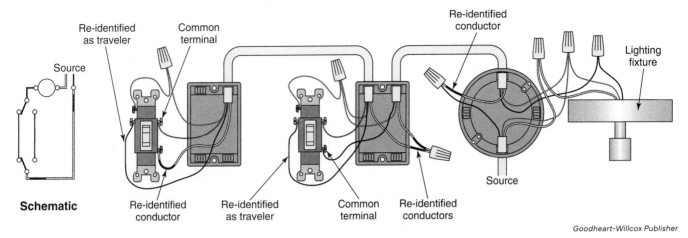

Goodheart-Willcox Publisher

Figure 13-7. Schematic and wiring diagram of a three-way switched circuit with the power at the lighting outlet location. Note the re-identified white conductor as an ungrounded conductor and the re-identified black conductor as a traveler.

traveler to the first switch. Depending on the position of the second switch, power then leaves the switch on the common conductor, energizes the lighting load, and returns to its source on the white, grounded conductor of the supply feed.

A third wiring scheme for a three-way switched circuit is the *double dead-end*, where the power and the three-conductor switch cables are installed at the lighting outlet location, and both three-wire conductors dead-end in their respective boxes, **Figure 13-8**. As the power is supplied to the circuit at the lighting outlet, there is no grounded conductor in the switched circuit, and the white conductors of the three-conductor are re-identified as ungrounded conductors.

Be the electron and trace the path of the current through the circuit. Power enters the circuit on the black conductor of the two-wire feed, which is spliced to one of the three-conductor cable's black conductor and serves as the common for the first switch. The position of the first switch

determines whether power travels through the lighting outlet box to the second switch on either the red or re-identified white conductor. The position of the second switch then determines whether power leaves the second switch and energizes the lighting load. Power returns to the source on the white conductor of the two-wire power feed.

13.3.3 Four-Way Switched Circuits

A *four-way switched circuit* is used to control a lighting load from three or more locations. Based on need, many four-way switches can be installed in the circuit if they are installed between two three-way switches. The four-way switch has four terminal screws, two for each pair of travelers. The common conductor is not connected to the four-way switch. The switching action of the four-way switch is shown in **Figure 13-9**.

Three-Way Switches, Power at Lighting Outlet, Double Dead-End

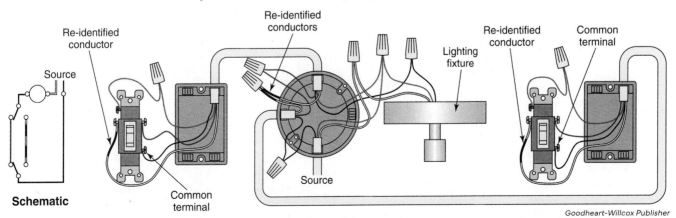

Goodheart-Willcox Publisher

Figure 13-8. A double dead-end three-way switched circuit. Note the re-identified white conductors.

Four-Way Switch

Travelers   Travelers

Travelers   Travelers

Schematic

Travelers to one
three-way switch

Travelers to
a second
three-way switch

**Switch position determines which
traveler is connected to common**

Goodheart-Willcox Publisher

Figure 13-9. The internal operation of a four-way switch. Note the crisscross action of the switch.

As with other switched circuits, the four-way switched circuit can receive its power at the first switch location or the lighting outlet location. As with the three-way switched circuit, three-conductor cables are installed between the switches.

At the switch location, when power is brought to the first switch, the black conductor of the two-wire feed is connected to the common terminal, **Figure 13-10.** Since the incoming power is at the switch, the white conductors throughout the circuit are grounded conductors and thus are not re-identified. Be the electron and follow the path of current through the circuit. The position of the first switch determines whether power travels on either the red or the black traveler conductor of the three-conductor cable to the four-way switch. Depending upon the position of the four-way switch, the power travels to the second three-way switch on either the red or the black conductor of the three-wire cable. Power leaves the second three-way switch on the black conductor of the two-conductor cable to the lighting outlet and returns to the source on the grounded, white conductor.

Four-Way Switch, Power at Switch

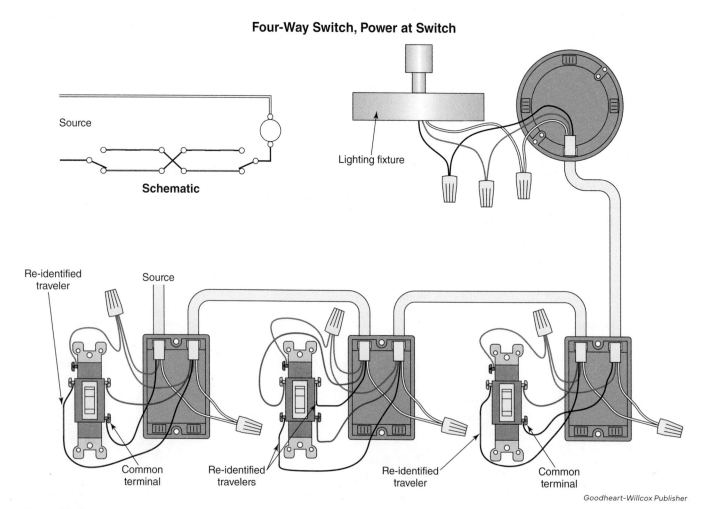

Source

Schematic

Lighting fixture

Re-identified
traveler

Source

Common
terminal

Re-identified
travelers

Re-identified
traveler

Common
terminal

Goodheart-Willcox Publisher

Figure 13-10. Schematic and wiring diagram of a four-way switched circuit with the power at the first switch.

A four-way switched circuit with power brought to the lighting outlet is wired in the same manner as the three-way circuit with power at the light. The white conductors of the three-conductor cables are re-identified as ungrounded conductors. See **Figure 13-11**. Ensure that the travelers are easily identified to trace or troubleshoot.

Be the electron and follow the path of current through the circuit. At the lighting outlet location, power enters the switched circuit on the ungrounded, black conductor of the two-conductor cable feed. The black conductor of the feed is connected to the re-identified white of the two-conductor switch loop to the first three-way switch. At the first switch box, the re-identified white conductor is connected to the black conductor of the three-conductor cable to become the common conductor. The common conductor bypasses the four-way switch without a connection and continues to the second three-way switch, where it is connected to the common terminal. Both the red and black traveler conductors are connected to the traveler terminals of the three-way switch. The black of the two-conductor switch loop is connected to the common terminal of the two-conductor switch loop on one end and to the luminaire on the other. Power is returned to the source on the white, grounded conductor of the power feed.

Four-Way Switch, Power at Lighting Outlet

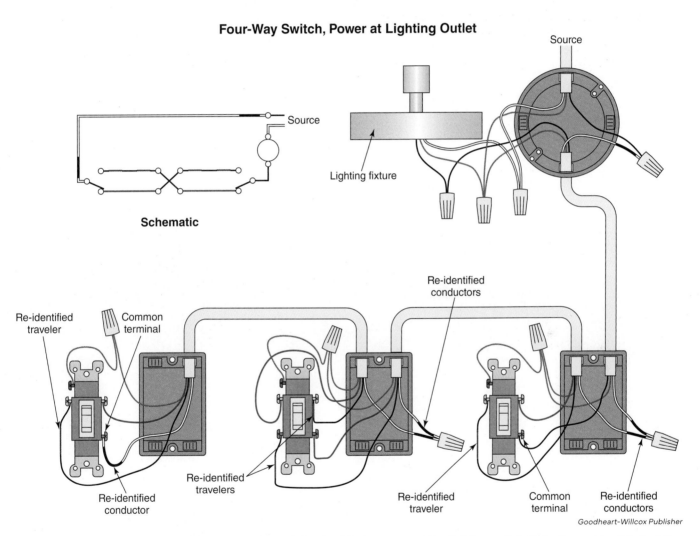

Goodheart-Willcox Publisher

Figure 13-11. Wiring diagram of a four-way switched circuit with the power at the lighting outlet. Note the re-identified white conductors and the re-identified traveler conductor.

Summary

- A switched circuit can receive its power at the switch location or the lighting outlet. Electrical current leaves its source, travels through the circuit, energizes a lighting load, and returns to its source.
- *Section 404.2(B)* does not permit switches to disconnect the grounded conductor. All switches are to be connected to the ungrounded conductor.
- A single-pole switched circuit controls a lighting load from a single location. When the switch is in the down or OFF position, the internal contacts are open, and no current can flow. When the switch is in the up or ON position, the internal contacts close, and current flows to the lighting outlet.
- Three-way switched circuits control a lighting load from two locations. A three-way switched circuit will always have two three-way switches with a three-conductor cable installed between them. A three-way switch has three terminal screws: two silver traveler terminals and one black common terminal screw. The power can enter the circuit at either of the two switches or at the lighting outlet location.
- Four-way switched circuits are used to control a lighting load from three or more locations. Four-way switches have four terminal screws, two for each pair of travelers. The common conductor is not connected to the four-way switch. As with other switched circuits, the four-way switched circuit can receive its power at the first switch location or the lighting outlet location.

Know and Understand

1. *True or False?* In an electrical circuit, the current flows from the source, energizes a load, and flows to ground to complete the circuit.
2. *True or False?* A switched lighting circuit can only receive its power at the lighting outlet location.
3. *True or False?* In a switched circuit, the white conductor is always a grounded conductor.
4. When using Type NM cable, electrical current is supplied to the load which conductor?
   A. Black, ungrounded conductor
   B. Red, ungrounded conductor
   C. White grounded conductor
   D. Either the black or red ungrounded conductor
5. When using Type NM cable, electrical current returns to its source on which conductor?
   A. Black, ungrounded conductor
   B. Red, ungrounded conductor
   C. White grounded conductor
   D. Either the black or red ungrounded conductor.
6. Single-pole switched circuits control a lighting load from _____ location(s).
   A. one   C. three
   B. two   D. four
7. Three-way switched circuits control a lighting load from _____ location(s).
   A. one   C. three
   B. two   D. four

8. Current enters and leaves a three-way or four-way switched circuit on the _____ conductor.
   A. traveler   C. grounded
   B. common   D. ungrounded
9. Four-way switched circuits control a lighting load from _____ locations.
   A. two   C. four
   B. three   D. three or more
10. Four-way switches are always installed _____.
    A. in pairs
    B. between two three-way switches
    C. with three-way switches between them
    D. to the common conductor

Apply and Analyze

1. Explain what must be done to the white conductor of a two-conductor switch loop.
2. A switch's terminal screws are rated for only one conductor. What must be done to connect more than one conductor to a terminal screw?

Critical Thinking

1. Although not required by the *Code*, explain why it is a best practice to identify the black traveler conductor in a three-way switched circuit.
2. Explain why it is the best practice to bring power to a switched circuit to the switch location rather than the lighting outlet location.

CHAPTER 14

Advanced Switched Circuits

CHAPTER OUTLINE

14.1 Receptacles as a Lighting Outlet

14.2 Single-Pole Switch Controlling Two Split Duplex Receptacles

14.3 Switch-Controlled Lighting Outlet and a Constant-Power Receptacle Downstream from the Light

14.4 Two Single-Pole Switches Separately Controlling Two Lights on a Single Feed

LEARNING OBJECTIVES

After completing this chapter, you will be able to:

- Explain how switch-controlled receptacles can serve as the required lighting outlet.
- Explain how a standard duplex receptacle can be "split" to provide switched power at one outlet and constant power at the other.
- Describe the installation and operation of a circuit with a single-pole switch controlling two split receptacles.
- Describe the installation and operation of a circuit with a switch-controlled light and a constant-power receptacle downstream from the light.
- Describe the installation and operation of a circuit with two single-pole switches separately controlling two lights on a single feed.
- Describe the installation and operation of a circuit with a constant power receptacle and single-pole switch-controlled light on a single feed.

TECHNICAL TERMS

constant-power receptacle
switch-controlled light

Introduction

Advanced switching circuits can be a bit more complicated than the basic switched circuits discussed in the previous chapter—standard single-pole, three-way, and four-way switched lighting circuits. Advanced switching circuits include switch-controlled receptacles, constant-power receptacles downstream from a switch-controlled light, or multiple switches controlling multiple lighting outlets or devices from a single feed. These types of circuits are common throughout residential wiring and should be thoroughly understood by residential electricians.

14.1 Receptacles as a Lighting Outlet

Recall that the *National Electrical Code* requires a switched lighting outlet in every habitable room in a dwelling, but the lighting outlet is not required to be an overhead luminaire. Instead, one or more switched receptacles are permitted to serve in place of lighting outlets in rooms other than kitchens, bathrooms, or laundry areas as per *Section 210.70(A)(1), Exception 1*. Chapter 10, *Electrical Devices*, introduced a split receptacle. These receptacles can be used in a switched circuit to provide switched power on one outlet to supply a luminaire and a constant power outlet for general use. Splitting a receptacle is accomplished by removing the tab between the two brass-colored terminal screws of an ordinary duplex receptacle, **Figure 14-1**. If necessary, both outlets of a duplex receptacle can be switch controlled, either as a unit or separately.

Goodheart-Willcox Publisher

Figure 14-1. Splitting a duplex receptacle using needle-nose pliers to remove the tab between the two ungrounded, hot terminal screws.

14.2 Single-Pole Switch Controlling Two Split Duplex Receptacles

As mentioned above, a switched receptacle with a lamp plugged into it can be used as the required lighting outlet in most rooms of a dwelling. A switched circuit with a single-pole switch controlling two split receptacles with the power source at the switch location is shown in **Figure 14-2**. This circuit can also be controlled by three-way or four-way switches and may contain as many split receptacles as needed.

In the illustration, note that a three-conductor cable is used from the switch to the first split receptacle and from receptacle to receptacle. Although not an *NEC* requirement, it is common practice and industry standard to use the red conductor for switched power and the black conductor for constant power. Following this scheme helps keep track of the function of each conductor.

Be the electron and follow the paths of current through the circuit. Starting at the switch location, notice the ungrounded black conductor of the two-conductor power source is pigtailed to provide power to the switched portion of the circuit as well as constant power to half of each receptacle. The red conductor of the three-conductor cable provides switched power to the other half of the receptacle. The white, grounded conductor provides the return path for both the switched power and the constant power. All the bare equipment grounding conductors are spliced together with a pigtail to each of the devices.

Single-Pole Switch Controlling Two Split Receptacles

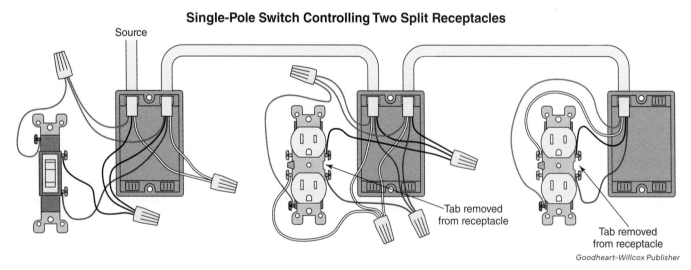

Goodheart-Willcox Publisher

Figure 14-2. A single-pole switch controls two split duplex receptacles. Note the tab removed from between the two ungrounded terminal screws. Be the electron and follow the paths of current through the circuit.

14.3 Switch-Controlled Lighting Outlet and a Constant-Power Receptacle Downstream from the Light

Like the single-pole switch controlling two split receptacles circuit, the wiring scheme for a switch-controlled light and a constant-power receptacle downstream from the light uses a three-conductor cable to deliver both constant power and switched power to the loads, **Figure 14-3**. A two-conductor cable is run from the lighting outlet to the downstream receptacle outlet. Note that a *switch-controlled light* is a lighting outlet that is controlled by a switch or combination of switches, while a *constant-power receptacle* is not controlled by a switch but always remains energized.

The switch location is wired in the same manner as the previous circuit—with the black conductor of the incoming power spliced to the black conductor of the three-conductor cable and a pigtail to provide power to the switch. The black conductors are tied together at the lighting outlet location, bypassing the luminaire to provide the downstream receptacle with constant power. Back at the switch location, the red conductor of the three-conductor cable is connected to the switch and provides switched power to the lighting outlet. The white, grounded conductor provides the return path for both the switched power of the luminaire and the constant power of the receptacle. All bare equipment grounding conductors are spliced together with a pigtail to each device.

14.4 Two Single-Pole Switches Separately Controlling Two Lights on a Single Feed

Figure 14-4 illustrates how a single power source can be "split" to provide power to different switched loads by splicing the incoming power to the required number of pigtails. In this example, there are two pigtails. There are many instances where a single feed is required to power two or more switched circuits, such as at an outside doorway where one switch controls the porch light and the other controls the floodlights or at a ceiling fan location where the fan and light are controlled by two separate switches.

Each of the pigtails supplies power to the switches. A two-conductor cable is run from each switch to their respective lighting loads. The black conductor of these cables is connected to the switch at one end and the lighting outlet on the other end. The white grounded conductors are spliced together in the switch box to provide a return path for the current from the luminaires. All bare equipment grounding conductors are spliced together with a pigtail to each device and luminaire. A double-gang box is used to house the two switches.

Be the electron and follow the path of current through the circuit. Power enters the switch location through the ungrounded black conductor, then flows through the pigtails to each of the two switches. If a switch is in the OFF position, the circuit is open, the flow of electrons is stopped, and light does not illuminate. If a switch is in the ON position,

Switch-Controlled Light and Constant-Power Receptacle

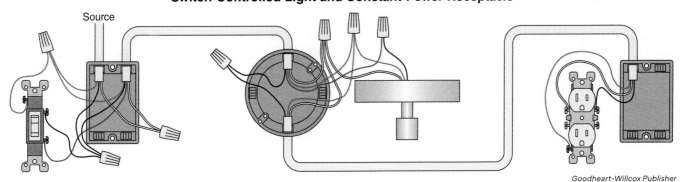

Goodheart-Willcox Publisher

Figure 14-3. Switch-controlled light and a constant-power receptacle downstream from the light. Be the electron and follow the paths of current through the circuit.

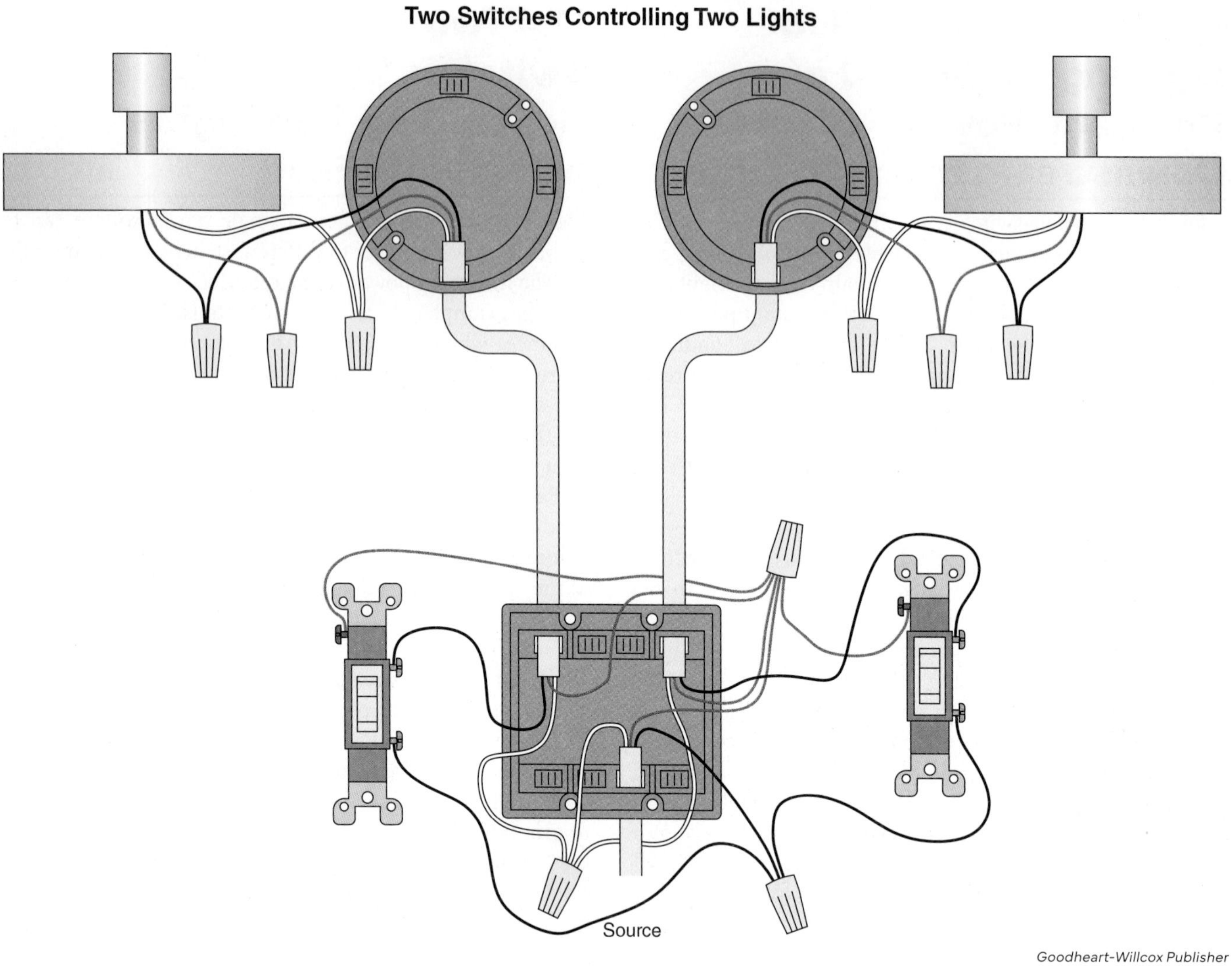

Two Switches Controlling Two Lights

Figure 14-4. Two single-pole switches are separately controlling two lights on a single feed. Be the electron and follow the paths of current through the circuit.

current flows through the switch into the ungrounded black conductor that feeds the light. Electroncs flow through the light, causing it to illuminate, and return to the switch box and then to the source through the grounded white conductor.

A slight modification of the previous circuit allows for a constant power receptacle to be housed in the same double-gang switch box as the switch, **Figure 14-5**. The incoming power source is spliced to two pigtails, one

for the switch and the other for the receptacle. A two-conductor cable is run from the switch location to the lighting outlet box. The white grounded conductors are tied together to provide a return path for the current. All bare equipment grounding conductors are spliced together with a pigtail to each device and luminaire. This type of circuit might be installed at the vanity of a bathroom to provide the required GFCI receptacle and a switched light over the vanity mirror.

Receptacle and Switch-Controlled Light

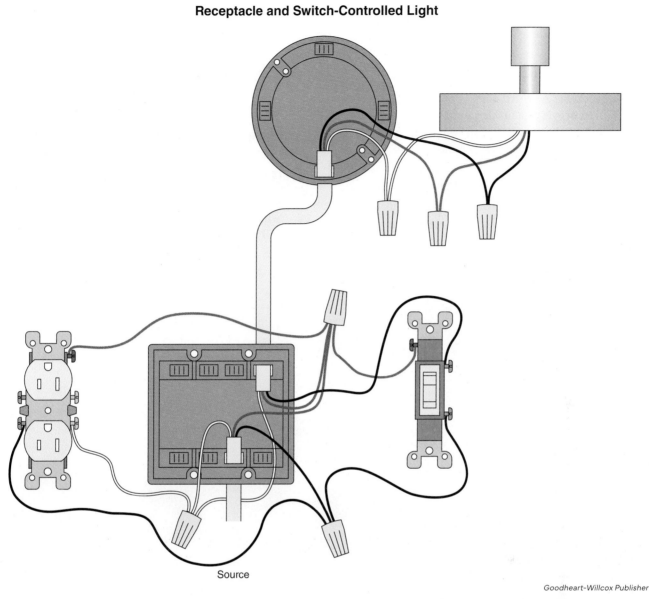

Source

Figure 14-5. A double-gang box with a constant power receptacle and single-pole switch-controlled light on a single feed. Be the electron and follow the paths of current through the circuit.

Summary

- Advanced switching circuits include switch-controlled receptacles, constant-power receptacles downstream from a switch-controlled light, or multiple switches controlling multiple lighting outlets or devices from a single feed.
- The *NEC* requires a switched lighting outlet in every habitable room in a dwelling, but the lighting outlet is not required to be an overhead luminaire. Instead, one or more switched receptacles are permitted to serve in place of lighting outlets in rooms other than the kitchen or bathroom.
- It is common practice and industry standard to use the red conductor for switched power and the black conductor as constant power.
- A switched circuit with a single-pole switch controlling two split receptacles with the power source at the switch location can be controlled by three-way or four-way switches and may contain as many split receptacles as needed.
- A switch-controlled light and a constant-power receptacle downstream from the light uses a three-conductor cable to deliver both constant power and switched power to the loads.
- A single power source can be "split" to provide power to different switched loads by splicing the incoming power to the required number of pigtails.

Know and Understand

1. *True or False?* A switch-controlled lighting outlet is required in every habitable room in a dwelling.
2. A common practice in wiring switching circuits is to use the _____ conductor as switched power.

   A. black   C. white

   B. red   D. bare

3. The _____ conductor delivers power to the load.

   A. black   C. white

   B. red   D. A or B

4. *True or False?* When a receptacle is split, only one outlet can be switched, and the other outlet must have constant power.
5. Electrical current is returned to the source on the _____ conductor.

   A. black, ungrounded

   B. red, ungrounded

   C. white, grounded

   D. bare equipment grounding

Apply and Analyze

1. Describe the process for splitting a receptacle.
2. How can a single source conductor be used to supply power to more than one device?

Critical Thinking

1. Other than the examples given, describe a scenario where you would use a circuit such as the one shown in **Figure 14-4**.

SECTION 5

The *National Electrical Code* and Construction Basics

Flegere/Shutterstock.com

CHAPTER 15
The *NEC*: Purpose, Scope, and Navigation

CHAPTER 16
Print Reading and Estimating

CHAPTER 17
Residential Framing

All electrical installations must be compliant with the rules and requirements of the *National Electrical Code (NEC)*, so residential electricians must understand and be able to navigate it. They also must also be proficient at reading blueprints and creating estimates for the cost of a job, as well as have knowledge of the basics of residential framing, either in new construction or remodeling work.

Chapter 15, *The NEC: Purpose, Scope, and Navigation*, presents an overview of the purpose and arrangement of chapters in the *NEC*, as well as the best way to navigate the contents. Chapter 16, *Print Reading and Estimating*, summarizes blueprint fundamentals and includes methods of estimating the cost of an electrical installation. The language of construction is contained in the blueprints of the project. Finally, Chapter 17, *Residential Framing*, introduces the terminology, materials, and methods commonly used in residential framing.

CHAPTER **15**

The *NEC*: Purpose, Scope, and Navigation

National Electrical Code®

International Electrical Code® Series

2023

SPARKING DISCUSSION ⚡

Review the Table of Contents for *Chapters 1–4* of the *NEC* and identify the articles you believe would be most useful when wiring a residential occupancy. Why do you think this is the case?

CHAPTER OUTLINE

15.1 Creating and Revising the *NEC*

15.2 Purpose and Scope of the *NEC*

15.3 Enforcement of the *NEC*

15.4 Chapter Arrangement of the *NEC*

15.5 Article Structure and Formatting in the *NEC*

15.6 Strategies for Navigating and Using the *NEC*

Introduction

The *National Electrical Code (NEC)* was introduced in Chapter 1, *Getting Started in the Industry*, and a variety of sections and articles have been referenced throughout this text. This chapter provides an in-depth exploration of the *NEC* and answers the following questions:

- What exactly is the *NEC*?
- What are the origins and purpose of the *NEC*?
- Who writes the *NEC*?
- Who enforces the requirements of the *NEC*?
- Why must we understand the arrangement of the *NEC*?

15.1 Creating and Revising the *NEC*

After the incandescent light bulb was invented, there were not many electricians who knew how to correctly install them. The commercial use of electricity was in its infancy, and dozens of electric companies competed for dominance even though the phenomenon of electricity was poorly understood by many, and electrical installation methods and materials lacked standardization. Electrical fires were becoming more and more common, causing many deaths and property damage. The insurance industry finally recognized a need to standardize the rules for electrical installations, and the first edition of the *National Electrical Code* was published in 1897 by the National Board of Fire Underwriters. Now, the *National Electrical Code* is published by the **National Fire Protection Association (NFPA)**, a trade association that publishes hundreds of industry codes and standards. The NFPA aims to minimize death, injury, and property losses due to fire, electrical, and related hazards, and the *NEC* is often referred to as *NFPA 70, National Electrical Code*.

To reflect industry changes, such as newly approved materials or equipment, the *NEC* is reviewed and updated every three years by a team of authors divided into 18 Code-Making Panels. **Code-Making Panels (CMPs)** are made up of representatives from the electrical industry, including electrical utilities, manufacturers, research and testing laboratories, electricians, inspectors, and insurers. Each CMP has 15-25 members and is responsible for several articles of the *Code*. For example, in 2020, CMP-1 had 23 members who oversaw *Article 90, Article 100, Article 110, Chapter 9, Table 10, Annex A, Annex H, Annex I,* and *Annex J.* The *NEC* is a "consensus standard," meaning that proposals for additions, deletions, or changes must go through a review process.

The latest edition of the *NEC* is the 2023 edition, to which this text is aligned, but the process for revising the next edition begins upon publication of the current edition.

PRO TIP   **Developing Standards**

The NFPA has a four-stage process for developing standards:

- **Public input.** Anyone can submit proposals for changes to the *NEC*. A form is included in the back of the *NEC*, or submittals can be made online. A technical committee then votes on the public inputs and produces a First Draft Report.
- **Public comment.** Anyone can then submit a public comment on the First Draft Report, and a Second Draft Report is developed based on public comments.
- **NFPA annual technical meeting.** The NFPA annual technical meeting is held each June to provide an opportunity for debate and discussion of the Second Draft Report. Anyone may offer input at the technical meeting, but voting on measures at the NFPA Technical Meeting is limited to NFPA members.
- **Council appeals and issuance of standards.** Appeals help ensure that all the procedures and NFPA rules were followed. Once all appeals are satisfied, the Standards Council vote on the proposal. Their decision is final and subject only to limited review by the NFPA Board of Directors.

15.2 Purpose and Scope of the *NEC*

The *National Electrical Code* is a safety standard and, as stated in *Section 90.2(A)*, its purpose is "*the practical safeguarding of persons and property from the hazards arising from the use of electricity.*" While not intended as a how-to manual for untrained persons, a trained electrician must utilize the *NEC* for proper design and installation practices. An electrical installation that complies with

Section 90.1(B) and is adequately maintained may not necessarily be "*efficient, convenient, or adequate for good service or future expansion,*" but it may be essentially free of any hazards and thus compliant.

It is important to note that the *NEC* only sets the minimum standard for electrical work. Oftentimes, job specifications have additional requirements that must be followed that are more stringent than the *NEC*. Installing electrical work to the *NEC* minimum is not an indicator of a good, quality electrical job.

As outlined in *Article 90.2*, the scope of the *NEC* coverage includes the installation and removal of electrical equipment in residential, commercial, and industrial locations, whether public or private. It does not cover electrical installations on ships, railway cars, aircraft, or automobiles other than recreational vehicles and mobile homes. It also does not apply to electrical utility installations, such as service drops, service laterals, and metering equipment, or installations that are an integral part of a generating plant, substation, or control center. The electric utility's offices, warehouses, machine shops, and similar locations are subject to the requirements of the *NEC*.

15.3 Enforcement of the *NEC*

The goal of the *National Electrical Code* is to be adopted as a legal document by state and/or local governmental bodies that oversee electrical installations under their jurisdiction. Local jurisdictions can adopt the *NEC* in its entirety, with specific additions or exceptions. Although rare, the *NEC* may not be adapted at all. The responsibility of oversight is usually that of the local building and inspections department's electrical inspector, or some municipalities may choose to contract a private inspection company for this task. The electrical inspector, known as the **authority having jurisdiction (AHJ)**, is responsible for enforcing the requirements of the *NEC*. The AHJ is also responsible for interpreting the rules of the *Code*, approving equipment and materials used in electrical installations, and granting special permissions addressed throughout the *NEC*. The AHJ also has the rarely used authority to waive certain sections of the *Code* or to suggest alternate methods that are *NEC* compliant.

Rules in the *NEC* are distinguished as either mandatory or permissive. **Mandatory rules** are actions that are specifically required or specifically prohibited, and they are identified by the terms "shall" and "shall not." **Permissive rules** identify actions that are allowed but not required. Permissive rules are identified by the phrase "shall not be required" or "shall be permitted."

Mandatory rules and permissive rules may have *Exceptions*. **Exceptions**, or rules concerning alternate methods or materials, immediately follow the general rules they modify and are formatted in italics. If there is more than one exception, they are numbered, **Figure 15-1.**

(A) General. Branch-circuit conductors shall have an ampacity not less than the larger of the following and comply with **110.14(C)** for equipment terminations:

(1) Where a branch circuit supplies continuous loads or any combination of continuous and noncontinuous loads, the minimum branch-circuit conductor size shall have an ampacity not less than the noncontinuous load plus 125 percent of the continuous load in accordance with **310.14.**

Exception to (1): If the assembly, including the overcurrent devices protecting the branch circuits, is listed for operation at 100 percent of its rating, the ampacity of the branch-circuit conductors shall be permitted to be not less than the sum of the continuous load plus the noncontinuous load in accordance with 110.14(C).

(2) The minimum branch-circuit conductor size shall have an ampacity not less than the maximum load to be served after the application of any adjustment or correction factors in accordance with **310.15.**

Exception to (1) and (2): Where a portion of a branch circuit is connected at both its supply and load ends to separately installed pressure connections as covered in 110.14(C)(2), an allowable ampacity in accordance with 310.15 not less than the sum of the continuous load plus the noncontinuous load shall be permitted. No portion of a branch circuit installed under this exception shall extend into an enclosure containing either the branch-circuit supply or the branch-circuit load terminations.

Reproduced with permission of NFPA from NFPA 70, National Electrical Code, 2023 edition. Copyright © 2022, National Fire Protection Association. For a full copy of the NFPA 70, please go to www.nfpa.org

Figure 15-1. *Section 210.19(A) Exception No. 1 and Exception No. 2.*

Exceptions are enforceable and are used when the main *NEC* text may not apply or where a different method can be used as an alternative.

Informational notes, on the other hand, are explanatory material that help interpret the intent of a rule, such as references to other codes, standards, or other sections of the *NEC*. Informational notes are not enforceable by the AHJ. These notes are generally text, formatted in a smaller font size but can also be presented as a table or illustration.

15.4 **Chapter Arrangement of the *NEC***

Recall from Chapter 1, *Getting Started in the Industry* that the *NEC* is divided into an introduction, nine chapters, and informative annexes, **Figure 15-2.** The beginning chapters of the *NEC* apply to all general electrical installations. *Chapter 1* (the 100s) is called the *General Chapter* and consists of two articles covering terminology that directly applies to residential buildings—such as ampacity, bathroom, branch circuits, cabinet, cooking unit, dwelling unit, feeder, ground-fault circuit interrupter, kitchen, service, general-use snap switch, and voltage—and general requirements for electrical installations, including examination and use of electrical conductors and equipment.

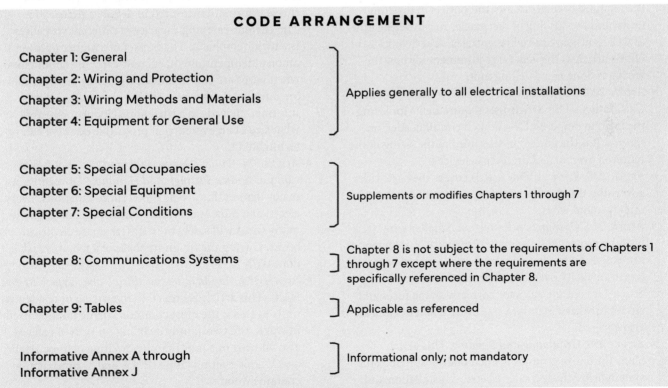

CODE ARRANGEMENT

Chapter 1: General
Chapter 2: Wiring and Protection
Chapter 3: Wiring Methods and Materials
Chapter 4: Equipment for General Use
— Applies generally to all electrical installations

Chapter 5: Special Occupancies
Chapter 6: Special Equipment
Chapter 7: Special Conditions
— Supplements or modifies Chapters 1 through 7

Chapter 8: Communications Systems
— Chapter 8 is not subject to the requirements of Chapters 1 through 7 except where the requirements are specifically referenced in Chapter 8.

Chapter 9: Tables
— Applicable as referenced

Informative Annex A through Informative Annex J
— Informational only; not mandatory

Reproduced with permission of NFPA from NFPA 70, National Electrical Code, 2023 edition. Copyright © 2022, National Fire Protection Association. For a full copy of the NFPA 70, please go to www.nfpa.org

Figure 15-2. Figure 90.3 of the *National Electrical Code* details the arrangement of the chapters of the *NEC*.

While this may seem like an intimidating amount of information, a residential electrician can briefly review the material and identify items concerning residential installations. Some examples include conductor type, how wire size is measured, mechanical execution of work, temperature limitations of a conductor, terminal connection torque, and working spaces of equipment.

Chapter 2, Wiring and Protection of the *NEC* (the 200s) is one of the most critical chapters for all electricians to study and utilize in their careers. It is essential to install electrical equipment and wiring sized correctly. Most requirements in this chapter are determined before the electrician arrives at the jobsite. Requirements for the branch circuits, electrical load calculations, feeder and service size, overcurrent protection (fuses or circuit breakers), grounding, and bonding are all planned before the installation process.

Key articles for the residential electrician found in *Chapter 2* include the following:

- *Article 200, Use and Identification of the Grounded Conductors.* Grounded conductors are most often the neutral conductor, and this article covers their coloring requirements.
- *Article 210, Branch Circuits Not Over 1000 Volts ac, 1500 Volts dc, Nominal.* This is one of the most used articles when wiring a dwelling unit and covers items such as when a ground-fault circuit interrupter (GFCI) or an arc-fault circuit interrupter (AFCI) is required, the required branch circuits and outlets in dwelling units, and the spacing of receptacles in a dwelling unit, as well as other required receptacles. A key component of this article is the general requirement for *how* the wiring is done in a dwelling unit.
- *Article 220, Branch-Circuit, Feeder, Service and Load Calculations.* This article lists requirements for sizing the load, in amperes, for a branch circuit, feeder, or service. It is the guide for determining the sizing of the required overcurrent protective devices.
- *Article 230, Services.* This article covers the basic rules governing the installation of services with overhead or underground wires.
- *Article 240, Overcurrent Protection.* Smaller wire must be protected by a smaller ampacity overcurrent protective device. Larger wires are allowed to have a larger ampacity overcurrent protective device. This article lays out the rules for correctly sizing fuses and circuit breakers. Standard size overcurrent devices are also covered.
- *Article 250, Grounding and Bonding.* This article plans the entire grounding and bonding system. This system helps create a safer electrical installation and is just as important in a small dwelling as it is in a large commercial or industrial installation. This is one

of the longer and more intense articles in the *NEC*. Fortunately, dwelling units do not have as many items to ground and bond, but it is still crucial to do them without error.

Chapter 3, Wiring Methods and Materials (the 300s) covers wiring methods and materials and consists of 51 articles. The information in this chapter is used to install and build the electrical portion of a job, but many of these articles and the wiring methods listed in them are not used in residential work. However, some of the *NEC* rules in the 300s are extremely important to the residential electrician.

- *Article 300, General Requirements for Wiring Methods and Materials.* Everyday installation methods are often found in this part of the *NEC*. Some items from this article used in residential installations include protecting cables from physical damage, requirements for buried cables and raceways, securing and supporting various wiring methods, conductor continuity, and the length of free conductors at junction boxes.
- *Article 310, Conductors for General Wiring.* The conducting material for wire shall be aluminum, copper-clad aluminum, or copper. Nearly all conductors used in branch circuits are copper. The feeders and service entrance cable can be copper or aluminum. One of the major differences in wire is in the insulating material protecting the conductor. All this is information covered in detail in *Article 310*. The current-carrying capacity of different size wires is a major emphasis. The impact of a higher or lower ambient temperature on current-carrying conductors when wires are run in the same raceway or when cables are bundled is a key concept in this article. All this information leads to a final ampacity of a wire and what size of an overcurrent protective device it can be terminated to.
- *Article 314, Outlet, Device, Pull, and Junction Boxes; Conduit Bodies; Fittings; and Handhole Enclosures.* A major item in this article is box fill calculations. Every electrician must learn the method of calculating how many wires will safely fit in different size electrical boxes. During the rough-in phase of a job, this task is done daily.
- *Article 334, Nonmetallic-Sheathed Cable: Type NM and NMC.* This article is crucial to the wiring of residential units as this is the most common wiring method used in single and two-family dwelling units. NM cable is also allowed in a multi-family dwelling of Type III, IV, and V construction. Type III, IV, and V construction contains wood.
- Other materials used by residential electricians are covered in *Article 320, Armored Cable: Type AC,*

Article 330, Metal Clad Cable: Type MC, Article 338, Service-Entrance Cable: Type SE and USE, and *Article 340, Underground Feeder and Branch-Circuit Cable: Type UF.*

Chapter 4, Equipment for General Use (the 400s) covers equipment for general use. The key word is *use*. The articles in this chapter address wiring or devices that distribute electricity to items that use or utilize electricity. Some *NEC* articles important in wiring a dwelling unit include the following:

- *Article 400, Flexible Cords and Flexible Cables* and *Article 402, Fixture Wires.* Flexible cords and fixture wires can be used in many luminaires in a residential occupancy. Cord, cord sets, or cables are not allowed for fixed (permanent) wiring in a building. This includes dwelling units.
- *Article 404, Switches* and *Article 406, Receptacles, Cord Connectors, and Attachment Plugs (Caps).* These articles provide guidance for requirements on switch and receptacle use and selection.
- *Article 408, Switchboards, Switchgear, and Panelboards.* Every dwelling will have a panelboard to distribute electricity to utilization items. This article gives an overview of what is required for a panel, switchboards, and switchgear.
- *Article 410, Luminaires, Lampholders, and Lamps.* The correct luminaire needs to be installed in the correct environment. Is the environment wet, dry, corrosive? Is it a bath or shower area? Depending on weight, extra support might be needed.
- *Article 422, Appliances.* This article covers the rules for connecting appliances to a branch circuit. A dwelling unit can have many appliances, including a stove, storage-type water heater, central vacuum, waste disposer, dishwasher, trash compactor, range hood with or without a microwave, and ceiling fans.
- *Article 430, Motors, Motor Circuits, and Controllers.* The electrician working on a dwelling unit will need some familiarity with this article. Although most motors in a dwelling unit are of the fractional HP variety, some dwelling units will have sump or sewage pumps.
- *Article 440, Air-Conditioning and Refrigerating Equipment.* Some dwelling units will have air conditioning units installed. This article will give wire and overcurrent protective device sizing and disconnect location guidance.

Chapters 5–7 in the *NEC* apply to special occupancies, special equipment, and special conditions. *Chapter 5, Special Occupancies* (the 500s) covers many different types of occupancies that have special wiring requirements above and beyond those listed in the general Chapters 1–4.

- *Article 550, Mobile Homes, Manufactured Homes, and Mobile Home Parks.* There are times when an electrician is called to repair or supplement work on mobile homes or in mobile home parks. This article should be referenced for this type of work.

Chapter 6, Special Equipment (the 600s) mostly covers special equipment that is not used in residential wiring except for *Article 690, Solar Photovoltaic (PV) Systems* and *Article 694, Wind Electric Systems. Chapter 7, Special Conditions* (the 700s) covers special conditions that apply to commercial, industrial, and institutional electrical work.

Chapter 8, Communications Systems (the 800s) is considered a standalone chapter because it covers communication systems that are not subject to the requirements of *Chapters 1–7* unless specifically referenced. The *NEC* requirements apply to many different types of communication circuits used in dwelling units, including internet and cable television. Information on how to properly bond and ground is of importance to electricians working in a residential setting. Finally, *Chapter 9, Tables* includes reference tables that contain useful information compiled from the requirements of the previous chapters.

Informative Annexes A–J are for reference only and are not enforceable requirements. The annexes include sample ampacity calculations, product standards, torque requirements for nuts and bolts, load calculations for dwelling and non-dwelling units, and Americans with Disabilities Act (ADA) information which can apply to multi-family dwelling units.

PRO TIP — *Informative Annex D*

Informative Annex D contains examples of some of the many calculations required by the *NEC*. The samples in *Annex D* illustrate the process of calculating the size of an electrical service or a feeder for dwelling and non-dwelling units. Knowing the location of this information and how to use it is a valuable tool in mastering residential wiring. Many municipalities require a dwelling unit load calculation to make certain the electrician is installing a service of adequate size.

15.5 Article Structure and Formatting in the *NEC*

Understanding how the *National Electrical Code* is structured will make navigating its contents much easier, especially if you know where to and where *not* to search for relevant information.

The *NEC* is written in an outline format that arranges topics from the most general to most specific. Each chapter is divided into articles. An *article* covers general topics and are designated with a three-digit number, such as *Article 210* covering branch circuits. The exception to this is *Article 90, Introduction*, which does not include specific installation requirements and is meant as an introductory article.

Most articles are divided into *parts*. Parts divide articles into more manageable sub-topics. These are designated by a Roman numeral, such as *Article 210, Part II, Branch-Circuit Ratings*. Parts are further divided into sections. A *section* contains the text of the *Code* where the requirements are specified. They are designated by a *dot* following the article number, as in *Section 210.52* (read as *two ten dot fifty-two*). Sections can be further divided into *subdivisions*. There are no more than three subdivisions of a section. The first-level subdivision uses an uppercase letter in parentheses, *250.21(A)*, for example. The second-level subdivision uses a number in parentheses, *250.21(A)(3)*. The third-level subdivision uses a lowercase letter in parentheses, as in *Section 250.21(A)(3)(a)*.

The outline of the *NEC* may require time to familiarize, but it is also designed in a logical and intentional way which makes navigation easier once better understood. The National Fire Protection Association (NFPA) adopts a common numbering system used in the *NEC*. For example, in *Wiring Materials Articles 320-399* any section that ends ".10" will specify Uses Permitted, or allowed wiring and materials installations, whereas a section that ends ".12" specifies Uses Not Permitted, or what is not allowed. Additionally, sections that end ".2" include definitions of terms unique to that article. Definitions for terms used in more than one article are found in *Article 100*. Refer to **Figure 15-3** for the common numbering system for wire, cable, and raceway articles found in the back of the *NEC* Handbook.

If changes were made to the content of an article in the latest *NEC* revision, special formatting is used as an indication, **Figure 15-4**. Changes to the content are indicated with gray shading, or a gray triangle, within each section to highlight the revisions. A figure with the entire caption in gray shading indicates a change to a preexisting figure.

New sections, tables, and figures are indicated by a bold and italic *N* with gray shading to their left. A bold and italic *N* with gray shading next to an article title indicates a new article. A bullet point (•) is used between the paragraphs to indicate where one or more paragraphs have been deleted. Bracketed text ([text]) references other NFPA documents or contains text that has been extracted from other NFPA documents, however these are not typically used in articles relating to dwelling units.

Articles 320–396
Common Numbering Format

Section	Title
Part I	General
Part II	Installation
Part III	Construction Specifications

Articles 320–362

Section	Title
3XX.01	Scope
3XX.2	Definitions
3XX.03	Reconditioned Equipment
3XX.06	Listing Requirements
3XX.10	Uses Permitted
3XX.12	Uses Not Permitted
3XX.14	Dissimilar Metals
3XX.15	Exposed Work
3XX.17	Through or Parallel to Framing Members
3XX.20	Size
3XX.22	Number of Conductors
3XX.23	In Accessible Attics
3XX.24	Bends-How Made, Bending Radius
3XX.26	Bends-Number in One Run
3XX.28	Reaming and Threading
3XX.30	Securing and Supporting
3XX.40	Boxes and Fittings
3XX.42	Couplings and Connectors
3XX.44	Expansion Fittings
3XX.46	Bushings
3XX.48	Joints
3XX.56	Splices and Taps
3XX.60	Grounding
3XX.80	Ampacity
3XX.100	Construction
3XX.104	Conductors
3XX.108	Equipment Grounding Conductor
3XX.112	Insulation
3XX.120	Marking
3XX.130	Standard Lengths
3XX.140	Conductors and Cable
3XX.140	Conductor Fill

Reproduced with permission of NFPA from NFPA 70, National Electrical Code, 2023 edition. Copyright © 2022, National Fire Protection Association. For a full copy of the NFPA 70, please go to www.nfpa.org

Figure 15-3. Common Numbering System for Wire, Cable, and Raceway Articles from the back of the 2023 *NEC* Handbook.

NEC Revisions and Extracted Text

SHADED TEXT	Shading behind text is used to indicate revisions from the previous issue of the *Code*. For example, see Article 100 where the definition of device had been revised from the previous edition of the *NEC*.
N	An italic *N* with shading is used to indicate new material. For example, see *110.21(A)(2)* where new information about reconditioned equipment has been added to the 2017 edition of the *NEC*.
•	A bullet is used to indicate that one or more paragraphs have been deleted. For example, see *314.71* where an Exception concerning terminal housings has been deleted form the 2017 edition of the *NEC*.
[EXTRACTED TEXT]	Brackets are used to denote extracted text from another document that appears in the *Code*. For example, see *517.35(A)* which was extracted from *Health Care Facilities Code*, *NFPA 99-2012, Section 4.4.1.1.4*.

Reproduced with permission of NFPA from NFPA 70, National Electrical Code, 2023 edition. Copyright © 2022, National Fire Protection Association. For a full copy of the NFPA 70, please go to www.nfpa.org

Figure 15-4. *NEC* Revisions and Extracted Text.

15.6 Strategies for Navigating and Using the *NEC*

Finding information on a particular topic in the *NEC* can be accomplished by understanding how the *Code* is structured. With this foundation, a solid plan to navigate the *NEC* can be developed. Unfortunately, there are no shortcuts to learning or experience, but some strategies can be employed to help the new *NEC* user to find the correct information.

First, form a question that needs to be answered. Then, identify the most significant key word(s), phrase(s), or concept(s) in the question and review the Table of Contents for the most appropriate article(s). Recall the chapter arrangement when trying to answer a specific type of question—*Chapter 1* for general information, *Chapter 2* for design or planning, *Chapter 3* for installation, or *Chapter 4* for an equipment question. It is possible the question may be more specific to *Chapters 5–8*, but when dealing with dwelling units, code question answers are generally found in *Chapters 1–4*. After choosing an article in the Table of Contents, further narrow down to relevant parts where the answer is likely to be found. This should result in only searching a few pages.

For example, you need to answer the question, "In general, when laying out receptacle outlets, what is the maximum spacing allowed between them in a bedroom or living room?" The key words used are *receptacle* and *spacing*, and the major concept is *laying out*. Laying out is part of planning, and planning is covered in *Chapter 2, Wiring and Protection*. When reviewing the 200s, *Article 210, Branch Circuits Not Over 1000 Volts ac, 1500 Volts dc, Nominal* is relevant. When further reviewing the article outline, *Part III, "Required Outlets"* would likely have the answer. *Part III* is less than four pages, and a quick scan can identify the following:

From *Section 210.52(A)(1)*: *"Receptacles shall be installed such that no point...of any wall space is more than 1.8 m (6 ft) from a receptacle outlet."*

PRO TIP   **Read Carefully**

Finding the correct information is meaningless if it is misinterpreted or misapplied. Consider the example question of receptacle spacing. Some *Code* users may quickly read *Section 210.52(A)(1)* and conclude the maximum distance between receptacles is *always* six feet. This is incorrect, however. A careful reading of the requirement that any wall space cannot be more than six feet from a receptacle actually means that there shall be no more than 12 feet *between* receptacles.

It may be tempting to start a search in the Index. When searching for a topic or key word within a textbook, beginning with the Index may be a quick and easy solution. However, the *NEC* is *not* a textbook, and searching the Index should *not* be your first approach. A key word will not always be helpful when found in the Index and can yield over 50 choices, which can seem daunting to new users. The starting point for all *NEC* searches should be the Table of Contents.

Consider the following question: *"What is the minimum burial depth for a 120-volt, 20-amp GFCI protected circuit to an outdoor residential lamp post?"* The key words here are *minimum burial depth*, *GFCI*, and *outdoor*. *Minimum burial depth* will not be found in the Index. *GFCI* and *outdoor* will not lead to an acceptable answer. However, an experienced user will know to review *Chapter 3* (the 300s) for an answer. Installation of a wiring method is going to occur—something is being buried—and installation practices are covered in this area. *Article 300, General Requirements for Wiring Methods and Materials* is the key to finding the burial depth. Turning to *Section 300.5(A)*, minimum cover requirements, will lead to *Table 300.5(A)*. The answer to the question can be found in the first row of Column 4: 12″.

Many users of the *NEC* find it helpful to mark up their codebook with highlighters and margin notes to make it easier to find the information in the future. Pre-printed, self-adhesive tabs are available from multiple sources and can be placed on key pages and articles, **Figure 15-5**, to easier find specific information in the *NEC*. Some tab publishers color-code their tabs by subject, which makes the navigation of the *Code* faster and easier.

Goodheart-Willcox Publisher

Figure 15-5. A copy of the *NEC* with color-coded page tabs for easy reference.

Summary

- The *National Electrical Code* is published by NFPA. Teams of representatives from the electrical industry author the revisions to the *NEC* every three years.
- The NFPA has a four-stage process for developing standards: public input, public comment, NFPA annual technical meeting, and council appeals and issuance of standards.
- As stated in *Section 90.1(B)*, the purpose of the *NEC* is "*the practical safeguarding of persons and property from the hazards arising from the use of electricity.*"
- The scope of the *NEC* coverage includes the installation and removal of electrical equipment in residential, commercial, and industrial locations, whether public or private.
- The *NEC* is designed and intended to be adopted as a legal document by state and/or local governmental bodies that oversee electrical installations under their jurisdiction. The authority having jurisdiction (AHJ) is responsible for enforcing the requirements of the *NEC* at the local level of cities, towns, villages, or counties.
- Mandatory rules are actions that are specifically required or prohibited. Permissive rules identify actions that are allowed but not required.
- *Exceptions* immediately follow the general rules they modify and are formatted in italics. *Informational notes* are explanatory material that helps in interpreting the intent of a rule. *Informational notes* are not enforceable by the AHJ.
- The *NEC* is divided into an introduction, nine chapters, and informative annexes. *Chapters 1* through *4* will be most useful for the residential electrician.
- The *NEC* is written in an outline format that arranges topics from the most general to the most specific.
- Revised content is identified with marginal notations and editorial markings throughout the *Code*, including gray shading, bullets, and bracketed text.
- When looking for information in the *NEC*, first form a question, then identify the key word or concept, and determine which article would most likely contain the answer.

Know and Understand

1. The *NEC* is updated every _____ years.
   - A. two
   - B. three
   - C. four
   - D. five

2. *True or False?* The *NEC* is often used as a how-to manual for laypersons.

3. *True or False?* An electrical installation in compliance with the *NEC* will be essentially free from hazards.

4. *True or False?* The *NEC* ensures that all electrical installations are efficient, convenient, and adequate for good service.

5. *True or False?* The *NEC* covers electrical installations in ships, railway cars, and underground mines.

6. Who is responsible for enforcing the requirements of the *NEC*?
   - A. The electrical contractor
   - B. The AHJ
   - C. The NFPA
   - D. Both the AHJ and the NFPA

7. The terms *shall* and *shall not* identify which of the following in the *NEC*?
   - A. Permissive rules
   - B. Mandatory rules
   - C. Exceptions
   - D. Informational notes

8. The terms *shall not be required* or *shall be permitted* identify which of the following in the *NEC*?
   - A. Permissive rules
   - B. Mandatory rules
   - C. Exceptions
   - D. Informational notes

9. What are enforceable rules that concern alternate methods or materials?
   - A. Permissive rules
   - B. Mandatory rules
   - C. Exceptions
   - D. Informational notes

10. Which of the following contains explanatory, unenforceable content in the *NEC*?
    - A. Permissive rules
    - B. Mandatory rules
    - C. Exceptions
    - D. Informational notes

11. How many chapters are included in the *NEC*?
    - A. Five
    - B. Seven
    - C. Nine
    - D. Eleven

12. *True or False?* The *Informative Annexes* of the *NEC* are enforceable by the electrical inspector.

13. What format arranges topics from the most general to the most specific?
    - A. Subdivision format
    - B. Article arrangement
    - C. Outline format
    - D. Funnel format

14. Which of the following indicates where a change has been made in the *NEC*?
    - A. The use of brackets
    - B. A shaded, italic N
    - C. A bullet point
    - D. Shaded text

15. Which of the following is a marginal notation that indicates deleted paragraphs from the previous edition of the *NEC*?
    - A. Brackets
    - B. A shaded, italic N
    - C. A bullet point
    - D. Shaded text

16. Which of the following is a marginal notation that indicates the addition of new material in the *NEC*?
    - A. Brackets
    - B. A shaded, italic N
    - C. A bullet point
    - D. Shaded text

17. Which of the following indicates a reference to other NFPA standards?
    - A. The use of brackets
    - B. A shaded, italic N
    - C. A bullet point
    - D. Shaded text

18. *True or False?* It is considered best practice to begin searching a topic in the Index when looking for information in the *NEC*.

Apply and Analyze

Use your copy of the *NEC* to answer the following.

1. According to *Article 90*, what is the purpose of the *NEC*?

2. List five examples from *Article 90.2* that are covered by the *NEC*.

3. What article of the *NEC* covers overcurrent protection?

4. What article of the *NEC* covers non-metallic sheathed cable?

5. What article of the *NEC* covers switches?

6. *Section* _____ of the *NEC* requires non-metallic-sheathed cable, Type NM, to be secured and supported at intervals not exceeding _____.

Critical Thinking

1. The *NEC* is adopted and utilized in different ways around the country. Discuss how this is done in your area. Who is the AHJ? What edition of the *NEC* is currently being enforced? Does your state or local governmental body make any changes or adjustments to the adopted *NEC*?

2. The *NEC* is a comprehensive code. Not every article will apply to the wiring of dwelling units. With the knowledge gained from covering this unit, discuss which articles are most applicable to the residential world and why this is the case.

3. Discuss the advantages and disadvantages of using the Table of Contents versus the Index as a first resource when trying to find specific *NEC* references and information.

CHAPTER 16

Print Reading and Estimating

Ground Picture/Shutterstock.com

LEARNING OBJECTIVES

After completing this chapter, you will be able to:

- Explain a scale of a blueprint.
- Differentiate between the architects' scale and the engineers' scale.
- Explain how architectural drawings are created.
- Describe types of blueprint layouts, including a floor plan, interior and exterior elevation drawings, a section, and an architectural detail drawing.
- Explain the significance of reference letters *A, C, S, E,* and *M* on a blueprint.
- Identify a section cutting line on a blueprint.
- Define the term *schedule* in reference to a set of building plans.
- Describe the purpose of plan specifications.
- Identify electrical symbols used in construction drawings.
- Explain how electrical plans are developed.
- Differentiate between the detailed estimating process and the approximate estimating process.

TECHNICAL TERMS

approximate estimate
architects' scale
architectural drawing
architectural details
architectural section
blueprint
change order
detailed estimate
easement
electrical plan
elevation
engineers' scale
floor plan
home run
incidental
legend
orthographic projection
request for information (RFI)
scale
schedule
shrinkage
site plan
specifications
structural plans
take-off
title block

Introduction

Building plans, familiarly known as **blueprints** or *prints*, are used in the construction process and for the planning and installation of the electrical system. They are also a critical reference for estimating the project's cost. A residential electrician must have the skills to read and understand a set of blueprints. Several components of a basic blueprint must be considered and understood. This chapter will explain how blueprints are arranged, discuss the various types of plans, introduce the unique language of symbols used on blueprints, and show how blueprints aid in estimating project costs.

16.1 Blueprint Scaling and Types of Scales

Residential blueprints are not drawn to the actual size of the house. Blueprints are drawn at a reduced scale so they can fit on a manageably-sized sheet of paper. A blueprint's **scale** refers to the relationship of the drawing size to the actual size of the building. In architecture, the scale of the drawing is often expressed as a fraction of an inch being equal to one foot, such as 1/4″ = 1′-0″, where each 1/4″ on the plan translates to 1′ of the finished building. A blueprint's drawing scale is indicated on each sheet of the building plans.

The term *scale* also applies to the instrument used to make measurements of the plan. Two common scales are the architects' scale and the engineers' scale. An **architects' scale** is used on plans for buildings and structures and includes scales such as 1/4″ = 1′-0″, 1/8″ = 1′-0″, 3/8″ = 1′, and 1/2″ = 1″. An **engineers' scale** is used with larger-scaled land planning projects and compares one inch on the drawing to tens or hundreds or more actual feet. For instance, common scales include 1″ = 1′, 1″ = 10′-0″, or 1″ = 100′-0″. Most scales are triangular-shaped and read plans of almost any scale, **Figure 16-1**. These tools are special rulers calibrated to allow a user to measure plans in a scale, thus eliminating the need to convert from common measurements to scale.

16.2 Blueprint Layouts

Blueprints convey a great deal of information in a limited space. Traditional blueprints are made of a sturdy paper but still can be damaged or lost. Digital blueprints are becoming more dominant in the construction world as portable computers and tablets become more powerful and reliable. It is an easy task to print the needed parts of a large set of blueprints. Every aspect of the construction process is included in the blueprints, whether digitized or a hard copy, and it takes several pages of prints to show all aspects of the project clearly. Most prints are drawn in a two-dimensional style known as an orthographic projection. **Orthographic projection** is a method of rendering a three-dimensional object in only two dimensions where each view of the building is drawn separately. This generally results in five views of a single-story building—a top view and one for each side.

16.2.1 Floor Plans

The type of orthographic projection most people are familiar with is the floor plan. A **floor plan** shows the layout of the house as viewed from above, **Figure 16-2**, as if the roof was removed. Floor plans are made by assuming a horizontal cut 42″ to 48″ above the floor to show the walls, windows, and doors. Items located above this cut line, such as upper cabinets in the kitchen, will be drawn with dashed lines indicating these items are hidden. Each level of the building, including the basement, will have a floor plan. All floor plans in the set should be drawn to the same scale, and the outlines of the upper floors will align with those of the lower floor.

Another type of orthographic projection is called the elevation. An **elevation** shows the vertical view of the building's exterior sides or an interior feature such as a built-in bookcase or kitchen cabinets. See **Figure 16-3**.

All on-site trades use the same floor plan, but more likely, there will be a floor plan with unique mechanical details for each trade. This is true for commercial, industrial, and institutional construction and often applies to residential plans. Each trade or phase of the building process is given its own reference letter, such as *E* for electrical, *S* for structural, and *M* for mechanical (HVAC and plumbing). A sheet number and a reference letter for the trade or phase of work depicted are included for quick identification. For instance, *E1* is the first page in the electrical plans.

The most basic floor plans and elevations are called the **architectural drawings** and are identified by the letter *A*. Architectural drawings make up the majority of the set of prints and tie together all other prints in the set, including floor plans, elevations, and any architectural sections, architectural details, and schedules. Other floor plans use the architectural drawings with relevant mechanicals, such as plumbing, HVAC, or electrical symbols inserted.

In a set of plans, a thick borderline at the edges of the page usually contains some form of a title block. A **title block** contains information such as the project name or owner, the name of the architect or architecture firm, revision number, if any, and required seals or stamps of approval. It is usually located on the right side of the prints so that it remains visible when plans are rolled up. Refer to **Figure 16-4** for key components of a title block.

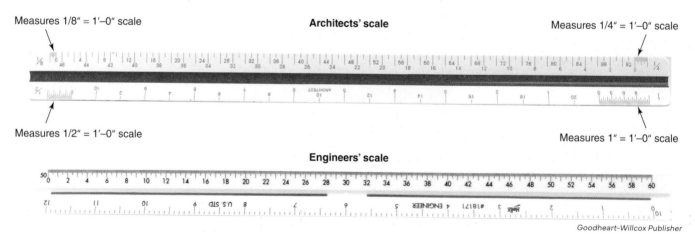

Goodheart-Willcox Publisher

Figure 16-1. Triangular architects' scales contain many common scales used on blueprints such as 1/4″ = 1′ and 1/2″ = 1′. Notice that each edge of the architects' scale has two scales, one read from left to right and one read from right to left.

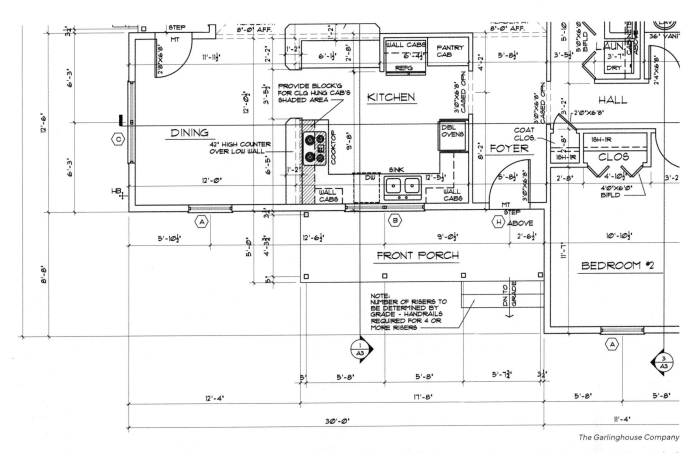

The Garlinghouse Company

Figure 16-2. A floor plan provides basic size and location information for the walls, doors, windows, and other components needed to build a house. The portion of a floor plan shown here describes the dining room, kitchen, foyer, and front porch for a house.

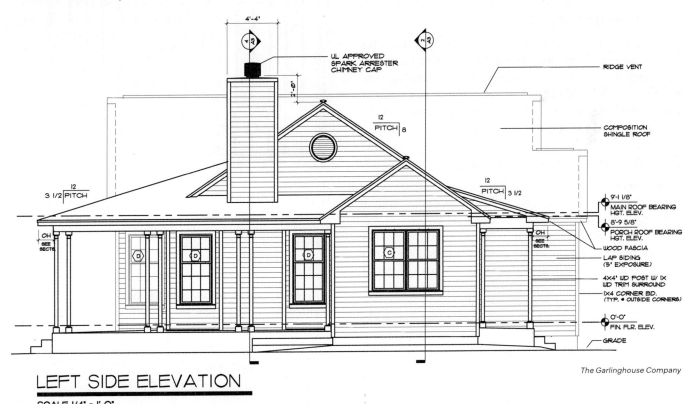

The Garlinghouse Company

LEFT SIDE ELEVATION
SCALE 1/4" = 1'-0"

Figure 16-3. An elevation drawing depicting the exterior features of a house. There may also be interior elevations that depict the layout of a built-in bookcase or the cabinetry layout of a kitchen.

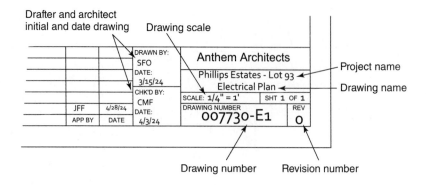

Figure 16-4. Title block for an electrical plan.

16.2.2 Site Plans

The first sheet in the set of plans is typically the site plan and may be identified by reference letter *C* for civil plans. A *site plan* is an aerial view of the property. It shows the property boundaries and how the building is situated on the lot. Features such as streets, trees, utilities, neighboring properties, and easements may also be shown. An *easement* is the right of one party to cross or access another's property for a specific use. The most common easements are for utilities, such as power lines, water mains, and sewers. However, a landlocked property owner with no road frontage might be given an easement to use a portion of the neighbor's property for a driveway.

16.2.3 Structural Plans

Structural plans, reference letter *S*, generally follow civil plans. These plans are used as a guide to install the building foundation and any concrete, wooden, or steel structural members, such as the size and spacing of framing members. **Figure 16-5** shows a detail of the structural plan for the floor framing of a new home.

16.2.4 Sections and Details

Architectural sections, referred to as *sections*, represent a cutaway view of the building. Floor plans are a type of sectional drawing, but most sections show vertical elements of the structure, such as stairways or walls. See **Figure 16-6**. Sectional drawings are usually drawn at a larger scale than the floor plan to allow for greater detail. Ideally, a section of a drawing will also indicate wall finish types and the thickness of each wall type. Examples include drywall, wood, and tile. This information is important to the electrician to ensure the rough-in phase of the job is done correctly. Sections can be taken from floor plans or elevations and identified by section cutting lines. Section cutting lines, sometimes called *cutting-plane lines*, usually contain arrow points to indicate the direction of the view and reference a page and drawing number where the sectional drawing can be found.

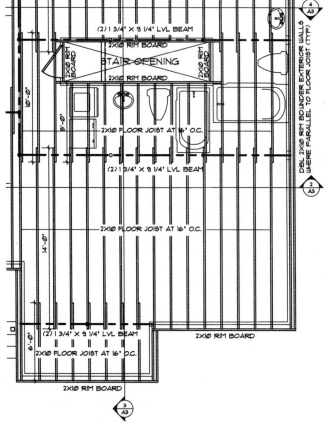

The Garlinghouse Company

Figure 16-5. Detail of a floor framing plan showing the size and spacing of the floor joists.

Architectural details, referred to as *details*, also show an enlarged view of a portion of the plan to provide greater clarity. Many sectional drawings are also detail drawings.

16.2.5 Schedules

A *schedule* is a table that lists specific materials used in the building process. For example, there are door and window schedules that give the details of which type of door or window goes into a particular opening. See **Figure 16-7**.

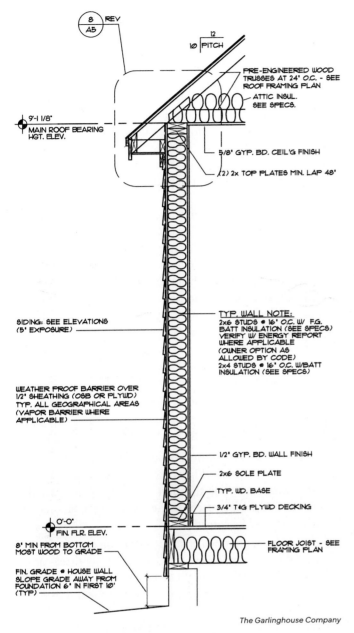

SIDING: SEE ELEVATIONS
(5' EXPOSURE)

WEATHER PROOF BARRIER OVER
1/2' SHEATHING (OSB OR PLYWD)
TYP. ALL GEOGRAPHICAL AREAS
(VAPOR BARRIER WHERE
APPLICABLE)

8
A5   REV

12
10 PITCH

PRE-ENGINEERED WOOD
TRUSSES AT 24' O.C. - SEE
ROOF FRAMING PLAN

ATTIC INSUL.
SEE SPECS.

9'-1 1/8'
MAIN ROOF BEARING
HGT. ELEV.

5/8' GYP. BD. CEIL'G FINISH

(2) 2x TOP PLATES MIN. LAP 48'

TYP. WALL NOTE:
2x6 STUDS @ 16' O.C. W/ F.G.
BATT INSULATION (SEE SPECS)
VERIFY W/ ENERGY REPORT
WHERE APPLICABLE
(OWNER OPTION AS
ALLOWED BY CODE)
2x4 STUDS @ 16' O.C. W/BATT
INSULATION (SEE SPECS)

1/2' GYP. BD. WALL FINISH

2x6 SOLE PLATE

TYP. WD. BASE

3/4' T&G PLYWD DECKING

O'-O'
FIN. FLR. ELEV.

8' MIN FROM BOTTOM
MOST WOOD TO GRADE

FIN. GRADE @ HOUSE WALL
SLOPE GRADE AWAY FROM
FOUNDATION 6' IN FIRST 10'
(TYP)

FLOOR JOIST - SEE
FRAMING PLAN

The Garlinghouse Company

Figure 16-6. Example of an architectural section of a wall. The drawing is at a larger scale than the floor plan to show the elements in greater detail.

WINDOW SCHEDULE

WINDOWS SHALL BE ANDERSEN 'NARROWLINE' '200 SERIES DOUBLE HUNG WINDOWS

MARK	MANUF.	MODEL #	UNIT DIMENSION	REMARKS
(A)	ANDERSEN	3056	3'-1 5/8' X 5'-9 1/4'	EGRESS
(B)		2-2842	5'-7 1/4' X 4'-5 1/4'	
(C)		2-2846	5'-7 1/4' X 4'-9 1/4'	
(D)		2856	2'-9 5/8' X 5'-9 1/4'	
(E)		2-2856	5'-7 1/4' X 5'-9 1/4'	
(F)		20310	2'-1 5/8' X 4'-1 1/4'	
(G)		6010	5'-11 1/4' X 1'-0'	DOOR TRANSOM
(H)		3010	3'-0' X 1'-0'	DOOR TRANSOM
(J)		2810	2'-8' X 1'-0'	DOOR TRANSOM
(K)	▼	2817	2'-7 5/8' X 1'-7 1/4'	BASEMENT/UTILITY WINDOW

NOTE: ADD 1/2' TO WINDOW HEIGHT AND WIDTH TO DETERMINE ROUGH OPENING DIMENSIONS

The Garlinghouse Company

Figure 16-7. A window schedule.

Each door and window opening shown on the plans has a symbol that is cross-referenced to the door and window types listed on the schedule. There are also room finish schedules to indicate the floor and wall finishes for each room, such as carpet type, wood floors, tiling, and paint colors.

Depending on the size and overall cost of a project, a dwelling unit may not have a dedicated lighting schedule. The new homeowner works directly with the general contractor or the electrical contractor to design the lighting layout and choose the light fixtures and ceiling fans. However, residential electricians should pay particular attention to the lighting schedule if one is available, as it details the type and brand of luminaire to be installed at each lighting outlet. Multi-family and single-family units designed to be sold by the builder or developer are likely to have a common lighting schedule.

16.2.6 Specifications

Specifications, also called *specs*, are written instructions that define the scope of work to be completed and the materials used in the construction process. Specifications for large commercial jobs can be hundreds of pages long and assembled in a binder or book form. The specifications for a residential construction project are typically less extensive and are included in the plans, usually on the first page. Some trade-specific specifications are included in the form of notes in that trade's plan. For example, an electrical specification in commercial and industrial work may require all 120-volt circuits to be rated at 20 amps. This is generally not done in residential occupancies.

Contractors use plans and specifications together in the estimation or bidding process. Specifications for a custom home may call for high-end, designer devices, hard-to-get high-end material, limited work hours, restricted parking areas, specific waste removal requirements, and possibly special insurance, while a speculation-built home built to be sold by a developer may specify standard, less expensive devices. An electrician who does not read the specifications prior to submitting a bid can be then held responsible for installing items more expensive than anticipated. During the construction process, specifications can aid in settling disputes such as this. When there is a conflict between the plans and the specifications, the specifications take precedence. On smaller jobs, such as single- and two-family dwellings, it is crucial to read every note on the prints. These notes can act as a specification for installation and material requirements.

16.2.7 Requests for Information (RFIs) and Revisions

At times, despite everyone's best efforts, an aspect of the blueprints or specifications might be omitted or be unclear. In these instances, a ***request for information (RFI)*** is submitted in writing to the general contractor or architect. RFIs can be issued during the bidding process, but they are more likely to arise after the work has begun. On rare occasions, an RFI can lead to a partial or a total shutdown of the job until the matter is resolved. Even a minor request can lead to delays and extra expenses. Once the request for information has been satisfied, a change order may be requested if the new information leads to an increase in the cost of the job. A *change order* is a revision in the agreed-upon cost of an installation that must be agreed on by all parties and approved in writing.

It is not uncommon for plans to undergo changes or revisions. For example, a wall may be added or removed, or the location of a switch, receptacle, or luminaire may be changed. Revisions are tracked by the date of the revision, and all affected parties are typically given a revised set of plans. Revisions to a drawing are indicated by a cloud-shaped figure around the change, **Figure 16-8**, and include notes relating to the change or changes made.

16.3 **Electrical Plans and Symbols**

Electrical plans are floor plans that use a set of symbols to convey the locations of switches, receptacles, lighting outlets, and other aspects of the electrical system, **Figure 16-9**. A *legend* is included on the electrical plan, and it identifies the symbols used on the plans. **Figure 16-10** shows many symbols used on residential electrical plans. In residential construction, the electrical designer or contractor usually

develops the electrical prints, but sometimes it is the job of the architect or engineer.

On an electrical plan, curved dashed lines are drawn from each switch location to the lighting outlet or outlets controlled by that switch, **Figure 16-11**. It is up to the electrician to determine the routing of the conductors for each circuit. At times, the real-world conditions of the jobsite make it impossible to follow the plans exactly, and an RFI might be issued. It is a good idea to make the blueprints and specifications available to all the electricians on the job site. More attention to these project details will result in fewer mistakes.

Although the electrician is primarily concerned with the electrical plans, they should also consult the plans of other trades, particularly those for plumbing, heating, and air conditioning. These plans often contain information of interest to the electrician, such as the location and amperage draw of the furnace and air conditioning unit, sump pumps, sewage pumps, and well pumps. Also, appliances may be installed by another trade, such as the water heater, but an electrician is responsible for the electrical connections and must know the electrical requirements. If there is a dedicated electrical plan, these items may not be shown there. They would be shown on the mechanical prints.

16.3.1 **Designing an Electrical Plan**

If the electrical contractor is responsible for designing an electrical plan, they should always review the floor plan and perform a mental walk-through of the residence. In this mental walk-through, each of the following locations and questions should be considered:

- At the front door, what is the view upon entry?
- Are there multiple ways into a room? If so, three- or four-way switching might be needed.
- What is the likely traffic pattern from room to room?
- What ways do the doors open/swing? This helps in choosing switch locations for lighting switches and outlets.

The first step in planning residential electrical design is to locate the breaker panel. The panel should be located so it is accessible but not obtrusive or unsightly. A short feeder is desirable and cost-effective. When possible, installing the electrical panel near the kitchen area is desired. The kitchen could have six to ten or more dedicated circuits for the small appliance branch circuit (two or more), microwave, dishwasher, garbage disposal, refrigerator, oven/range, and an instant water heater. In general, the utility company will determine where the service drop will be, thus the location for the service equipment. For curb appeal, it is best if the service can be in the back or on the side of the house. Ideally, all utilities should be installed where they are the least likely to be seen by the general public.

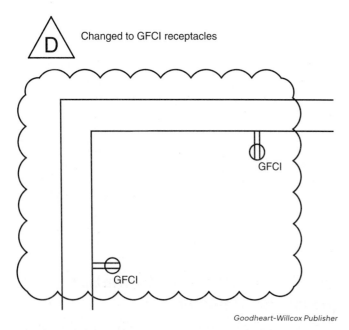

Goodheart-Willcox Publisher

Figure 16-8. A cloud on a blueprint indicates a revision from a previous plan.

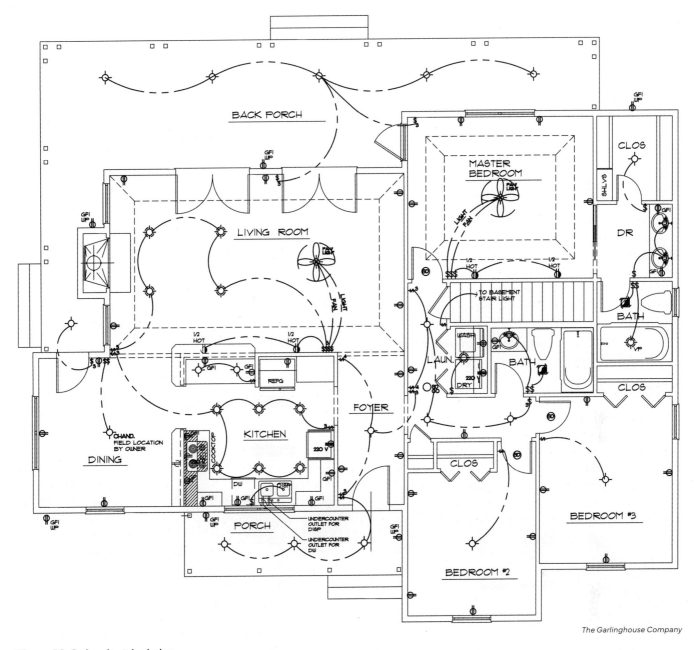

The Garlinghouse Company

Figure 16-9. An electrical plan.

16.4 **Estimating Project Costs Using Blueprints**

Blueprints are an effective tool for estimating the cost of installing the electrical system. Most estimation methods can be categorized as either a detailed estimate or an approximate estimate. A *detailed estimate* involves counting and pricing every item to be supplied and installed and knowing with reasonable certainty how long the project will take. These are more time-consuming but result in a very accurate estimate. Detailed estimates are often used in the bidding process, and the electrician who bids the lowest price will generally be awarded a contract to

perform the work. An *approximate estimate* is created by an electrician using their experience to arrive at a price based on similar, previous jobs. These are much easier to make but do not typically have the same accuracy of a detailed estimate.

16.4.1 **Detailed Estimates**

The detailed estimating process always begins with take-offs. *Take-offs* refer to the counting of major items that are supplied and installed by the electrical contractor, such as wires, boxes, devices, disconnects, breakers, and service equipment. Normally, the electrical contractor does not supply the luminaires for a residence. The general contractor or

Electrical Symbols

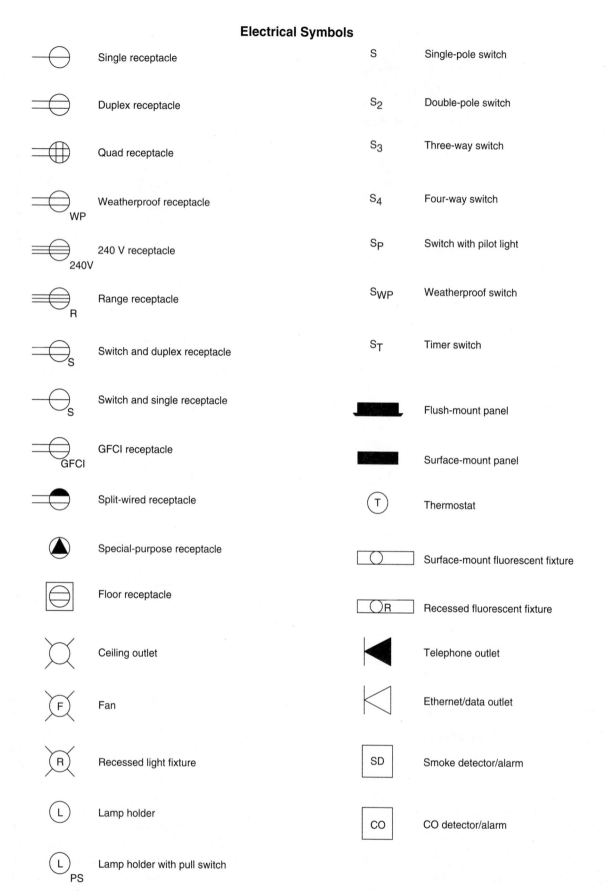

Symbol	Name	Symbol	Name
Single receptacle		S	Single-pole switch
Duplex receptacle		S_2	Double-pole switch
Quad receptacle		S_3	Three-way switch
Weatherproof receptacle WP		S_4	Four-way switch
240 V receptacle 240V		S_P	Switch with pilot light
Range receptacle R		S_{WP}	Weatherproof switch
Switch and duplex receptacle S		S_T	Timer switch
Switch and single receptacle S			Flush-mount panel
GFCI receptacle GFCI			Surface-mount panel
Split-wired receptacle		T	Thermostat
Special-purpose receptacle			Surface-mount fluorescent fixture
Floor receptacle		R	Recessed fluorescent fixture
Ceiling outlet			Telephone outlet
F Fan			Ethernet/data outlet
R Recessed light fixture		SD	Smoke detector/alarm
L Lamp holder		CO	CO detector/alarm
L PS Lamp holder with pull switch			

Figure 16-10. Various symbols are used on an electrical plan.

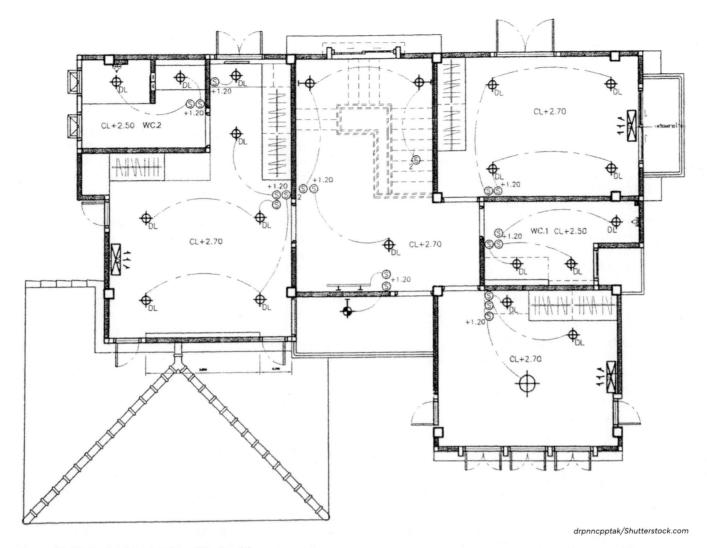

drpnncpptak/Shutterstock.com

Figure 16-11. Basic electric plan with circuiting.

the homeowner selects and provides fixtures for the electrician to install. There are a few exceptions to this for recessed lighting housings, new LED lighting, and bathroom ventilation/light combinations that are installed during the rough-in phase. Other possible exceptions include crawl space, basement, and attic lighting.

Performing Take-Offs

When performing take-offs, it is helpful to use a highlighter to keep track of the items that have been counted. The count should always be accurate. Highlighting items helps ensure everything is counted, and nothing gets counted more than once. It should be noted that one count could include several individual items. For example, when counting a receptacle outlet, you are actually counting one box, one receptacle, and one cover plate.

To estimate the number of conductors used, measure the runs with an architect's scale. Measuring the horizontal dimensions of the walls in each room gives a close estimate of the wire needed. The length of wire from the first outlet back to the electrical panel, where all the branch circuits originate, is referred to as the *home run*. Never forget to include the length of the home run conductors when calculating the overall length needed for wire in a job. Add 10′ to 20′ of wire to each circuit to account for vertical drops down from the attic or up from the crawlspace or basement to the switch or receptacle locations. Occasionally, a job may be more complex with higher ceilings or unique construction, such as a log cabin. In instances like this, more material and labor will likely be required.

Once all the items have been counted and priced, it is prudent to add a reasonable percentage—perhaps within a range of 3% to 5%—to the cost of materials to account for incidentals and shrinkage. *Incidentals* are small items that are impractical to charge for individually, such as staples, wire nuts, and cable clamps. *Shrinkage* refers to any item that

CODE APPLICATION

Adhering Electrical Plans to the *NEC*

Scenario: An electrician needs to locate the distribution panel in a readily accessible location in a dwelling. Consider the relevant codes and determine your solution:

Section 240.24(D)
Overcurrent protective devices shall not be located in the vicinity of easily ignitable material, such as in clothes closets.

Section 240.24(E)
Overcurrent devices, other than supplementary overcurrent protection, shall not be located in bathrooms, showering facilities, or locker rooms with showering facilities.

Section 240.24(F)
Overcurrent protective devices shall not be located over the steps over a stairway.

The *NEC* prohibits the installation of overcurrent devices, and thus the distribution panel, in certain areas of the dwelling.

Solution: The *Code* specifies that you do not locate the distribution panel in a clothes closet, bathroom, or over the steps of a stairway. Good design suggests a panel should be readily accessible but not obtrusive. If a basement is available and provides a dry environment, it is generally the best place to install a service panel. This is especially true in northern climates. If a basement is not available, a good choice for the panel location is a back hall, laundry room, mechanical room, or garage.

After locating the floor plan, next, design the circuitry room by room, drawing the receptacles, switches, lighting fixtures, and ceiling fans. Some electrical contractors will attempt to provide a branch circuit for each room. A different way to handle this is to evenly divide the load among the required general lighting (and receptacle) loads. Most electrical contractors will provide more than the minimum required branch circuits and divide the load accordingly. The number of required general lighting and receptacle circuits is based on the square footage area of the house. *NEC* rules for required receptacles and receptacle spacing and required lighting outlets will be discussed in later chapters.

Next, indicate which luminaire is controlled by each switch. Special-purpose outlets, also called *individual branch circuits*, are those circuits that supply only one load, such as the dryer, water heater, or stove. Some of these items will not appear on the electrical plans, such as the air conditioner, the heater, and the water heater, as other trades determine the locations for these appliances.

When planning the electrical installation, make sure nothing is overlooked. Are the outside lights and receptacles appropriately placed? Many local and state codes will require electrically interconnected smoke detectors and possibly carbon monoxide detectors. Check with the local AHJ concerning the requirements in your area.

has been purchased but is no longer useable, including wire scraps at the end of the reel and items that are lost, broken, or damaged. Always remember to include sales tax. The sum of the cost of the installed materials, equipment, incidentals, shrinkage, and sales tax make up the total materials cost.

Labor pricing can be complicated in a business. This is because the cost of labor includes much more than the employee's hourly wages. A business owner is also responsible for employee payroll deductions, such as state and federal payroll tax withholding, insurance, workers' compensation, and likely retirement contributions, vacation, and sick pay. This will yield a rate for estimating labor. Depending on the job, the final labor rate will likely be a composite or a combination of different labor rates for employees with different levels of skill and experience.

To estimate the labor costs for a project, break the project into specific tasks. Consider the house room-by-room, and estimate how long each task may take. For example, how long would it take to rough-in one of the bedrooms? How long would it be to trim out the kitchen? When the time it takes to perform all tasks has been determined, multiply the total hours by the hourly or composite labor rate to arrive at the total labor costs. Using this detailed estimating method,

the total cost of the project will be the total labor costs plus the total materials cost plus a reasonable profit.

Detailed estimates can be very time-consuming, and many times there is no guarantee that the estimator will be awarded the job. However, detailed estimating is an important skill to develop as it will often be required on the job. Nearly every job will have unforeseen or extra work, and knowing how to estimate this cost is a valuable skill. **Figure 16-12** shows the basic detail bid sheet for a small house.

16.4.2 Approximate Estimates

Using the approximate estimate method, the electrician provides a rough estimate based on some aspects of the building. This estimate can be based on many factors, such as a portion of the total project budget, a dollar amount per square foot of floor space, or a price based on the number of drops or openings. When counting drops or openings, a price is determined for all the materials and time needed to install each switch, receptacle, luminaire, or other outlet. Approximate estimates are often used by builders and homeowners to determine if a project is worth pursuing.

Electrical Bid Sheet

Project Name:
Project Number:
Project Location:
Date:

Item	Qty	Mat'ls (Each)	Mat'ls (Total)	Labor	Total
Receptacles/Boxes					
Lighting Fixtures					
Cable/Conduit/Conductors					
Service, Circuit Breakers					
Appliance Circuits					
Other					
Totals					

Goodheart-Willcox Publisher

Figure 16-12. Detail bid form.

A contractor with enough experience working on dwelling units that are similar in size and complexity can often generate a competitive bid taking into account the number of openings and the number of appliances that operate on 240-volts. With this basic information, an effective bid can be generated. Often, allowances for other lighting can be easily added. Specialty luminaires costs can be determined by customer preference. Specialty equipment like a hot tub or heated flooring can always be added to the base bid.

Summary

- The blueprint's scale refers to the relationship of the drawing size to the actual size of the building. It is often expressed as a fraction of an inch is equal to one foot.
- The architects' scale is used on plans for buildings and structures such as homes and retail buildings. The engineers' scale is used with large-scale land planning projects such as highway projects and residential subdivision plans.
- Most prints are drawn using orthographic projection, a method of rendering a three-dimensional object in only two dimensions where each view of the building is drawn separately.
- A floor plan shows the layout of the house as viewed from above. An elevation shows the vertical view of the building's exterior sides or an interior feature such as a built-in bookcase or kitchen cabinet. A site plan shows an aerial view of the property. A section represents a cutaway view into the building, mostly showing vertical elements of the structure, such as stairways or walls. Details show an enlarged view of a portion of the plan to provide greater clarity.
- Each trade or phase of the building process is given its own reference letter, such as *E* for electrical, *S* for structural, and *M* for mechanical or HVAC to include plumbing. The most basic floor plans and elevations are called the architectural drawings that are identified by the letter *A*.
- Section cutting lines usually contain arrow points to indicate the direction of the view and reference a page and drawing number where the sectional drawing can be found.
- A schedule is a table that lists specific materials used in the building process. Electricians may have a luminaire schedule that lists the types of fixtures installed in specific locations throughout the house.
- Specifications are written instructions that define the scope of work to be completed and the materials used in the construction process.
- Electrical prints use the floor plan and a set of symbols defined in a legend to convey the locations of switches, receptacles, lighting outlets, and other aspects of the electrical system.
- The electrical plans show the locations of the devices and electrical outlets, but it is up to the electrician to determine the routing of the conductors for each circuit.
- A detailed estimate is more time-consuming but results in a very accurate estimate. An approximate estimate is much faster but may not have the accuracy of a detailed estimate. The cost of the installed materials and incidentals, shrinkage, and sales tax make up the total materials cost. Using the approximate estimate, the electrician will provide a rough estimate based on some aspect of the building, such as a dollar amount per square foot of floor space or a price based on the number of drops or openings.

Know and Understand

1. The relationship between the drawing size on a blueprint to the actual size of the building is called a(n) _____.
   A. detail
   B. scale
   C. ratiometric
   D. orthographic projection

2. A(n) _____ is a measuring instrument used on plans for buildings and structures.
   A. tape measure
   B. ruler
   C. architects' scale
   D. engineers' scale

3. The type of plan that shows the layout of the interior of a building as viewed from above is a(n) _____.
   A. floor plan
   B. detail
   C. elevation
   D. section

4. A _____ is an enlarged view of a portion of the plan to provide greater clarity.
   A. floor plan
   B. detail
   C. elevation
   D. structure plan

5. A cut-away view showing the interior of the building is called a _____.
   A. floor plan
   B. detail
   C. elevation
   D. section

6. A view on the plan that shows the vertical features or the exterior or interior of a building is known as the _____.
   A. floor plan
   B. detail
   C. elevation
   D. section

7. *True or False?* In print reading, a schedule refers to the timeline of the construction process.

8. Written instructions that define the scope of the work to be done and are used in conjunction with the blueprints are the _____.
   A. details
   B. sections
   C. schedules
   D. specifications

9. *True or False?* All of the items requiring an electrical connection will always be shown on the electrical plans.

10. The process of counting material on the blueprint that is supplied and installed by the electrical contractor is known as _____.
    A. shrinkage
    B. home run
    C. take-offs
    D. approximate estimate

11. *True or False?* Approximate estimates are often used to judge if a project is worth pursuing.

Apply and Analyze

1. What type of scale would most likely be used on the plans for the construction of a highway interchange?

2. Explain what is meant if a drawing is scaled to 1/4″ = 1′-0″.

3. How many floor plans would be included for a two-story home with a basement?

4. Which type of plans would the carpenters use to know the size and spacing of the lumber used for the floor joists?

5. Which type of plans make up the majority of the set and serve as the basis for each trade's floor plans?

6. Draw the symbols for the following electrical components.
   A. Single-pole switch
   B. Three-way switch
   C. Duplex receptacle
   D. 240-volt receptacle
   E. GFCI receptacle
   F. Split-wired/switched receptacle
   G. Surface-mounted luminaire
   H. Recessed luminaire
   I. Recessed fluorescent fixture

7. How is a change from a previous plan indicated on a revised set of plans?

8. Explain why an employee's total labor costs are greater than their hourly pay rate.

Critical Thinking

1. Explain how an ordinary tape measure could be used on the jobsite to measure scaled drawings.

2. What is the actual length of a wall that measures 4″ on a plan at a scale of 1/8″ = 1′-0″?

3. Why would the electrical utility need an easement on a homeowner's property?

CHAPTER **17** | Residential Framing

UfaBizPhoto/Shutterstock.com

Introduction

Wood framing is the most common type of residential construction method. As a building material, wood is a prevalent, versatile, and renewable resource. It is also strong in relation to its weight and easy to work with using common hand and power tools. Homes made of wood can have many appealing architectural details and features. This chapter introduces different types of wood products and their uses, discusses the different types of framing methods, and identifies some of the terms common in residential construction.

Building techniques are widely known by skilled tradespersons, including carpenters, electricians, sheet metal workers, and plumbers. During the construction process, every trade will be impacted by the framing process. Electricians will drill through and attach wiring and receptacles to all the framing done by the carpenters. When this is done, care must be taken not to drill holes too large for a piece of lumber or drill and notch wood in the wrong place, as this may impact the structural integrity. When attaching electrical equipment to framing members, the framing structure must be adequate. The work practices of one trade will likely impact another. It is essential not only to master the knowledge to be a competent electrician but also to understand other parts of construction that impact what they do.

17.1 Lumber Basics

Most homes are built using lumber. *Lumber* refers to wood that has been cut to specific dimensions for use in construction. It is mainly used for structural purposes and serves as the framework for the entire home. Lumber can be made of either hardwood or softwood. *Hardwood* generally comes from slow-growing deciduous trees, such as oaks or maples, that lose their leaves in the winter. This wood is strong, very durable, and difficult at times to use for construction. When attaching an electrical box to hardwood, either by drilling through or cutting in, electricians must use drill bits and cutting blades designed to not as easily damage this type of wood. *Softwood* comes from faster-growing evergreen trees such as pine and fir. This wood is softer and easier to cut for construction. Most of the wood used in residential framing is softwood. Electricians are able to use drill bits and saw blades that will cut and drill this type of wood in a faster and more efficient way.

The strength of the wood is a result of its grain. *Grain* refers to the direction the wood cell fibers grow. Different types of wood have different wood grains. Straight-grained wood is very strong and is used in construction. The expert carpenter can determine the type of wood from its grain pattern. Sawing a length of board in the direction of the grain is called *ripping*. Sawing a length of board across the grain is a *crosscut*.

17.1.1 Sizing Lumber

In the early 1900s, as lumber technology and better milling developed, a need for standardization of lumber sizes across the United States was deemed necessary. It would create additional labor and delays in building a house to have part of the lumber arriving from a mill in Georgia, another shipment from Minnesota, and the last load from Oregon if each mill was cutting its final product to a slightly different size.

Lumber is sized and referred to by its cross-sectional dimensions, or the two-dimensional area of the lumber. For example, a lumber size of 2×4, pronounced *2-by-4*, means it has a cross-sectional dimension of approximately 2″ by 4″. This is considered its nominal size. However, these dimensions do not represent the actual measurements of the lumber. A 2×4 actually measures 1-1/2″ by 3-1/2″. Other nominal measurements are listed in **Figure 17-1**. The larger the lumber, the greater the weight it can support.

17.1.2 Lumber Treatment

Although wood is a strong building material, it quickly degrades when exposed to the elements if not adequately treated or preserved in some manner. There are several ways of protecting wood, including painting and pressure treating. All wood that is to be exposed to the weather or in

Lumber Terms and Actual Measurements

Nominal Size	Actual Size in Inches
2 × 4	1-1/2 × 3-1/2
2 × 6	1-1/2 × 5-1/2
2 × 8	1-1/2 × 7-1/4
2 × 10	1-1/2 × 9-1/4
2 × 12	1-1/2 × 11-1/4
4 × 4	3-1/2 × 3-1/2
6 × 6	5-1/2 × 5-1/2

Goodheart-Willcox Publisher

Figure 17-1. Lumber is typically referred to by its nominal size rather than its actual size in inches.

contact with concrete should be pressure treated. *Pressure treated wood* is lumber processed under pressure with a water/copper solution that acts as a preservative to slow rot and has antifungal and insecticide properties. Pressure treated wood has a green or yellow hue and may still be wet from the manufacturing process when purchased.

SAFETY NOTE   Pressure Treated Lumber

Tradespeople need to be safety conscious when installing and disposing of treated lumber. It will contain varying degrees of copper and arsenic. Do not burn treated lumber. When the cellulose burns off, the copper or any other metal present in the wood will remain in the ashes at high and possibly dangerous toxic concentrations.

17.1.3 Types of Lumber

The most common types of lumber used in residential construction, plywood and oriented strand board (OSB), are forms of engineered wood products. *Engineered wood products* are fabricated building materials that enhance the natural strength of wood through the manufacturing process. Many engineered wood products repurpose scraps and otherwise unusable chips and sawdust into usable products, contributing to fewer trees being cut.

Plywood is made of several thin, individual sheets, called *veneers*, that are peeled from a log using a giant lathe, then cross-stacked according to their grain, and finally glued together under high pressure. Cross-stacking the grain gives plywood greater strength that resists bending or breaking. *Oriented strand board (OSB)* is made from various sizes of wood flakes or chips glued together under high pressure giving it a variegated appearance. It is common to use OSB to sheath, or cover, the outside wall or possibly as a subfloor of a house. This type of material is not as expensive as traditional plywood. Some AHJs do not like to see OSB used as support for anything electrical. It may lack the strength of plywood. Wood screws, if over tightened, will not hold properly.

Engineered wood framing members such as rafters, joists, and beams are widespread in residential construction and can span greater distances without support. Three common engineered wood framing members are laminated veneer lumber (LVL), the I-joist, and the open web truss. *Laminated veneer lumber (LVL)* is a manufactured wooden beam made of thin veneers with the grain oriented in the long direction and glued under high pressure, **Figure 17-2**. The *I-joist*, characterized by its I-shaped profile, is constructed with top and bottom chords of traditional lumber and a webbing of OSB, **Figure 17-3**. An *open web truss* is similar to the I-joist in that it is constructed with top and bottom chords, but the webbing is made of vertical and diagonal wood members secured by metal nail plates, **Figure 17-4**. Two common types of open web trusses are the floor truss and the roof truss.

credit: https://www.weyerhaeuser.com/woodproducts/engineered-lumber/microllam-lvl/microllam-lvl-beams/

Figure 17-2. Microllam LVL beam with the grain going the way of the long dimensions. In general, do not drill this type of a beam as this may weaken the structure. Route the electrical wiring above or below.

TFoxFoto/Shutterstock.com

Figure 17-3. Residential wooden I-joist.

George__GL/Shutterstock.com

Figure 17-4. This is an example of an open web floor truss. There is plenty of room to run the wires. Take care not to run cable too close to the metal in the trusses.

PRO TIP **Routing Electrical Conductors**

Drilling holes in an LVL for the routing of electrical conductors should be avoided. If holes are drilled in the wrong spot or are too large, they can severely weaken the beam. Route conductors around the LVL. Some manufacturers allow some drilling with size and location limitations, such as the middle third of the depth and the middle third of the overall span. Consult the specifications before any drilling is to occur. The structure can be weakened if the installation requirements are not followed.

The vertical and diagonal members of open web trusses should never be removed or altered. Also, the metal nail plates of an open web truss can damage the outer sheathing of NM cable. Avoid having the NM cable cross the sharp edge of the metal nail plate.

The upper and lower chords of an I-joist or web truss should never be drilled or notched. The OSB webbing of an I-joist is more forgiving and may be cut or drilled according to the manufacturer's instructions for routing conductors, piping, or ductwork. See **Figure 17-5**. Often, prepunched holes can be removed with a hammer to allow the routing of cables. A good carpenter will make sure if they have to shorten an OSB I-joist, that upon installation, the prepunched holes will still line up to facilitate use by plumbers and electricians.

Headers, beams, and studs also have requirements for proper hole drilling. The requirement is that the middle third of a beam or header depth is the area to drill. This hole should be no larger than 1/3 the depth of the beam. Any drilled hole placed next to another should be placed at twice the diameter of the largest hole. Holes drilled in studs are limited by the size of the stud. An example would be no hole larger than 1-3/8″ for a stud with an outside dimension of 3-1/2″ or 2-3/16″ for a board measuring 5-1/2″. No hole should be drilled within 5/8″ of the outer edge.

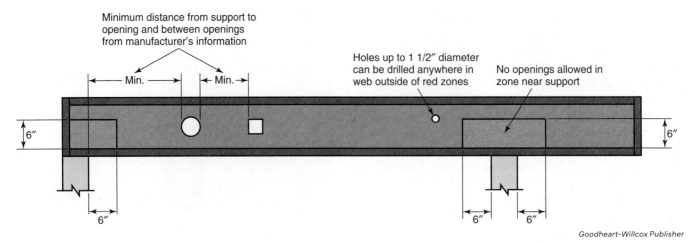

Goodheart-Willcox Publisher

Figure 17-5. Allowable hole sizes for I-joists. Check with the specific manufacturer to make sure the holes are not oversized and the location of each hole is not too close to the edge or the next hole.

17.2 **Framing Basics**

Vertical framing members used to build walls are called *studs*. These are typically 2×4 or 2×6. Horizontal framing members that support floors and ceilings are called *joists*. Floor joists must support a great deal of weight and are usually 2×8 or larger depending on the *span*, or distance between the joist's supports. The greater the distance, the larger the lumber size needed for the span. Long spans may be supported at intervals by posts. *Rafters* are diagonal structural members of a roof that support the roof decking. Rafters are typically 2×4 or 2×6.

17.3 **Framing Methods**

The three methods of residential framing are balloon framing, post-and-beam framing, and platform framing. How a residence is framed may have an impact on the installation of the electrical system. Therefore, residential electricians should become familiar with each construction method and understand the benefits and drawbacks associated with them.

Balloon framing was the standard construction method for homes built before the 1940s but is uncommon today. Balloon-framed houses are typically two-story houses. In *balloon framing*, the studs for the exterior walls are continuous from the top of the building foundation to the ceiling of the highest floor, **Figure 17-6**. The space between the studs creates a convenient chase, or clear path, from the basement to the attic, where electrical conductors, piping, and ductwork can be installed. The second-floor floor joists rest on what is called a *ribbon* or *ledger* that is notched into the wall studs. A ribbon or ledger board is a long horizontal board that is intentionally notched in the vertical studs with the intent of supporting the upper floor. When an electrician is called to update or add wiring in an old house

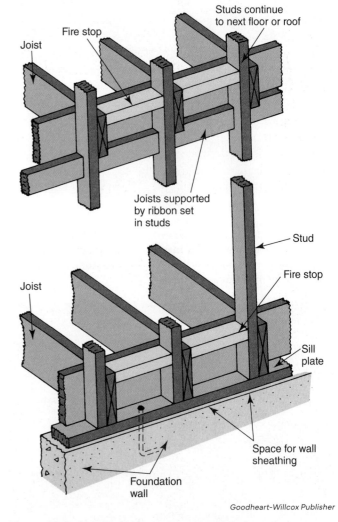

Goodheart-Willcox Publisher

Figure 17-6. Detail of balloon framing. Note how the upper floor joists are supported by the ribbon that has been notched into the studs.

knowing a chase exists to pull in or fish wire can save time. In other types of construction, it is more difficult to get new wiring to the second floor. Often, interior walls may have to be opened to help with this task.

Post-and-beam framing uses larger framing members that can be spaced farther apart than in balloon or platform framing, **Figure 17-7**. The vertical framing members are the posts, and the horizontal framing members are the beams. They both often remain visible as an architectural feature when the house is completed.

Homes built in the post-and-beam style have spacious, open floor plans and large windows, sometimes from floor to ceiling. This can be more costly to wire. Routing wires to areas without walls can be difficult and may require extra time. Receptacles are required in kitchen islands and may be required in the floor near large floor-to-ceiling windows, or the homeowner may want them in open areas where they will likely have furniture. High ceilings are a special challenge for installing lighting. These all add to the cost of wiring houses that utilize post-and-beam construction.

Platform framing is by far the most common house framing method used today. With *platform framing*, also called *western framing*, the first floor is framed on top of the foundation, and the subfloor is installed on floor joists. This becomes the platform on which the walls are assembled and raised. If there is a second floor, the floor joists and subfloor are installed and become the platform for the second-story walls, **Figure 17-8**. Platform framing uses more wood than balloon framing or post-and-beam framing, but the pieces of wood used can be shorter. As a result, fewer carpenters are needed to frame a house using the platform method. Although not cost-effective, a single carpenter could likely do most of the work alone.

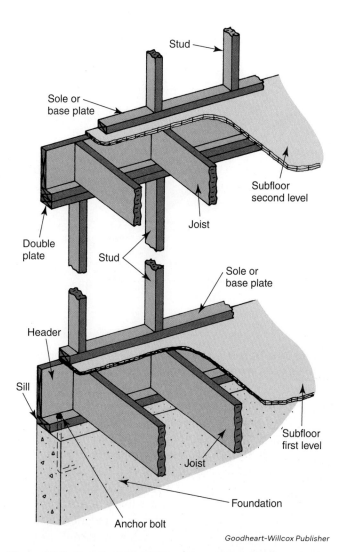

Goodheart-Willcox Publisher

Figure 17-8. Detail of the platform framing method.

stevecoleimages/iStock/Getty Images Plus via Getty Images

Figure 17-7. Typical post-and-beam framing system. Notice the absence of wall studs at regular intervals.

The remainder of this chapter will focus on different aspects of platform framing, although some terms will also apply to other framing methods. The wiring of a house built with platform framing is generally straightforward. High ceilings, kitchen islands, large open areas, and some large and tall window areas could be encountered in high-end dwelling units. This type of construction method, when wired with NM cable, is conducive to an easy wiring job.

17.3.1 Floor Framing

There are many homes built on concrete slabs, but most homes have foundation walls made of poured concrete or concrete blocks. The foundation walls support the floor framing, as shown in **Figure 17-9**. A *sill plate* is usually a 2×6 and attached to the top of the foundation wall with anchor bolts cast into the concrete. This is where the foundation ends, and the framing begins. A *header*, also called a *rim joist* or *band joist*, is attached to the sill plate and is

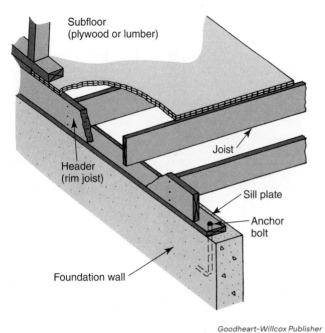

Goodheart-Willcox Publisher

Figure 17-9. Detailed view of the typical floor framing method used in residential construction.

used to secure the floor joists. The header is the same size as the floor joists being used. The floor joists can be solid lumber or an engineered wood product such as an I-joist or an open web joist. Floor joists are typically run parallel to one another and spaced at 12″ to 16″ intervals. The floor joists of a second floor will also serve as the ceiling joists for the first floor. The electrician is likely to run the service conductors through the basement or foundation header to the main panel. In warmer regions this panel could be in the

garage or a convenient and unobtrusive location. If a basement is available, the panel will likely be installed there.

The floor joists are topped with a plywood or OSB subfloor. A **subfloor** is the lowermost floor layer and will serve as the floor during construction. A finished floor with carpet, hardwood, tile, or other covers the subfloor.

17.3.2 **Wall Framing**

With platform framing, the pieces of lumber that make up the wall are hammered together horizontally on the floor and then tipped up into a vertical position and secured. A **sole plate**, or *bottom plate*, is the portion of the wall frame that attaches to the subfloor, **Figure 17-10.** Studs are the bulk of the wall framing and support the **top plate** or double plate. A top plate is the uppermost portion of the wall frame and supports the ceiling joists.

Walls are either load bearing or nonload bearing. A **load-bearing wall** supports the weight of the building above it and has some form of structural support below it, such as a foundation wall or a beam. As a rule, all exterior walls are load-bearing. A nonload-bearing wall does not support any weight from above and does not need any structural support below. Interior walls that are perpendicular to the floor and ceiling joists are often load-bearing walls. Interior walls that are parallel to the floor and ceiling joists are often nonload-bearing walls.

If an opening, such as a door or a window, is necessary for a load-bearing wall, a header is used to distribute the weight of the building around a window or door opening. The headers can be constructed of two 2×6s or larger with a 1/2″ spacer between them, making the header the same

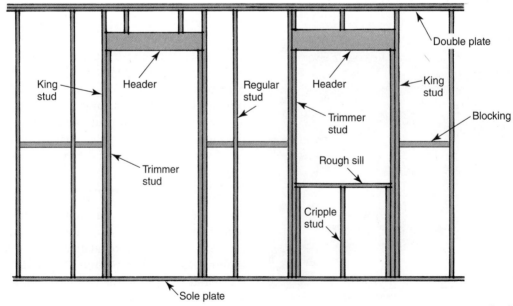

Goodheart-Willcox Publisher

Figure 17-10. Detail of typical wall framing parts and terms.

thickness as the 2×4 walls. The width of the opening determines the size of the header. Headers are also available as an engineered wood product. Nonload-bearing walls also require a header. An opening in a nonload-bearing wall uses a smaller header than would be needed for a load-bearing wall.

The header is installed between two studs, called **king studs**, that run the entire length from the sole plate to the top plate. **Trimmer studs**, also called *jack studs*, support the header and are nailed to the inside of the king stud. The jack studs run from the top plate to the bottom of the header. The header is nailed to the king studs and the trimmer studs.

If the opening is for a window, a **rough sill** serves as the bottom support of the window's rough opening. The studs that support the rough sill are called **cripples**. Electricians need to have a basic understanding of common construction practices when working side-by-side with other trades to ensure accurate communication on a jobsite.

17.3.3 Roof Framing

Residential electricians must understand basic terminology for roof framing in case they are needed for elements located in attics. The **rise** is the height of the roof and applies to the vertical distance between the top plate of the wall and the ridge beam at the peak of the roof. **Run** is the horizontal distance from the wall supporting the rafter to the center of the ridge board and is one-half the length of the span. The roof's **pitch**, or slope, is an indication of the steepness of the roof and is the relationship between the number of inches of rise to every 12″ of run. For example, a roof with a 6/12 pitch will rise 6″ for every 12″ of run, **Figure 17-11**.

Other roof features are illustrated in **Figure 17-12**. The **valley** is the inside corner where two sloping roofs meet. In the northern climates this can be a problem for snow and ice accumulation. Electricians may be asked to install electrical heat tape to help with this problem. A **dormer** is a roofed structure that projects out from a sloped roof. A dormer allows for installing a window on a sloped roof, lets light into the attic, and increases the usable space in the attic. The **eave** is the portion of the roof that extends past and overhangs the outer walls of the house. See **Figure 17-13**. The underside of the eave is called the **soffit**; some houses will have recessed lights installed in this area. The front of the eave is the **facia**. Rain gutters are usually attached to the facia. If installed, electricians are responsible for installing heat tape in downspouts.

Valley   Dormer

Rick's Photography/Shutterstock.com

Figure 17-12. Dormers and valleys are common elements of roof design and construction.

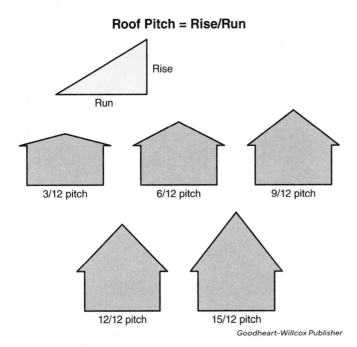

Roof Pitch = Rise/Run

Rise

Run

3/12 pitch   6/12 pitch   9/12 pitch

12/12 pitch   15/12 pitch

Goodheart-Willcox Publisher

Figure 17-11. Typical roofing pitches.

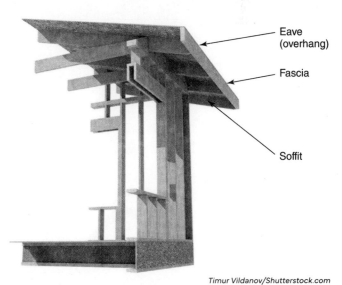

Eave (overhang)

Fascia

Soffit

Timur Vildanov/Shutterstock.com

Figure 17-13. The roof eave (overhang) has a vertical fascia on the end and a horizontal soffit below.

Roof framing is accomplished through either stick-built framing or truss framing. **Stick-built roof** frames, **Figure 17-14**, are constructed by the carpenters onsite using individual sticks of lumber or engineered wood products as the rafters. The rafters are attached to a **ridge beam** at the peak of the roof. A notch called a **bird's mouth** is made in the rafter, where the diagonal rafters meet the vertical walls. This notch allows the rafter to sit flush on the top plate of the wall. **Collar ties** connect the opposing rafters to add strength to the roof frame. Due to cost considerations, this method is not as widely used as in years past.

Roof trusses, **Figure 17-15**, are engineered roofing frames assembled off-site and craned into place on the job-site. Roof trusses are designed to have longer unsupported spans than can be achieved with stick-built roofs. Truss-framed roofs are characterized by the triangular cross-bracing of the webbing. This cross bracing can compromise the usable attic storage space. The ridge beam is usually installed below the top chords of a truss. Roof trusses made off-site and installed with the help of a crane is a great time saver. This is the best, least expensive, and most efficient way to top off a house.

Whichever framing method is used, the shape of the roof determines the architectural style of the house, **Figure 17-16**. The most common roof style is the gable roof that slopes in two directions. A hip roof slopes in all directions. A Dutch hip roof is a hybrid of the gable and the hip roof. The flat roof has little or no slope, and the shed roof slopes in one direction. The A-frame roof is an exaggerated gable roof where the roof is two of the outer walls of the house. The butterfly roof is an inverted gable roof where the roof slopes inward to the center of the house. The gambrel and mansard roofs use a double slope to provide more usable space in the attic. The gambrel is a gable roof with two different slopes, and the mansard is a hip roof with two different slopes. The most common residential roof types are hip and gable. As an electrician, it is useful to have knowledge beyond the electrical trade. It will help make a better, well-rounded tradesperson.

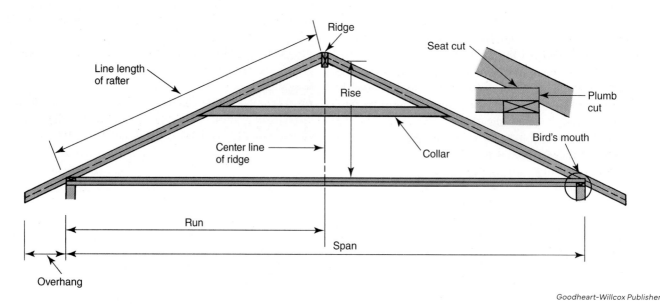

Goodheart-Willcox Publisher

Figure 17-14. Stick-built roof frame.

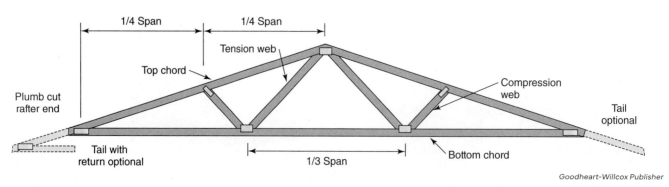

Goodheart-Willcox Publisher

Figure 17-15. Typical roofing truss.

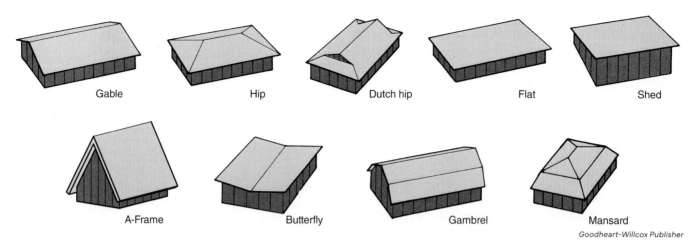

Figure 17-16. Different types of roofing styles.

Goodheart-Willcox Publisher

Summary

- As a building material, wood is a prevalent, versatile, and renewable resource. It is also strong in relation to its weight and easy to work with using common hand and power tools.
- Hardwood generally comes from slow-growing deciduous trees, such as oaks or maples, that lose their leaves in the winter. Softwood comes from faster-growing evergreen trees such as pine and fir.
- Lumber is sized and referred to by its cross-sectional dimensions.
- The two most common types of lumber used in residential construction are plywood and OSB. Plywood is made of several thin, individual sheets called veneers that are peeled from a log using a giant lathe and cross-stacked according to their grain. OSB (oriented strand board) is made from various sizes of wood flakes or chips glued together under high pressure. Engineered wood products are fabricated building materials that enhance the natural strength of wood through the manufacturing process.
- The three methods of residential framing are balloon framing, post-and-beam framing, and platform framing. In balloon framing, the studs for the exterior walls are continuous from the top of the building foundation to the ceiling of the highest floor. Post-and-beam framing uses larger framing members that can be spaced further apart than in balloon or platform framing. With platform framing, the first floor is framed on top of the foundation, and the subfloor is installed on floor joists. This becomes the platform on which the walls are assembled and raised.
- Most homes have foundation walls made of poured concrete or concrete blocks. The foundation walls support the floor framing and the entire weight of the house. The subfloor is the lowermost floor layer and will serve as the floor during construction. A finished floor of carpet, hardwood, or tile will cover the subfloor. Horizontal framing members are called joists and support floors and ceilings. Floor joists must support a great deal of weight and are usually 2×8 or larger.
- Using platform framing, the pieces of lumber that will make up the wall are hammered together horizontally on the floor and then tipped up into a vertical position and secured. Vertical framing members used to build walls, called studs, are typically 2×4 or 2×6.
- A load-bearing wall supports the weight of the building above it and will have some form of structural support below it, such as a foundation wall or a beam. Nonload-bearing walls require a smaller header.
- The pitch, or slope, of the roof is an indication of the steepness of the roof and is the relationship of the number of vertical inches of rise to every horizontal 12″ of run.
- The valley is the inside corner where two sloping roofs meet. A dormer is a roofed structure that projects out from a sloped roof. Dormers increase the usable space of an attic and serve as a source of light. The eave is the portion of the roof that extends past and overhangs the outer walls of the house.
- Roof framing is accomplished by stick-built framing or truss framing. Stick-built roof frames are constructed by the carpenters onsite using individual sticks of lumber or engineered wood products. Roof trusses are engineered roofing frames assembled off-site and craned into place on the jobsite. They are designed to have longer unsupported spans than can be achieved with stick-built roofs.

Know and Understand

1. Wood that has been cut to specific dimensions for use in construction is called _____.
   - A. timber
   - B. lumber
   - C. OSB
   - D. I-joist

2. *True or False?* Softwood comes from fast-growing evergreen trees such as spruce and fir.

3. *True or False?* The majority of house framing is done with hardwoods such as oak or hickory.

4. *True or False?* The nominal size of lumber refers to its actual size in inches.

5. *True or False?* Plywood and OSB are types of engineered wood products.

6. What is the benefit of using engineered wood products over natural wood for floor and ceiling joists?
   - A. They enhance the natural strength of the wood.
   - B. They can span greater distances without support.
   - C. They increase sustainability by using wood that may otherwise be rejected.
   - D. All of these are benefits of engineered wood products.

7. A wooden beam made of many thin layers of wood with the grain oriented to the long dimension is called a(n) _____.
   - A. LVL
   - B. I-Joist
   - C. open web truss
   - D. header

8. A framing member used for floor joist of roof rafters that has a top and bottom chord of solid wood and webbing of OSB is called a(n) _____.
   - A. LVL
   - B. I-Joist
   - C. open web truss
   - D. header

9. A framing member used mostly for floor joists that top and bottom chords of solid wood and vertical and diagonal solid wood webbing is called a(n) _____.
   - A. LVL
   - B. I-Joist
   - C. open web truss
   - D. header

10. Vertical framing members used to build walls are called _____.
    - A. joists
    - B. studs
    - C. headers
    - D. rafters

11. Horizontal framing members that support floors or ceilings are called _____.
    - A. joists
    - B. studs
    - C. headers
    - D. rafters

12. The diagonal framing members of a roof are called _____.
    - A. joists
    - B. studs
    - C. headers
    - D. rafters

13. _____ is the most common method of house framing used today.
    - A. Balloon
    - B. Post-and-beam
    - C. Platform
    - D. Open

14. _____ framing uses less wood than other methods and results in an open floor plan with large floor to ceiling windows.
    - A. Balloon
    - B. Post-and-beam
    - C. Platform
    - D. Open

15. In _____ framing, the outer wall studs run continuously from the top of the foundation to the ceiling of the highest level.
    - A. Balloon
    - B. Post-and-beam
    - C. Platform
    - D. Open

16. *True or False?* In floor framing, the header will be the same size as the floor joists.

17. The portion of the wall frame that rests on the subfloor is the _____.
    - A. sole plate
    - B. double plate
    - C. blocking
    - D. header

18. A framing member used between studs to distribute the weight of the building around a door or window opening is called a _____.
    - A. king stud
    - B. trimmer stud
    - C. header
    - D. rough sill

19. *True or False?* A header is not required on a nonload-bearing wall.

20. Which of the following is the steepest roof?
    - A. 4/12
    - B. 6/12
    - C. 10/12
    - D. 12/12

Apply and Analyze

1. What is the wood type that comes from slow-growing deciduous trees?
2. What are the actual dimensions of a 2×10 floor joist?
3. A wall that supports the weight of the building above it is called a(n) _____.
4. Which two factors determine the steepness of a roof?
5. What is the primary benefit of the gambrel and mansard roof?
6. What is the underside of the eave called?
7. What is the purpose of a dormer?

Critical Thinking

1. List five benefits that wood construction has over other methods.
2. Explain how engineered wood products are considered to increase ecological sustainability.

Justin_Krug/Shutterstock.com

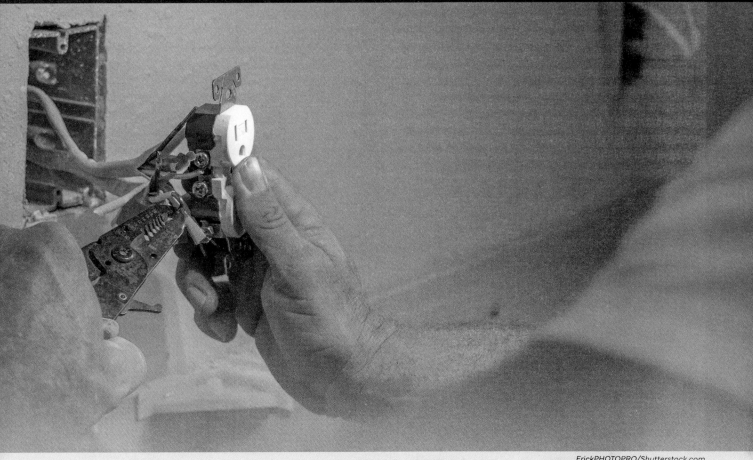

CHAPTER 18
Branch Circuit/
Receptacle Requirements

CHAPTER 19
Conductor Sizing and
Ampacity Adjustments

CHAPTER 20
Service Calculations

While the previous chapter introduced the student to the *National Electrical Code*'s scope and purpose, this section takes a deeper dive. Chapter 18, *Branch Circuits/Receptacle Requirements*, sets forth the *NEC* requirements for branch circuits in a residence. Therefore, required outlets, receptacle spacing, and overcurrent protection are underscored. Chapter 19, *Conductor Sizing and Ampacity Adjustments*, covers the *NEC* requirements for conductor sizing. One of the essential skills of the electrician is the ability to properly size conductors for their intended use.

Chapter 20, *Service Calculations*, covers the vital skill of calculating the size of the electrical service for a dwelling. This chapter summarizes the difference between connected and calculated loads and gives examples of service calculations outlined in the *NEC*.

CHAPTER 18

Branch Circuit/Receptacle Requirements

Ungvar/Shutterstock.com

LEARNING OBJECTIVES

After completing this chapter, you will be able to:

- Define the term branch circuit.
- Calculate the minimum number of branch circuits required in a dwelling.
- Identify locations in a dwelling where lighting outlets are required.
- Summarize the ratings of branch circuits and conductor sizes required.
- Summarize the spacing requirements for receptacle outlets throughout a dwelling.
- Identify where receptacle outlets are required throughout a dwelling.
- Summarize the requirements for GFCI and AFCI protection throughout a dwelling.

TECHNICAL TERMS

accessible, as applied to equipment

branch circuit

dwelling unit

general-purpose branch circuit

individual branch circuit

readily accessible

small appliance branch circuit

Introduction

Recall that a **branch circuit** is defined as the conductors between the final overcurrent protection device and the outlets being served. The branch circuits in a dwelling power all the lights, receptacles, and dedicated appliances.

> **PRO TIP**   **Dwellings**
>
> According to the *NEC*, a **dwelling unit** is a *"single unit, providing complete and independent living facilities for one or more persons, including permanent provisions for living, sleeping, cooking and sanitation."* This chapter, as well as many others, will refer to dwellings in relation to *NEC* requirements and calculations. A dwelling can be one-family, two-family, or multifamily.

A single-family dwelling may have 30 to 40 or more branch circuits depending upon its square footage and the number of circuits needed for appliances and equipment. Some examples of appliances and equipment include a dishwasher, refrigerator, microwave, garbage disposal, air conditioner, furnace, sump pump, well pump, water heater, and electric vehicle charger.

This chapter covers branch circuit and receptacle requirements and ratings, as well as how to calculate the minimum number of branch circuits required in a dwelling. This chapter also covers lighting outlet spacing requirements and certain requirements for arc-fault protection, ground-fault protection, or both.

18.1 Branch Circuit Requirements

Dwelling units are wired differently than commercial and industrial installations. It is common practice to have both lighting and receptacles on the same branch circuit called a *"Branch Circuit, General-Purpose"* in the *NEC*. A **general-purpose branch circuit** supplies two or more general-purpose receptacles or outlets for lighting and appliances.

An **individual branch circuit** is a circuit that supplies only a single load. Some examples of individual branch circuits in dwelling units include a single 240-volt branch circuit for electric ranges that require a 40- or 50-amp receptacle and electric dryers that require a 30-amp receptacle. Sump pumps and sewage pumps may have a single receptacle, rated 15 or 20 amp, or a disconnect switch, possibly the breaker, if directly wired. Other individual branch circuits include the central air conditioner, furnace, and electric water heater. Each of these is not cord-and-plug connected.

The following sections cover branch circuit requirements for small appliance, laundry, bathroom, and garage branch circuits found in a dwelling.

18.1.1 Small Appliance Branch Circuit Requirements

A **small appliance branch circuit**, referenced in *Section 210.11(C)(1)*, requires a minimum of two 20-amp circuits. It is not uncommon to install more than the required two circuits. The small appliance branch circuit is intended to supply receptacles in the kitchen, pantry, dining room, or similar areas of a dwelling unit. *Section 210.52(B)(2), Exception 1* allows a specific appliance to be supplied from an individual branch circuit rated 15 amps or greater. This would cover appliances such as the range, dishwasher, microwave, and garbage disposal.

Section 210.52(B)(2) does not allow the small appliance branch circuit to feed any other outlets except for an electric clock or supplementary equipment that is part of a gas range. The receptacles for the kitchen countertops must be fed from a minimum of at least two small appliance branch circuits per *Section 210.52(B)(3)*. This is done just in case power is lost to part of the countertop receptacles and to divide the load of the small plug-in appliances. The rules for the minimum spacing between receptacles are covered in *Section 210.52(C)*.

18.1.2 Laundry Branch Circuit Requirements

Section 210.11(C)(2) covers laundry branch circuit requirements and states, *"At least one additional 20-ampere branch circuit shall be provided to supply the laundry receptacle outlet(s) required by Section 210.52(F). This circuit shall have no other outlets."* A laundry branch circuit can consist of a receptacle installed for only the washer or a number of receptacles installed in a laundry area. According to *Section 210.50(C)*, receptacle outlets in dwelling units for an appliance *"shall be installed within 6' of the intended location of the appliance."*

18.1.3 Bathroom Branch Circuit Requirements

Section 210.11(C)(3) covers bathroom branch circuit requirements. The basic requirement for a bathroom branch circuit is a minimum of one 20-amp branch circuit to power receptacle outlets in one or more bathrooms. This circuit is not allowed to power any other receptacles outside of the bathrooms or lighting outlets. An inexpensive option is to provide one GFCI receptacle or breaker to feed the receptacles in two or more bathrooms. However, this has the potential for overloading or drawing too much power in this circuit.

Common small appliances plugged into the GFCI-protected receptacle, as covered in *Section 210.8(D)*, include hair dryers and curling irons. The amperage draw of two hair dryers and a curling iron could exceed the 20 amp rating of the circuit. If the overload persists for too long, the OCPD will open.

Section 210.11(C)(3) Exception covers another commonly used option to provide a branch circuit to the bathroom. This method allows for a 20-amp branch circuit to supply all the outlets, lighting, and receptacles in a single bathroom. The exception directs the *NEC* user to *Section 210.23(B)(1)–(2)* for more specific information.

18.1.4 Garage Branch Circuit Requirements

Section 210.11(C)(4) covers garage branch circuit requirements. The requirement for a garage circuit is straightforward. If the dwelling unit has an attached garage, it must be provided with a minimum of one 20-amp branch circuit. Note this is the minimum, and more can be provided if desired. If the garage is detached, the electrician is not required to provide power.

If the garage can fit more than one vehicle, at least one garage receptacle outlet must be provided in each vehicle bay. This receptacle must be no higher than 5-1/2' above the floor. An exception, per *Section 210.52(G)(1)*, states that detached garages in a multifamily dwelling shall not require a receptacle outlet in each bay.

18.2 Branch Circuit Conductor Sizing

Section 210.18 states that the lowest rating for a general-purpose branch circuit in a dwelling is 10 amps. According to *Table 310.16* and the note referencing *Section 240.4(D)*, a

15-amp branch circuit requires size 14 AWG conductors. Some residential electricians install 20-amp branch circuits for general lighting and receptacles, and these 20-amp circuits require size 12 AWG conductors. The kitchen small-appliance branch circuits, the laundry circuit, the bathroom circuit, and the garage circuit are all required to be rated at 20 amps per *Section 210.11(C)(1)–(4)*. The disadvantage of working with 12 AWG NM-cable is the higher cost of material, and its larger size makes it harder to make up electrical connections and neatly install in a device box.

Conductors for individual branch circuits are sized to the single load of the circuit. For example, a 30-amp circuit, such as the water heater or clothes dryer, will use minimum size 10 AWG conductors. A kitchen range is usually supplied by size 8 AWG conductors and protected by a 40-amp circuit breaker. Some ranges will require a larger OCPD, resulting in a larger size wire. This could be 6 AWG conductors protected by a 50-amp circuit breaker. An electrician should consult with a heating and air conditioning contractor to determine the size of the air conditioning unit for the HVAC system.

Article 440, Air–Conditioning and Refrigerating Equipment will provide information on sizing various air conditioning units. The general practice for residential units is to review the nameplate for guidance and sizing of the wires leading to the unit and for sizing the required OCPD. Sometimes a smaller wire rated less than the ampacity of the OCPD is allowed. The larger OCPD is in place for the higher starting or *in-rush current*. Once the unit is running, extra current will not be drawn, and the smaller wire is adequate to handle the running current. Air conditioning units are not typically installed during the rough-in phase, but power must be run to their locations.

18.3 Calculating the Number of Branch Circuits for General-Purpose Receptacles and Outlets

This section covers how to calculate the number of circuits—usually 15 amp but can be 20 amp—that are required in each dwelling. In a residence, there are many circuits and outlets to consider. Some electricians provide a branch circuit for each room in a dwelling, which can be a requirement in some municipalities. The *NEC* also requires specific circuits in certain locations, such as the small kitchen appliance, bathroom, and laundry branch circuits. Regardless, a minimum number of branch circuits must exist in a household. According to *Section 220.41*, "the minimum unit load shall not be less than 3 volt-amps per square foot." However, installing electrical to the *NEC* minimum will not result in a quality job, just an adequate job.

This minimum can be calculated based on the total square feet of the house and the rating of the branch circuit. This calculation results in the *minimum* number of general lighting and receptacle circuits required. For instance, *Section 220.41* requires a unit load of 3 volt-amps per square foot of living space for general lighting and receptacles. Under this rule, a 2500 square foot dwelling would require 7500 volt-amps. To calculate how many volt-amps are required, use the following equation:

$$3 \text{ VA} \times 2500 \text{ ft}^2 = 7500 \text{ VA}$$

Recall from Part 1, *Basic Knowledge and Skills* that by utilizing the power wheel for Ohm's law and Watt's law, **Figure 18-1**, you can calculate the total amps required for the lights and receptacles by dividing the total volt-amps, also called *power* or *watts* in dwelling units, by 120 volts nominal per *220.5*:

$$7500 \text{ VA} \div 120 \text{ V} = 62.5 \text{ A}$$

If the branch circuit OCPD, usually a circuit breaker in residential work, is rated at 15 amps, the minimum allowed would be five branch circuits at 15 amps each. This is determined by the following calculation:

$$62.5 \text{ A} \div 15 \text{ A} = 4.167 \text{ A}$$

This number would be rounded up to five as there can be no partial circuits. If the branch circuits are rated at 20 amps, four branch circuits are required.

$$62.5 \text{ A} \div 20 \text{ A} = 3.125 \text{ A}$$

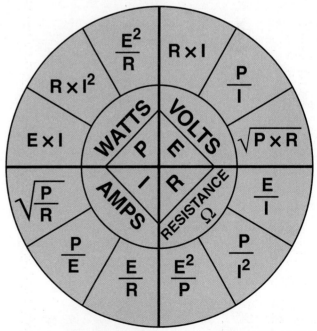

Goodheart-Willcox Publisher

Figure 18-1. Use formulas from the power wheel to calculate unknown electrical quantities.

The point of using the 3 volt-amperes per square foot is to make sure, no matter the size of the dwelling unit, a minimum amount of power is provided for the general-purpose lights and receptacles. Most electricians would likely provide eight or more circuits to cover all the general-purpose lighting and receptacle loads.

18.4 Lighting Location Requirements

There are several lighting requirements outlined in the *NEC. Section 210.70(A)(1)* requires at least one lighting outlet controlled by a wall switch to be installed in every habitable room, kitchen, and bathroom. There are two exceptions to this requirement:

- One or more receptacle outlets are controlled by a wall switch instead of a dedicated lighting outlet in rooms other than a kitchen or bathroom. This allows table lamps or floor lamps plugged into the switch-controlled receptacle to serve as the required lighting outlets. Further, it attempts to clarify that kitchen and bathrooms will need a dedicated opening strictly for the installation of a luminaire. In kitchens, this will indicate an overhead fixture for practical purposes. For bathrooms, the luminaire could be installed on a wall or in the ceiling.

- Lighting outlets are controlled by occupancy sensors rather than wall switches. This assumes that the occupancy sensor is in addition to the wall switch or located in a typical wall switch location and has a manual override that allows the sensor to operate as a wall switch. Although allowed, this type of installation is not typical.

Additional required lighting outlets are listed in *Section 210.70(A)(2)*. This section outlines requirements for a wall-switch-controlled lighting outlet—not receptacle—for hallways, stairways, attached garages, and detached garages supplied with electric power. It also states at least one wall-switch-controlled lighting outlet shall be installed to provide illumination for the exterior of outdoor entrances with grade-level access, **Figure 18-2**. In addition, there shall be a wall switch at each floor level and landing for interior stairways with six or more risers and an entryway.

Zach Wright/Shutterstock.com

Figure 18-2. A wall-switched exterior lighting outlet is required at all entrances to a house.

Section 210.70(C) requires at least one lighting outlet—not receptacle—to be installed in attics, crawl spaces, basements, and utility rooms where these spaces are used for storage or include equipment that might require servicing.

Lighting load ampacity calculations are calculated with the general receptacle load from *Section 220.14*.

18.5 **Required Spacing for Receptacle Outlets**

Article 210 of the *NEC* covers the minimum requirements and spacing for dwelling unit receptacles. This includes general use, such as in the living room, bedrooms, recreation rooms, home offices, and dens. Also included are the required receptacles in the kitchen, dining room, pantry, and breakfast nook areas. Some specific appliances, like a dryer or range, will require their own receptacle. A laundry circuit and the bathroom circuits are allowed to serve only receptacles in those areas. Mounting heights for receptacles

are not specified by the *NEC* but mounting at a convenient height for the intended use should be done.

18.5.1 **General-Use Receptacle Spacing**

Article 210 Part III addresses required outlets for branch circuits, and *Section 210.52* provides the requirements for 15- and 20-amp, 125-volt receptacles for general use. The first rule for receptacle spacing requires that no point on a wall can be more than 6′ from a receptacle outlet. This translates to a total distance of 12′ between receptacles, **Figure 18-3**. For the purposes of the *Code*, a wall space includes any wall space 2′ or more in width, a fixed panel in a wall such as the fixed portion of a sliding glass door, and any space provided by a room divider such as a counter, railing, or half-wall. A floor receptacle installed within 18″ of the wall can be counted as a required wall receptacle. Even though there is no theoretical restriction on the maximum number of receptacles permitted on a residential branch circuit, *Section 210.11(B)* requires the load to be evenly proportioned among the required branch circuits.

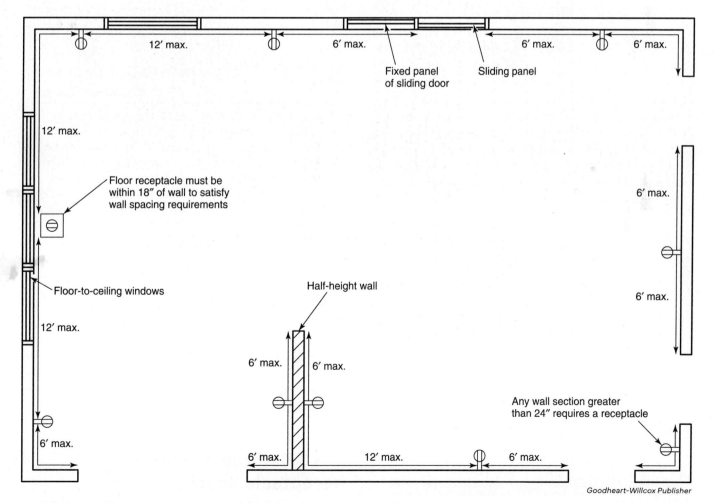

Goodheart-Willcox Publisher

Figure 18-3. Depiction of the *six-foot rule* of receptacle placement, which results in a maximum spacing between receptacles of 12′.

Consistent Receptacle and Switch Heights

The *NEC* does not specify a height above the floor for general-use receptacles, although 16″ to 20″ to the top of the box is common. For visual consistency, all receptacle boxes should be at the same height. Using a scrap 2×4 or piece of PVC pipe cut to length to create a homemade jig can speed the installation process by eliminating the need to measure and mark the stud for each receptacle outlet box. The jig can be used to mark a stud or to position a box for installation.

A dwelling unit will not have as many switches as receptacles. Using a jig is not necessary when laying out and marking the switch heights. A tape measure would be a better choice. Switch heights will generally vary between 42″ and 52″ measured to the top.

18.5.2 Kitchen Countertop Receptacle Spacing

Kitchen countertop receptacles are served by a minimum of two small appliance circuits required by *Section 210.11(C)(1)*. In addition, *Section 210.52(C)(1)* requires a receptacle outlet on each counter space that is twelve inches wide or wider and spaced such that no point along the countertop is more than 2′ from a receptacle outlet, resulting in a total distance of 4′ between countertop receptacles. See **Figure 18-4**.

In the 2023 *NEC*, a receptacle outlet is optional for an island countertop space or peninsular countertop space that measures 12″ by 24″ or greater. If a receptacle is installed in an island countertop, the receptacle assembly must be listed for that purpose. These are often called *pop-up* receptacles. These are receptacles that can be flush with the surface when not in use, but by pressing down on the receptacle, they will pop-up to a usable level, **Figure 18-5**. If a receptacle is not installed in an island, power must be provided to a junction box in the island base to allow for the future addition of a receptacle.

Goodheart-Willcox Publisher

Figure 18-5. A pop-up countertop receptacle must be listed for that use.

Receptacles on Island Cabinet Bases

Prior editions of the *Code* required that a receptacle be installed in island and peninsular countertops. These prior editions also allowed the receptacle to be installed on the cabinet base, as long the countertop extended more than 6″ beyond its support base and the receptacle was located no more than 12″ below the countertop. The 2023 *NEC* does not allow receptacles to the be installed on the cabinet base.

The *NEC* does not specify a vertical distance requirement for how much above the countertop receptacle outlets should be. However, outlets must be located less than 20″ above the countertop to be considered a countertop receptacle. Most upper cabinets in a kitchen are installed 18″ above the countertops. The top of the kitchen counter is 36″ above the floor, and most countertops have a 4″ to 6″ backsplash. Therefore, mounting the outlet box 46″ to 48″ above the floor puts the receptacle outlet a couple of inches above the backsplash, **Figure 18-6**.

18.5.3 Bathroom Receptacle Spacing

Per *Section 210.52(D)*, each bathroom shall have at least one receptacle outlet installed within 3′ of the outside edge of

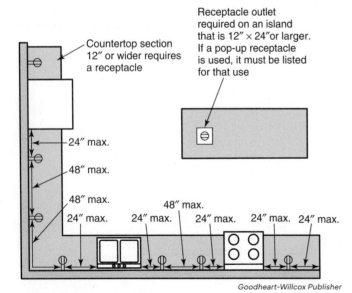

Goodheart-Willcox Publisher

Figure 18-4. A typical kitchen countertop receptacle layout.

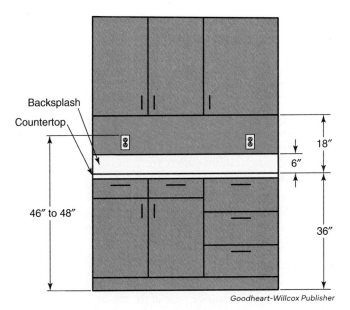

Goodheart-Willcox Publisher

Figure 18-6. Kitchen countertop receptacle installed above the backsplash.

the sink or basin. The receptacle shall be on a wall adjacent to the basin or basin countertop or in the base cabinet that supports the basin. A bathroom countertop with two sink basins can share a single receptacle if the receptacle is within 3′ of the edge of each basin. More likely, each basin must have its own receptacle outlet, **Figure 18-7.**

A bathroom countertop measures 30″ from the floor and must have a 4″ to 6″ backsplash. Mounting an outlet box 38″ to 42″ above the floor puts the outlet a few inches above the backsplash. It is wise to check prints to see if a standard cabinet or a cabinet that is a little bit taller is to be installed. Also, check if a tile finish will be applied over the drywall. If this is the case, an adjustable box should be used during the rough-in phase of the job to ensure the outer edge of the box will be within 1/4″ of the tile.

18.5.4 Outdoor Receptacle Spacing

Section 210.52(E) outlines the requirements for outdoor receptacle outlets for dwellings. A single-family dwelling shall have a minimum of two outdoor readily accessible receptacles installed no more than 6′ 6″ above grade level at the dwelling's front and rear. What is considered the front and rear of a dwelling unit can be left open to interpretation, but most reasonable inspectors will allow a receptacle in the side of a dwelling unit in the front half or the back half to count as a required receptacle. *Readily accessible* is defined in *Article 100* as "*capable of being reached quickly for operation renewal, or inspections without requiring those to whom ready access is requisite to take actions such as to use tools (other than keys), to climb over or under, to remove obstacles or to resort to portable ladders, and so forth.*"

A receptacle outlet is also required in an accessible location for the servicing of heating and air-conditioning equipment in *Section 210.63*. *Accessible, as applied to equipment*, is defined in *Article 100* as capable of being reached for operation, renewal, and inspection. If one of the required outdoor outlets is within 25′ of the heating and air equipment, it can count as the required outlet.

A receptacle outlet shall also be installed at each balcony, deck, or porch that is accessible from inside the dwelling. The receptacle must be accessible and not located more than 6′ 6″ above the porch, deck, or balcony surface as referenced

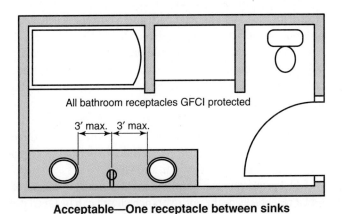

Goodheart-Willcox Publisher

Figure 18-7. Two options for installing a receptacle outlet in a twin-basin bathroom vanity.

in *Section 210.52(E)(3)*. Most outdoor receptacle outlets are installed 18″ to 24″ above grade level, putting them above the splash zone of falling rain and above a normal snow fall.

18.5.5 Laundry Receptacle Outlet Spacing

Section 210.52(F) requires at least one receptacle outlet to be installed for the use of laundry equipment. A receptacle for this equipment must be installed within 6′ of the intended location of the washing machine as required in *Section 210.50(C)*. This is the same receptacle outlet that is required for the service calculation in *Section 220.52(B)*.

18.5.6 Spacing Requirements for Basements, Garages, and Accessory Buildings

Garages, accessory buildings, and unfinished portions of basements are not subject to the spacing rules of general-use receptacles. In an attached garage and in each detached garage that is supplied with electric power, *Section 210.52(G)(1)* requires the installation of at least one receptacle outlet for each vehicle bay. The receptacle shall not be installed higher than 5′ 6″ above the floor. Per *Section 210.52(G)(2)*, at least one receptacle outlet shall be installed in an accessory building, such as a workshop or shed supplied with electric power. *Section 210.52(G)(3)* also requires at least one receptacle outlet must be installed in each separate unfinished portion of a basement. Note that finished basements are to conform to the spacing requirements of general-use receptacle outlets. Finished basements would be classified as a bedroom, recreation room, family room, or a similar room or area as referenced in *Section 210.52(A)*.

18.5.7 Hallway and Foyer Receptacle Outlet Spacing

Receptacle outlets for hallways and foyers are covered in *Section 210.52(H)* and *Section 210.52(I)*, respectively. A hallway that is 10′ or more in length, measured along the center line, shall have at least one receptacle outlet regardless of the length. Hallways shorter than 10′ are not required to have a receptacle. An entry foyer with an area of 60 ft² or more, and not part of a hallway, shall have a receptacle outlet on each wall space that is 3′ or more in width.

18.5.8 Appliance Outlet Spacing

Receptacle outlets for a dedicated appliance such as a clothes dryer, washing machine, or kitchen range shall be installed within 6′ of the intended location of the appliance. Appliances rated 240 volts, such as ranges and clothes dryers, will install easier in a two-gang box to accommodate the larger-sized receptacle and conductors.

18.6 Ground-Fault Circuit Protection Requirements

The *NEC* rules concerning ground-fault and arc-fault protection of branch circuit outlets are found in *Article 210*, and it is beneficial to reference these rules during the rough-in phase. Ground-fault circuit interrupter (GFCI) protection requirements for dwellings can be found in *Section 210.8(A)(1)–(12)*, which states that all 125-volt and 250-volt receptacles be installed in the following locations:

- Bathrooms
- Bathtubs or shower stalls where the receptacle is installed within 6′ of the outside edge of the tub or shower stall
- Sinks where the receptacles are installed within 6′ of the top inside edge of the basin
- Kitchens
- Laundry areas
- Garages and accessory buildings with a floor at or below-grade level and limited to storage areas, work areas, and the like.
- Basements with an exception for permanently installed fire and security alarm systems
- Crawlspaces at or below-grade level, including crawl space lighting per *Section 210.8(C)*
- Boathouses
- Outdoors with limited exceptions for snow melting and de-icing equipment
- Any indoor damp or wet locations
- Areas with sinks and permanent provisions for food preparation, beverage preparation, or cooking

The feature common to all these locations is the presence or potential presence of water. Water will generally contain minerals that are conductive to electricity. Standing in water while contacting a hot or ungrounded conductor will result in a shock. By providing the required GFCI protection, in the event electricity leaves its intended path and a current is detected to ground, possibly through a person, the GFCI will open, de-energizing the circuit. This should prevent an electric shock or electrocution (death) of anyone in contact with the ungrounded conductors. GFCIs will trip when any path to ground is established if the current to ground exceeds 6 mA. This can happen if an extension cord with a "nick" in the hot wire is plugged into a GFCI-protected circuit and the "nick" allows a ground fault to occur. Review the *Article 100* definitions related to ground fault, ground-fault current path, and GFCI.

GFCI protection can be provided at the outlet by means of a GFCI receptacle, or the entire branch circuit can be GFCI protected using a GCFI circuit breaker in the distribution panel.

18.7 Arc-Fault Circuit Protection Requirements

An arc-fault circuit interrupter (AFCI) is a device designed to protect an electrical circuit from arcing faults. For the purposes of a dwelling, these devices are used to protect many of the branch circuits.

Arcs have a unique electronic signature. When an arc is sensed, the arc-fault circuit breaker or receptacle opens the circuit. The reason for requiring these in dwelling units is to reduce or eliminate arcs, which can cause fires. Two instances where this can happen are a drywall screw penetrating NM-cable, creating an arc inside the wall, or a cracked or partially shorted cord to a lamp. One of the common items creating arcs is malfunctioning holiday lights.

The *NEC* requires that all 120-volt, single-phase, 15- and 20-amp branch circuits supplying outlets or devices in the following areas must be AFCI-protected by any of the means described in *Section 210.12(A)(1)–(6)*:

- Dwelling unit kitchens and bedrooms
- Family, dining, and living rooms
- Parlors, dens, and sunrooms
- Libraries and recreation rooms
- Closets, hallways, and laundry areas

This covers most of the circuits used in a dwelling unit with the exception of bathrooms, unfinished basements, attics, garages, and specialty branch circuits exceeding 20 amps. The easiest and most common method of providing AFCI protection in a dwelling unit, as mentioned in *Section 210.12(A)(1)*, is to install a listed combination-type AFCI circuit breaker in the distribution panel to provide protection to the entire branch circuit. The other options for complying with the *NEC* requirements for arc-fault protection are more involved and costly.

Note that GFCI protection is provided for receptacles, whereas AFCI protection is provided for outlets and devices. There is also some overlap, and some outlets may require both AFCI and GFCI protection. For example, a kitchen countertop receptacle outlet is required to be GFCI protected and to provide arc-fault protection. This protection is usually provided in one of two ways: installing a combination AFCI breaker in the distribution panel and a GFCI receptacle at the countertop outlet or using a dual-purpose GFCI/AFCI circuit breaker that provides both GFCI and AFCI protection.

Summary

- A branch circuit is defined as the conductors between the final overcurrent protection device and the outlets being served. The branch circuits in a dwelling power all the lights, receptacles, and dedicated appliances.
- The minimum number of branch circuits can be calculated based on the total square feet of the house and the rating of the branch circuit. *Section 220.14* requires a unit load of 3 volt-amps per square foot of living space for general lighting and receptacles. Ohm's law must be used for the calculations.
- At least one lighting outlet controlled by a wall switch should be installed in every habitable room, kitchen, and bathroom in a dwelling.
- The lowest rating for a general-purpose branch circuit in a dwelling is 15 amps. *Section 210.18* states that conductors for individual branch circuits that supply only a single load are sized to that load.
- *Section 210.52(A)(1)* states that no point on a wall can be more than 6′ from a receptacle outlet. This translates to a total distance of 12′ between receptacles. There are no restrictions on the maximum number of receptacle outlets permitted on each branch circuit, but the electrical load must be evenly divided between the required branch circuits.
- The *NEC* does not specify a height above the floor for general-use receptacles. Typical mounting heights are 16″ to 20″ to the top of a receptacle box.
- Kitchen countertop receptacles are served by a minimum of two small appliance circuits. A receptacle outlet is required on each counter space 12′ wide or wider, and they should be spaced no more than 2′ from a receptacle outlet. This results in a total distance of 4′ between countertop receptacles.
- Each bathroom shall have at least one receptacle outlet installed within 3′ of the outside edge of the sink or basin, per *Section 210.52(D)*.
- *Section 210.52(E)(2)* requires an accessible receptacle outlet shall be installed at each balcony, deck, or porch. Garages, accessory buildings, and unfinished portions of basements are not subject to the spacing rules of general-use receptacles.
- A hallway longer than 10′ must have at least one receptacle outlet. An entry foyer larger than 60 ft, and not part of a hallway, requires a receptacle outlet on each wall space that is wider than 3′.

- Receptacle outlets for a dedicated appliance, such as a clothes dryer, washing machine, or kitchen range, shall be installed within 6′ of the appliance location.
- GFCI protection can be provided at the outlet by means of a GFCI receptacle, or the entire branch circuit can be GFCI protected using a GCFI circuit breaker in the distribution panel.
- The most common method of providing AFCI protection in a dwelling is to install a listed combination-type arc-fault circuit interrupter circuit breaker in the distribution panel to provide protection to the entire branch circuit.

Know and Understand

1. The *NEC* requires a unit load of _____ per square foot of living space to supply general lighting and receptacles.
   A. 2 volt-amps
   B. 3 volt-amps
   C. 5 volt-amps
   D. 8 volt-amps

2. *True or False?* Every habitable room, kitchen, and bathroom in a dwelling is required to have at least one wall-switch controlled overhead lighting outlet.

3. For interior stairways with _____ or more risers require a wall switch at each floor level.
   A. three
   B. four
   C. five
   D. six

4. What is the lowest rating for a branch circuit according to the *NEC*?
   A. 5 amps
   B. 10 amps
   C. 15 amps
   D. 30 amps

5. The *NEC* requires a receptacle on any wall space that is _____ or wider.
   A. 1/2′
   B. 1′
   C. 1-1/2′
   D. 2′

6. *True or False?* The fixed portion of a sliding glass door is considered to be a wall space for the purposes of the receptacle layout in a dwelling.

7. *True or False?* The *NEC* specifies a height of 12″ to 20″ above the floor for general-use receptacles in a dwelling.

8. In a dwelling, each kitchen countertop space _____ or more requires a receptacle outlet.
   A. 12″
   B. 18″
   C. 2′
   D. 4′

9. A kitchen island measures 24″ by 48″. How many receptacles does the *NEC* require for an island of this size?
   A. One
   B. Two
   C. Three
   D. None

10. Kitchen countertop receptacles shall be located not more than _____ above the countertop surface.
    A. 6″
    B. 12″
    C. 18″
    D. 20″

11. The *NEC* requires a receptacle outlet to be installed within _____ of the outside edge of a bathroom sink.
    A. 12″
    B. 18″
    C. 24″
    D. 36″

12. A single-family dwelling is required to have a minimum of _____ outdoor receptacle outlet(s).
    A. one
    B. two
    C. three
    D. zero

13. *True or False?* A duplex receptacle outlet in a two-car garage will satisfy the *NEC* requirement for a receptacle outlet to be provided for each vehicle bay.

14. *True or False?* An accessory building to a dwelling, such as a shed, workshop, or detached garage, is required to be provided with electric power.

15. In a dwelling, the *NEC* requires at least one receptacle outlet to be installed in hallways that measure _____ or more in length.
    A. 6′
    B. 10′
    C. 12′
    D. 15′

Apply and Analyze

1. How many 15-amp branch circuits would be required for a 1,500 square foot dwelling?

2. Explain the two exceptions to the requirement in *Section 210.70(A)(1)*, which requires at least one lighting outlet controlled by a wall switch to be installed in every habitable room, kitchen, and bathroom in a dwelling.

3. What determines whether attics, crawl spaces, basements, or utility rooms require a lighting outlet?

4. What is the feature common to all locations in a dwelling that are required to have GFCI protection?

5. Describe two methods for providing both AFCI and GFCI protection to a branch circuit receptacle outlet.

Critical Thinking

1. Considering the *NEC* is a safety standard, why is a receptacle outlet required on outdoor balconies, porches, and decks?

2. A wall in a dwelling begins at a doorway and continues for 6′ before making a 90-degree turn. It then continues for 21′ and makes another 90-degree turn and continues for another 4′ until it reaches a second doorway. What is the minimum number of receptacles required for the 31′ of wall space?

CHAPTER 19

Conductor Sizing and Ampacity Adjustments

Praphan Jampala/Shutterstock.com

CHAPTER OUTLINE

LEARNING OBJECTIVES

After completing this chapter, you will be able to:

- Summarize the conditions that determine the ampacity of a conductor.
- Select the appropriate ampacity correction factor for ambient temperature from *Table 310.15(B)(1)(1)*.
- Calculate the corrected ampacity of a conductor based on the ambient temperature of its use.
- Select the appropriate ampacity adjustment factor for more than 3-current carrying conductors from *Table 310.15(C)(1)*.
- Define voltage drop as it applies to extended lengths of conductors.
- Explain the role of the current draw in the voltage drop on a circuit.
- Calculate the voltage drop of a particular sized conductor of a given length.
- Calculate the voltage drop as a percentage of the source voltage.
- Calculate the increased wire size required to compensate for voltage drop in a circuit.

TECHNICAL TERMS

ampacity
ampacity adjustment
derate
power loss
voltage drop

Introduction

Conductors must be sized correctly to ensure proper and safe residential wiring. If a conductor is too small for the current it is drawing, it can overheat and start a fire. Cost increases as conductor size increases, so using a conductor larger than required adds unnecessary cost to a project. However, an appropriate size increase in conductors to accommodate for voltage drop can save money overall. Wire that is undersized will have unnecessary heating, or power losses, as shown in the Ohm's law formula:

$$P = I^2 \times R$$

Power loss is measured in watts and can be reduced by an increase in wire size. The cost incurred by a larger conductor sized to the load will be recouped over time. This is especially true if the conductor is near the allowable ampacity of the overcurrent protection device (OCPD).

All conductors have an ampacity rating that cannot be exceeded. It is the electrician's responsibility to determine the proper ampacity for a select conductor by referring to the *NEC* and making separate calculations based on a conductor's conditions of use. Conductor conditions may include considerations to temperature, number of current-carrying conductors, and voltage drop over a given distance. The size of the conductor may need to be increased when the load is a great distance from the distribution panel. This chapter will explain how to size conductors under common installation conditions.

19.1 Ampacity Adjustments

Ampacity adjustments will be used throughout an electrician's career. The overall goal of the **ampacity adjustment** process is to ensure the wire that reaches all the panels and final utilization devices is adequately sized for all the factors that will impact the installation. This includes the voltage needed by the panels and the utilization equipment. For example, the number of current-carrying conductors run in the same raceway, or the number of cables bundled together need to be adjusted for more than 3-current carrying conductors and higher ambient temperatures.

In residential occupancies, the situation most often encountered is the bundling of electrical cables such as NM, UF, SE, and SER. Bundling can be described as the routing together of the cables. If too many cables containing current-carrying conductors are bundled together, extra heating could, depending on the amperage draw, occur in the bundle. The bundling or grouping together of conductors makes it more difficult for the heat to disperse.

Occasionally ampacity adjustments are needed for longer runs of wire in a dwelling unit. This will not occur in the typical dwelling unit with cable runs of 100′ or less.

However, longer cable runs for separate buildings, such as garages or workshops, will need to account for voltage drop.

Ampacity and voltage drop are linked. Recall that **ampacity** is the current-carrying ability of a conductor. It is defined in *Article 100* as "*the maximum current, in amperes, that a conductor can carry continuously under the conditions of use without exceeding its temperature rating.*" Conditions of use refers specifically to the ambient temperature surrounding the conductor and the number of current-carrying conductors contained in a cable, cables bundled together, conduit, or buried together in the earth.

Allowable ampacities for a given conductor size are shown in *Table 310.16* of the *NEC*, excerpted in **Figure 19-1**. Refer to Chapter 8, *Electrical Conductors and Connectors* for a full review of this table. Read the table's heading and notes to effectively and correctly apply the table to your work. The table heading clarifies the ampacities listed are valid only if there are three or fewer current-carrying conductors in a raceway (conduit), cable assembly, or directly buried in the earth. Note 1 of the table further explains the ampacities listed are based on an ambient temperature of 30°C (86°F). If the conditions of use are different than those described in the table, adjustments to the conductor's ampacity must be made.

Detail of Table 310.16 showing conductor sizes typically used in residential wiring.

Size AWG or kcmil	Temperature Rating of Conductor [See Table 310.4(1)]						Size AWG or kcmil
	60°C (140°F)	75°C (167°F)	90°C (194°F)	60°C (140°F)	75°C (167°F)	90°C (194°F)	
	Types TW, UF	Types RHW, THHW, THW, THWN, XHHW, USE, ZW	Types TBS, SA, SIS, FEP, FEPB, MI, PFA, RHH, RHW-2, THHN, THHW, THW-2, THWN-2, USE-2, XHH, XHHW, XHHW-2, XHWN, XHWN-2, XHHN, Z, ZW-2	Types TW, UF	Types RHW, THHW, THW, THWN, XHHW, XHWN, USE	Types TBS, SA, SIS, THHN, THHW, THW-2, THWN-2, RHH, RHW-2, USE-2, XHH, XHHW, XHHW-2, XHWN, XHWN-2, XHHN	
	COPPER			ALUMINUM OR COPPER-CLAD ALUMINUM			
18*	—	—	14	—	—	—	—
16*	—	—	18	—	—	—	—
14*	15	20	25	—	—	—	—
12*	20	25	30	15	20	25	12*
10*	30	35	40	25	30	35	10*
8	40	50	55	35	40	45	8
6	55	65	75	40	50	55	6
4	70	85	95	55	65	75	4
3	85	100	115	65	75	85	3
2	95	115	130	75	90	100	2
1	110	130	145	85	100	115	1
1/0	125	150	170	100	120	135	1/0
2/0	145	175	195	115	135	150	2/0
3/0	165	200	225	130	155	175	3/0
4/0	195	230	260	150	180	205	4/0

Notes:
1. Section 310.15(B) shall be referenced for ampacity correction factors where the ambient temperature is other than 30°C (86°F).
2. Section 310.15(C)(1) shall be referenced for more than three current-carrying conductors.
3. Section 310.16 shall be referenced for conditions of use.
*Section 240.4(D) shall be referenced for conductor overcurrent protection limitations, except as modified elsewhere in the *Code*.

Figure 19-1. *Table 310.16*

Correction factors based on ambient temperature are commonly used in residential wiring because it is not unusual in many areas, such as an attic, to exceed temperatures of 50°C (120°F) in the summer months. *Table 310.15(B)(1)(1)* lists correction factors used to adjust the allowable ampacity for ambient temperatures over 30°C (86°F), **Figure 19-2**. The table is divided into five columns. The outermost columns list the ambient temperatures to be corrected with degrees Celsius on the left and the equivalent degrees Fahrenheit on the right. The center three columns list the correction factors based on the temperature rating of the conductor. The correction factor *must* be chosen from the same temperature rating as the conductor. Recall the temperature rating is the maximum temperature the conductor can be exposed to before its insulation degrades.

To correct a conductor's ampacity, multiply its normal ampacity, as shown in *Table 310.16*, by the correction factor chosen from *Table 310.15(B)(1)(1)*.

To properly **derate** a conductor, the electrician must know the type of insulation on the wire. *Table 310.16* lists many different insulation types. NM and NMC cable, per *Section 334.112*, must contain conductors with 90-degree centigrade insulation. NM cable utilizes THHN wire as listed in *Section 310.4(1)*, *Conductor Construction and Applications*. Ampacity derating can start from the highest listed ampacity. When completing final termination on a receptacle, switch, or an OCPD, typically a circuit breaker, reference

the asterisk beneath *Table 310.16* to determine the maximum allowable ampacity for small conductors. Most of the wire installed in a dwelling unit will be governed by the small conductor rules. Every electrician needs to memorize the basic rules of small wire maximum OCPD sizing: 14 AWG at 15 amps, 12 AWG at 20 amps, and 10 AWG at 30 amps.

NM cable, as referenced in *Article 334.80*, reminds the electrician that the 90°C (194°F) rating is allowed to be used for ampacity adjustment and correction. Once the adjustments are made the final calculated ampacity cannot exceed the value listed for the 60°C (140°F) rated conductor. Per *Section 110.14(C)(1)(a)(1)–(4)*, the 60°C (140°F) column is used to determine the termination provisions for ampacity for devices and OCPDs. For type UF and NM cable the 60°C (140°F) rating shall be used for all ampacity calculations. The following sections will provide a few examples to calculate ampacity adjustments.

PRO TIP   Calculations for Conductors

When performing any calculations concerning conductors, if not stated otherwise, the conductor sizes given in the *NEC* apply to copper. Where the conductors are aluminum or copper-clad aluminum the size shall be changed to accommodate a given ampacity, a larger size wire that is not copper will be needed.

Table 310.15(B)(1)(1) Ambient Temperature Correction Factors Based on 30°C (86°F)

For ambient temperatures other than 30°C (86°F), multiply the ampacities specified in the ampacity tables by the appropriate correction factor shown below.

Ambient Temperature (°C)	Temperature Rating of Conductor			Ambient Temperature (°F)
	60°C	75°C	90°C	
10 or less	1.29	1.20	1.15	50 or less
11–15	1.22	1.15	1.12	51–59
16–20	1.15	1.11	1.08	60–68
21–25	1.08	1.05	1.04	69–77
26–30	1.00	1.00	1.00	78–86
31–35	0.91	0.94	0.96	87–95
36–40	0.82	0.88	0.91	96–104
41–45	0.71	0.82	0.87	105–113
46–50	0.58	0.75	0.82	114–122
51–55	0.41	0.67	0.76	123–131
56–60	—	0.58	0.71	132–140
61–65	—	0.47	0.65	141–149
66–70	—	0.33	0.58	150–158
71–75	—	—	0.50	159–167
76–80	—	—	0.41	168–176
81–85	—	—	0.29	177–185

Note: Table 310.15(B)(1)(1) shall be used with Table 310.16 and Table 310.17 as required.

Figure 19-2. *Table 310.15(B)(1)(1)*, Ambient temperature correction factors.

19.1.1 Conductor Size Selection

Although the purpose of ampacities tables is to correct and adjust the ampacities of conductors according to their conditions of use, these tables can also be applied to select the properly sized conductors for a particular installation when the current drawn by the load is known. These calculations are made by dividing the known load current by the correction factor or adjustment factor.

Example: What sized conductor is required for an 18-amp noncontinuous load in an ambient temperature of 54°C (129°F) if using a conductor with THWN-2 insulation? According to the asterisk associated with *Table 310.16* the maximum amperage a 12 AWG conductor is allowed to draw is 20. The chosen conductor must be capable of drawing this amount of current, or more.

The correction factor from the 90°C column in *Table 310.15(B)(1)(1)* is 0.76 for THWN-2 conductors. Therefore, the value 18 amps, divided by 0.76 equals in the following equation:

$$18 \text{ A} \div 0.76 = 23.68 \text{ A}$$

The final amperage is 23.68 amps. This means the conductor must be sized to this revised ampacity, which in this case is still a 12 AWG with Type THHN insulation. Per *Section 310.16*, the allowable ampacity of this conductor is 30 amps before the application of any adjustment factors. After adjustment for the ambient temperature the adjusted ampacity is above the minimum requirement of 20 amps.

The above calculation shows the importance of a proper understanding of the use and application of *Table 310.16*, *Table 310.15(B)(1)(1)*, and, in other calculations, *Table 310.15(C)(1)*. In this example, the electrician needs to calculate using the high ambient temperature. An inexperienced electrician would not have the experience to consider this higher ambient temperature and could potentially undersize the conductors.

19.1.2 Temperature Corrections to Ampacity Rating

Recall that to correct a conductor's ampacity, multiply its normal ampacity, as shown in *Table 310.16*, by the correction factor chosen from *Table 310.15(B)(1)(1)*.

Example 1: What is the corrected ampacity for a copper 8-3 AWG NM cable with THHN insulation used to feed a 240-volt electric range in a dwelling unit? This cable is routed in an attic with a summertime ambient temperature of 55°C (130°F). What is the maximum size OCPD (circuit breaker) this cable can be terminated on?

The first step is to determine the normal ampacity of the conductor using *Table 310.16*, which is 55 amps. Next, select the correction factor from the appropriate column of temperature correction factors in *Table 310.15(B)(1)(1)*

for conductors in an ambient temperature of 55°C (131°F). In this example, the correction factor is 0.76. Multiply the normal ampacity by the correction factor using the following equation:

$$55 \text{ A} \times 0.76 = 41.8 \text{ A}$$

Therefore, the corrected ampacity is 41.8 amps.

Section 334.80 states the final calculated ampacity cannot exceed 40 amps. Even though the conductors, after derating, can handle 41.8 amps. The largest size OCPD as listed in *Table 240.6* is a 40 amp.

Example 2: What is the corrected ampacity for an aluminum SER cable containing three, 1 AWG THWN-2 conductors, two of which are current carrying, used in an ambient temperature of 50°C (122°F)?

The ampacity from the 90-degree column of *Table 310.16* for this conductor is 115 amps. The correction factor from the 90°C column of *Table 310.15(B)(1)(1)* is 0.82. To calculate the corrected ampacity, use the following equation:

$$115 \text{ A} \times 0.82 = 94.3 \text{ A}$$

Therefore, this leads to a corrected ampacity of 94.3 amps.

Example 3: When used as a feeder cable to a sub panel, what is the largest size OCPD that could be utilized?

Refer to *Section 240.4(B)* which states that OCPDs rated 800 amps or less can be rounded up to the next higher OCPD from *Table 240.6(A)* provided the conductors are not part of a multi-wire branch circuit supplying receptacles for cord-and-plug connected loads, the ampacity does not correspond to a standard size fuse or circuit breaker, and the next higher OCPD does not exceed 800 amps.

The corrected ampacity is 100 amp OCPD.

PRO TIP

Single-Phase Dwelling Services and Feeders

Table 310.12 lists acceptable conductor sizes for the feeders and services. Compare this table to *Table 310.16* to see the different ampacities for the same size conductors. A typical example would be a 100-amp service in a non-dwelling unit would require a 3/0 AWG copper conductor from the 75°C column.

Table 310.12 will allow a conductor one size smaller to be installed on the OCPD, a 4 AWG copper will work. Also, a 2/0 AWG copper conductor is capable of handling 200 amps in a residential setting. The same 200 amps will require a 3/0 AWG in a non-dwelling application. The reason behind the allowing of a smaller size conductor is dwelling units seldom draw anywhere near the maximum allowed current for any length of time. The reduction in wire size will also make termination of conductors easier, and the cost to purchase will be slightly less.

19.1.3 Ampacity Adjustments for More Than Three Current-Carrying Conductors

Ampacities must be adjusted when there are more than three current-carrying conductors in a raceway, cable, multiple cables without maintaining spacing, or directly buried in the earth. See **Figure 19-3**. The footnote to the table references *Section 310.15(E)–(F)* clarifies that a neutral conductor that carries unbalanced current of other conductors of the same circuit is not counted, nor is the grounding or bonding conductor when applying the adjustment factors. The table further notes that conductors that cannot be energized at the same time are not counted. All other conductors, including spare conductors that may be energized later, are counted.

Table 310.15(C)(1) contains only two columns. The left column lists the number of conductors in increasing ranges. The right column lists the percentages by which the conductor's normal ampacity is adjusted.

Example 1: What is the adjusted ampacity of six 12-2 NM cables and two 12-3 NM cables ran in contact with thermal insulation without maintaining spacing in a dwelling unit? The cables are bundled.

As with making temperature corrections to the conductor's ampacity, the first step is to determine the normal ampacity using *Table 310.16*, which is 25 amps. *Table 310.15(C)(1)* lists an adjustment factor of 50% for 10-20 conductors. Each 12-2 conductor contains a hot and a neutral that are current-carrying. The two 12-3 conductors also contain only two current-carrying conductors per *Section 310.15(E)(1)*. The neutral conductor only carries the unbalanced current from the two hot or ungrounded conductors

and does not count as current carrying. In Example 1, there are a total of 16 current-carrying conductors.

$$30 \text{ A} \times 0.5 \ (50\%) = 15 \text{ A}$$

Therefore, the adjusted ampacity for the 8 cable and 16 current-carrying conductors is 15 amps. *Section 240.4(B)* will allow the ungrounded conductors to be terminated on a 15-amp circuit breaker.

Example 2: A 1 5/8″ hole is drilled in the sill plate between the first floor and the attic. The attic is an unconditioned space, meaning it is not heated or cooled. This hole will house seven 12-2 NM cables containing 14 current-carrying conductors and will be sealed with the thermal insulation, caulk, or sealing foam. Once in the attic, these cables will be routed to the kitchen appliances and receptacles. The maximum ambient temperature in the attic is 60°C (140°F) in August. Given this information, what is the maximum size circuit breaker this 12 AWG wire can be terminated on? For this problem you will do a *double derate*.

The conductors have a normal maximum ampacity of 30 amps, from the 90°C column as shown in *Table 310.16 referenced from Section 334.80*. *Table 310.15(C)(1)* has an adjustment factor of 50% for 10-20 conductors routed or bundled together. The second part of the derating will use the temperature adjustment form *Table 310.15(B)(1)(1)* based on 30°C (86°F). An ambient temperature 60°C (140°F) will result in an additional adjustment of 71% or 0.71. To derive the final allowable ampacity is a simple matter of taking the maximum amperage allowable, 30 amps multiplied by 50% for the bundling multiplied 71% for the ambient temperature adjustment.

$$30 \text{ A} \times 0.5 \times 0.71 = 10.65 \text{ A}$$

This results in an adjusted ampacity of 10.65 amps. Next, refer to *Section 240.4(B)* and round up to the next size larger OCPD, which is 15 amps per *Table 240.6*. Normally a 12 AWG wire will be allowed on a 20-amp OCPD, but not in this case. What could be done to allow the use of a 20-amp OCPD in this case?

Simply drill three holes through the sill plate to limit the amount of NM cables with no more than three cables or six current-carrying conductors in each running through the holes sealed with the thermal insulation, caulk, or sealing foam. The *double derate* formula would now be set up as follows:

$$30 \text{ A} \times 0.8 \times 0.71 = 17.04 \text{ A}$$

This can be rounded up to a 20-amp OCPD.

19.2 Voltage Drop Calculations

Electrical circuits that extend long distances, typically 100′ or greater, can experience voltage drop, which means the voltage at the load is less than the voltage at the source. *Voltage drop* is the loss of voltage in a circuit due to the

Table 310.15(C)(1) Adjustment Factors for More Than Three Current-Carrying Conductors

Number of Conductors*	Percent of Values in Table 310.16 Through Table 310.19 as Adjusted for Ambient Temperature if Necessary
4–6	80
7–9	70
10–20	50
21–30	45
31–40	40
41 and above	35

*Number of conductors is the total number of conductors in the raceway or cable, including spare conductors. The count shall be adjusted in accordance with 310.15(E) and (F). The count shall not include conductors that are connected to electrical components that cannot be simultaneously energized.

Figure 19-3. *Table 310.15(C)(1),* Adjustment factors for more than three current-carrying conductors.

internal wire resistance combined with the current being drawn by the load. Voltage drop can be minimized on longer runs by increasing the size of the conductors resulting in less resistance. Power or heat loss is also reduced. The Ohm's law formula for power or watts is P equals I squared times R. This loss can be sensed or felt in the form of heat on a wire too small for the load. For example, a 14 AWG conductor that is run a great distance, perhaps 200′, and drawing the full 15 amp maximum.

SAFETY NOTE   **Warm Wires**

The escaping heat or power, measured in watts, can be felt by the touch. Warmer wires are a good indicator of too great of a current on the conductor, wire length too long for a given current, or a loose connection.

On long runs of conductors, this drop in voltage can be enough to impact the efficient operation of certain loads, notably motor loads. Under-voltage is a leading cause of motor failure. Resistive loads, such as incandescent light bulbs and electric baseboard heaters, are less affected by voltage drop and will operate at a reduced output, meaning the light bulb will be dimmer, and the heater will not heat as much. A general rule is that most devices that use electricity will operate within a parameter of plus or minus 10 percent of the equipment's voltage rating. This would be a range of 108 to 132 volts for many household items.

Increasing wire size can prove to be cost effective if the power or heat loss, also called the I squared times E losses, are decreased. If the amperage draw is high enough and the voltage drop is great enough, payback for buying a larger size conductor could be as soon as two or three years.

The voltage drop formulas can be used in more than one way. Listed beneath are four variations:

Formula 1: VD = $2 \times K \times I \times L$ / cmil
Formula 2: VD (E) = $I \times R$ (R total both ways)
Formula 3: Wire size in cmil = $2 \times K \times I \times L$ / VD
Formula 4: Wire length = VD $\times$ cmil / $2 \times K \times I$

where
VD = voltage drop expressed in the actual volts dropped, such as 3% of 120 volts is 3.6 volts

2 = constant to double the length for current to the load and returning to the source, for single phase loads. For three-phase loads, replace 2 with 1.732 (square root of 3)

K = direct current resistance at 75°C (167°F) for a 1000 kcmil conductor, 1000′ long

I = Actual current of the load (in amps)

R = Resistance

L = one-way length of the conductors from the source to the load

cmil = the cross-sectional area of the conductors from Table 8, Chapter 9

The first two voltage drop formulas will give an answer of the voltage that will be dropped knowing the current draw (I), length or resistance, and the cmil of the chosen wire. These parameters are combined with the known length of the wire L, the constant (K) for copper or aluminum, and the multiplier of 2 for a single-phase circuit.

The second formula for voltage drop replaces the constant (K) with the actual resistance of the given length of wire. This resistance can be found in Table 8, Conductor Properties, Chapter 9 of the NEC. Look at the direct current resistance at 75°C, uncoated copper to get the resistance per 1000′. Using a ratio, determine the resistance of the length of conductor used in this formula.

The third formula will allow the electrician to determine the size of the conductor needed to minimize voltage drop over a known distance. Decide the amount of actual volts that are willing to be dropped, often 3% of the voltage at the source or supply. Identify the current needed to operate electrical equipment the conductors are feeding, add the multiplier of 2 for single-phase, choose the needed constant (K) for aluminum or copper, and then plug in the numbers. The answer will be in cmil. Review Table 8, Chapter 9 to determine where this number will fall. Reference the wire size in AWG or cmil.

The last formula is seldom used. It will allow the user to determine the maximum length of a conductor knowing the amperage draw, wire size, and type. If the wire type (aluminum or copper) and amperage draw are known, plug in the constant (K), the cmil size of the wire, and the multiplier of 2.

In most dwelling units, voltage drop is not a consistent problem. It is seldom that a particular circuit will draw near the maximum allowed amperage and create a voltage drop issue because the conductors feeding the load are too small. The example below is a residential question concerning a feeder to an outbuilding. It will illustrate why distance, wire size, and ampacity of the load all need to be considered when designing and installing properly working electrical installations.

Example 1: A detached garage with a workshop area is located 250′ from the main electrical service of a single-family dwelling. The dwelling has a 200 amp, 240/120-volt single-phase service, a 100 amp, 240/120-volt single-phase sub-panel is to be installed in the garage/workshop. The estimator at shop has ordered a 2-2-2-4 AWG, XLPE aluminum cable rated for direct burial in the earth. This cable contains 2-ungrounded conductors, 1 neutral (grounded), and one smaller size grounding conductor. Table 310-16 lists the ampacity from the 75°C column as 90 amps. The OCPD is listed at 75°C to allow a 90-amp breaker. This is one option. Another option for residential feeders is to utilize Table 310.12 with an allowable ampacity of 100 amps for 2 AWG aluminum conductors.

Given this scenario, answer the following questions: Consider a 2-2-2-4 AWG aluminum cable. What is the voltage drop and is it within the recommended maximum of 3% for feeders?

In this scenario, we will first use Formula 1:

$$VD = 2 \times K \times I \times L / \text{cmil}$$
$$2 \times 21.2\ \Omega \times 100\ A \times 250' / 66{,}360 \text{ circular mils} = 15.97\ V$$
$$15.97/240 = X/100 = 6.65\% \text{ voltage drop}$$

This exceeds the recommended 3% voltage drop at a full 100 amps current draw. Using Formula 2, we can calculate the following:

$$VD\ (E) = I \times R \text{ in } \Omega\ (R \text{ total both ways of the conductors})$$
$$100\ A \times (0.319 \times 0.5) = 15.95\ V$$
$$250' \text{ of wire each way} \times 2 = 500'$$

Thus, 500' is 50% of 1000'. The *Table 8, Chapter 9* resistance of 1000' of 2 AWG aluminum is 0.319 Ω. 0.319 × 0.5 (50%) = 0.1595. This means that 0.1595 Ω × 100 A = 15.95 V dropped.

This is an alternate way to determine voltage drop resulting in a very similar answer. Consider the following: what is the minimum size aluminum conductor needed to not exceed the recommended 3% voltage drop? For this, we will use Formula 3.

Wire size in cmil = $2 \times K \times I \times L$ / VD

$$2 \times 21.2 \times 100 \times 250' / 7.2\text{-volts (3\% VD)} = 147{,}222 \text{ cmil}$$

The next larger size wire above 147,222 cmil as listed in *Table 8, Chapter 9* is a 3/0 AWG aluminum at 167,800 cmil conductor. This size conductor limits the voltage drop to no more than 3%.

Consider an electrician is given a 2-2-2-4 aluminum cable to install. What is the maximum length this conductor can be run and still deliver a full 100 amps to the sub-panel? Use Formula 4 to calculate.

Wire length = VD × cmil / $2 \times K \times I$
$$7.2 \times 66{,}360 \text{ cmil} / 2 \times 21.2 \times 110\ A = 112.69'$$

At this distance, the panel can operate 100 A without a voltage drop exceeding 3%.

Everything covered in this section has dealt with seeing what happens to a 100-amp load under different conditions of use. The question that needs to be asked is, how many amps is the realistic maximum this panel would draw at any given time? More than likely that amount would rarely exceed 50 amps.

Example 2: Given a 2-2-2-4 AWG aluminum cable, what is the voltage drop and is it within the recommended maximum of 3% for feeders?

$$2 \times 21.2\ \Omega \times 50\ A \times 250' / 66{,}360 \text{ circular mils} = 7.99\ V$$
$$7.99/240 = X/100 = 3.33\% \text{ voltage drop.}$$

Remember, voltage drop is just a recommendation. Is the cost of upsizing the wire worth the additional expense? The 2-2-2-4 aluminum cable on a 100-amp OCPD for a dwelling unit is a code-compliant installation.

Note that the source voltage is not part of the formula, but it can be used to determine the percentage of the voltage drop, which in this example is 6.65%. Voltage drop is not an enforceable rule in dwelling units in the *NEC*, but instead it is an equipment performance issue and not considered a safety issue. It is mentioned only in Informational Notes of *Section 210.19(A)* and *Section 215.2(A)(2)* where it is recommended that voltage drop not exceed 3% for branch circuits and 5% for the combined branch circuit and feeder circuits. See **Figure 19-4**.

Another way to consider voltage drop is the resulting voltage at the load or the difference between the source voltage and the amount of voltage drop calculated. In Example 1, this is 240 volts minus 15.97 volts equals 224.03 volts. In Example 2, 240 volts minus 7.99 volts equals 232.01 volts. In both instances, the voltage is within 10%. The electrical consuming devices will likely work as designed.

Table 8 Conductor Properties

Size (AWG or kcmil)	Area mm^2	Area Circular mils
18	0.823	1620
18	0.823	1620
16	1.31	2580
16	1.31	2580
14	2.08	4110
14	2.08	4110
12	3.31	6530
12	3.31	6530
10	5.261	10380
10	5.261	10380
8	8.367	16510
8	8.367	16510

Reproduced with permission of NFPA from NFPA 70, National Electrical Code, 2023 edition. Copyright © 2022, National Fire Protection Association. For a full copy of the NFPA 70, please go to www.nfpa.org

Figure 19-4. Detail of *Table 8, Chapter 9* showing the circular mil area of conductors.

Summary

- Ampacity is the current-carrying ability of a conductor and is defined in *Article 100* as "*the maximum current, in amperes, that a conductor can carry continuously under the conditions of use without exceeding its temperature rating.*"
- For the purposes of determining the ampacity of a conductor, its conditions of use refers specifically to the ambient temperature surrounding the conductor and the number of conductors contained in a raceway, cable, or directly buried in the earth.
- *Table 310.16* lists the allowable ampacities of conductors based on an ambient temperature of 30°C (86°F) where there are no more than three current-carrying conductors in a raceway, cable, or directly buried in the earth.
- *Table 310.15(B)(1)(1)* lists correction factors used to adjust the allowable ampacity for ambient temperatures ranging from 10°C (50°F) and less to a maximum of 85°C (185°F).
- The correction factor must be chosen from the same temperature rating as the conductor temperature rating.
- To correct a conductor's ampacity, multiply its normal ampacity as shown in *Table 310.16* by the correction factor chosen from *Table 310.15(B)(1)(1)* or *Table 310.15(C)(1)*.
- *Table 310.15(C)(1)* is used to adjust ampacities when there are more than three current-carrying conductors in a raceway, cable, or directly buried in the earth.
- Voltage drop is the loss of voltage in a circuit due to the internal wire resistance of the conductor in conjunction with the amperage draw.
- The amount of voltage drop is determined by the length and size of the conductor, the conductor's resistance, and the current draw of the load.
- Voltage drop, with the exception of fire pumps, is not an enforceable rule in the *NEC*, as it is an equipment performance issue not considered a safety issue.
- Voltage drop can be minimized by installing a larger conductor size.

Know and Understand

1. According to *Table 310.16*, what is the allowable ampacity of a 14 AWG copper conductor with Type THW-2 insulation?
   A. 15 amps
   B. 20 amps
   C. 25 amps
   D. Cannot be determined.

2. What is the temperature rating of a conductor with Type THHN insulation?
   A. 30°C   C. 75°C
   B. 60°C   D. 90°C

3. According to *Table 310.16*, what is the allowable ampacity of an aluminum 4/0 USE conductor?
   A. 150 amps   C. 205 amps
   B. 180 amps   D. 230 amps

4. *True or False?* The ambient temperature correction factors are the same for copper conductors and aluminum or copper-clad conductors.

5. The correction factor for a conductor with a 75°C temperature rating in an ambient temperature of 100°F is _____.
   A. 0.88   C. 0.87
   B. 0.82   D. 0.91

6. The correction factor for a conductor with a 90°C temperature rating in an ambient temperature of 60°F is _____.
   A. 0.58
   B. 0.71
   C. 1.08
   D. A conductor cannot be used at this temperature.

7. *True or False?* If the conductor type is not mentioned, it is assumed to be made of copper.

8. The adjustment factor for 8 current-carrying conductors in the same raceway is _____.
   A. 80%   C. 50%
   B. 70%   D. 45%

9. *True or False?* Resistive loads are greatly affected by voltage drop and may be damaged if operated at an under-voltage.

10. Which of the following are factors that lead to voltage drop in an electrical circuit?
    A. Long lengths of conductors in the circuit
    B. High wire resistance
    C. The cross-sectional area of the conductors
    D. All of these.

Apply and Analyze

1. What is meant by a conductor's condition of use as applied to conductor ampacity?

2. What is the corrected ampacity for a 2/0 aluminum USE conductor in an ambient temperature of 125°F?

3. Calculate the adjusted ampacity for nine 10 AWG conductors with type RHW-2 insulation.

4. What is the revised ampacity for eight 6 AWG conductors with type ZW-2 insulation in an ambient temperature of 39°C?

5. What type of load is most affected by voltage drop?

6. How can voltage drop in an electrical circuit be minimized?

7. Calculate the voltage drop using aluminum THHN 3/0 AWG conductors for a 240-volt, 120-amp load that is 600′ from the power source.

8. What is the percentage of voltage drop in the previous circuit?

9. A customer would like a decorative fountain in their pond that is 375′ from their home. The fountain pump runs on 120 volts and draws 9 amps of current. What size copper conductors are required to properly wire this circuit? The amount of voltage drop should be 3% or less.

Critical Thinking

1. Why is the voltage drop on a conductor not an enforceable rule of the *NEC*?

2. When making ampacity corrections based on ambient temperature, why does the ampacity increase at temperatures above 86°F?

CHAPTER 20

Service Calculations

Chase Grimm/Shutterstock.com

Introduction

A residential electrician is responsible for installing and sizing all electrical service equipment, including the correct sizing of the main breaker or fuse and conductors, which are part of a service calculation. A ***service calculation*** determines the amount of power a dwelling requires. This involves completing a series of calculations to add up all the electrical loads in a dwelling unit to determine the service size, rated in amps.

All calculations must be accurate to size the service conductors and equipment appropriately. A service that is too small can experience problems such as circuit breakers tripping repeatedly or conductors overheating causing a fire. A service that is too large costs more to install, but it is generally not considered hazardous. A properly sized electrical system is the most desirable since it is the most cost-efficient and provides years of trouble-free service. This chapter will focus on providing you with the tools to complete accurate service calculations while on the job.

20.1 Connected Loads, Calculated Loads, and Demand Factors

You must understand the meaning of connected loads and calculated loads before making a service calculation. Recall that an electrical load is the component in an electrical system that is connected to the power source, such as a panelboard, and uses electrical power. Loads can be small and draw a small amount of power, like a light bulb, or large, like for an entire dwelling.

For any loads that are nearly 100% resistive, it is expressed in watts or kilowatts, such as the heating elements referenced in *Section 220.54* (electric clothes dryers) or *Section 220.55* (electric cooking appliances). The units VA or kVA are different units of measure than the watt or kilowatt. The major portion of VA is the wattage component; however, VA results in a higher number because it factors in the inductive reactance of certain connected loads such as motors and electronics. The *NEC* permits treating watts and VA as equivalent for service calculations.

Calculating Values on a Nameplate

Recall that VA stands for volt-amps, and it is the product of the applied voltage multiplied by the current draw of an electrical appliance. A product's nameplate provides the voltage and the amps or the voltage and the wattage. If wattage is given, the current can be calculated using the formula:

$$I = P/E$$

where

I = current measured in amperes (A)
E = voltage measured in volts (V)
P = power in watts (W)

A *connected load* is the consumption of all appliances or electrical equipment operating and drawing power collectively. It is determined by the nameplate ratings of each component. A *calculated load* is less than the connected load and is based on the power being used at a point in time. The calculated load takes the total connected load for a residence and applies reductions, called demands, to arrive at the calculated load.

In *NEC Article 100* a **demand factor** is defined as "*the ratio of the maximum demand of a system, or part of a system, to the total connected load of a system or the part of a system under consideration.*" This permitted decrease in the connected load recognizes an electrical load rarely draws the maximum amperage for which it is rated. This knowledge can help an installer understand why a 200-amp main breaker on a panel allows 42 openings for circuit breakers containing a mix of 15-, 20-, 30-, 40- or 50-amp circuit breakers. Each individual breaker has the potential to draw amperage near its rated ampacity, but when reviewing the total connected load at any given point in time, the total draw from all the circuit breakers will not exceed the rating of the 200-amp main circuit breaker size.

20.2 Methods of Calculation for Existing Dwellings

The *NEC* provides one method of calculation for an existing dwelling. This is used to determine if the existing service has enough capacity to serve additional loads. Perhaps a homeowner wishes to add an electric mini-split system for heating and cooling, a hot tub, or a swimming pool or to construct an addition to the house. This calculation method considers all the existing loads and applies a demand factor to determine if the additional load will exceed the rating of the existing service. Existing dwelling service calculations will be covered in Chapter 39, *Service Upgrades*.

20.3 Methods of Calculation for New Dwellings

The *NEC* includes two methods of service calculations for a new dwelling: the general method and the optional method. The **general method** involves applying several demand factors to the total connected load to arrive at the calculated load. The **optional method** applies a single demand factor to the total connected load to determine the calculated load. You will likely produce two different service sizes when using both methods for the same residence. Both are considered valid, and in a competitive bidding situation, many electricians perform both methods and choose the lower, less costly service size.

The optional method from *Section 220.80* often results in a smaller size service. The most-used main breaker panelboards in one- and two-family dwelling units are 100 amp and 200 amp. Very large dwellings or multi-family units may require a 400 amp or larger service. The larger the service, the more costly the equipment and installation costs—although cost differences between values like 100 amp and 200 amp are not that significant. If conductors are larger, a panel is more costly and, if used, larger conduit adds to the price.

It is also important to note that if using the general or optional method, and the service size calculation is greater than 100 amp, it is common practice to install a 200-amp service. Other sizes noted in *Table 240.6(A)* that are greater than 100 amp but not commonly used include 110, 125, 150, and 175 amp.

CODE APPLICATION

Connected Loads

Scenario: An electric cooking range has a connected load of 12 kW. This represents the maximum power draw if all the stovetop burners and the oven were all on their highest setting.

Table 220.55: *NEC 220.55* allows the use of *Table 220.55* to reduce the maximum demand from a listed 12 kW rating to 8 kW.

Solution: The *NEC* recognizes that this scenario is unlikely, and even when it might occur, the heating elements cycle on and off, and the connected load is rarely reached. Therefore, per *Table 220.55*, a calculated load of 8 kW is permitted by the *NEC* to be used for the service calculation.

20.3.1 The General Method

The general method involves nine steps of calculations to arrive at the service size. These steps are scattered throughout *Article 220* and are summarized in worksheet form in **Figure 20-1**.

> **PRO TIP   General Method Guidelines**
>
> It is also useful to note that if the general method is chosen, and the calculated service size is 200 amp or slightly higher, it would be worth the time to recalculate to see if the optional method will result in a calculation that would allow the use of a 200-amp service. This is still a safe and compliant installation and save the customer money.

Example: Calculate the minimum service size for a dwelling with the following loads. The electrical system is 120/240 volts:

- 1000 ft² of living area
- 4080 VA of air conditioning (17-amp, 240-volt compressor motor)
- 6 kW heating
- 12 kW cooking range
- 6 kW clothes dryer
- 4.5 kW water heater
- 1150 VA food waste disposer (9.8-amp, 120-volt motor)
- 1200 VA dishwasher
- 1650 VA well pump (6.9 amps, 240 volts)

The first step of a service calculation is to calculate the living area of the dwelling. *Section 220.11* instructs to calculate the floor area using the outside dimensions of the dwelling. This does not include open porches, garages, or unused or unfinished spaces that are not adaptable for future use. In this example, the floor area is given. The lighting load includes the general-use receptacles in the dwelling.

$$1000 \text{ ft}^2 \times 3 \text{ VA} = 3000 \text{ VA}$$

Next, add the two small appliance circuits and the laundry circuit. *NEC 220.52(A)* and *(B)* require 1500 VA for each small appliance and laundry circuit.

$$3 \text{ VA} \times 1500 = 4500 \text{ VA}$$
$$3000 \text{ VA} + 4500 \text{ VA} = 7500 \text{ VA}$$

For Step 3, the *NEC* in *Section 220.42* permits a demand factor for the general lighting load of a residence:

- The first 3000 VA is at 100%.
- 3001 to 120,000 VA is at 35%.
- Any remainder over 120,000 VA is at 25%

The General Method of Residential Service Calculation

Step	*NEC* Article	Calculation	Value
1	220.41	Lighting load. 3 VA per square foot of living space.	
2	220.52	Two small appliance circuits and one laundry circuit at 1500 VA per circuit. Each additional appliance branch circuit results in an additional 1500 VA to be added to the basic calculation.	
		Total Steps 1 and 2.	
3	Table 220.42(A)	Apply the general lighting demand to the total of Steps 1 and 2. 1st 3000 VA at 100%. 3001 to 120,000 VA (next 117,000 VA) at 35%. Remainder over 120,000 at 25%.	
4	220.60	Noncoincident loads. Omit the *smaller* of the heating and air conditioning loads.	
5	220.54	Clothes dryers. Use the nameplate rating or 5kW, whichever is greater.	
6	Table 220.55	Cooking equipment with demand factor applied; minimum is 8000 VA.	
7	220.53	Fastened-in-place appliances. 75% demand permitted for four or more.	
8	220.14(C)	The largest motor in full-load current increased by 25%.	
		Total Steps 3–8.	
9	Table 310.12	Divide the total VA by the source voltage to determine the amp-rating of the service. Use the next higher standard OCPD size from *Table 240.6(A)*, and size the service conductors using *Table 310.12*. Size the grounding electrode conductor using *Table 250.66*.	

Goodheart-Willcox Publisher

Figure 20-1. General method service calculation worksheet.

We will apply the general lighting demand to the total of Steps 1 and 2, which was 7500 VA. The first 3000 VA of this total is 100%, giving us the full 3000 VA. The remaining 4500 is applied at 35%, which is 1575 VA.

In Step 4, compare the heating load against the air conditioning load. The *NEC* recognizes these two loads as being **noncoincident loads**, meaning it is unlikely that they will both be running at the same time. Therefore, it is permitted to drop the smaller of the two loads, as per *NEC 220.60*. In this example, the heating load is 6000 watts, and the air conditioning load is 4080. Therefore, the air conditioning load is permitted to be dropped from the calculation.

Step 5 is to add the dryer load to the calculation, which is 6000 W. In Step 6, the cooking equipment is added to the calculation. When employing the general method, use *Table 220.55* to apply the demands for the electric range. See **Figure 20-2**. This table can be quite confusing to apply. However, with only one 12 kW range in our example, using the table is rather simple.

The table is divided into four columns: the leftmost column is the number of appliances. Columns A, B, and C have demand factors for different appliances based on their kW rating:

- Column A is for appliances with a rating less than 3-1/2 kW.
- Column B is for appliances rated between 3-1/2 kW and 8-3/4 kW.
- Column C is used for all appliances rated not more than 12 kW.

The demand factors in columns A and B are percentage multipliers, while the demand factor in Column C is expressed in kW. There is one 12 kW range in the example calculation, and *Table 220.55* permits a maximum demand of 8 kW, or 8000 W.

Fastened-in-place or fixed appliances are included in Step 7 of the service calculation. These are appliances other than the clothes dryer and cooking equipment that are connected to the dwelling by means of plumbing or some form of mechanical connection. Some examples include a water heater, dishwasher, disposal, built-in microwave, and possibly a refrigerator with a built-in icemaker and water dispenser. If there are four or more fastened-in-place appliances, *NEC Section 220.53* permits a demand factor of 75% to be applied to the total connected load of the appliances.

Some other loads referenced in *Section 220.14* to consider in the calculations (based on the ampere rating) include specific appliances or loads such as an electrical vehicle charger, air compressor, central vacuum, hot tub, and RV receptacles. In the example, there are four such appliances:

- Water heater at 4500 W
- Food waste disposer in the sink at 1150 VA
- Dishwasher at 1200 VA
- Well pump at 1650 VA

$$4500 \text{ W} + 1150 \text{ VA} + 1200 \text{ VA} + 1650 \text{ VA} = 8500 \text{ VA}$$
$$8500 \text{ VA} \times 75\% = 6375 \text{ VA}$$

In Step 8, increasing the largest motor in full-load current by 25% has a minimal impact on the calculation. It is included because the *NEC* requires conductors that serve motor loads to have an ampacity of not less than 125% of the motor's full-load current. This extra capacity is necessary to withstand the increased current flow when the motor is started, which can be four to six times the motor's full load running current. In the example, the largest motor in full-load current is the disposer motor. Therefore, the air conditioning motor was omitted from the calculation in Step 4.

If the heating load not containing a motor is used in this calculation, some municipalities still may require the motor load from the air conditioner to be used as the largest motor for the service calculation. Steps 3 through 8 are then added to arrive at the total calculated load.

$$1150 \text{ VA} \times 25\% = 288 \text{ VA}$$
$$3000 \text{ VA} + 4575 \text{ VA} + 6000 \text{ W} + 6000 \text{ W} + 8000 \text{ W} + 6375$$
$$\text{VA} + 288 \text{ VA} = 31{,}238 \text{ VA}$$

Table 220.55 Demand Factors and Loads for Household Electric Ranges, Wall-Mounted Ovens, Counter-Mounted Cooking Units, and Other Household Cooking Appliances over 1¾ kW Rating (Column C to be used in all cases except as otherwise permitted in Note 3.)

Number of Appliances	Demand Factor (%) (See Notes)		
	Column A (Less than 3½ kW Rating)	Column B (3½ kW through 8¾ kW Rating)	Column C Maximum Demand (kW) (See Notes) (Not over 12 kW Rating)
1	80	80	8
2	75	65	11
3	70	55	14
4	66	50	17
5	62	45	20

Figure 20-2. Detail of *Table 220.55* that outlines the demand factors for household cooking equipment.

The amp-rating of the service is determined by dividing the total calculated load by the source voltage of 240 volts, which results in 130 amps in the example.

$$31,238 \text{ VA} \div 240 \text{ V} = 130 \text{ A}$$

However, 130 amps is not a standard circuit breaker size listed in *NEC Table 240.6(A)*, and it is permitted to use the next higher standard size, which is 150 amps.

As seen in this example, the general method calculation results in a service size of 150 amps which requires a 1 AWG copper or 2/0 aluminum service conductor, per *Table 310.12*, and a 6 AWG copper grounding electrode conductor from *Table 250.66*. See **Figure 20-3** and **Figure 20-4**. The service conductors and grounding electrode conductors will be discussed in Section 8, *Service Installation*.

The minimum service size for a dwelling allowed by the *NEC* is 100 amps, per *NEC 230.79(C)*. See **Figure 20-5**. Although the example calculation resulted in a 150-amp service, a 200-amp service is more common, which would not cost substantially more, and allows for future expansion.

20.3.2 The Optional Method

The rules for the optional method for a dwelling unit are found in *NEC 220.82*. Using this method, the service or feeder conductors must have an ampacity of 100 amps or more. This method divides the loads into two groups: the general load and the heating and air conditioning load. The general load is comprised of the general lighting load, the small appliance loads, and laundry loads, as well as the nameplate rating of all fastened-in-place appliances, cooking equipment, clothes dryers, water heaters, and any permanently connected motors not associated with any appliance.

Table 310.12 Single-Phase Dwelling Services and Feeders

Service or Feeder Rating (Amperes)	Conductor (AWG or kcmil)	
	Copper	Aluminum or Copper-Clad Aluminum
100	4	2
110	3	1
125	2	1/0
150	1	2/0
175	1/0	3/0
200	2/0	4/0
225	3/0	250
250	4/0	300
300	250	350
350	350	500
400	400	600

Note: If no adjustment or correction factors are required, this table shall be permitted to be applied.

Figure 20-3. *Table 310.12* of the *NEC* is used to size residential service conductors.

Table 250.66 Grounding Electrode Conductor for Alternating-Current Systems

Size of Largest Ungrounded Conductor or Equivalent Area for Parallel Conductors (AWG/kcmil)		Size of Grounding Electrode Conductor (AWG/kcmil)	
Copper	Aluminum or Copper-Clad Aluminum	Copper	Aluminum or Copper-Clad Aluminum
2 or smaller	1/0 or smaller	8	6
1 or 1/0	2/0 or 3/0	6	4
2/0 or 3/0	4/0 or 250	4	2
Over 3/0 through 350	Over 250 through 500	2	1/0
Over 350 through 600	Over 500 through 900	1/0	3/0
Over 600 through 1100	Over 900 through 1750	2/0	4/0
Over 1100	Over 1750	3/0	250

Figure 20-4. *Table 250.66* is used for sizing the grounding electrode conductor.

The heating and air conditioning load is calculated as the largest of the following six options:

1. 100% of the nameplate rating of the air conditioner.
2. 100% of the rating of the heat pump when used without supplemental electric heating.
3. 100% of the heat pump and 65% of the supplemental heat. If the heat pump is prevented from running concurrently with the supplemental heat, the heat pump need not be added.
4. 65% of the nameplate rating of the electric heating if there are less than four heating units under separate control.
5. 40% of the nameplate rating of the electric heating if there are four or more heating units under separate control.
6. 100% of the rating of electric thermal storage systems where the usual load is expected to operate for three hours or more.

These rules of *NEC 220.82* are summarized in worksheet form in **Figure 20-6**.

Example: Calculate the minimum service size for a dwelling with the following loads using the optional method. The electrical system is 120/240 volts. This is the same criteria as was used for the general method example.

- 1000 square feet of living area
- 4080 VA of air conditioning (17-amp, 240-volt compressor motor)

The General Method of Residential Service Calculation

Step	NEC Article	Calculation	Value
1	220.41	Lighting load. 3 VA per square foot of living space. 1000 ft² x 3 VA = 3000 VA	3000 VA
2	220.52	Two small appliance circuits and one laundry circuit at 1500 VA per circuit. 3 x 1500 = 4500 VA	4500 VA
		Total Steps 1 and 2.	7500 VA
3	Table 220.42(A)	Apply the general lighting demand to the total of Steps 1 and 2. 1st 3000 VA at 100%. 3001 to 120,000 VA (next 117,000 VA) at 35%. Remainder over 120,000 at 25%.	3000 VA 1575 VA
4	220.60	Noncoincident loads. Omit the *smaller* of the heating and air conditioning loads.	6000 W
5	220.54	Clothes dryers. Use the nameplate rating or 5 kW, whichever is greater.	6000 W
6	Table 220.55	Cooking equipment with demand factor applied.	8000 W
7	220.53	Fastened-in-place appliances. 75% demand permitted for four or more. Water heater at 4500 W Food disposer at 1150 VA Dishwasher at 1200 VA Well Pump at 1650 VA 8500 VA x 75% = 6375 VA	6375 VA
8	220.14(C)	The largest motor in full-load current increased by 25%. 1150 VA x 25% = 288 VA	288 VA
		Total Steps 3–8.	31,238 VA
9	Table 310.12	Divide the total VA by the source voltage to determine the amp-rating of the service. 31,238 VA ÷ 240 V = 130 Amps Use the next higher standard OCPD size from *Table 240.6(A)*. 150 Amps Size the service conductors using *Table 310.12*. 1 AWG Copper or 2/0 Aluminum Size the grounding electrode conductor using *Table 250.66*. 6 AWG copper Grounding Electrode Conductor	

Goodheart-Willcox Publisher

Figure 20-5. A completed worksheet from the in-text example using the general method.

- 6 kW heating
- 12 kW cooking range
- 6 kW clothes dryer
- 4.5 kW water heater
- 1150 VA food waste disposer (9.8-amp, 120-volt motor)
- 1200 VA dishwasher
- 1650 VA well pump (6.9 amps, 240 volts)

When applying the worksheet, note that the first two steps are the same regardless of the calculation method used.

First, calculate the lighting load, which is 3 VA per square foot of living space.

$$1000 \text{ ft}^2 \times 3 \text{ VA} = 3000 \text{ VA}$$

Second, calculate the two small appliance circuits and one laundry circuit at 1500 VA per circuit.

$$3 \times 1500 = 4500 \text{ VA}$$

Step 3 adds all general load appliances at their *nameplate* value:

- Food waste disposal at 1150 VA
- Dishwasher at 1200 VA
- Range at 12,000 VA
- Clothes dryer at 6000 VA
- Water heater at 4500 VA

$$1150 \text{ VA} + 1200 \text{ VA} + 12,000 \text{ VA} + 6000 \text{ VA} + 4500 \text{ VA} = 24,850 \text{ VA}$$

The Optional Method of Residential Service Calculation

Step	NEC Article	Calculation	Value
1	220.82(B)(1)	Lighting load. 3 VA per square foot of living space.	
2	220.82(B)(2)	Two small appliance circuits and one laundry circuit at 1500 VA per circuit.	
3	220.82(B)(3)	Calculate ALL appliances at nameplate rating (excluding HVAC).	
4	220.82(B)(4)	Nameplate rating of motors not included in Step 3.	
		Total Steps 1–4 to arrive at the general load.	
5	220.82(B)	Apply the general load demand factor: 100% of the first 10 kVA, and 40% of the remainder.	
6	220.80(C)	Heating and air conditioning load.	
		Total Steps 6 and 7.	
7		Divide the total VA by the source voltage to determine the amp-rating of the service. Use the next higher standard OCPD size from Table 240.6(A), and size the service conductors using Table 310.12. Size the grounding electrode conductor using Table 250.66.	

Goodheart-Willcox Publisher

Figure 20-6. The optional method service calculation worksheet.

The well pump motor is added in Step 4, and the results are totaled.

3000 VA + 4500 VA + 24,850 + 1650 VA = 34,000 VA

The demand factor is applied to the general load in Step 5, which is 100% of the first 10 kVA and 40% of the remainder.

10,000 VA × 100% = 10,000 VA
24,000 VA × 40% = 9600 VA
10,000 VA + 9600 VA = 19600 VA

The heating and air conditioning load is added in Step 6. Like the general method, the smaller of the heating or air-conditioning load is omitted from the calculation. In this example, the 6000 watts of heat is greater than 4080 VA of air conditioning.

Steps 5 and 6 are added to arrive at the total calculated load. The result is then divided by the source voltage in Step 7 to arrive at the amp rating of the service.

19,600 VA + 6000 VA = 25,600 VA
25,600 VA ÷ 240 V = 107 A

Using the optional method of calculation resulted in a minimum service size of 110 amps, which requires a 3 AWG copper or a 1 AWG aluminum service entrance conductor from Table 310.12, Single-Phase Dwelling Services and Feeders, and an 8 AWG grounding electrode conductor from Table 250.66. See **Figure 20-7**.

Although the calculation in this example yielded a 110-amp service, this is an unusual service size in actual practice. Other available choices are a 125, 150, or 175 amp; however, a more common residential service size choice is 200 amps.

20.3.3 Household Cooking Equipment and Table 220.55

The service calculation adds all the electrical loads in the dwelling to determine an adequate service size for that dwelling. Table 220.55 is used to determine the load of household cooking equipment when calculating the service size. The NEC permits reductions of specific loads to account for load diversity. **Load diversity** means all the connected loads are not energized simultaneously, and many loads cycle off and on throughout the day. Table 220.55 recognizes load diversity by allowing for a reduced kW rating than the actual rating of the cooking appliance. The heading of the table clarifies that it is intended to be used for cooking appliances rated over 1–3/4 kW (1750 W).

The table is divided into four columns, as shown in **Figure 20-8**. The leftmost column is the number of appliances in the dwelling, from 1 appliance to 61 and over. It must be understood this table is primarily used for multi-family dwellings such as apartment complexes, which can have 61 or more units fed from a single service. Most likely, this service will not be single-phase 120/240-volt, but rather a 120/208-volt three-phase. A single-family dwelling kitchen, however, typically has either only one cooking range with an oven and a cooktop combined or a wall oven and a counter-mounted cooktop.

The three remaining columns contain demand factors for the cooking appliances:

- Column A is used for appliances with a rating of less than 3–1/2 kW and is typically applied to cooktops.
- Column B is used for appliances with a rating of 3–1/2 kW to 8–3/4 kW and is typically applied to wall ovens. The demand factors listed for columns A and B are percentages of the nameplate rating of the appliance.

The Optional Method of Residential Service Calculation

Step	*NEC* Article	Calculation	Value
1	220.82(B)(1)	Lighting load. 3 VA per square foot of living space. 1000 ft² x 3 VA = 3000 VA	3000 VA
2	220.82(B)(2)	Two small appliance circuits and one laundry circuit at 1500 VA per circuit. 3 x 1500 VA = 4500 VA	4500 VA
3	220.82(B)(3)	Calculate ALL appliances at nameplate rating (excluding HVAC). Food waste disposer 1150 VA Dishwasher 1200 VA Range 12,000 VA Clothes Dryer 6000 VA Water Heater <u>4500 VA</u> 24,850 VA	24,850 VA
4	220.82(B)(4)	Nameplate rating of motors not included in Step 3. Well Pump 1650 VA	1650 VA
		Total Steps 1–4 to arrive at the general load.	<u>34,000 VA</u>
5	220.82(B)	Apply the general load demand factor: 100% of the first 10 kVA, and 40% of the remainder 10,000 @ 100% = 10,000 24,000 @ 40% = <u>9600</u> 19,600	19,600 VA
6	220.80(C)	Heating and air conditioning load. Electric heat 6000 W	6000 W
		Total Steps 6 and 7.	25,600 VA
7		Divide the total VA by the source voltage to determine the amp-rating of the service. 25,600 VA ÷ 240 V = 107 Amps Use the next higher standard OCPD size from *Table 240.6(A)*. 110 Amps Size the service conductors using *Table 310.12*. 3 AWG Copper or 1 AWG Aluminum Size the grounding electrode conductor using *Table 250.66*. 8 AWG copper Grounding Electrode Conductor	

Goodheart-Willcox Publisher

Figure 20-7. A completed worksheet from the in-text example using the optional method.

- Column C is used for appliances rated not over 12 kW and is typically applied to ranges. The values listed in column C are the maximum demand ratings in kW for the number of appliances.

In addition to the four columns, there are five notes to the table that aid in its application. The heading of the table clarifies that column C shall be used in all cases except as permitted in Note 3.

Using Column C

Using Column C for appliances with a rating of 12 kW or greater is straightforward. Simply apply the maximum demand to the number of appliances. The maximum demand for one appliance rated not over 12 kW is 8 kW, the maximum for three appliances not over 12 kW is 14 kW, and the maximum demand for 25 appliances not over 12 kW is 40 kW.

The application of column C is more complicated for appliances rated over 12 kW. Note 1 or Note 2 must be applied. Note 1 is for ranges of all the same rating that are

rated over 12 kW but not more than 27 kW. The note states that the maximum demand in column C shall be increased by 5% for each additional kW rating, or major fraction, by which the appliance exceeds 12 kW.

Work through the following examples to better practice using the chart and column C.

Example 1: What is the maximum demand used for the service calculation for a 14 kW range in a dwelling?

The first thing to notice is that 14 kW is greater than 12 kW by 2 kW. Therefore, the maximum rating in Column C shall be increased by 2 × 5%, or 10%. The column C maximum demand for one appliance, 8 kW, is increased by 10%, resulting in a revised maximum demand of 8.8 kW.

Example 2: What is the maximum demand used for the service calculation for three 16 kW ranges?

The value 16 kW is greater than 12 kW by 4 kW. Thus, 4 kW × 5% = 20%, and the maximum demand in column C for three appliances is 14 kW. This results in a revised maximum demand of 16.8 kW for all three ranges.

Table 220.55 Demand Factors and Loads for Household Electric Ranges, Wall-Mounted Ovens, Counter-Mounted Cooking Units, and Other Household Cooking Appliances over 1¾ kW Rating (Column C to be used in all cases except as otherwise permitted in Note 3.)

| Number of Appliances | Demand Factor (%) (See Notes) | | Column C Maximum Demand (kW) (See Notes) (Not over 12 kW Rating) |
	Column A (Less than 3½ kW Rating)	Column B (3½ kW through 8¾ kW Rating)	
1	80	80	8
2	75	65	11
3	70	55	14
4	66	50	17
5	62	45	20
6	59	43	21
24	31	26	39
25	30	26	40
26–30	30	24	15 kW + 1 kW for each range
31–40	30	22	
41–50	30	20	25 kW + ¾ kW for each range
51–60	30	18	
61 and over	30	16	

Notes:
1. *Over 12 kW through 27 kW ranges all of same rating.* For ranges individually rated more than 12 kW but not more than 27 kW, the maximum demand in Column C shall be increased 5 percent for each additional kilowatt of rating or major fraction thereof by which the rating of individual ranges exceeds 12 kW.
2. *Over 8¾ kW through 27 kW ranges of unequal ratings.* For ranges individually rated more than 8¾ kW and of different ratings, but none exceeding 27 kW, an average value of rating shall be calculated by adding together the ratings of all ranges to obtain the total connected load (using 12 kW for any range rated less than 12 kW) and dividing by the total number of ranges. Then the maximum demand in Column C shall be increased 5 percent for each kilowatt or major fraction thereof by which this average value exceeds 12 kW.
3. *Over 1¾ kW through 8¾ kW.* In lieu of the method provided in Column C, adding the nameplate ratings of all household cooking appliances rated more than 1¾ kW but not more than 8¾ kW and multiplying the sum by the demand factors specified in Column A or Column B for the given number of appliances shall be permitted. Where the rating of cooking appliances falls under both Column A and Column B, the demand factors for each column shall be applied to the appliances for that column, and the results added together.
4. Calculating the branch-circuit load for one range in accordance with Table 220.55 shall be permitted.
5. The branch-circuit load for one wall-mounted oven or one counter-mounted cooking unit shall be the nameplate rating of the appliance.
6. The branch-circuit load for a counter-mounted cooking unit and not more than two wall-mounted ovens, all supplied from a single branch circuit and located in the same room, shall be calculated by adding the nameplate rating of the individual appliances and treating this total as equivalent to one range.

Reproduced with permission of NFPA from NFPA 70, National Electrical Code, 2023 edition. Copyright © 2022, National Fire Protection Association. For a full copy of the NFPA 70, please go to www.nfpa.org

Figure 20-8 *Table 220.55* outlines the allowable demand factors for household cooking appliances.

This is quite a reduction and illustrates the impact of load diversity.

Note 2 is similar to Note 1 in its application. However, it is used for multiple ranges of unequal values. It states that an average value of ranges rated 8–3/4 kW through 27 kW (using 12 kW for any ranges less than 12 kW) can be calculated by adding together the ratings of the individual ranges to obtain the total connected load and dividing the total by the number of ranges. Then the maximum demand in column C shall be increased in the same manner as used in Note 1.

Example 3: What is the maximum demand used for a service calculation for one 10 kW range, one 12 kW range, one 14 kW range, and one 16 kW range for a multi-family dwelling?

Adding the individual ratings of the ranges results in a total of 54 kW (12 kW + 12 kW + 14 kW + 16 kW). Then, 54 kW is divided by four resulting in an average value of 13.5 kW, which rounds up to 14kW. The calculation is then performed as if there are four 14 kW ranges: 14 kW is greater than 12 kW by 2 kW. Therefore, the maximum demand for four ranges listed in column C, 17 kW, is increased by 10%. This results in a revised maximum demand of 18.7 kW. Again, this is quite a reduction.

Using Columns A and B

Using columns A and B, apply a percentage reduction to the nameplate rating of the appliance. Using column A, one appliance rated 3 kW can be reduced by 80%, which results in a revised demand of 2.4 kW (3 kW × 0.8 = 2.4 kW). Column B values are calculated in the same manner. Two appliances rated at 8 kW each have a revised demand of 65% applied to the sum of the two appliances: 65% of 16 kW is 10.4 kW.

Note 3 of *Table 220.55* provides an alternative to using column C for appliances rated more than 1–3/4 kW but not more than 8–3/4 kW, and it allows the nameplate ratings to be added together to multiply the sum by the demand factors of column A or column B. This derives the given number of appliances. The note further states that when the rating of the appliances falls under both columns A and B, the demand factors for each column shall be applied to the appliances of that column, and the results are added together.

Column B Example: Calculate the maximum demand for four wall ovens rated 4.5 kW, 5 kW, 7.5 kW, and 8 kW.

First, determine the combined rating of all appliances, which is 4.5 kW + 5 kW + 7.5 kW + 8 kW = 25 kW.

Then, locate the demand factor for four appliances under column B, which is 50%. Thus, 25kW × 50% = 12.5 kW.

Columns A and B Example: What is the service calculation value for 1–3 kW cooktop and 1–8.5 kW wall oven?

Determine the demand factor for one appliance in each column, which in this case is 80% for each. So, 3 kW × 80% = 2.4 kW and 8.5 kW × 80% = 6.8 kW. Then apply it to the nameplate value of each appliance and add the results: 2.4 kW + 6.8 kW = 9.2 kW.

For appliances that fit in either Column B or Column C, calculate for both columns and choose the lowest result.

20.3.4 Household Electric Clothes Dryers and *Table 220.54*

Multi-family dwelling units often include a receptacle for an electric dryer. The *NEC* permits reductions of multiple electric dryer loads to account for load diversity. *Table 220.54* provides a method utilizing percentages to accomplish this. See **Figure 20-9**. The more electric dryers, the smaller the percent demand factor. This demand factor drops from 100% for up to four dryers to a maximum percentage demand factor of 25% for 43 and over. The *NEC* allows this reduction because the likelihood of most dryers operating at the same time is remote.

Example 1: Consider the installation of 10 dryers in a 10-unit condominium dwelling unit with each receptacle rated for a maximum of 5000 watts per *Table 220.54*. What is the total demand in kW to be included in the service calculation?

Table 220.54 allows a 50% demand factor for this problem. Therefore, 5000 W × 10 dryers = 50,000 W. So, 50,000 × 0.5 (50%) = 25,000 W, or 25 kW, is included in the service calculation for the dryer load.

Table 220.54 Demand Factors for Household Electric Clothes Dryers

Number of Dryers	Demand Factor (%)
1–4	100
5	85
6	75
7	65
8	60
9	55
10	50
11	47
12–23	47% minus 1% for each dryer exceeding 11
24–42	35% minus 0.5% for each dryer exceeding 23
43 and over	25%

Reproduced with permission of NFPA from NFPA 70, National Electrical Code, 2023 edition. Copyright © 2022, National Fire Protection Association. For a full copy of the NFPA 70, please go to www.nfpa.org

Figure 20-9 *Table 220.54* outlines the allowable demand factors for household electric clothes dryers.

Example 2: Consider the installation of 40 dryers in a 40-unit condominium dwelling unit with each receptacle rated for a maximum of 5000 watts per *Table 220.54*. What is the total demand in kW to be included in the service calculation?

Table 220.54 allows a 35% demand factor minus 0.5% for each dryer exceeding 23. In this case, 40 dryers exceed 23 by 17. Therefore, 17 × 0.5 (.5%) = 8.5%, and 35% – 8.5% = 26.5% as final demand factor. This means there is a 40 × 5000 W per dryer = 200,000 W. Thus, 200,000 W × 0.265 (26.5%) = 53,300 W, or 53.3 kW, is included in the service calculation.

Summary

- The service calculation is a series of calculations that adds all the electrical loads in a house to determine the service size, rated in amps.
- The electrical load is the component in an electrical system that is connected to and uses electrical power.
- The connected load is the nameplate rating of an appliance or piece of electrical equipment and the actual amount of consumed power.
- The calculated load is a reduction in the connected load using rules of the *NEC*.
- A demand is an *NEC* permitted decrease in the connected load.
- The *NEC* includes two methods of service calculations for a new dwelling, the general method and the optional method, and one method for an existing dwelling.
- Using both calculation methods for the same residence will likely result in two different service sizes. Both are considered valid.
- The first step of a service calculation is to calculate the living area of the dwelling.
- Noncoincident loads are electrical loads that are unlikely to be energized simultaneously, such as a heating unit and an air conditioner.
- The amp-rating of the service is determined by dividing the total calculated load by the source voltage.
- The minimum service size for a dwelling allowed by the *NEC* is 100 amps.
- The general method of calculation involves nine steps of calculations scattered throughout *Article 220*.
- The rules for the optional method of calculation for a dwelling unit are found in *Section 220.82*.
- *Table 220.55* is used to determine the load of household cooking equipment when calculating the service size.
- The *NEC* permits reductions of specific loads to account for load diversity.
- *Table 220.55* recognizes load diversity by allowing for a reduced kW rating than the actual rating of the cooking appliance.

Know and Understand

1. The component in an electrical system that consumes electrical power is the _____.
   A. source
   B. load
   C. conductor
   D. the overcurrent device

2. To determine the current draw of an appliance when the voltage and the wattage are given, use the formula _____.
   A. $I = E \times P$
   B. $I = E/P$
   C. $I = E + P$
   D. $I = P/E$

3. The nameplate rating of an appliance represents its _____.
   A. connected load
   B. calculated load
   C. demand factor
   D. noncoincident load

4. Which of the following is not included when computing the floor area of a dwelling for the purposes of a service calculation?
   A. Open porches
   B. Garages
   C. Crawl spaces
   D. All of these.

5. *True or False?* General use receptacles are not included in the general lighting load of a dwelling.

6. When using the general method of calculation, a 75% demand factor can be applied to fastened in place appliances if there are _____ or more.
   A. two
   B. three
   C. four
   D. six

7. What table in the *NEC* is used for sizing dwelling service conductors?
   A. *Table 240.6(A)*
   B. *Table 310.12*
   C. *Table 250.66*
   D. *Table 310.16*

8. What table of the *NEC* is used for sizing the grounding electrode conductor?
   A. *Table 240.6(A)*
   B. *Table 310.12*
   C. *Table 250.66*
   D. *Table 310.16*

9. Standard sizes of overcurrent devices are found in _____.
   A. *Table 240.6(A)*
   B. *Table 310.12*
   C. *Table 250.66*
   D. *Table 310.16*

10. *True or False?* Electricians are required to perform both the general method and the Optional Method and use the higher of the two results.

11. Which table in the *NEC* is used to calculate the service demand for household cooking equipment?
    A. *Table 8, Chapter 9*
    B. *Table 310.16*
    C. *Table 310.16(A)*
    D. *Table 220.55*

12. Which column of *Table 220.55* is typically used for cooktops?
    A. Column A
    B. Column B
    C. Column C
    D. None of the above.

13. What is the demand on the service of a duplex with 2—10 kW ranges?
    A. 20 kW
    B. 16 kW
    C. 11 kW
    D. 17 kW

14. Column C shall be used in all cases except as permitted in Note _____.
    A. 1
    B. 2
    C. 3
    D. 4

15. What is the demand factor for six wall ovens rated at 8 kW?
    A. 80%
    B. 70%
    C. 59%
    D. 43%

Apply and Analyze

1. Two or more appliances that are not expected to run simultaneously are known as _____.

2. The minimum Residential service size allowed by the *NEC* is _____.

3. For the purposes of service calculations, the small appliance and laundry loads are factored at _____ VA each.

4. A dwelling with outside dimensions of 32′ by 45′ would have a lighting load of _____ VA.

5. If using the Optional Method, include _____ percent of the nameplate rating of the electric heat if there are four or more separately controlled units.

6. The concept that assumes all of the electrical loads in a household will not be energized at the same time and that many loads automatically cycle off and on is known as _____.

7. What is the demand on the service for two wall ovens in a dwelling if one oven is rated at 8.5 kW and the other is rated 6 kW?

8. What is the demand on the service for three ranges rated 8 kW, 10 kW, and 12 kW?

Critical Thinking

1. What is the major difference between the general method and the optional method of service calculations?
2. Calculate the service size for the following residence using the general method worksheet format.
   - 60′ × 35′ of usable living space
   - 10 kW of heating
   - 12 kW of air conditioning with two 22-amp, 240-volt compressor motors
   - 10 kW electric range
   - 7 kW clothes dryer
   - 8 kW water heater
   - 960 VA attic ventilation fan with an 8-amp, 120-volt motor
   - 1500 VA dishwasher
   - 1920 VA pool pump with an 8-amp, 240-volt motor

3. Calculate the service size for the following residence using the optional method worksheet format.
   - 60′ × 35′ of usable living space
   - 10 kW of heating
   - 12 kW of air conditioning with two 22-amp, 240-volt compressor motors
   - 10 kW electric range
   - 7 kW clothes dryer
   - 8 kW water heater
   - 960 VA attic ventilation fan with an 8-amp, 120-volt motor
   - 1500 VA dishwasher
   - 1920 VA pool pump with an 8-amp, 240-volt motor

SECTION **7**

Rough-in

Rachid Jalayanadeja/Shutterstock.com

CHAPTER 21
Temporary Power

CHAPTER 22
Outlet Box Installation

CHAPTER 23
Conductor Routing
and Installation

House wiring is typically performed in two stages: rough-in and trim-out. This section covers the rough-in stage, and Section 9, *Trim Out* covers the trim-out stage. During the rough-in, electricians perform all work that is concealed behind the finished walls. Device boxes are installed, conductors are run from box to box, and the electrical contractor to establishes the temporary power for all other trades.

Chapter 21, *Temporary Power*, discusses how to provide temporary power to the worksite for both overhead and underground services. Chapter 22, *Outlet Box Installation*, outlines the receptacle outlets and lighting outlets that the *NEC* requires. Chapter 23, *Conductor Routing and Installation*, addresses conductor routing and installation, emphasizing the *NEC*'s proper securing requirements for the cables.

CHAPTER 21

Temporary Power

LeeAnn White/Shutterstock.com

CHAPTER OUTLINE

21.1 Requirements of Temporary Power

21.2 Overhead Temporary Services

21.3 Underground Temporary Services

Introduction

On nearly every construction project, the electrician is one of the first trades on site. It is their responsibility to provide a source of power, called **temporary power**, that is used by all the trades. This power is used until the project has reached a state of completion where permanent power can be established. This requires an electrician to install a temporary service consisting of a meter enclosure, panelboard, often called a *panel*, with overcurrent protection, a grounding electrode, and utility-approved methods to support the equipment. After this is completed and the dwelling unit is ready, temporary lighting and power would be brought to the inside.

There are *NEC* requirements for temporary power on a jobsite that an electrician must be familiar with and be able to implement. This chapter will discuss these requirements for a residential construction site for both overhead and underground temporary service installation.

21.1 Requirements of Temporary Power

Temporary service must be adequately sized and durable enough to withstand construction site use. Electrical wire run overhead or draped around the jobsite to and from the temporary service is subject to damage from construction workers and their machinery. Whatever the season, severe weather is another consideration. It is likely that every temporary service will risk damage.

Temporary service is provided through a temporary power pole, often called a *temp pole* or *power pole*, providing electrical power to a panel rated for an outdoor environment. *Table 110.28* lists enclosure ratings acceptable for this type of environment. Typically, a 3R enclosure for the panelboard is used. A panelboard rated at 100 amps is the most common size used for a single or duplex dwelling unit temporary service. However, larger apartment or condominium complexes could have 200-amp or 400-amp temporary services. The required meter enclosure and panelboard can be fed either through **overhead temporary service**, from a utility pole-mounted transformer, or **underground temporary service**, from a pad-mounted transformer. Often, the type of permanent power a dwelling will have determines the type of temporary pole that will be used: a dwelling permanently fed from an overhead service uses an overhead temporary pole, and one with permanent underground service uses an underground temporary pole.

Temporary wiring installations are covered in *NEC Article 590*. Per *590.6(A)(1)*, all receptacles not part of the permanent wiring that are single-phase, 15-, 20- and 30-amp and are in use by personnel shall have ground-fault circuit-interrupter protection.

Although not required by the *NEC*, temporary panels should provide power to a minimum of two ground-fault protected circuits that provide power to receptacles used for power tools. Depending on the size of the construction projects, more circuits would be needed. In addition, *Section 590.4(D)(1)* prohibits the use of receptacles on a circuit used for temporary lighting. Although not specifically required in the *NEC*, it is wise to provide a GFCI protected circuit for lighting branch circuits.

PRO TIP   **Temporary Wiring**

Section 590.1, Temporary Installations states, *"The provisions of this article apply to temporary electric power and lighting installations."* The scope of this article covers all temporary power, including but not limited to dwelling units, commercial and industrial installations, remodeling in an existing structure, holiday lighting, art fairs, and the electricity created by a generator for temporary use.

The dangers of electricity used on a temporary basis can be greater than a fully completed installation. Construction sites are constantly changing and may have some unsafe conditions. Other than the requirements for GFCI protection, the rules of wiring methods are more lenient. Branch circuits and feeders are allowed to be temporarily installed with minimal restrictions concerning routing and protection of wires and regardless of the type of job, commercial, residential, or industrial, NM, NMC, and Type SE cable allowed.

An electrical meter is installed on a temporary pole with a meter socket, allowing incoming power to pass through the load center in the panelboard. The load center is responsible for then distributing power for temporary lighting and temporary electrical outlets for needed tools and equipment within a dwelling unit under construction. The utility company may provide the meter socket or the temporary power pole, and the electrician then assembles the necessary components to house the circuit breakers and receptacles. Another method is to purchase an outdoor-rated load center which includes the meter socket with the breakers and receptacles preinstalled. Refer to **Figure 21-1**.

Upon installation, the temporary service may be subject to inspection by the AHJ before the serving utility energizes it. In some municipalities, the utility makes the connection without an AHJ inspection if it meets its installation requirements. Sometimes, areas with limited inspection require an affidavit signed by the electrician or the electrical contractor before the energization of a temporary service.

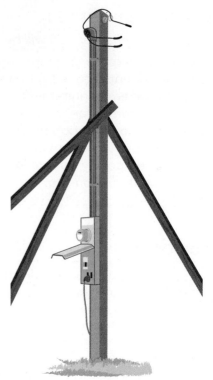

Goodheart-Willcox Publisher

Figure 21-1. A typical overhead temporary power pole installation.

PRO TIP   **Rating Enclosures**

The National Electrical Manufacturers Association (NEMA) rates enclosures based on potential exposure to live parts and for the conditions of use. Enclosures for most residential outdoor must be a NEMA 3R rating. Refer to *Table 110.28*. This rating indicates the enclosure provides protection against the entry of rain, sleet, and snow. In contrast, an enclosure with a NEMA 1 rating provides no protection against the weather and is restricted to indoor use. There are nearly two dozen different ratings for enclosures that are used throughout the electrical industry, protecting against an array of hazards, including hose-down, corrosive agents, and temporary or prolonged submersion.

21.2 Overhead Temporary Services

An overhead temporary power pole supports power lines and a temporary ***service mast***, which serves as a transfer point from the serving utility's overhead power distribution system to a residence. The serving utility provides the construction specifications for a temporary service mast onsite. Most utilities require a minimum 6 × 6 post or a pole with a minimum diameter of 5″ at the top. The mast must be able to withstand the weight of the service drop conductors.

Service drop conductors are overhead conductors that drop from the utility transformer to the service point, **Figure 21-2**. The *service point* marks the location where the utility's wiring ends and the customer's wiring begins. It is typically at the top of the pole at the weather head.

Figure 21-3 is a sample of a structure the utility may expect an electrician to install. All wood used for the service mast should be pressure treated. Three diagonal support braces connected to 2 × 4 stakes driven to a minimum depth of 2′ help to stabilize the mast. The rear brace should align with the service drop. The bottom end of the service mast or structure should extend at least 3′ into the ground. Some overhead service clearances note to include staying at least 18′ above parking areas, alleys, and public streets and roads that are subject to truck traffic and staying 12′ over residential property and driveways not subject to truck traffic including 12′ above sidewalks and areas subject to only pedestrian traffic.

For overhead poles, a meter socket is typically mounted 3′-6″ to 5′-6″ to the bottom of the enclosure, above grade. The service-entrance conductors, conductors that run from an exterior point of attachment to the utility pole and to an electric panel, are often Type SEU cables. An SEU (Service Entrance, U-shaped) cable contains two different phase conductors (electrically hot) and a concentric neutral. It does not contain a ground wire. This type of cable can leave the meter socket and be used as the conductor that attaches to the utilities service drop. Often this cable is more economical than using a raceway to leave the meter socket. Individual conductors installed in a raceway are sturdier and physically protected. It is slightly more costly in material and labor to

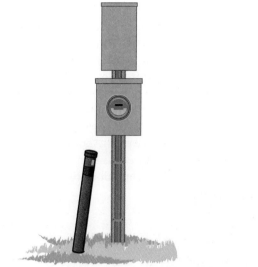

Goodheart-Willcox Publisher

Figure 21-3. A typical underground temporary power pole installation.

do this, but it is a better installation. Check with your local utility for specific requirements.

Some utilities may require a raceway with individual conductors. The metal parts and the neutral conductor for the service need to be grounded to earth. This is accomplished by means of a single driven ground rod with 8′ in contact with earth if the electrician can prove an electrical resistance to ground or earth of 25 ohms or less. A supplemental grounding electrode will be required if this is not

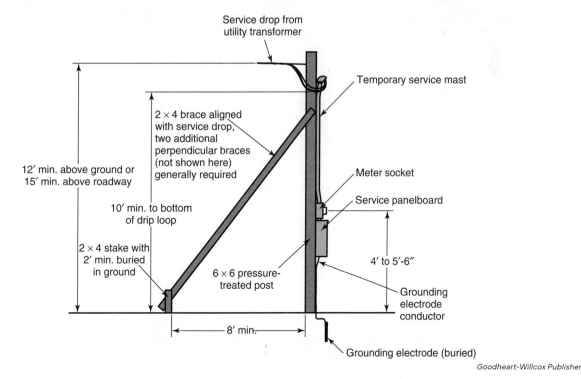

Goodheart-Willcox Publisher

Figure 21-2. Overhead temporary power pole installation requirements are specified by the utility.

the case. A second driven ground rod is the easiest way to do this. Review *Section 250.52(A)(4)* and *Section 250.53(A)(2) Exception* for the *NEC* references.

Also, a minimum 6 AWG grounding electrode conductor (GEC) is required to run from the ground rod(s) to the main temporary panel or in some cases, the meter enclosure. If physical damage to the GEC, as referenced in *Section 250.64(B)(1)–(2)*, is expected to occur, a larger size wire, a 4 AWG, could be used. Another option is to protect the wire with one of the raceways mentioned *Section 250.53(A)(2)* such as a rigid metal conduit or rigid nonmetallic conduit, schedule 80.

21.3 Underground Temporary Services

Most utilities require a 4 × 4 post for underground temporary service poles. The post can be smaller than that required for an overhead temporary service because it does not have to support the weight of a service drop. It also does not require extra bracing that an overhead service requires.

The service post should be installed near the pad-mounted transformer. The closer to the pad-mounted transformer, the less digging for the service lateral. At least 3′ of the support post should be in the ground. Underground service poles have utility-provided conductors that supply an underground service called *service laterals*. Some utilities require a minimum 2-1/2″ conduit is to be provided from the meter socket to a depth below grade where a 45° elbow is installed, **Figure 21-4**. The conduit provides protection to the service lateral conductors from physical damage. Other utilities do not have this requirement and may allow a smaller 2″ conduit without a

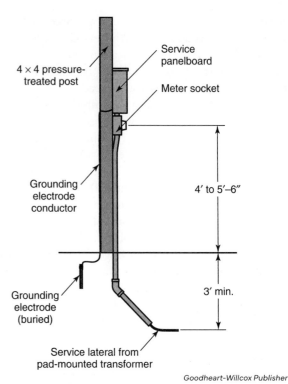

Goodheart-Willcox Publisher

Figure 21-4. Underground temporary power pole installation requirements are specified by the utility.

45° elbow. However, the support post burial requirements would likely be similar. Another option is to purchase a meter enclosure rated for direct burial. This will eliminate the need for a raceway entering the ground.

The meter socket should be 3′-6″ to 5′-6″ above grade to the bottom of the enclosure. The grounding electrode and grounding electrode conductor rules are the same for all services, following the rules for an overhead service is the best practice.

Summary

- Temporary power is provided on construction sites by the electrician for all trades to use in the construction process.
- Temporary services are typically rated 100 amps. Larger temporary services may be required for multi-family dwellings, possibly 200 to 400 amps.
- Temporary services can be either overhead or underground. An overhead temporary service will be fed from a utility pole-mounted transformer, and an underground temporary service will be fed from a pad-mounted transformer.
- It is recommended that temporary power poles be provided with a minimum of two ground-fault circuit interrupter protected circuits.
- For an overhead temporary service, most utilities will require a minimum 6 × 6 post or a pole with a minimum diameter of 5″ at the top.
- The bottom end of the service support mast should extend at least 3′ into the ground.
- The overhead service drop should have a clearance of not less than 18′ above driveways, alleys, public streets, and roads that are subject to truck traffic and 12′ above residential driveways, sidewalks, and areas subject to only pedestrian traffic.
- The meter socket should be 3′-6″ to 5′-6″ above grade to the bottom of the enclosure.
- The temporary service is grounded by means of an 8′ driven ground rod and a minimum 6 AWG grounding electrode conductor with a resistance to ground of 25 ohms or less. Another option is to drive or install a second ground rod.
- Most utilities require a minimum 4 × 4 post for an underground temporary service with a minimum of 3′ driven into the ground.
- The service laterals are the utility-installed conductors underground between the electric supply system and the service point.

Know and Understand

1. *True or False?* Temporary power poles do not require a ground rod or grounding electrode conductor.
2. *True or False?* All temporary power receptacles are required to have arc-fault protection.
3. A panelboard rated at _____ is the most common size used for a single dwelling unit temporary service.

   A. 30 amps
   B. 100 amps
   C. 200 amps
   D. 400 amps

4. Although not required by the *NEC*, it is good practice to provide a temporary service with a minimum of _____ GFCI branch circuits.

   A. two
   B. three
   C. four
   D. six

5. An enclosure that is exposed to the outdoor elements must have a NEMA rating of _____.

   A. 1
   B. 3R
   C. 4X
   D. 12

6. The mast for an overhead temporary service must be of sufficient strength to withstand _____.

   A. any weather condition imposed
   B. the weight of the service drop
   C. an automobile impact
   D. rot and degradation from contact with the soil

7. The _____ marks the point between the utility wiring and the premise wiring.

   A. temporary power pole
   B. service drop
   C. service point
   D. service lateral

8. The overhead conductors from the serving utility to the service point are called the service _____ conductors.

   A. entrance
   B. drop
   C. point
   D. lateral

9. The underground conductors between the electric supply system and the service point are called the service _____ conductors.

   A. entrance
   B. drop
   C. point
   D. lateral

10. *True or False?* The mast for an underground service does not have to be buried as deep as a mast for an overhead service.

Apply and Analyze

1. What determines if the temporary service will be overhead or underground?
2. Why does the overhead temporary pole need to be a larger size than an underground temporary pole?
3. When is the temporary service removed?

Critical Thinking

1. What is the minimum length of an overhead temporary power pole where the service drop is above a road subject to truck traffic?
2. Why does the *NEC* prohibit the use of receptacles on any branch circuit that supplies temporary lighting?

Chad Robertson Media/Shutterstock.com

CHAPTER OUTLINE

LEARNING OBJECTIVES

After completing this chapter, you will be able to:

- Explain the proper method of installation for outlet boxes.
- Identify the maximum number of conductors permitted in a particular box size using *Table 314.16(A)* if all the conductors are the same size.
- Calculate the required volume allowance for a device or junction box.
- List the volume allowances and basis for the various items that contribute to box fill.
- Calculate the volume allowance of conductors using *Table 314.16(B)(1)*.

TECHNICAL TERMS

box extender
box fill calculation
cut-in box
internal clamp
yoke

Introduction

All residential electrical installations begin with rough-in. This first stage in rough-in involves box layout and installation according to the electrical plan for the project. It is necessary to note, however, that not all projects come with an electrical plan. In these instances, the electrician is responsible for determining the placement of the receptacle and the lighting outlets. In a dwelling unit, the placement of the receptacles is driven by the *NEC. Article 210, Part III* provides guidance for the minimum required spacing. Recall that the *NEC* only lists the *minimum* requirements. A good electrical installation will include thoughtful placement of every lighting outlet and receptacle, and this type of job will exceed the underachieving code minimum.

The most common outlet boxes in new residential construction are non-metallic nail-on device boxes and outlet boxes, **Figure 22-1**. You should understand how these boxes work and how they are installed during the rough-in phase. This chapter will provide an overview of the proper installation methods for these boxes. It will also outline how to perform box fill calculations, which requires determining the total volume of conductors, devices, and fittings in a box. Recommendations for what sizes of boxes are best to use for common installations will also be covered.

Two-gang

Three-gang

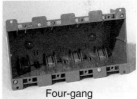
Four-gang

Goodheart-Willcox Publisher

Figure 22-1. Examples of single-gang, two-gang, three-gang, and four-gang non-metallic nail-on boxes.

22.1 Box Installation

During the rough-in stage of a job, a consistent mounting height for all the device boxes should be chosen. Depending on job specifications or the electrician's choice, receptacle outlets are installed as low as 12″ or as high as 20″ to the top of the box. The higher the receptacle is installed, the easier it is to reach down and plug in utilization equipment. Switches could be as low as 42″ to the top or as great as 54″. If the drywall is mounted horizontally, try to avoid 48″ to the top. This is the top of the lower piece of drywall, and some joint compound may be inadvertently pushed into the installed box by the drywallers.

Receptacles and switches installed in countertops need to be installed with care. The height of a typical kitchen countertop is 36″ and may have a 4″ backsplash. In this case, the bottom of the device box should be no lower than 42″ to the bottom or 46″ to the top.

Some bathroom sinks have a lower countertop at 32″. In this case, make sure any device box is mounted no lower than 38″ to the bottom or 42″ to the top. It is best practice to check with the carpenter or kitchen and bath designer to confirm counter heights and if tile or a backsplash will be installed.

When installing tile, use boxes that are still adjustable after the drywall and tile are installed. This type of box is also useful near wood surfaces or other combustible materials.

22.1.1 Nail-on Box Installation

Non-metallic nail-on boxes have *stand-off ridges* that position the box forward from the framing stud to which it is nailed, **Figure 22-2**. This stand-off permits the sheetrock or other wall finish to be flush with the front edge of the box. Device boxes should never be installed with the front edge of the box flush with the stud. The stand-off is 1/2″, the depth of the typical drywall used in dwelling units. It should be noted drywall is also available in 3/8″ and 5/8″ thicknesses.

Stand-off ridge

Goodheart-Willcox Publisher

Figure 22-2. A non-metallic nail-on box with a stand-off ridge for framing.

> **PRO TIP**　　　　　　　**Sizing Boxes**
>
> For single-gang boxes, it is good practice to install a larger box than what is required. The cost difference in box sizes is minimal, and the larger box eases installation and speed of trimming boxes. Larger and deeper devices can cause problems. It is convenient to have the extra space a deeper box offers. Two-gang, three-gang, and four-gang boxes generally do not have issues with device installation.

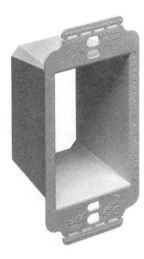

Arlington Industries, Inc

Figure 22-3. Box extender.

Section 314.20 allows the front edge of the box to be recessed no more than 1/4″ from the finished surface if the wall finish is of noncombustible material such as tile, gypsum, or plaster. When tile is installed on drywall, the box would likely be recessed more than 1/4″ and would require the addition of a *box extender*, or an adjustable box allowing the box's front edge to be within the required 1/4″, **Figure 22-3**. When installing a box extender, it must be compatible with the box being extended. One manufacturer's box extender may not fit with another brand.

After the drywall taping and mudding is completed, it is not uncommon to have gaps or spacing greater than 1/8″ by the edge of the box. An electrician must consider repairing noncombustible surfaces that surround incomplete or broken boxes per *Section 314.21*. A smaller gap will ensure the receptacle or switch faceplate will cover the entire opening,

and if installed flush with the drywall, the ears of the receptacle will rest snugly against the box. If the wall surface is wood or other combustible material, the box edge shall extend to or project past the finished surface.

> **SAFETY NOTE**　　　**Flush-Mounted Boxes**
>
> Under no circumstances is the box, box extender, or a mud ring allowed to be recessed from the combustible surface. A flush-mounted box reduces the likelihood of a shorted-out receptacle igniting a fire on the combustible material.

22.1.2 Old Work Box Installation

Old work boxes, also called *cut-in boxes*, are used to add a receptacle or a switch to an existing finish wall. Often, the process of cutting into an existing wall is difficult. Extra care and planning are essential.

The easiest part of this process is the cutting in of the box; the hardest part is routing the required wire to the new box. When cutting in the box, make sure you land between two vertical studs. Lightly use a pencil to trace the minimum size opening needed. If the opening is cut too large, a major patching job will be required. It is always better to have made the opening a little too small; it can be made larger if needed.

Additional care and patience must be taken in older houses that have lathe and plaster walls. Using a battery-operated multitool cutter on a slow speed is helpful. Plaster will crack, and the wood lathe can pull away from the plaster. The key to a successful job is careful and precise measurements from above or below the desired location for the device box. You do not want the hole you drill for the wire to come up in the living room floor or bedroom ceiling. It is a common best practice to measure twice and drill once.

Also, the outside walls of a house will likely be filled with insulation. It is always harder to fish, or route, wires through insulation, but it can be done. Note that NM cable needs to reach a place that will provide electricity to the switch or receptacle. A fish tape is often pushed through the newly drilled hole and needs to be brought out the newly cut hole for the old work box. Once you can reach the fish tape, strip the NM cable back about 8″ and cut off two of the conductors, bend the remaining conductor in half, and attach it to the fish tape. Add electrical tape to smooth the edges and pull up or down the desired length of wire. Terminate both ends of the wire with the power off, then energize. Oftentimes, fishing wires requires two people.

22.2 Box Fill Calculations

Performing a *box fill calculation*, or identifying the proper box size and maximum fill limitations to provide adequate space for the conductors and devices they contain, is an essential task for the residential electrician. Device boxes and junction boxes have maximum fill limitations and must be sized to provide adequate space for the conductors and devices they contain. The residential electrician must determine the appropriately sized box for each application. If a box is crowded with too many conductors, it may be difficult to install the switch or receptacle, cause a ground-fault between the equipment grounding conductor and an energized device terminal, or reduce the airflow around the conductors, leading to overheating.

Section 314.16 has provisions for calculating the maximum box fill for a given application. However, even when following the *NEC* rules, there are times when a device box will not comfortably fit a receptacle. This can occur when a receptacle with greater depth than normal is installed. Some examples of receptacles with a greater depth include those capable of charging cellphones with a USB charger, a ground-fault circuit interrupter (GFCI), and an arc-fault circuit interrupter (AFCI) type installed in a single-gang box. Even if the box is correctly sized per the *NEC*, these receptacles still may not be easy to install.

Occasionally metal boxes are used in dwelling units. The volume of these boxes is shown in *Table 314.16(A)*, **Figure 22-4**. This table shows the maximum number of conductors allowed in standard metal box sizes if all the conductors are the same size. The table contains three columns. The leftmost column lists the box sizes divided by type. The center column lists the minimum volume of each box in cm³, cubic centimeters, and in³, cubic inches. The rightmost column lists the maximum number of conductors permitted to be installed, arranged by AWG sizes 18–6.

PRO TIP   Non-Metallic Box Calculations

Unlike the metal boxes listed in *Table 314.16(A)*, non-metallic boxes do not have a stand-alone table. *Section 314.16(A)(2)* states the volume of non-metallic boxes shall be *"durably and legibly marked"* by the manufacturer. Non-metallic boxes will also commonly include a list of the maximum number of same size conductors that can be installed. For example, an 18 in³ box would be internally stamped with the following information about the installation of the same size wire: 9/14 AWG, 8/12 AWG, 7/10 AWG. This type of box is designed for NM cable up to a maximum size of 10-3. Although the *NEC* allows larger conductors in any box they will fit in, an electrician must always follow the listing rule from the manufacturer and not exceed those specifications.

Different sized wires can be mixed in the same size box. *Table 314.16(B)(1)* assigns a cubic centimeter value and a cubic inch value to wires in the range of 18 AWG to 6 AWG. Using these two tables to identify how many and what size conductors are entering the box, you can calculate the volume of the box occupied by wires.

Additional items and the volume allowance required per conductor are listed in *Sections 314.16(B)(2)–(5)*. The additional items used to determine box fill include internal

Table 314.16(A) Metal Boxes

Box Trade Size			Minimum Volume		Maximum Number of Conductors* (arranged by AWG size)						
mm	in.		cm³	in.³	18	16	14	12	10	8	6
100 × 32	(4 × 1¼)	round/octagonal	205	12.5	8	7	6	5	5	5	2
100 × 38	(4 × 1½)	round/octagonal	254	15.5	10	8	7	6	6	5	3
100 × 54	(4 × 2⅛)	round/octagonal	353	21.5	14	12	10	9	8	7	4
100 × 32	(4 × 1¼)	square	295	18.0	12	10	9	8	7	6	3
100 × 38	(4 × 1½)	square	344	21.0	14	12	10	9	8	7	4
100 × 54	(4 × 2⅛)	square	497	30.3	20	17	15	13	12	10	6
120 × 32	(4¹¹⁄₁₆ × 1¼)	square	418	25.5	17	14	12	11	10	8	5
120 × 38	(4¹¹⁄₁₆ × 1½)	square	484	29.5	19	16	14	13	11	9	5
120 × 54	(4¹¹⁄₁₆ × 2⅛)	square	689	42.0	28	24	21	18	16	14	8
75 × 50 × 38	(3 × 2 × 1½)	device	123	7.5	5	4	3	3	3	2	1
75 × 50 × 50	(3 × 2 × 2)	device	164	10.0	6	5	5	4	4	3	2
75 × 50 × 57	(3 × 2 × 2¼)	device	172	10.5	7	6	5	4	4	3	2
75 × 50 × 65	(3 × 2 × 2½)	device	205	12.5	8	7	6	5	5	4	2
75 × 50 × 70	(3 × 2 × 2¾)	device	230	14.0	9	8	7	6	5	4	2
75 × 50 × 90	(3 × 2 × 3½)	device	295	18.0	12	10	9	8	7	6	3
100 × 54 × 38	(4 × 2⅛ × 1½)	device	169	10.3	6	5	5	4	4	3	2
100 × 54 × 48	(4 × 2⅛ × 1⅞)	device	213	13.0	8	7	6	5	5	4	2
100 × 54 × 54	(4 × 2⅛ × 2⅛)	device	238	14.5	9	8	7	6	5	4	2
95 × 50 × 65	(3¾ × 2 × 2½)	masonry box/gang	230	14.0	9	8	7	6	5	4	2
95 × 50 × 90	(3¾ × 2 × 3½)	masonry box/gang	344	21.0	14	12	10	9	8	7	4
min. 44.5 depth	FS — single cover/gang (1¾)		221	13.5	9	7	6	6	5	4	2
min. 60.3 depth	FD — single cover/gang (2⅜)		295	18.0	12	10	9	8	7	6	3
min. 44.5 depth	FS — multiple cover/gang (1¾)		295	18.0	12	10	9	8	7	6	3
min. 60.3 depth	FD — multiple cover/gang (2⅜)		395	24.0	16	13	12	10	9	8	4

*Where no volume allowances are required by 314.16(B)(2) through (B)(5).

Figure 22-4. *Table 314.16(A)*, Allowable conductor fill for metal boxes.

clamp fill (excluding small fittings such as locknuts and bushings), support fittings commonly called *fixture studs* or *hickeys*, device or equipment fill, and equipment grounding conductor (EGC) fill.

An **internal clamp** may be present in some metal boxes. It is used to secure the wire to a box to prevent the conductor from moving. All internal clamps result in a *single* volume allowance in cubic inches based on the required volume allowance of the largest conductor entering the box.

Fixture studs and hickeys are threaded metal with a nut or threaded cap designed to support a light fixture. They are used to support heavier luminaires, like a chandelier. A single volume allowance of the largest conductor in the box shall be added for a hickey and another single volume allowance for a fixture stud.

Devices and equipment fill includes switches and receptacles. These devices will be allocated a double volume allowance based on the size of the wire attached to the device. Calculate the double volume allowance for each yoke or strap. A **yoke** is the structural frame of a switch, receptacle, or similar device. A yoke is normally attached to the outlet box

with two screws: one at the top of the box and one at the bottom. Each yoke or strap can be visualized as a single opening intended to hold a device. A two-gang box would contain two yokes or straps, a three-gang box would have three, and so on.

Receptacles for electric dryers and ranges will not easily fit in a single-gang box under one yoke or strap. If the device or utilization equipment is wider than 2″ a double volume allowance shall be provided for each gang or yoke the device must use. A range or dryer receptacle could count as one device if installed in a roomier 4″ square box with a single-gang mud ring. A 2 1/8″ deep box is best. However, if this larger receptacle were to be installed in a two-gang box, the double allowance would apply to each of the gangs. These devices are designed to fit under a single yoke with a mud ring or an appropriate two-gang box.

The *NEC* allows a single volume allowance for the initial four EGC's but after that "*a 1/4 volume allowance shall be made for each additional equipment grounding conductor that enter the box,*" per *Section 314.16(B)(5)*. This volume allowance is based on the largest equipment grounding conductor entering the box.

22.2.1 Calculating Box Fill Using *Table 314.16(A)*

The following examples outline how to calculate the box size with *Table 314.16(A)*.

Example 1: What is the maximum number of 12 AWG conductors permitted in a metal device box that measures $3'' \times 2'' \times 2$-$1/2''$?

To determine the answer, find the row for that box size where it meets the 12 AWG column to determine if this box can have a maximum of five 12 AWG conductors.

Note that the table can also be used to select a box size based on the number of conductors present.

Example 2: What is the minimum size metal device box required for seven 14 AWG conductors?

Following the 14 AWG column down to the device box division where seven conductors are specified shows that a $3'' \times 2'' \times 2$-$3/4''$ is the minimum size box of 14.0 in³. A larger box could also be used if conditions permit.

22.2.2 Calculating Box Fill Using *Table 314.16(B)(1)*

Table 314.16(B)(1) must be used in conjunction with *Table 314.16(A)*. The maximum number of conductors shown in *Table 314.16(A)* is based on the volume allowed for each of the conductors as determined by *Table 314.16(B)(1)*, **Figure 22-5**. This table is used to calculate the minimum box size when different sizes of conductors are used.

Example: What depth 4″ square box is required for six 14 AWG and four 12 AWG conductors?

Table 314.16(B)(1) shows a volume allowance of 2.00 in³ for each 14 AWG and 2.25 in³ for each 12 AWG, resulting in a total minimum volume required of 21 in³.

$$14 \text{ AWG, } 6 \times 2.00 \text{ in}^3 = 12 \text{ in}^3$$
$$12 \text{ AWG, } 4 \times 2.25 \text{ in}^3 = 9 \text{ in}^3$$
$$12 \text{ in}^3 + 9 \text{ in}^3 = 21 \text{ in}^3$$

Then refer to *Table 314.16(A)*, which shows the square 4″ × 1-1/2″ deep box has a minimum volume of 21 in³.

Table 314.16(B)(1) Volume Allowance Required per Conductor

Size of Conductor (AWG)	Free Space Within Box for Each Conductor	
	cm³	in.³
18	24.6	1.50
16	28.7	1.75
14	32.8	2.00
12	36.9	2.25
10	41.0	2.50
8	49.2	3.00
6	81.9	5.00

Figure 22-5. *NEC Table 314.16(B)(1)*, Volume allowance required per conductor.

22.2.3 Calculating Box Fill for Various Types of Electrical Boxes

Junction boxes are often used as a distribution point where no device, such as a switch or receptacle, is installed. These boxes contain only wires and conductors. However, most boxes in dwelling units are used to house devices or support luminaires. Calculating the box size under these conditions is a bit more complicated because the electrician must consider other items besides the conductor size. Anything that takes up space in the box must be accounted for. This includes internal cable clamps where the clamping mechanism is inside the box, luminaire mounting studs, and devices as detailed in *Sections 314.16(B)(2)–(5)*. The requirements are summarized in **Figure 22-6**.

Items Contributing to Box Fill Calculations

Items in the Box	Volume Allowance	Based upon [refer to NEC Table 310.16(B)(1)]
Conductors that originate outside the box	One for each conductor	Actual conductor size
Conductors less than 12″ in length that pass through the box without a splice or connection	One for each conductor	Actual conductor size
Conductors greater than 12″ in length that are looped (for future use) and pass through the box without a splice or connection	Two for each conductor	Actual conductor size
Conductors that originate in the box and do not leave the box (pigtails)	None	Not applicable
Internal cable clamps, one or more	One only	Largest-sized conductor installed in the box
Support Fittings (luminaire stud or hickey)	One for each type present in the box	Largest-sized conductor installed in the box
Devices or equipment supported by a yoke or mounting strap	Two for each yoke or strap	Largest conductor attached to the device or equipment
Equipment grounding conductors, four or fewer	One only	Largest-sized equipment grounding conductor installed in the box
Equipment grounding conductors, additional greater than four	¼ for each	Largest-sized equipment grounding conductor installed in the box

Goodheart-Willcox Publisher

Figure 22-6. Summary of items contributing to box fill calculations.

Note that any number of internal cable clamps, luminaire support fittings, and up to four equipment grounding conductors have only one volume allowance, each based on the largest-sized conductor in the box. Devices or equipment supported by a yoke have two volume allowances based on the largest conductor connected to the device. Also, pigtails that may be used to connect more than one conductor to a terminal screw are not counted, nor is the twist-on wire plug type connector.

Care must be taken when counting the conductors of Type NM cable. For example, for the purposes of box fill calculations, two Type NM 14-2 with ground cables is a total of five conductors—the two white and two black insulated conductors each count once, and the two bare equipment grounding conductors count once for both present.

Example 1: What size metal device box is required for a duplex receptacle and two Type NM 12-2 with ground cables?

The box contains two internal cable clamps, as shown in **Figure 22-7.**

$$12 \text{ AWG insulated conductors: } 4 \times 2.25 = 9 \text{ in}^3$$
$$\text{Internal clamps: } 1 \times 2.25 = 2.25 \text{ in}^3$$
$$\text{Receptacle: } 2 \times 2.25 = 4.5 \text{ in}^3$$
$$\text{Bare equipment grounds: } 1 \times 2.25 = 2.25 \text{ in}^3$$
$$9 + 2.25 + 4.5 + 2.25 = 18.00 \text{ in}^3$$

The calculation results in a minimum capacity of 18.00 in³, which requires a device box sized 3″ × 2″ × 3-1/2″.

Recall from Chapter 7, *Electrical Boxes*, how metal device boxes can be ganged to accommodate more than one device.

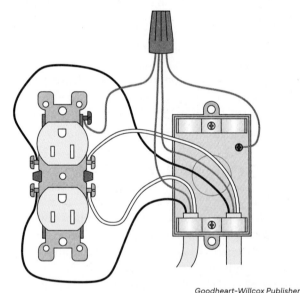

Goodheart-Willcox Publisher

Figure 22-7. Box fill example with a duplex receptacle connected to 12 AWG conductors of Type NM cable.

When calculating the box fill for multiganged metal boxes, the minimum volume listed in *Table 314.16(A)* is multiplied by the number of boxes used. The following example illustrates this point.

Example 2: What size metal device boxes are required to be ganged together containing one single-pole switch and one duplex receptacle as shown in **Figure 22-8**?

The switch is connected to a Type NM 14-2 with ground wire, and the receptacle is connected to two Type NM 12-2

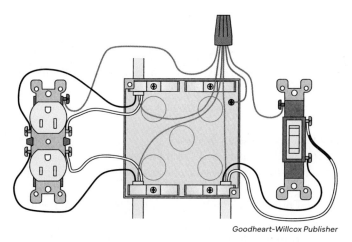

Goodheart-Willcox Publisher

Figure 22-8. Box fill example with a duplex receptacle connected to 12 AWG conductors and a single-pole switch connected to two 14 AWG conductors.

with ground wires. Each metal box will contain two internal cable clamps.

$$14 \text{ AWG insulated conductors: } 2 \times 2.00 = 4.00 \text{ in}^3$$
$$12 \text{ AWG insulated conductors: } 4 \times 2.25 = 9.00 \text{ in}^3$$
$$\text{Internal clamps: } 1 \times 2.25 = 2.25 \text{ in}^3$$
$$\text{Receptacle: } 2 \times 2.25 = 4.50 \text{ in}^3$$
$$\text{Switch: } 2 \times 2.00 = 4.00 \text{ in}^3$$
$$\text{Bare equipment grounds: } 1 \times 2.25 = 2.25 \text{ in}^3$$
$$4.00 + 9.00 + 2.25 + 4.50 + 4.00 + 2.25 = 26.00 \text{ in}^3$$

The minimum capacity required for this example is 26.00 in³. Two boxes sized 3″ × 2″ × 2 3/4″ would provide 14 in³ each for a total of 28 in³.

An *NEC* table for common size non-metallic boxes does not exist. Non-metallic box manufacturers make many different sizes and types. Single-gang non-metallic nail-on boxes range in size from 18 cubic inch up to 24 cubic inch.

Example 3: A 14-3 NM cable is a home run to a single-gang non-metallic device box on the second-floor hallway of a single-family dwelling. Two 14-2 cables and a duplex receptacle are also to be installed. What is the minimum size non-metallic box that could be used?

$$14 \text{ AWG insulated conductors: } 7 \times 2.00 = 14 \text{ in}^3$$
$$\text{Receptacle: } 2 \times 2.00 = 4.0 \text{ in}^3$$
$$\text{Bare equipment grounds: } 1 \times 2.00 = 2.00 \text{ in}^3$$
$$14.00 + 4.00 + 2.00 = 20.00 \text{ in}^3$$

A 20.3 cubic inch or 20.5 cubic inch box will work.

Example 4: A 12-3 NM cable is a home run to a single-gang non-metallic box beneath the kitchen sink of a single-family dwelling. In addition to the 12-3 cable, one 12-2 cable and a GFCI receptacle is to be installed. What is the minimum size non-metallic box that could be used?

$$12 \text{ AWG insulated conductors: } 5 \times 2.25 = 11.25 \text{ in}^3$$
$$\text{Receptacle: } 2 \times 2.25 = 4.50 \text{ in}^3$$
$$\text{Bare equipment grounds: } 1 \times 2.25 = 2.25 \text{ in}^3$$
$$11.25 + 4.50 + 2.25 = 18.00 \text{ in}^3$$

An 18 cubic inch box is allowable per the *NEC*. However, it would be difficult to install a deep GFCI and the required wire connections for the conductors. A larger box would result in a quicker and easier installation.

Example 5: Two 14-2 NM cables are installed in a switch box with two 14-3 NM cables to feed two separate ceiling fan and light combinations. What is the minimum size two-gang non-metallic box that could be used? Each of the two switches is designed to provide power to a fan/light combo.

$$14 \text{ AWG insulated conductors: } 10 \times 2.00 = 20 \text{ in}^3$$
$$2 \text{ switches: } 4 \times 2.00 = 8.0 \text{ in}^3$$
$$\text{Bare equipment grounds: } 1 \times 2.00 = 2.00 \text{ in}^3$$
$$20.00 + 8.00 + 2.00 = 30.00 \text{ in}^3$$

This is the minimum required. A common size two-gang nail-on non-metallic box is 35 in³.

Example 6: One 14-2 NM cable is installed in a ceiling box feeding a ceiling fan with a light. What is the minimum size lighting device box that could be used?

$$14 \text{ AWG insulated conductors: } 3 \times 2.00 = 6.00 \text{ in}^3$$
$$\text{Bare equipment grounds: } 1 \times 2.00 = 2.00 \text{ in}^3$$
$$\text{Four 16 AWG fixture wires with}$$
$$\text{a ground from a domed canopy}$$
$$\text{from the fan/light: } 0 \times 2.00 = 0.00 \text{ in}^3$$
$$6.00 + 2.00 = 8.00 \text{ in}^3$$

The 14 AWG wires will not occupy much space, but more needs to be considered. *Section 314.16(B)(1) Exception* mentions canopies used to enclose wires from a domed luminaire or a similar canopy, such as a fan. If there are four conductors or less, and they are smaller than a 14 AWG, they do not need to be counted. Also, the EGC from the fan/light does not need to be counted. However, if the wires dropping from the fan/light are 14 AWG or larger and they exceed four, they would need to be counted.

Summary

- Proper installation of the nail-on box will ensure the installed device is flush with the finished wall surface. All non-metallic nail-on boxes have stand-off ridges that position the box forward from the framing stud to which it is nailed. Boxes adjustable in depth to match the finished surface are also available.
- Device boxes and junction boxes have maximum fill limitations and must be sized to provide adequate space for the conductors and devices they contain.
- *Section 314.16* has provisions for calculating the maximum box fill for a given application.
- The volume of metal boxes is shown in *Table 314.16(A)*. The *NEC* does not have a dedicated table indicating the volume of nonmetallic boxes. Nonmetallic boxes are required to have their volumes durably and legibly marked by the manufacturer.
- *Table 314.16(A)* lists the box's minimum volume and the maximum number of conductors allowed in standard metal box sizes if all the conductors are of the same size.
- *Table 314.16(B)(1)* is used to calculate the minimum box size when different sizes of conductors are used. This table is used for both metal and nonmetallic boxes.
- The volume allowances for items in a device or junction box are detailed in *Sections 314.16(B)(2)–(5)*.
- Conductors that originate outside the box and terminate inside the box count one volume allowance based on the conductor's actual size.
- Conductors that originate in the box and don't leave the box, like pigtails, have no volume allowance when computing box fill.

Know and Understand

1. In a wall finished with non-combustible material, the front edge of the box should be recessed no more than _____ finished wall.
   - A. 1/8″
   - B. 1/4″
   - C. 1/2″
   - D. 3/4″

2. What is the maximum number of 12 AWG conductors permitted in a square 4″ × 2-1/8″ metal box?
   - A. 8
   - B. 9
   - C. 13
   - D. 15

3. What depth is required for six 14 AWG conductors in a 2″ × 3″ metal device box?
   - A. 1—1/2″
   - B. 2″
   - C. 2—1/4″
   - D. 2—1/2″

4. When computing box fill, what is the cubic inch volume allowance required for a 10 AWG conductor?
   - A. 2.25 in^3
   - B. 2.50 in^3
   - C. 3.00 in^3
   - D. 5.00 in^3

5. When computing box fill, which of the following items do not have volume allowances in the device box?
   - A. Conductors that originate outside the box and terminate inside the box.
   - B. Conductors greater than 12″ in length that are looped or pass through the box without a splice or connection.
   - C. Devices or equipment that are supported by a yoke or mounting strap.
   - D. Conductors that originate in the box and do not leave the box.

Apply and Analyze

1. What is the first stage in rough-in?
2. Describe the purpose of the stand-off tabs on a nonmetallic outlet box.
3. A two-gang non-metallic box is stamped with a cubic inch capacity of 35 in^3. Is this box of sufficient capacity to house two NM cables Type 14-2 with ground, two NM cable Type 14-3 with ground, and two three-way switches?
4. For the purposes of box fill calculations, what is the volume allowance for eight equipment grounding conductors if four are 14 AWG and four are 12 AWG?

Critical Thinking

1. What are the consequences of using an improperly sized box?

Conductor Routing and Installation

ARENA Creative/Shutterstock.com

SPARKING DISCUSSION

Why is it necessary to understand the *NEC* requirements for conductor installation?

Introduction

After the outlet boxes are in place, the next step of the rough-in is to determine the circuit layout and to drill holes in the framing members for cable installation. This chapter serves to apply foundational knowledge of electrical conductors, as well as AWG sizing and ampacity, introduced in Chapter 8, *Electrical Conductors and Connectors*, to the installation of electrical conductors during the rough-in stage.

While this chapter will focus exclusively on installing Type NM cable, the primary type used in residential wiring, some installations will require EMT, MC, or AC cable. This will typically occur if a local code prohibits the use of NM cable or if the building construction type is I or II, as referenced in *Annex E* of the *NEC*. It is key to understand the *Code* and its guidelines for using and installing NM cable.

23.1 Circuit Layout

Electricians must understand the circuit layout of the residence they are working at and identify the location to drill holes. Before drilling, it is best practice to install all the required boxes, determine the location of the home runs, and then divide the boxes fed from each home run into its own circuit. Recall that a home run is the length of conductor from the panel to the first outlet in the room. Make sure not to accidentally tie one home run and the boxes on that branch circuit with another home run and the boxes on it.

When possible, it is best to drill holes at a consistent height, typically at waist height and at least 1-1/4″ from the edge of the joist or stud. This is done between device boxes, including around corners. Every dwelling constructed with wood studs will require many drilled holes for the wiring. Doing this at waist height is easier than bending or reaching to drill. It is easier to pull wire while standing on the ground as opposed to working off a ladder. Sometimes drilling will be required in ceiling joists. Typically, this is the case for wires, luminaires, and smoke detectors installed in basement ceilings and ceiling joists in the upper floors.

Section 300.4(A)(1) requires the edge of the hole to be no less than 1-1/4″ from the nearest edge of the framing member. When drilling into a 2 × 4 stud with an outside nominal dimension of 3-1/2″, the largest wood drill bit that can be comfortably used without installing nail plates is 7/8″. A 1-1/8″ bit is used in larger studs and joists to allow for the installation of more cables. If this distance cannot be maintained, or if the framing member is notched rather than drilled, *Section 300.4(A)(2)* states to use a **nail plate**, or steel plate, that is at least 1/16″ thick, to protect the cables from nails or screws, **Figure 23-1**.

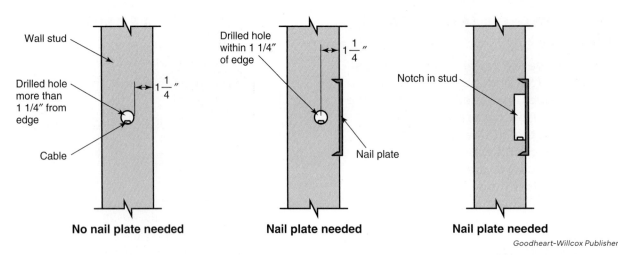

No nail plate needed   **Nail plate needed**   **Nail plate needed**

Goodheart-Willcox Publisher

Figure 23-1. Metal nail plates not less than 1/16″ are required where the cable is less than 1-1/4″ from the edge of the stud. The nail plates prevent screws and nails from damaging the conductors.

Drywall installers should use 1-1/4″ screws. If this is the case, the length of a screw should never reach the depth to penetrate the NM cable. However, if the hole is drilled too close to the edge of a stud or joist, the likelihood of drilling through the wire will increase. This can cause a short circuit or a ground fault. This will probably not be discovered until after the circuit is initially energized. Once drywall is installed, it is not an easy task to find or replace a damaged cable.

23.2 Installing and Securing the Cable

After all holes are drilled, you can then prepare for cable installation. *Section 110.12* requires all work to be executed in a neat manner and enforces that one must take pride in the installation of their work. As it relates to cable installation, this means competent electricians should:

- Run cables such that they are parallel and perpendicular to the framing members.
- Properly secure and support all cables that are used.
- Ensure all equipment is securely fastened and is installed level and plumb.
- Try to avoid twists in cables laying on a flat surface.

Select the correct cable for each portion of the circuits. *Section 210.11(C)* requires that the installed branch circuits are 15 amp, with 14 AWG conductors, and 20 amp with 12 AWG conductors. The 15-amp circuits will generally cover the branch circuits for lighting and receptacles in the living room, bedrooms, recreation room, hallways, and foyers. Required 20-amp circuits include those installed for bathrooms, kitchen and laundry areas, and the garage.

23.2.1 Type NM Cable Classifications

The requirements for Type NM cable use and installation are covered in *Article 334. Section 334.2* defines NM cable as, "*a factory assembly of two or more insulated conductors enclosed within an overall nonmetallic jacket.*" There are additional classifications for the types of non-metallic-sheathed cable:

- **Type NM.** Regular, everyday NM cable. Commonly referred to as *Romex*.
- **Type NMC.** NM cable with a corrosion-resistant sheathing.
- **Type NMS.** Cable that includes signaling, data, and communication conductors in addition to the insulated power conductors.

Neither Type NMC nor Type NMS are commercially available. Type NM cable is generally permitted in one- and two-family dwellings, including attached or detached garages, storage buildings, and multi-family dwellings of wood construction. Some jurisdictions may not allow this as a wiring method. Check with your local inspector for special rules in your area. It can be installed exposed or concealed in normally dry locations, including being fished through the voids in masonry block walls. Type NM and Type NMS cables cannot be embedded in masonry, concrete, or plaster.

23.2.2 Type NM Cable Installation

For new installations, NM cable is sold in 1000′ reels and 250′ coils. The 250′ coils are usually contained in a plastic wrap. When using a 250′ coil, it is best to loop off a needed length of wire from the outside of the coil and walk the loop out to help flatten the conductor and ease the installation. This method will result in fewer twists in the wire when stapling on a flat surface. Another option is to invest in a wire spool dispenser, **Figure 23-2.**

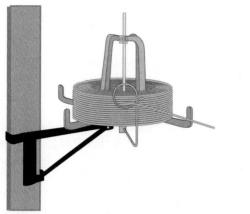

Goodheart-Willcox Publisher

Figure 23-2. An example of a dispenser for electrical cable.

When installing and estimating cable length, be sure to label each home run at the service panel end and at the outlet box end. Recall that a home run is the length of cable from the distribution panel to the first outlet box in the circuit. Use a permanent marker on the outer sheathing on the NM cable. The label might read, "living room and front foyer," or "north bedrooms." A simple labeling of "home run" is sufficient at the device box at the end of the pull of cable.

Before securing the cable, estimate the cable length needed. Estimating the amount of wire and cutting it from the reel or coil before installing it can lead to wasted cable if the estimated length is too long or too short. At the distribution panel, pull enough extra cable to make up all the required connections. A length of cable equal to the length of the panel is generally sufficient. Adding a little bit more than this is a good precaution to ensure adequate cable length.

When installing cable, you must understand the *NEC* requirements related to this task. *Section 334.30* requires NM cable to be supported by staples, cable ties, straps, or hangers at intervals no greater than 4-1/2″. They must be secured within 12″ of every cable entry into an outlet box, junction box, cabinet, or fitting. *Section 314.17(B)(2) Exception* reduces this distance to 8″ for a single-gang nonmetallic box if there is no clamping mechanism at the box.

The general 12″ distance for securing a conductor from a box requires a clamping method per *Section 300.17(B)(2)* at the point where the cable enters. Also, at least 1/4″ of the sheath of the cable must enter the box. Horizontal runs of cable through bored holes or notches are considered supported and secured per *Section 334.30(A)*, where the support does not exceed 4-1/2″. *Section 334.30* states that flat cables are not to be stapled on the edge, **Figure 23-3**.

Take extra caution to not damage the conductors by driving the staple too tight into the wood. The cables should be snug under the staple.

23.2.3 Type NM Cable Runs

An exposed run of cable is cable that is not covered by a finished wall and remains visible. This causes the cable to be subject to physical damage. Exposed cable must be protected by metallic conduit or by PVC with a thicker wall, such as Schedule 80 PVC. This type of PVC is identified for areas that could experience physical damage. Schedule 40 PVC is not rated for this. Where the exposed cable emerges from the floor, the protection shall extend no less than 6″ above the floor per *Section 334.15(B)*, **Figure 23-4**. Protection may extend higher if needed.

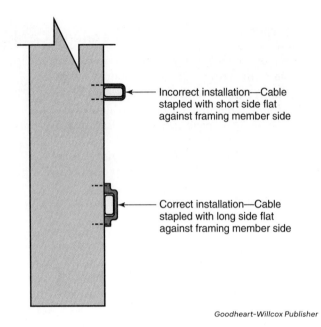

Goodheart-Willcox Publisher

Figure 23-3. NM cable must be stapled so its long side is flat against the framing member.

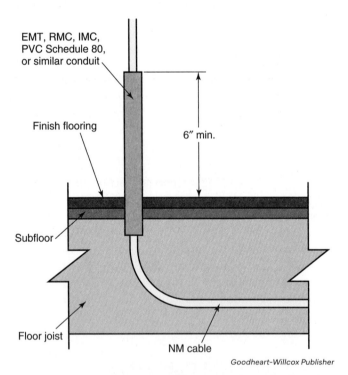

Goodheart-Willcox Publisher

Figure 23-4. When exposed cable passes through a floor, it must be protected by conduit that extends at least 6″ above the floor.

Exposed runs of NM cable are allowed in one- and two-family dwelling in a drop ceiling to feed luminaires or smoke detectors installed in the ceiling.

In unfinished basements and crawlspaces where the cables are installed at right angles to the joists, cables no smaller than 6 AWG two-conductor cable (6-2 NM with ground) or

8 AWG three-conductor cable (8-3 NM with ground) are permitted to be secured directly to the bottom edges of joists. Smaller cables must be installed in bored holes or on running boards, **Figure 23-5**. A *running board* is a board attached to the bottom edges of joists that provides an attachment surface for NM cable. Any size cable can be installed on the face of the joist if properly secured and supported.

Cables run in accessible attics must comply with *Section 334.23* and its reference to *Section 320.23*. *Section 320.23(1)* requires cable protection by using guard strips or running boards if the cable is installed on the tops of floor joists or within 7′ above the floor or floor joists on the face of rafters or studs. If the attic space has no permanent stairs or ladders, the cables shall be protected by guard strips within 6″ of the scuttle hole access, **Figure 23-6**. No protection is required for cables installed parallel to the sides or on the face of the floor joist or framing member per *Section 320.23(2)*. Cables run through bored holes in attic joists, rafters, and studs are considered protected. Do not drill pre-manufactured rafters without permission from the manufacturer. The structural integrity could be compromised by doing this. This could result in a very expensive problem to fix.

To prevent damage to NM cables, *Section 334.24* requires the bending radius shall have a curve not less than five times the diameter of the cable. NM cable can be folded over, bent, and stretched to a point of damaging the internal copper wire.

NM Cable in Unfinished Basements

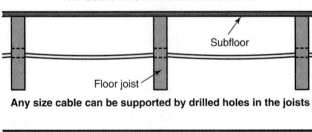

Any size cable can be supported by drilled holes in the joists

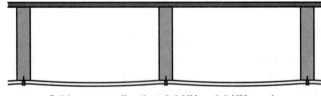

Cables no smaller than 6-2 NM or 8-3 NM can be attached to the bottom edges of the joists

Any size cable can be attached to a running board attached to the bottom edges of the joists

Goodheart-Willcox Publisher

Figure 23-5. The support method used in an unfinished basement or crawlspace depends in part on the size of the NM cable.

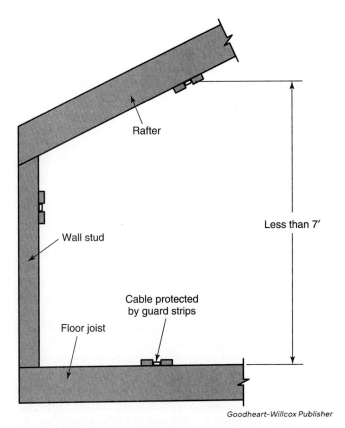

Figure 23-6. When cable is installed perpendicular to floor joists, wall studs, or rafters in an attic area, guard strips are required in areas within 7' of the floor.

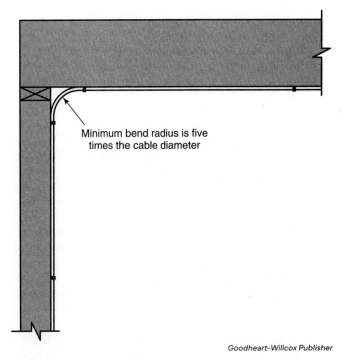

Goodheart-Willcox Publisher

Figure 23-7. The required bending radius of Type NM cable is five times the cable diameter.

The *NEC* addresses this by limiting how tight of a bend can be made while still maintaining the integrity of the wire in the NM cable. A typical 12-2 NM cable has a diameter of about 1/2″, which results in a bending radius of 2-1/2″, or a little more than the diameter of a soda can, **Figure 23-7.**

An example of a typical *NEC* violation is leaving a horizontal cable run through a drilled hole and dropping down to a device box. If a loop is not made as the cable changes direction, there is a good chance a *NEC* violation will occur. It is possible the cable could be bent to the point the wires inside could be weakened, creating a potential failure point in the circuit.

23.3 Conductor Prepping for Device Installation

Conductors must be prepped at the outlet box during the rough-in stage. This involves removing the outer sheathing, stripping, and preparing the individual conductors for connection to a device, such as a receptacle or a switch. Completing this task will allow for a smoother trim-out, and it also will alert the installer to any missing runs of cable. It is much easier to add a length of cable during rough-in than to discover during trim-out that a cable run is missing.

When removing the outer sheathing, be sure to leave at least 1/4″ of sheathing extending into the box beyond any clamping mechanism per *Section 314.17(B)–(C)*. *Section 300.14* requires that all conductors extend at least 6″ beyond the sheathing.

Additionally, *Section 300.14* states where the opening of an outlet box is less than 8″ in any dimension, the conductors must be long enough to extend at least 3″ outside the opening. The minimum conductor length of 6″ might not be long enough with the use of extension boxes. The extra length ensures that there is enough conductor for the make-up of joints and the installation of the devices, **Figure 23-8.**

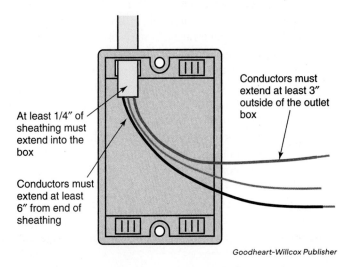

Goodheart-Willcox Publisher

Figure 23-8. Representation of allowable free conductors required at an outlet box.

PROCEDURE

Installing NM Cables in a Device Box

Complete the following steps to properly install receptacles and switches, also referred to as devices, in a junction box during the rough-in stage: "*Strip, stuff, staple, and make electrical connections.*"

1. **Strip**. After the NM cable is pulled and is long enough to reach the device box, the outer sheath of the cable must be removed to allow enough of the conductor portion of the cable to enter the box with at least 1/4" of the sheath exposed in the device box. Many tools are made for exclusively striping the sheath of the NM cable.

2. **Stuff**. After the cable sheath is stripped, the wire must be inserted into the device box. Remember to have at least 1/4" of the sheath extend inside the device box. It is a good practice to leave at least 8" of free conductor, sometimes more for larger boxes, to trim the wires to needed lengths and connect wires under a clamping device.

3. **Staple**. The NM cable must be stapled within 12" of a device or junction box. Single gang non-metallic boxes without an internal or external clamping method must have the cable stapled within 8". Two-gang boxes will be provided with an internal clamping method or a connector with a locknut that is considered external to the box. According to *Section 300.4(D)* cables cannot be stapled closer than 1-1/4" from the outside edge of the stud or joist. Manufacturers make items that will maintain this distance.

4. **Make electrical connections**. The ungrounded or *hot* conductors from the same *line*, L1-black or L2-red, must be jointed together, black to black and, if installed, red to red. A short piece of wire, usually 6" to 8", called a *tail*, must leave this joint to go to every receptacle and each switch that needs one. Some three-way switches do not require a hot wire but only need the two travelers to the other three-way switch and a wire to the device, usually a light or fan called a switch leg. All the grounded conductors, commonly called *neutral conductors*, must be jointed together. In residential wiring, these conductors will be white. Each receptacle must receive a white tail. Last, terminate all the EGCs together and provide a tail for every device in the junction box. In NM cable, this will be the bare copper conductor. The EGC may also be a wire with green insulation. This is the case for individual conductors in raceways.

5. **Neatly fold and bend the conductors back in the box**. Make sure to try to push them back to the point when the drywall is installed and openings are cut to accommodate the device boxes the electrical wires are not damaged. Identify each switch leg. If more than one switch is installed, identify each switch leg differently. An example would be a twist in the wire or two twists if there are two switches installed and so on.

Summary

- Consecutive holes should all be bored at the same height above the floor, making it much easier to pull the cable through the holes.
- The *NEC* requires the edge of the hole to be no less than 1-1/4" from the nearest edge of the framing member. If this distance cannot be maintained, or if the framing member is notched rather than drilled, a method of protection of the cables from the penetration of nails or screws shall be provided by using a steel plate that is at least 1/16" thick.
- The length of cable from the distribution panel to the first outlet box in the circuit is the home run.
- The *NEC* requires NM cable to be supported by staples, cable ties, straps, or hangers at intervals not to exceed 4-1/2" and to be secured within 12" of every cable entry into an outlet box, junction box, cabinet, or fitting. The exception being for non-metallic boxes that do not have an internal or external wire clamping method. The NM cables must be supported within 8" of this entry point.
- Horizontal runs of cable through bored holes or notches shall be considered to be supported and secured where the support does not exceed 4-1/2".
- The requirements concerning the use and installation of Type NM cable are covered in *NEC Article 334*.
- Unless prohibited by local codes, type NM cable is permitted in one and two-family dwellings, including attached or detached garages and storage buildings. It can be installed exposed or concealed in normally dry locations, including being fished through the voids in masonry block walls.
- Exposed cables that are subject to physical damage shall be protected by metallic conduit or by Schedule 80 PVC.

- The *NEC* requires the protection of cables by the use of guard strips or running boards if the cable is installed on the tops of floor joists or within 7″ above the floor or floor joists on the face of rafters or studs. If the attic space has no permanent stairs or ladders, the cables shall be protected by guard strips within 6″ of the scuttle hole access.
- To prevent damage to the conductors, the bending radius of Type NM cable shall have a curve not less than five times the diameter of the cable.
- Prepping the conductors during the rough-in makes for a quicker trim-out and alerts the installer to any missing runs of cable.

Know and Understand

1. Who determines the routing of the cables during the rough-in stage?
   A. The electrician
   B. The homeowner
   C. The architect
   D. The general contractor

2. Type NM cable shall be supported and secured by approved means at intervals not exceeding _____.
   A. 1″
   B. 4.5″
   C. 6″
   D. 12″

3. Which Article of the *NEC* addresses the installation of Type NM cable?
   A. *Article 332*
   B. *Article 340*
   C. *Article 338*
   D. *Article 334*

4. *True or False?* Type NM cable is permitted to be embedded in masonry, concrete, or plaster.

5. What is the first step in the process of installing cables?
   A. Run cables to be parallel and perpendicular to the framing members.
   B. Label home run conductors.
   C. Properly secure and support all cables that are used.
   D. Drill holes in framing members at waist height.

6. The smallest cable allowed to be installed directly to the bottom of the floor joists in an unfinished basement or crawlspace is _____.
   A. 8-3
   B. 6-3
   C. 10-3
   D. 14-2

7. *True or False?* A length of wood installed across joists or rafters to protect cables from physical damage is known as a running board.

8. The bending radius of Type NM cable shall not be less than _____ the diameter of the cable.
   A. five
   B. eight
   C. ten
   D. twelve

9. The *NEC* requires at least _____ inches of free conductor to extend beyond the opening of the outlet box.
   A. two
   B. three
   C. five
   D. six

Apply and Analyze

1. What determines the number of receptacle outlets required in a dwelling? Given a 2400-square-foot dwelling and knowing that each square foot is calculated at 3 VA (volt-amps), what is the minimum number of 15-amp branch circuits required? If using a 20-amp branch circuit, how many circuits would be required?

2. When the edge of a bored hole is less than 1-1/4″ from the edge of the stud, what is to be done to protect a cable from physical damage from a nail or screw?

3. Explain why consecutive holes drilled horizontally in framing studs should all be at the same height.

4. What are the requirements for securing a cable at a single-gang nonmetallic device box where there is no clamping mechanism at the box, and where can this be found in the *NEC*?

5. List some of the methods of protecting exposed Type NM cable from physical damage. Where are these requirements found in the *NEC*?

6. Explain the benefits of prepping the conductors for device installation during the rough-in phase of construction.

Critical Thinking

1. The *NEC* requires all work to be performed in a neat manner. Explain what this means in relation to a rough-in and provide examples.

2. Discuss different ways to accomplish the "*stripping, stuffing, and securing*" of NM cable. If possible, practice different methods of terminating cable and their conductors and stuffing the box to help make adding of switches and receptacles (the trimming) go smoothly.

SECTION **8**

Service Installation

Noel V. Baebler/Shutterstock.com

CHAPTER 24
Service Equipment

CHAPTER 25
Service Installation

The heart of the residential electrical system is the service. Therefore, all the electricity that powers a dwelling must flow through the service equipment. An understanding of the components that make up a service, as well as the *NEC* requirements for service installations, is critical for the residential electrician.

Chapter 24, *Service Equipment*, introduces the basics of electrical service equipment, including the meter socket and service disconnecting means. Also discussed are the differences between three- and four-wire services and grounding residential electrical systems. Chapter 25, *Service Installation*, covers the *NEC* requirements of the service installation, overhead and underground service connections, and rules for grounding and bonding residential electrical systems.

Service Equipment

SPARKING DISCUSSION

Why must the residential electrician understand the components that make up the electrical service for a dwelling?

LEARNING OBJECTIVES

After completing this chapter, you will be able to:

- Describe the purpose of the electrical service.
- Identify the two major components of the electrical service.
- Explain the difference between service-entrance conductors and feeder conductors.
- Contrast three-wire and four-wire services.
- Explain the two reasons for grounding electrical systems.
- Identify the different types of grounding electrodes and their installation requirements.
- Explain the purpose of the grounding electrode conductor.
- Describe three common methods of residential system grounding.

TECHNICAL TERMS

combination meter socket/disconnect
concrete-encased electrode
feeder conductor
four-wire service
grounding electrode
grounding electrode conductor
main bonding jumper
meter socket
service
service-entrance conductor
supplemental grounding electrode
three-wire service
underground meter socket
utility meter

Introduction

The *NEC* defines a ***service*** as "*the conductors and equipment for connecting the serving utility to the wiring system of the premises served.*" An electrical service is an essential component for all electrical systems and is comprised of service equipment and conductors. Service equipment provides the pathway for power to be distributed from the electrical utility to the customer's distribution panelboard and branch circuits.

The first component of service equipment in a residential electrical service is the underground or overhead service conductor. These conductors bring the utility connection into the utility metering equipment, which is usually installed on an exterior wall of the dwelling. Once the overhead or underground service conductors are routed through the utility meter, they become service-entrance conductors and terminate in the service disconnecting means. The service disconnecting means provides a convenient place for emergency responders and homeowners to de-energize the dwelling's electrical supply. Once the conductors have passed through the service disconnecting means, they transition into feeder conductors that supply power to the distribution panelboard. The panelboard bussing and circuit breakers then supply power to the loads through the branch circuit conductors.

This chapter will discuss the equipment and conductors that make up an electrical service, as well as the installation methods and requirements by the *NEC* for providing such service.

24.1 Meter Socket and Service Disconnecting Means

Two major components of the service are the utility meter socket and the service disconnecting means. A ***utility meter*** records the amount of electrical usage associated with a dwelling and is used as the basis for the customer's utility bill. The utility owns and provides the meter, but in most services, the customer owns the meter socket, the enclosure the meter is installed in.

Meter sockets come in several configurations, **Figure 24-1.** An overhead service often contains a simple *meter socket* that encloses the service conductors and has a set of terminals for the termination of the service conductors and service-entrance conductors. The meter socket provides a connection point for the meter to be plugged into the socket, which allows the current to flow from the service conductor terminals through the meter into the service-entrance terminals, eventually powering the loads within the dwelling. **Underground meter sockets** are often in a pedestal configuration that provides a means to route the underground utility wiring from below grade to the meter terminals.

A meter socket and service disconnecting means can be built from individual components or be provided as a preassembled combination meter socket/disconnect. The **combination meter socket/disconnects** are more expensive but are much faster to install. They also have a neater finished appearance than one built from individual components. Note this textbook primarily focuses on services built with the combination meter socket/disconnect.

Recall that service disconnecting means is a piece of service equipment that disconnects the dwelling from the utility supply. A service disconnecting means can be located outside with the metering equipment, or as a main breaker in the distribution panelboard. *Section 230.70(A)(1) requires the disconnecting means to be installed in "a readily accessible location either outside a building or structure or inside nearest the point of entrance of the service conductors."* The nearest point of entrance is directly on the inside of the wall where the conductors emerge into the structure, **Figure 24-2.** This requirement limits the length of the conductor that is permitted to be run on the inside of the dwelling. Service conductors and service-entrance conductors are protected from short circuit overcurrents by the utility at the utility transformer. It is typical for a utility transformer to serve more than one dwelling, so the overcurrent protection is often higher than it would be for a single dwelling in accordance with the *NEC*. A service-entrance conductor that is run directly from the

Simple meter socket

Underground meter socket (with pedestal)

Combination meter socket and disconnect

Goodheart-Willcox Publisher

Figure 24-1. A variety of meter socket configurations including a simple meter socket, underground meter socket, and combination meter socket/disconnect.

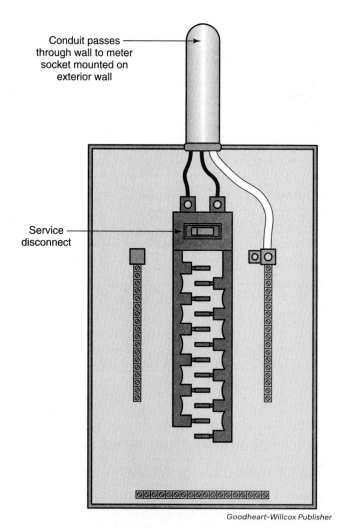

Conduit passes through wall to meter socket mounted on exterior wall

Service disconnect

Goodheart-Willcox Publisher

Figure 24-2. If not located outside with the metering equipment, the *NEC* requires the service disconnecting means to be located nearest the point of entry of the service-entrance conductors. In this illustration, the disconnecting means is in a panel located in the basement.

meter to the main breaker in the distribution panel has limited overcurrent protection. This can become dangerous if the conductors are cut or pierced by accident, resulting in a fault that could lead to significant damage, like a fire. Some municipalities limit the length of the inside run of service-entrance conductors for this very reason.

Locating the service disconnecting means on the outside of the house with the metering equipment also gives firefighters a convenient location to shut off electrical power in the event of a fire. Firefighters do not spray water on a building that is on fire if it is still energized. If they cannot readily disconnect power, emergency responders must call for the utility to disconnect the power at the transformer, which can consume critical time.

24.2 Service-Entrance Conductors and Feeder Conductors

Service-entrance conductors are the conductors between the terminals of the service equipment—the metering device and service disconnecting means—and a point typically outside the structure where a connection is made to the utility's overhead service drop or underground service lateral. Service-entrance conductors are a group of three conductors: two insulated ungrounded, hot, conductors and a bare or insulated grounded neutral conductor. This group of conductors does not include a grounding conductor. Service-entrance conductors used in residential wiring are typically Type USE or Type SEU cables or single conductors installed in a conduit.

The service disconnecting means becomes the demarcation point between service-entrance conductors and feeder conductors. The conductors on the utility side of the service disconnecting means are connected to the utility transformer without additional overcurrent protection. Due to the limited overcurrent protection, the installation practices outlined by the *NEC* are more restrictive for service-entrance conductors. The overcurrent protection within the service disconnecting means provides additional protection for the conductors on load side, so the *NEC* installation standards are less restrictive, and the conductors are defined by a new term, feeder conductor.

Service conductors must not be confused with *feeder conductors*, often simply called *feeders*, which are the circuit conductors between the service equipment and the final branch circuit overcurrent protection device. Feeder conductors have four individual conductors: two insulated ungrounded conductors, an insulated grounded neutral conductor, and an insulated or bare equipment grounding conductor. Those used in residential wiring are typically Type SER cable or individual conductors installed in a raceway.

24.3 Three-Wire Services and Four-Wire Services

Distribution panelboards are considered service equipment if the panelboard contains a main circuit breaker that is used as the service disconnecting means. A distribution panelboard that is considered service equipment is fed by service-entrance conductors and is known as a ***three-wire service*** because the service-entrance conductor group does not contain a grounding conductor. The grounded neutral conductor terminal bus bar and the equipment grounding conductor terminal bus bar in the panelboard are bonded together and to the panel enclosure with the main bonding jumper. Based on the *NEC*, a ***main bonding jumper*** is the "*connection between the grounded circuit conductor and the equipment grounding conductor, or the supply-side bonding jumper, or both, at the service.*" In residential wiring, this is usually a green screw or a copper strap that makes the bonding connection between the bus bar and the panel enclosure, **Figure 24-3**. With a

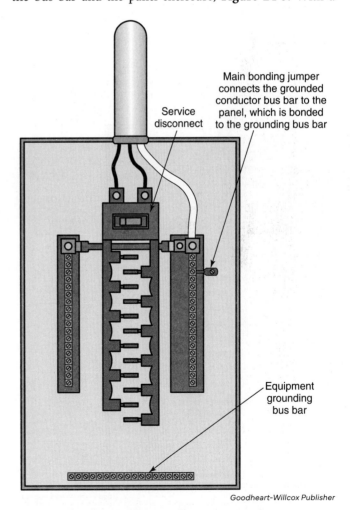

Service disconnect

Main bonding jumper connects the grounded conductor bus bar to the panel, which is bonded to the grounding bus bar

Equipment grounding bus bar

Goodheart-Willcox Publisher

Figure 24-3. A typical three-wire service. Note the grounded neutral bar and the equipment grounding bar are bonded together and to the panel enclosure by means of the main bonding jumper. The service disconnecting means is the main breaker in the distribution panel.

three-wire service, the branch circuit grounded conductors and the branch circuit equipment grounding conductors are intermixed at the terminal bus bars.

A distribution panelboard that does not contain the service disconnecting means is not considered service equipment. The panelboard is fed from a feeder conductor and is known as a **four-wire service** because the feeder contains a fourth conductor, an equipment grounding conductor. *Section 250.24(C)* only allows a main bonding jumper to be installed in one location, the service disconnect. A main bonding jumper is not used in a distribution panel that is fed by a four-conductor feeder because it is located outside at the disconnecting means. This panelboard contains a grounded conductor terminal bus bar that is isolated from the equipment grounding conductor terminal bus bar and panel enclosure by a nonmetallic standoff. The equipment grounding terminal bus bar is attached directly to the panel enclosure, **Figure 24-4**. The branch circuit grounded conductors and the branch circuit equipment grounding conductors are separated and isolated from one another.

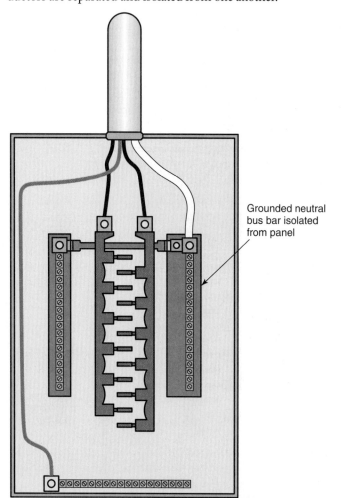

Grounded neutral bus bar isolated from panel

Goodheart-Willcox Publisher

Figure 24-4. A typical four-wire service. Note the grounded neutral bar is isolated from the panel enclosure, and the equipment grounding bus bar is directly connected to the enclosure. The service disconnecting means is located outside with the metering equipment.

24.4 **System Grounding and Grounding Electrodes**

Electrical systems are intentionally connected to the earth. This intentional connection is known as system grounding. *Section 250.4(A)(1)* provides two reasons for the grounding of the electrical system: to limit the voltage that might be imposed by lightning, surges from the utility line, or unintentional contact with higher voltage lines, and to stabilize the voltage to earth during normal operation. Note that system grounding plays no role in the operation of the overcurrent devices such as fuses and circuit breakers.

System grounding is accomplished by making a low-impedance connection to the earth. The electrical system must be connected to a conductive object, known as a **grounding electrode**, that has a direct connection to the earth. The *NEC* identifies several different types of permitted electrodes in *Section 250.52(A)(1)–(8)*. These include the following:

- A metal underground water pipe in direct contact with the earth for at least 10′.
- Any metal support structure in contact with the earth for no less than 10′ vertically.
- A concrete-encased electrode.
- A ground ring made of a bare copper conductor sized 2 AWG or larger in direct contact with the earth at least 20′ deep and encircling the building.
- Rod electrodes at least 5/8″ and pipe electrodes of trade size 3/4 or larger.
- Other listed electrodes shall be permitted.
- Plate electrodes of at least two square feet in direct contact with the earth at least 20′ deep, made of coated ferrous conductive metallic material at least 1/4″ in thickness, or solid, uncoated nonferrous metals at least 0.06″ in thickness.
- Other local underground systems or structures such as piping systems and underground tanks.

A dwelling unit may contain several grounding electrodes as defined by the *NEC*. *Section 250.50* requires that all electrodes present shall be bonded together to form the building's grounding electrode system, **Figure 24-5**. The electrician is not free to pick and choose which electrodes are used to create the system ground; if they are present, they must be connected. Utilizing more electrodes reduces the impedance in the ground connection and provides higher reliability of the grounding system.

24.4.1 **The Grounding Electrode Conductor**

A **grounding electrode conductor** connects the grounded service conductor to the grounding electrode. This conductor can be copper or aluminum but must be copper if it terminates within 18″ of the earth. It must be a continuous

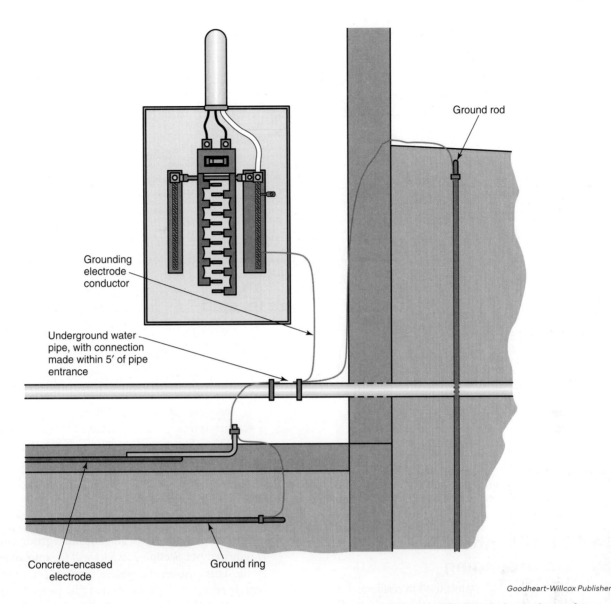

Ground rod

Grounding
electrode
conductor

Underground water
pipe, with connection
made within 5′ of pipe
entrance

Concrete-encased
electrode

Ground ring

Goodheart-Willcox Publisher

Figure 24-5. Examples of grounding electrodes. All grounding electrodes that are present shall be bonded together to form the grounding electrode system.

conductor without a splice or joint unless that splice is by irreversible connection like that made by compression connections or exothermic welding. The grounding electrode conductor is often a single bare conductor but is permitted to be insulated.

The *NEC* does not contain specific color requirements for grounding electrode conductor insulation. If insulated, the insulation is typically green or black. Some authorities having jurisdiction require the insulation of a grounding electrode conductor to be removed in a panelboard where it is terminated or to be identified with green tape. The grounding electrode conductor can be installed in a raceway. If the raceway is made of ferrous metal, the raceway must be bonded to the grounding electrode conductor at each end of the raceway to reduce the choking effect produced by the ferrous metal.

The grounding electrode conductor is sized in accordance with *Table 250.66* and is based on the size of the service-entrance conductors feeding the service. There are several types of grounding electrodes that have reduced grounding electrode conductor sizing requirements due to the current carrying limitations of the electrode. One example of this limitation is the driven ground rod. *Section 250.66(A)* states that the maximum size grounding electrode conductor required to be installed for connection of a rod electrode is 6 AWG copper wire. *Section 250.24(A)(1)* requires the grounding electrode conductor connection to be made at "*any accessible point from the load end of the overhead service conductors, service drop, underground service conductors, or service lateral to the terminal or bus to which the grounded service conductor is connected at the service disconnecting means.*" **Figure 24-6** shows three conventional points where the grounding electrode conductor can be connected to the grounded service conductor.

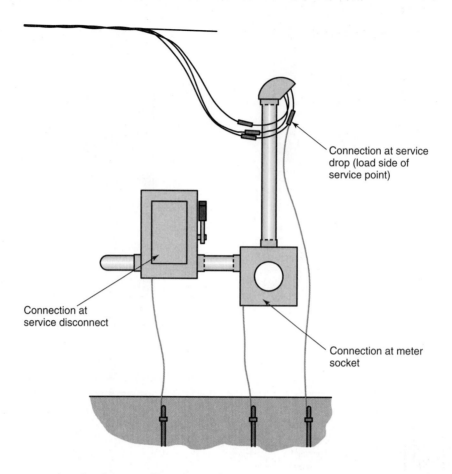

Connection at service drop (load side of service point)

Connection at service disconnect

Connection at meter socket

Goodheart-Willcox Publisher

Figure 24-6. Three common points for the grounding electrode conductor to be connected to the grounded service conductor. The most common is at the meter enclosure or at the service disconnecting means.

24.4.2 Residential System Grounding

The grounding electrodes most used in residential wiring are the underground metal water pipe, the rod electrode, and the concrete-encased electrode. This section will focus on the *NEC* requirements for installing each of these electrodes.

For metal water pipes, an additional grounding electrode is required per *Section 250.53(D)(2)* to supplement the electrode. This ensures the system remains grounded if a water pipe is replaced with a non-metallic pipe. A ***supplemental grounding electrode*** is also required for a single rod, pipe, or plate electrode, per *Section 250.53(A)(2)*. The exception to this is if the single rod, pipe, or plate electrode has a

---

CODE APPLICATION

Sizing a Grounding Electrode

Scenario: An electrician needs to size a typical 200-amp residential service.

Table 250.66: A grounding electrode conductor is sized using this table and is based on the size of the largest ungrounded service-entrance conductor used.

Table 310.16: This table covers the allowable ampacities of insulated conductors.

Section 250.66 (A): This section covers grounding connections to electrodes based on conductor size.

Section 250.66(B): The grounding electrode conductor connection to a concrete-encased grounding electrode does not need to be larger than 4 AWG copper.

Solution: A typical 200-amp residential service requires 4/0 aluminum or 2/0 copper ungrounded conductor based on *Table 310.16*. According to *Table 250.66*, this requires a 4 AWG copper grounding electrode conductor. However, *Section 250.66(A)* allows for no larger than a 6 AWG copper grounding electrode conductor if the connection is to a single or multiple rod, pipe, or plate electrodes, or any combination of these. *Section 250.66(B)* permits the grounding electrode conductor to be no larger than a 4 AWG copper when connected to a concrete-encased electrode.

resistance to earth of 25 ohms or less, and the supplemental electrode shall not be required. In both instances, the supplemental grounding electrode will most likely be a rod-type electrode. To meet the requirements of the *NEC*, multiple rod, pipe, or plate electrodes shall be installed at least 6' apart, **Figure 24-7**.

The installation requirements of a pipe or rod electrode can be found in *Section 250.53(A)(4)*. This electrode must be installed where 8' of length is in direct contact with the earth. If rock or other material is encountered, and this depth cannot be achieved, it is permitted to drive the electrode at an angle of up to 45°. As a last resort, the electrode shall be permitted to be buried horizontally at a depth of at least 30", **Figure 24-8**. Unless protected from physical damage, the upper end of the electrode shall be flush with or below the grade level.

For ***concrete-encased electrodes***, *NEC 250.52(A)(3)* outlines that either 20' of electrically conductive reinforcement bars of at least 1/2" in diameter or a bare copper conductor at least 4 AWG, encased in at least two inches of concrete that is in direct contact with the earth. This electrode is also known as the *Ufer ground*, named after its developer Herbert G. Ufer. A concrete-encased electrode uses the conductivity of concrete, which is superior to most types of soil, especially in dry soil regions, and it provides the best system grounding with the lowest impedance to earth than other types of electrodes. For this reason, some municipalities require them for all new construction. Many times, the

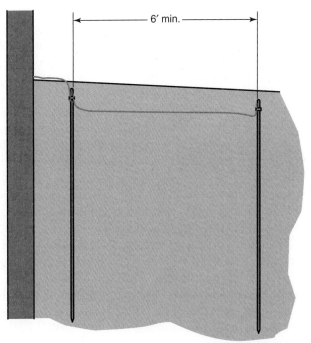

Goodheart-Willcox Publisher

Figure 24-7. Multiple rod-type electrodes must be installed at least 6' apart with 8' in direct contact with the earth.

footing of the home's foundation is used as the concrete-encased electrode, and the concrete installer provides an upturned piece of reinforcement bar for the connection of the grounding electrode conductor.

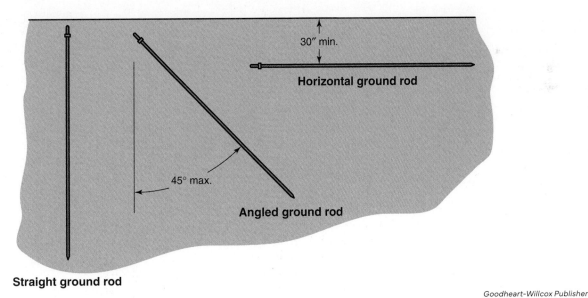

Goodheart-Willcox Publisher

Figure 24-8. Eight feet of electrode shall be in contact with the soil. If the electrode cannot be driven vertically to the proper depth, it shall be permitted to drive it at up to a 45-degree angle or to be buried at a depth not less than 30".

Summary

- The *NEC* defines a service as "*the conductors and equipment connecting the serving utility to the wiring system of the premises served.*" An electrical service is an essential component for all electrical systems and is comprised of service equipment and conductors.
- The two major components of the service are the utility meter socket and the service disconnecting means.
- Service-entrance conductors are defined as the conductors between the terminals of the service equipment (the metering device and disconnecting means) and a point usually outside the building where a connection is made to the overhead service drop or to the underground service laterals.
- A feeder is defined as the circuit conductors between the service equipment and the final branch circuit overcurrent device.
- A distribution panel that is fed by service-entrance conductors is known as a three-wire service. With a three-wire service, the branch circuit grounded conductors and the branch circuit equipment grounding conductors are intermixed at the terminal bus bars.
- A panel that is fed from a feeder is known as a four-wire service. After the service disconnecting means, the branch circuit grounded conductors and the branch circuit equipment grounding conductors are forever separated and isolated from one another.
- The *NEC* provides two reasons for the grounding of the electrical system: to limit the voltage that might be imposed by lightning, surges from the utility line, or unintentional contact with higher voltage lines and to stabilize the voltage to earth during normal operation.
- System grounding is accomplished via the grounding electrode and the grounding electrode conductor.
- The grounding electrodes most used in residential wiring are the underground metal water pipe, the concrete-encased electrode, and the rod electrode.
- The rod or pipe electrode must be installed so that 8′ of length is in direct contact with the earth. A concrete-encased electrode uses the conductivity of concrete, which is superior to most types of soil, especially in dry soil regions.

Know and Understand

1. *True or False?* The *NEC* requires disconnects to be installed on the outside of the residence.
2. The conductors between the terminals of the service equipment and a point where a connection is made to the utility supply are the _____.
   A. feeder conductors
   B. service-entrance conductors
   C. branch circuits
   D. service drop
3. The conductors between the service equipment and the final branch circuit overcurrent device are the _____.
   A. feeder conductors
   B. service-entrance conductors
   C. branch circuit
   D. service drop
4. A feeder conductor is usually of cable type _____.
   A. USE
   B. SEU
   C. SER
   D. UF
5. The _____ connects the grounded circuit conductor to the equipment grounding conductor at the service.
   A. grounding electrode
   B. grounding electrode conductor
   C. ground ring
   D. main bonding jumper
6. The branch circuit grounded conductors and the branch circuit equipment grounding conductors are forever isolated from one another after _____.
   A. the utility meter
   B. the service disconnecting means
   C. the main bonding jumper
   D. both the service disconnecting means and main bonding jumper
7. *True or False?* Electrical systems are grounded to facilitate the operation of the systems overcurrent devices.
8. A rod type grounding electrode shall be at least _____ in diameter and _____ in length.
   A. 3/4″; 6′
   B. 5/8″; 8′
   C. 1/2″; 8′
   D. 1/2″; 6′
9. *True or False?* All grounding electrodes that are present shall be bonded together to form the grounding electrode system.
10. The grounding electrode conductor is sized according to *NEC Table* _____.
    A. *250.24*
    B. *250.122*
    C. *250.66*
    D. *310.16*

11. Where the sole connection of the grounding electrode conductor is to a rod, pipe, or plate electrode, the grounding electrode conductor does not need to be any larger than a _____.
    A. 6 AWG
    B. 4 AWG
    C. 2 AWG
    D. 4/0 AWG
12. A supplemental grounding electrode is not required if the single electrode has a resistance to earth of _____ or less.
    A. 10 ohms
    B. 25 ohms
    C. 50 ohms
    D. 100 ohms
13. Multiple grounding rod or pipe electrodes should be installed so there is a distance of at least _____ between them.
    A. 2′
    B. 3′
    C. 4′
    D. 6′
14. A concrete-encased electrode must have no less than _____ of electrically conductive material encased in no less than two inches of concrete in direct contact with the soil.
    A. 6′
    B. 10′
    C. 20′
    D. 100′

Apply and Analyze

1. Which component of the electrical service records the amount of electrical power used?
2. Why do some municipalities limit the length of the inside run of a three-wire service-entrance conductor?
3. What are the two reasons for grounding an electrical system?
4. What can be done if bedrock or other debris prevents the rod or pipe electrode from being driven vertically into the ground?
5. According to *Table 250.66*, what is the size of the grounding electrode conductor based on?

Critical Thinking

1. Explain the conditions that require the branch circuit equipment grounding conductors and the branch circuit grounded conductors to be isolated from one another.

Service Installation

Viktorus/Shutterstock.com

CHAPTER OUTLINE

25.1 Service Equipment

25.2 Service and Distribution Panelboard Locations

25.3 Overhead Service Clearance Requirements

25.4 Types of Overhead Services

    25.4.1 Cable Mast

    25.4.2 Conduit Mast

    25.4.3 Through-the-Roof Mast

25.5 Underground Services

25.6 Subpanels

25.7 Grounding and Bonding

Introduction

The previous chapter focused on the service equipment required for proper electrical service setup. The proper installation of this equipment is necessary to provide many years of safe and efficient use. It must be securely installed and safe from approach with any energized portions out of reach or enclosed.

Proper service installation ensures that the service is operationally safe and free of hazards, and any fault that might occur is cleared quickly and efficiently. As expected, the *NEC* has many rules concerning the installation of electrical services, which will be the primary focus of this chapter.

25.1 Service Equipment

Electrical service equipment consists of the meter socket and the service disconnecting means. If the service disconnecting means is located in the distribution panelboard, the panelboard is also considered service equipment. When using the combination meter/main disconnect, the distribution panel is connected to the service equipment via a 4-conductor feeder, **Figure 25-1**. When the service disconnecting means is the main breaker of the distribution panelboard, the distribution panel is connected with only three conductors as the grounding conductor would not be installed between the meter socket and the service disconnecting means. This piece of equipment houses the final overcurrent devices that protect the branch circuit conductors. Circuit breakers are required but vary based on the type of circuits. A single-pole circuit breaker is used for 120-volt branch circuits, such as general lighting and receptacles. A double-pole circuit breaker is used for 240-volt circuits, such as electric water heaters and electric ranges.

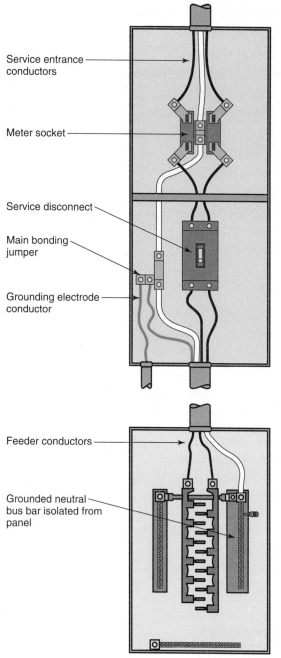

Service entrance conductors

Meter socket

Service disconnect

Main bonding jumper

Grounding electrode conductor

Feeder conductors

Grounded neutral bus bar isolated from panel

Goodheart-Willcox Publisher

Figure 25-1. The main bonding jumper for this system is installed in the service disconnecting means located on the outside of the dwelling. The distribution panelboard is fed from the service disconnect by four feeder conductors, including both a grounded conductor and an equipment grounding conductor. Note the absence of the main bonding jumper in the distribution panelboard. The grounded neutral bar is isolated from the panelboard enclosure, and the equipment grounding bus bar is directly connected to the enclosure. If the equipment grounding bus bar and the grounded neutral conductor bus bar were connected in the panelboard, a parallel path would be created, allowing unwanted current to flow on the equipment ground.

25.2 **Service and Distribution Panelboard Locations**

Like temporary service, an electrical service for a dwelling can be either underground or overhead. Underground services are typically fed from a pad-mounted transformer located in the yard. One transformer generally serves one or two homes. The overhead service is fed from a utility pole-mounted transformer that also serves one or two houses. The electrical utility determines the location of the service based on the location of their distribution transformer, and the service equipment is usually on the side of the house.

The distribution panelboard should be located in a convenient, readily accessible location. *Section 110.26* outlines the required working spaces for electrical equipment, including the distribution panelboard. See **Figure 25-2.** According to *Table 110.26(A)(1)*, the depth of the working space shall be a minimum of 3′. This distance shall be maintained for both surface-mounted and flush-mounted panels. The width of the working space shall not be less than 30″ or the width of the equipment, whichever is greater. This measurement does not intend for the equipment to be centered with 15″ to either side.

In all cases, however, the width of the working space shall allow for a 90-degree opening of hinged equipment doors,

Required Working Space Clearance

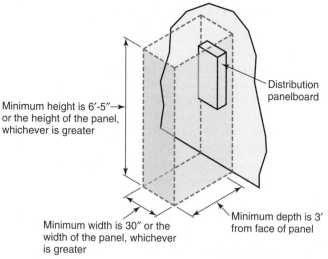

Minimum height is 6′-5″ or the height of the panel, whichever is greater

Distribution panelboard

Minimum depth is 3′ from face of panel

Minimum width is 30″ or the width of the panel, whichever is greater

Goodheart-Willcox Publisher

Figure 25-2. Clear working space is required for access to certain types of electrical equipment. A panelboard installed in a residence requires a minimum working depth of 3′, a minimum working width of 30″, and a minimum working height of 6′-6″.

as required by *Section 110.26(A)(2)*. *Section 110.26(A)(3)* requires the height of the working space to be 6′-5″ or the height of the equipment, whichever is greater.

Section 240.24(A) requires circuit breakers to be readily accessible with the center of the highest breaker handle no higher than 6′-7″ above the floor. *Section 240.24(D)* prohibits overcurrent protective devices, and thus the distribution panel, from being installed in the vicinity of easily ignitable materials, such as in a clothes closet. Nor can the panel be located in a bathroom, per *Section 240.24(E)*, or over the steps of a stairway, per *Section 240.24(F)*.

PRO TIP   **Line and Load**

The terms *line* and *load* are often used in electrical wiring, **Figure 25-3**. The line side of the equipment is the side connected to the electrical supply. The load side of the equipment is the output side of the equipment. As applied to meter sockets, the line side of the meter is for the connection of the incoming service entrance conductors, and the load side is for the connection of the outgoing conductors.

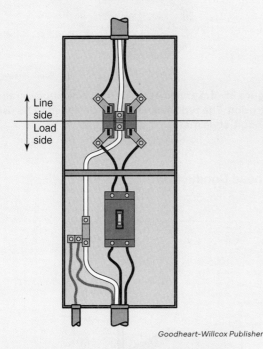

Goodheart-Willcox Publisher

Figure 25-3. The terminals on the top of this meter socket are receiving the electrical supply from the utility and would be considered the line side of this equipment. The load side of the equipment are the meter socket terminals connected to the conductors leaving the meter and going to the service disconnecting means breaker.

25.3 Overhead Service Clearance Requirements

Overhead services are fed from the utility by the service drop that originates from the distribution transformer that is located on top of the utility pole nearest the dwelling. There are many *NEC* clearance requirements that must be considered when installing overhead services.

Section 230.9(A) outlines minimum clearances for the installation of overhead service conductors. This includes various specifications for multiconductor service entrance conductors, overhead conductors above roofs, and weather heads. Multiconductor service entrance conductors without an overall jacket shall have a clearance of not less than 3′ from "*windows that are designed to be opened, doors, porches, balconies, ladders, stairs, fire escapes, or similar locations,*" **Figure 25-4**. The intent of the rule is to reduce the likelihood that someone would come in contact with the conductors by keeping open wiring out of reach. An exception allows conductors run above the top level of a window to be less than the 3′ requirement.

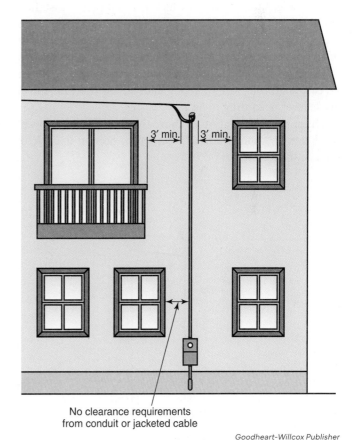

No clearance requirements from conduit or jacketed cable

Goodheart-Willcox Publisher

Figure 25-4. Overhead service conductors are required to have a minimum clearance of 3′ from building openings and points of access like windows, doors, and balconies.

Overhead conductors above roofs are subject to additional clearance requirements in *Section 230.24(A)*. The general rule requires a vertical clearance of at least 8′ above the roof surface. This distance shall be maintained for at least 3′ beyond the edge of the roof. This general requirement ensures that the conductors have enough clearance to be out of reach to a person standing on the roof surface. There are a few exceptions to this rule, including *Exception No. 2*, which reduces the required height to 3′ where the voltage between conductors does not exceed 300 volts and the roof has a slope of 4:12 or greater. When the roof pitch is 4:12 or greater, there is an assumption that persons would not be walking around on the roof surface and the likelihood of contact is reduced, thereby allowing a reduction in the clearance to 3′, **Figure 25-5**. The vertical distance is further reduced by *Exception No. 3* of *Section 230.24(A)* to not less than 18″. This rule applies when the voltage between conductors does not exceed 300 volts and no more than six feet of overhead service conductors, four feet horizontally, pass above the roof overhang, and they are terminated at a through-the-roof raceway.

Section 230.54(A) requires the use of a **service head**, commonly called *weather head*, which is a cap that is installed at the top of the service mast to prevent rain intrusion into the mast and service equipment. See **Figure 25-6**. The point of attachment for the utility's overhead service conductors shall be below the service head, per *Section 230.54(C)*. In no case shall the point of attachment be less than 10′ above grade level, according to *Section 230.26*.

Other vertical clearances above the final grade for overhead service conductors, as outlined in *Section 230.24(B)*, include the following:

- 10′ above areas or sidewalks accessible only to pedestrians where the voltage is less than 150 volts to ground.
- 12′ above residential property, including driveways and commercial areas not subject to truck traffic, where the voltage does not exceed 300 volts to ground.
- 15′ for areas listed in the 12′ classification where the voltage exceeds 300 volts to ground.
- 18′ over public streets, alleys, roads, and parking lots subject to truck traffic, as shown in **Figure 25-7**.

Note that these clearances are for *overhead service conductors*, which are installed by the electrical contractor, not the *overhead service drop conductors*, which are installed by the electrical utility. It is the location of the service point that determines if the conductors are overhead service conductors or overhead service drop conductors, **Figure 25-8**. Even though these clearances do not apply to the serving utility's conductors, they will have similar overhead clearances for their service drop conductors.

Goodheart-Willcox Publisher

Figure 25-6. Service head for the prevention of rain intrusion. The type shown here is designed to be used with a conduit-style mast.

Roof Clearance for Overhead Conductors

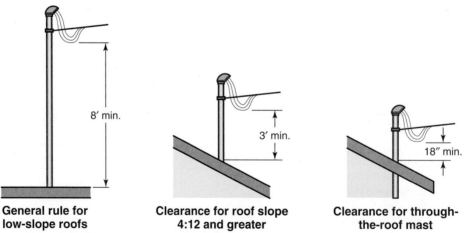

| General rule for low-slope roofs | Clearance for roof slope 4:12 and greater | Clearance for through-the-roof mast |

8′ min.   3′ min.   18″ min.

Goodheart-Willcox Publisher

Figure 25-5. Overhead service conductors must maintain clearance between the conductors and the roof. The general requirement is 8′ but can be reduced to 3′ if the slope of the roof is 4:12 or greater. If the service mast passes through the roof, the conductors passing over the overhang can be reduced to 18″ clearance as long as the horizontal distance does not exceed 4′ and the length of the conductor over the roof does not exceed 6′.

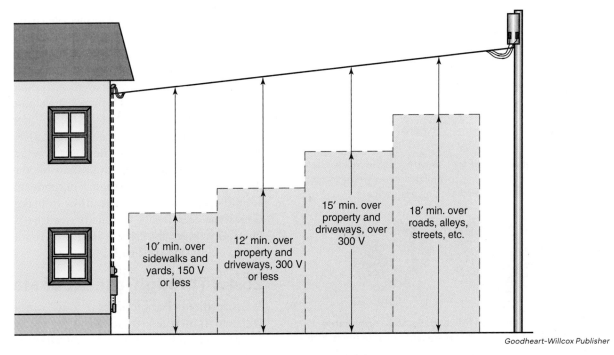

Goodheart-Willcox Publisher

Figure 25-7. Overhead service conductor clearances based on *Section 230.24(B)*.

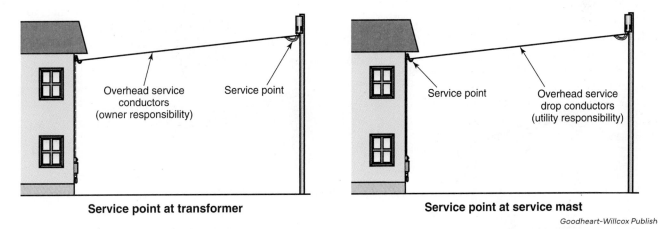

Service point at transformer   **Service point at service mast**

Goodheart-Willcox Publisher

Figure 25-8. The location of the service point determines if the overhead conductors are overhead service drop conductors or overhead service conductors. A service drop is under the purview of the electrical utility, whereas the overhead service conductors are installed and maintained by the customer.

25.4 Types of Overhead Services

There are three types of overhead services as determined by the type of service mast used: the cable mast, the conduit mast, and the through-the-roof mast. Each of these will be discussed in detail as well as *NEC* clearance requirements.

25.4.1 Cable Mast

A cable mast service is the simplest and least costly of the three types of overhead services. A *cable mast* uses a length of Type SEU cable attached directly to the building as the service mast. According to *Section 230.51(A)*, a cable must be secured within 12″ of the service head and at intervals not exceeding 30″. A service head is required for all overhead services by *Section 230.54(A)*. A service head used with a cable mast is shown in **Figure 25-9**.

A threaded hub and weather-proof cable connector must be used where the SEU cable enters the meter socket, as outlined in **Figure 25-10**. **Figure 25-11** shows a typical cable mast service where the feeder cable emerges from the back of the meter socket/disconnect and is fed through the wall to the distribution panel.

Figure 25-9. Service head for use with a cable mast. The service head is installed to prevent the intrusion of rainwater into the service mast and service equipment.

Goodheart-Willcox Publisher

Figure 25-10. A threaded hub and weather-tight cable connector for an overhead service with a cable mast. Note the rubber washer in the connector. This washer will squeeze around the cable when tightened to create a rain-proof seal.

25.4.2 Conduit Mast

A *conduit mast*, as its name implies, uses a length of conduit to protect the overhead service conductors or cables from weather and any physical damage. A threaded rigid metal conduit

(Type RMC) or a threaded intermediate conduit (Type IMC) is generally used in conjunction with a properly sized threaded hub at the meter socket/disconnect. Polyvinyl chloride conduit (Type PVC) can also be installed when the conduit is not used as a point of attachment for the overhead conductors. If EMT is used as the mast, a weather-tight compression-style connector must be used to secure the EMT to the threaded hub.

Any mast must be secured in accordance with the *NEC* requirements for its type. The point of attachment for the service drop should be no lower than 10′ above grade level per *Section 230.26*. No less than 3′ of conductor shall emerge from the service head for the utility to make their connection. Type SER feeder conductors can be protected from physical damage by using a length of conduit and an LB conduit body, **Figure 25-12.**

25.4.3 Through-the-Roof Mast

If the required clearances above grade cannot be achieved with a mast installed on the side of the house, it will be necessary to extend the service conductors through the roof. A *through-the-roof mast* is an electrical service in which a conduit extends through the roof and provides a point of attachment for the overhead service drop conductors, **Figure 25-13.** *Section 230.28(A)* requires the mast to be of sufficient strength to support the weight and strain of the service drop conductors. Type RMC or Type IMC raceways are typically used in these applications.

Figure 25-11. A typical cable mast type of service.

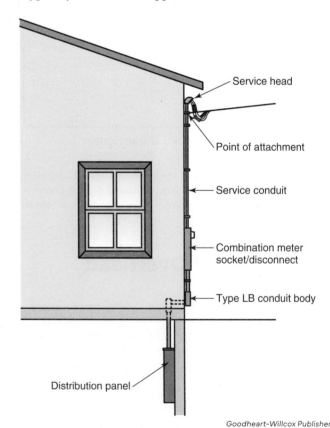

Figure 25-12. A typical conduit mast type of service.

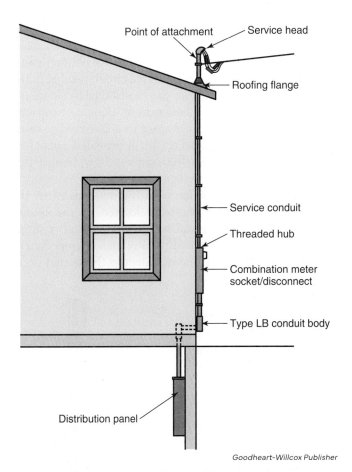

Goodheart-Willcox Publisher

Figure 25-13. An example of a through-the-roof type service mast.

The point of attachment for the service drop conductors is attached to the service mast just below the service head. *Section 230.26* requires the point of attachment to provide the minimum clearances outlined in *Section 230.24*. For residential through-the-roof installations, this would require a minimum clearance of 18″ for conductors passing over the overhang and at least 3′ for conductors over the main roof if the slope was 4:12 or greater. *Section 230.28* permits only the overhead service drop conductors to be attached to the service mast. No telephone or television cables are allowed. **Figure 25-15** shows a typical porcelain stand-off used as a point of attachment for a through-the-roof mast.

25.5 **Underground Services**

A house with an underground service is supplied by a set of underground service lateral conductors. These conductors are installed between the utility transformer and the service equipment. They are installed and maintained by the serving utility. The lateral conductors terminate at the service point or demarcation point between the utility and the customer wiring, typically located in the meter socket. The transformer is generally pad-mounted but, at times, may be pole-mounted, **Figure 25-16**. Underground service conductors are buried to a depth of not less than 30″, in accordance with *Table 300.5(A)*.

PROCEDURE

Installing a Mast Through-the-Roof Overhang

To install the mast, a hole must be cut through the roof overhang, which can pose a challenge.

1. With the meter socket/disconnect installed, use a plumb-bob or laser plumb to transfer the center point of the meter hub to the underside of the overhang.
2. Drill a pilot hole through the overhang at the center point marking.
3. Examine the overhang to ensure you have enough clearance to cut a hole for the conduit without cutting through structural framing supporting the roof.
4. Using a long pilot bit, extend the pilot hole from the underside of the overhang through the shingle side of the roof.
5. Using a hole saw, cut a hole just large enough to accommodate the conduit on the underside of the roof.
6. Using a slightly larger hole saw, cut a hole on the shingle side of the roof. Using the larger hole at the top will provide some play between the holes to ensure the conduit is plumb.
7. Install the conduit through the roof and secure it to the hub on the meter socket.
8. Cover the hole in the roof by installing a **roofing flange** or *boot*. The roofing flange is a type of seal that covers the hole in the roof and prevents rainwater intrusion into the roofing structure, **Figure 25-14**.
9. Install conductors through the mast.
10. Cap off the mast by installing the service weather head.

Goodheart-Willcox Publisher

Figure 25-14. A typical roofing flange for a through-the-roof service mast. The flange is installed over the mast, and a portion of the flange must be inserted below the asphalt shingles. Special roofing caulk is used to seal the flange. The roofing flange prevents rainwater from seeping into the attic.

The underground service is the least expensive and the fastest to install as there is no mast to build.

Goodheart-Willcox Publisher

Figure 25-15. A stand-off point of attachment for a through-the-roof service mast. The point of attachment is affixed to the mast and provides support for the overhead conductors terminating at the service. This version will fit masts from 1-1/4″ to 3″.

25.6 **Subpanels**

Recall that a subpanel is an additional distribution panel that is fed from an appropriately sized circuit breaker in the main distribution panel. They can be installed to reduce the lengths of branch circuit home runs or to increase capacity by adding additional breaker space to accommodate more circuits than the main panel originally provided. Subpanels are typically rated at 60 to 100 amps and can be installed with or without a main breaker.

A large, multistory home with the main panel in the basement might have a subpanel located on the second floor. It might be easier and less costly to install a single feeder to a subpanel for the upstairs branch circuits than to pull all home runs from the basement to the second floor.

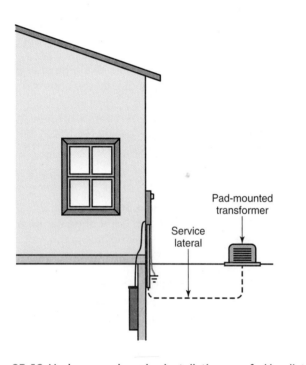

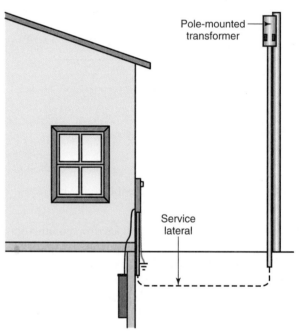

Goodheart-Willcox Publisher

Figure 25-16. Underground service installations are fed by distribution transformers. Most transformers are pad mounted, but some transformers are mounted on poles.

CODE APPLICATION

Section 300.5 Protection Against Damage

Scenario: A set of underground service conductors are run from the service point at a pad-mounted transformer to a meter socket mounted to the exterior wall of the home. What protections are provided to protect the underground service conductors from damage?

Section 300.5: *Section 300.5(A)* provides minimum cover requirements that help determine the minimum burial depth of the conductors. *Section 300.5(D)(1)* requires conductors emerging from grade to be protected from the minimum cover depth to a height of

8′ above grade. *Section 300.5(D)(3)* requires direct burial service conductors to be marked with a warning ribbon.

Solution: If these conductors are direct burial cables, the minimum cover in accordance with *Section 300.5* would be 24″. The requirements of *Section 300.5(D)(1)* would be met if the conductors of the system are installed in a conduit to protect them from damage. A warning ribbon will also be required to be installed in the trench at least 12″ above the direct burial conductors in accordance with *Section 300.5(D)(3)*.

A subpanel on the second floor also provides quicker access to the circuit breakers in an emergency.

Subpanels can be installed in workshops and detached garages. As with the main panel, when the service disconnect is located outside with the meter, the grounded conductors and the equipment grounding conductors are isolated from one another in the subpanel.

25.7 Grounding and Bonding

Electrical systems supplying residential occupancies are required to be grounded, or intentionally connected to the earth through the grounding system. This connection

---

PRO TIP

Intersystem Bonding Termination Device

Electrical systems are not the only type of system within a dwelling that may be required to be grounded. *Section 250.94* requires an *intersystem bonding connection*, which is a device for the interconnection of the grounding electrode conductors of other systems such as the telephone, cable or satellite TV, and broadband to the system ground of the dwelling service. The bonding together of all these systems minimizes the possibility of a potential difference between the equipment of different systems by bringing all systems to the same potential. The intersystem bonding connection is usually made at the service and utilizes the grounding electrode conductor, as shown in **Figure 25-17**.

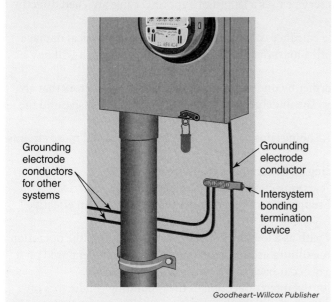

Grounding electrode conductors for other systems

Grounding electrode conductor

Intersystem bonding termination device

Goodheart-Willcox Publisher

Figure 25-17. This intersystem bonding termination device is connected to the grounding electrode conductor terminated in the meter socket. Notice the two bare conductors terminated on the bar provide a ground connection for other systems serving the dwelling.

---

provides stability for the electrical system by stabilizing the voltage and providing a pathway for excess voltage due to transients, such as lightning, to exit the system. The *NEC* recognizes two types of grounding: system grounding and equipment grounding. *System grounding* is used to connect a conductor of the electrical system to ground through one or more grounding electrodes.

Equipment grounding is used to connect the normally non-current carrying metal parts of the electrical equipment and raceways to ground. Equipment grounding is required for all electrical systems, even those systems that are not grounded. The intent of the equipment grounding system is to prevent electrical shock by limiting the voltage between the metallic components of the system and the earth.

Equipment grounding is not to be confused with *equipment bonding*. Bonding is an additional concept recognized by the *NEC* in which normally non-current-carrying metallic components are connected together to ensure a low impedance electrical path. This low-impedance path is an *effective ground-fault current path* because the low impedance allows a high level of current to flow during a ground-fault event. The high level of current will facilitate the opening of the overcurrent protection device, thereby ending the ground fault and eliminating the electrical hazard. According to *Section 250.4(A)(5),* an effective ground-fault current path must also be capable of carrying the maximum ground-fault current likely to be imposed upon it. This section requires the bonding connection to be substantial enough to carry the current required to trip the circuit breaker. The supply source in residential wiring is the utility distribution transformer. The low-impedance circuit making up the effective ground-fault current path is comprised of the bare equipment grounding conductors that are connected to the green grounding screw of devices and to the equipment grounding bus in the distribution panel.

When a ground fault occurs, the ungrounded, hot conductor makes contact with a normally non-current-carrying part of the equipment. Without proper grounding and bonding, that piece of equipment becomes energized at the line voltage value and can pose a risk to anyone in contact with it. Proper bonding through the effective ground-fault current path facilitates the operation of the overcurrent device to quickly open the circuit.

A key concept for understanding grounding and bonding is understanding the flow of current under a ground-fault condition. This example explains the current pathway for a ground fault occurring at a load currently plugged into an outlet. The ground fault can be the result of a faulty cord that has made contact with the metal frame of the machine. With an effective ground-fault current path, the following sequence describes the flow of fault current in the circuit:

1. Fault current flows from the ungrounded conductor onto the metal frame of the machine to the connection

---

of the equipment grounding conductor of the machine's cord and then to the grounding pin of the receptacle.

2. It then passes on the bare equipment grounding conductor of the branch circuit to the grounding bus bar of the distribution panelboard.

3. The equipment grounding conductor of the feeder provides a path for the fault current to flow from the grounding bus bar in the panelboard to the main disconnect, where the main bonding jumper connects the grounded conductors to the equipment grounding conductors.

4. It then flows through the main bonding jumper and onto the grounded neutral conductor from the main disconnect through the neutral lugs in the meter socket.

5. From the meter socket, the fault current leaves the house and travels on the service entrance grounded conductor back to the neutral point of the distribution transformer, where it comes rushing back into the dwelling on one of the ungrounded conductors and trips the circuit breaker in the distribution panelboard in the blink of an eye.

Once the overcurrent protection device opens due to a ground fault, the fault must be removed before resetting a circuit breaker or replacing a fuse. If the ground fault is not removed and an attempt to reset a breaker is made, the fault-current will take the same path and once again trip the overcurrent protection device. Repeated resetting of devices under fault conditions can lead to failure of the equipment, damage to conductors, and arcing that may lead to a fire.

Summary

- The electrical service for a dwelling can be either underground or overhead.
- The electrical utility determines the location of the service based on the location of their distribution transformer and will usually be on the side of the house.
- The distribution panel should be located in a convenient, readily accessible location.
- *Section 110.26* sets forth the required working spaces for electrical equipment, including the distribution panel.
- The width of the working space shall not be less than 30 inches or the width of the equipment, whichever is greater.
- The depth of the working space shall be a minimum of 3′ per *Table 110.26(A)(1)*.
- *Section 240.24(A)* requires circuit breakers to be readily accessible, with the center of the highest breaker no higher than 6′-7″ above the floor.
- There are three types of overhead services as determined by the type of service mast used: the cable mast, the conduit mast, and the through-the-roof mast.
- The cable mast is the simplest type of overhead service. This service uses a length of Type SEU cable attached directly to the building as the service mast.
- A conduit mast uses a length of conduit to protect the service entrance cables from the weather and physical damage.
- If the required clearances above grade cannot be achieved with a mast installed on the side of the house, it will be necessary to install a through-the-roof mast.
- A house with an underground service is fed from the transformer by underground service lateral conductors that are installed by and under the maintenance of the serving utility. The underground service is the least expensive and the fastest to install.
- Subpanels are sometimes installed to reduce the lengths of the home runs of the branch circuits or to add more circuits than the main panel is capable of.
- The *NEC* recognizes two types of grounding: system grounding and equipment grounding.
- Equipment bonding requires that the normally non-current-carrying conductive materials enclosing electrical conductors or equipment shall be connected to the electrical supply source in a manner that establishes an effective ground-fault current path.
- *Section 250.4(A)(5)* defines the effective ground-fault current path as a low-impedance circuit facilitating the operation of the overcurrent device and capable of safely carrying the maximum ground-fault current likely to be imposed on it from any point on the wiring system where a ground fault may occur back to the electrical supply source.
- *Section 250.94* requires an intersystem bonding connection for the interconnection of the grounding conductors of other systems, such as the telephone, cable or satellite TV, and broadband, to the system ground of the dwelling service.
- The intersystem bonding connection is usually made at the service and utilizes the grounding electrode conductor.

Know and Understand

1. What cable type is used for the feeder conductor when the disconnect is located outside with the meter?
   A. Type SEU
   B. Type USE
   C. Type SER
   D. All of these.

2. Who determines where the electrical service will be located on a dwelling?
   A. Electrician
   B. Utility
   C. AHJ
   D. Fire department

3. The depth of working space required for a residential distribution panelboard is _____.
   A. 1′
   B. 2′
   C. 3′
   D. 4′

4. *True or False?* The width of the working space for a residential panel shall be such that the hinged panel door can be opened to 18 degrees.

5. *True or False?* A clothes closet is a good location for the distribution panel.

6. Multiconductor service entrance conductors without an overall jacket shall have a clearance of not less than _____ from windows that are designed to be opened, doors, porches, balconies, ladders, stairs, fire escapes, or similar locations.
   A. 3′
   B. 4′
   C. 5′
   D. 10′

7. The general rule for overhead service conductors requires a vertical clearance of at least _____ above the roof.
   A. 2′
   B. 3′
   C. 4′
   D. 8′

8. The piece of the overhead service that prevents rainwater intrusion into the service mast is the _____.
   A. threaded hub
   B. service terminator
   C. weather protector
   D. service head

9. An overhead service conductor is installed where it crosses a residential driveway. It shall have a vertical clearance above grade of _____.
   A. 10′
   B. 12′
   C. 15′
   D. 18′

10. The _____ type of overhead service is the simplest and least expensive to install.
    A. through-the-roof mast
    B. conduit mast
    C. cable mast
    D. None of these.

11. *True or False?* All overhead services require a threaded hub where the mast enters the meter socket regardless of the type of mast installed.

12. The point of attachment for an overhead service drop should be no less than _____ above grade level.
    A. 10′
    B. 12′
    C. 15′
    D. 18′

13. The most common raceway used for a through-the-roof mast is _____.
    A. electrical metallic tubing, type EMT
    B. rigid metal conduit, type RMC
    C. intermediate metal conduit, type IMC
    D. Both RMC and IMC.

14. *True or False?* The through-the-roof service mast can be used as a point of attachment for the incoming telephone or CATV cables.

15. *True or False?* An underground service is always supplied by a pad-mounted transformer.

16. A distribution panelboard that is fed by a circuit breaker in another panelboard is called a(n) _____.
    A. auxiliary panel
    B. MLO panel
    C. back-up panel
    D. subpanel

17. A low-impedance circuit facilitating the operation of the overcurrent device and capable of safely carrying the maximum ground-fault current likely to be imposed on it from any point on the wiring system where a ground fault may occur back to the electrical supply source is known as a(n) _____.
    A. grounded conductor
    B. GFCI circuit
    C. AFCI circuit
    D. effective ground-fault current path

18. The supply source for a residential service is the _____.
    A. distribution transformer
    B. main distribution panel
    C. main service disconnect
    D. utility generating station

19. *True or False?* In the event of a ground fault, the fault current will use the metal enclosure of a piece of equipment as a path back to the source.

20. The piece of equipment that is used to bond the grounding electrode conductors from other systems, such as the communication system and the broadband system, to the system ground of the dwelling is the _____.
    A. main bonding jumper
    B. grounding electrode conductor
    C. intersystem bonding termination device
    D. grounding electrode

Apply and Analyze

1. A single-pole circuit breaker is used for _____-volt circuits, while a two-pole circuit breaker is used for _____-volt circuits.
2. The grounded conductors and the equipment grounding conductors shall be _____ after the service disconnect.
3. The height of the working space for electrical equipment shall be_____ or the height of the equipment, whichever is greater.
4. What are the three types of overhead services?
5. The vertical clearance for service drop conductors over a roof is _____ where the voltage between conductors does not exceed 300 volts and the roof has a slope of 4:12 or greater.
6. A cable mast must be secured within _____ and at intervals not to exceed _____.
7. What types of fittings must be used where the cable mast enters the meter socket enclosure?
8. What type of connector is used at the threaded hub when EMT is used as the conduit mast?
9. The piece of equipment used to prevent rainwater intrusion into the roofing structure is the _____.
10. The service conductors for an underground service shall be buried to a depth not less than _____.
11. What are the two types of grounding recognized by the *NEC*?
12. Why are the grounding electrode conductors of other systems such as telephone and cable TV bonded to the system ground of the dwelling?

Critical Thinking

1. Explain the difference between equipment grounding and equipment bonding.
2. Define the service point and explain how the location of the service point determines whether overhead conductors from the pole-mounted transformer to the point of attachment on the dwelling are overhead service conductors or overhead service drop conductors.

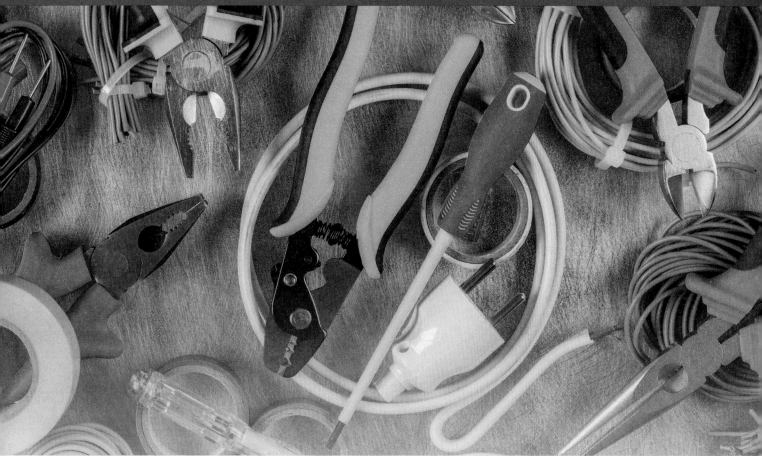

Flegere/Shutterstock.com

The trim-out stage of the electrical system installation occurs after the rough-in stage and involves installing devices, such as switches and receptacles, and making the electrical connections to major appliances, such as the furnace, air conditioning, and water heater. Luminaires are also installed, the distribution panel is wired, and then the system is energized and tested for proper operation.

Chapter 26, *Device and Appliance Trim Out*, covers proper device terminations and correct wiring of major appliance disconnecting means. Chapter 27, *Luminaire Trim Out*, covers the installation of luminaires during the trim-out stage. Chapter 28, *Distribution Panel Trim Out*, covers wiring the distribution panel with an emphasis on circuit breaker selection and installation. Chapter 29, *Branch Circuit Testing and Troubleshooting*, covers the testing of the electrical system once the trim-out is complete.

CHAPTER 26

Device and Appliance Trim Out

ungvar/Shutterstock.com

LEARNING OBJECTIVES

After completing this chapter, you will be able to:

- Explain the proper technique for making terminal loops for device connections.
- Discuss the correct method for terminating switches.
- Explain the correct procedure for terminating standard duplex receptacles.
- Outline the correct method for terminating GFCI receptacles.
- Explain the proper method of terminating range and dryer receptacles.
- Determine when a disconnecting means is required for a particular appliance.
- Discuss the correct wiring procedure for an appliance disconnecting means.

TECHNICAL TERMS

line termination
line/load configuration
line/line configuration
load termination
major appliance
split-wired
terminal loop
trim out

Introduction

After all the trades have completed their rough-in stages, the walls are closed in, and it is time for the electrical trim out. ***Trim out***, or *trimming*, refers to installing all devices and cover plates, connecting major appliances, installing the luminaires, and installing the breakers in the distribution panel. This phase is the most visible portion of the installation; thus, it is critical that the trim out be completed in a neat, diligent, careful, and professional manner.

An electrician must be able to perform trim out with care and accuracy. Many future problems in an electrical system can arise from a lack of attention during the trim-out stage, including loose terminations that lead to ground faults or short circuits. The proper mechanical torquing of bolts and screws should minimize or eliminate this hazard. This chapter will focus on the proper installation of devices and common household appliances, such as dishwashers, water heaters, ranges, dryers, furnaces, and central air conditioning units.

26.1 Proper Device Terminations

Any conductor to a device, whether a circuit breaker, switch, or receptacle, must always be properly terminated during the process of electrical trim out. Improper device terminations are a leading cause of electrical house fires. Loose connections cause an increase in resistance, which then causes heat. Therefore, it is crucial that every termination on every device in the dwelling be made correctly and properly torqued, **Figure 26-1**.

A correctly made device termination begins with the proper strip length of conductor insulation. About 3/4″ of insulation is removed from the end of the conductor. Removing too much insulation increases the likelihood of a short circuit or ground fault in the device box if the length of the bare, stripped "hot" or current carrying conductor makes contact with the neutral (grounded) or the equipment grounding conductor (EGC). Conversely, removing too little insulation may result in the insulation under the terminal screw, which hampers the connection to the conductor.

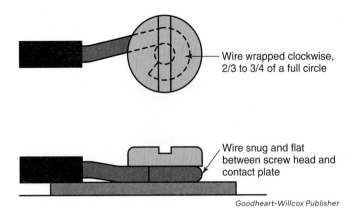

Wire wrapped clockwise, 2/3 to 3/4 of a full circle

Wire snug and flat between screw head and contact plate

Goodheart-Willcox Publisher

Figure 26-1. Proper installation of the terminal loop.

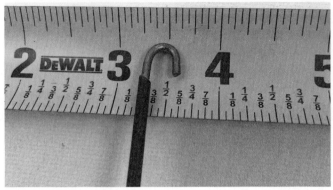

Goodheart-Willcox Publisher

Figure 26-3. A properly made terminal loop.

Wire strippers have a small hole that is used to form terminal loops in the conductors, **Figure 26-2.** A *terminal loop* is a small hook formed at the end of the conductor that wraps around the terminal screw of a device. The terminal loop should be a half-circle with the end of the bare conductor even with the insulation, **Figure 26-3.** This half-circle will allow the conductor to wrap two-thirds to three-quarters around the terminal screw with no overlap.

Once formed, the terminal loop should be wrapped around the terminal screw in a clockwise direction and squeezed close. The terminal screw should be tightened snugly. Wrapping the conductor clockwise around the screw ensures the conductor is pulled tighter around the screw post as the screw is tightened. Wrapping the conductor counterclockwise will result in the conductor being forced from under the screw as it is tightened.

When tightening the terminal screw, the manufacturer recommendation for torque of snap switches and receptacles is typically 12 to 14 inch-pounds. This is accomplished with a calibrated torque screwdriver. Manufactures generally allow only one conductor to be installed per terminal screw. It is not unusual, though, to have more than one conductor that needs to be terminated. Pigtails are used in these instances. A pigtail is a short length of conductor used when two or more spliced conductors need a connection to a device, **Figure 26-4.** Pigtails should be six to eight inches in length. Too short of a pigtail makes terminations difficult whereas too long of a pigtail will take up too much space in the device box.

26.2 Switch Terminations

Termination is the attachment of a wire to a terminal bar or a screw in equipment. According to *Article 100*, equipment is a general term intended to cover a multitude of items such as devices, appliances, luminaires, and machinery.

Goodheart-Willcox Publisher

Figure 26-2. Use the hole in the strippers to form the terminal loop.

Pigtails

Goodheart-Willcox Publisher

Figure 26-4. Pigtails are short lengths of conductors used to connect a device to spliced conductors.

Switches are devices installed in accordance with the rules of *Article 404*. In this article, *Section 404.2(A)* requires all three- and four-way switching circuits to be wired such that all switching is on the ungrounded circuit conductor. *Section 404.2(B)* clearly prohibits switching of the grounded circuit conductor with any switched circuit.

26.2.1 Single-Pole Switch Terminations

When installing a single-pole switch with the power at the switch location, two ungrounded conductors, typically black or red, should connect to the switch—one for the incoming power and one that leads to the luminaire. The white, grounded conductors are spliced together and not connected to the switch unless needed to operate a motion sensor. According to *Section 404.2(C)*, if a switch controls a lighting load and the loads are supplied by a grounded system, such as with a 120/240-volt system, the grounded (neutral) conductor must be installed at the switch location. This is to include circuits serving bathrooms, hallways, stairways, and habitable rooms or occupiable spaces. Receptacles controlled by a switch are exempt from this requirement.

If power is at the receptacle location, a switch loop must be brought to the switch location. A two-conductor cable with a re-identified white conductor provides constant power to the switch, while the black conductor provides switched power to the receptacle. The re-identification is typically accomplished by taping, painting, or other effective means, such as a permanent marker. Green is reserved for grounding, and white is the neutral conductor in dwelling units. These colors are not allowed for re-identifying the white as an ungrounded conductor. The specific requirements for the re-identification are found in *Section 200.7(C)(1)*.

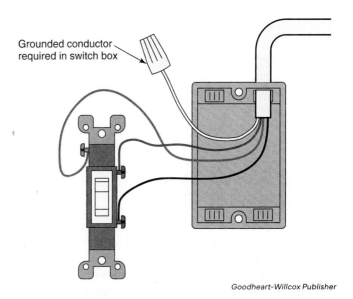

Goodheart-Willcox Publisher

Figure 26-5. Correct connection of a single-pole switch.

With a single-pole switch, it does not matter which ungrounded conductor is connected to either of the terminal screws, **Figure 26-5.** The switch merely opens and closes the circuit.

26.2.2 Three-Way Switch Terminations

When installing three-way switches, each of the two switch locations has one common conductor and two traveler conductors connected to the switch. The switches have three terminal screws each, one black or dark brown screw identified as the common terminal, and two silver or gold screws for the traveler conductors. For the circuit to function correctly, the common conductor must be connected to the common terminal screw, **Figure 26-6.** Recall from Chapter 13,

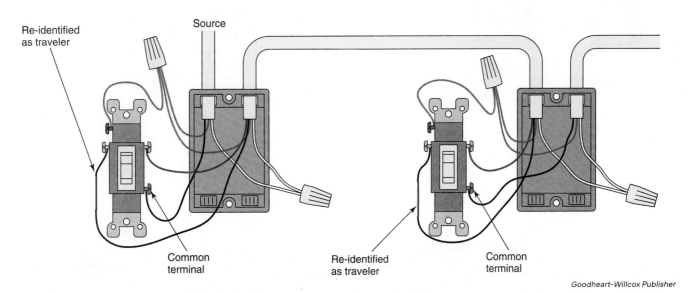

Goodheart-Willcox Publisher

Figure 26-6. Proper connection of a three-way switch. Note the common terminal.

Basic Switched Circuits, that the common conductor is the way into and out of a three-way switching circuit. Of the two three-way switches involved in controlling luminaires or receptacles, one of the common screws will have the hot or ungrounded conductor, and the other will have the switch leg to the controlled device, a luminaire, or a receptacle. Either traveler conductor can be connected to either traveler terminal screw.

26.2.3 Four-Way Switch Terminations

Four-way switches are always used between two three-way switches to control a lighting load from more than two locations. Four-way switches are connected only to the traveler conductors. The common conductor is not connected to a four-way switch. The four-way switch has four terminal screws, usually two black screws and two brass-colored screws. In general, the travelers from one cable are connected to the same color terminal screws, while those from the other cable are connected to the other set of terminal screws, **Figure 25-7.**

Regardless of the type of switch used, the bare or green equipment grounding conductor (EGC) must be connected to the hexagonal green grounding screw on the mounting yoke. Theoretically, there is no limit on the number of four-way switches that can be placed between the two three-way switches.

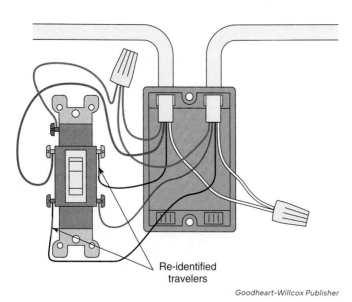

Re-identified travelers

Goodheart-Willcox Publisher

Figure 26-7. Proper connection of a typical four-way switch. Only the traveler conductors are connected to the switch.

26.3 **Receptacle Terminations**

Receptacles are addressed in *Article 406. Section 406.12(1)* requires all 15- and 20-amp, 125-volt receptacles in dwelling units, boathouses, mobile homes, and manufactured

homes to be of the tamper-resistant type. The exceptions to this, listed at the end of *Section 406.12*, include receptacles located more than 5-1/2′ above the floor, those that are part of a luminaire, a single or duplex receptacle located in a dedicated appliance space if the appliance is not easily moved (a microwave, refrigerator, washing machine), and non-grounding receptacles used for replacements. In addition, where installed in damp or wet locations, *Sections 406.9(A)–(B)* require receptacles listed as the weather-resistant type.

Recall from Chapter 10, *Electrical Devices*, that a standard duplex receptacle has five terminal screws: two brass-colored, two silver-colored, and one hexagonal green grounding screw. The silver terminal screws of a duplex receptacle are for the white grounded or neutral conductors, and the brass terminal screws are for the black ungrounded, hot conductors. As with switches, the green grounding screw is for the bare or green equipment grounding conductor. When there are two or more cables at the receptacle outlet location, it is a best practice to use pigtails to connect the receptacle, **Figure 26-8.**

Depending on the device, it is not a best practice to pass a current through to another device, receptacle to receptacle. Instead of an additional connection under an outlet screw, using wire nuts or their equivalent would provide a stronger connection. The use of a receptacle as a connection point can create problems if the screw connection becomes loose.

26.3.1 **Duplex Receptacle Terminations**

A standard duplex receptacle has a connecting tab between the terminal screws on either side. Removing the connecting

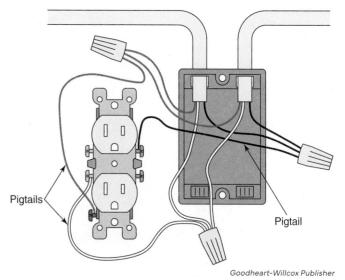

Pigtails

Pigtail

Goodheart-Willcox Publisher

Figure 26-8. Using pigtails to connect a receptacle. Although not an *NEC* requirement, it is a best practice that ensures the continuity of the circuit is not dependent on the device.

tabs between the ungrounded, hot conductor terminals allows the duplex receptacle to be *split-wired*, which means one-half of the receptacle outlet has constant power and the other half is switch-controlled, **Figure 26-9**. The tab between the white, grounded conductors is left intact. It is a best practice to switch the bottom half of the receptacle and have the top half receive constant power.

Before GFCI receptacles became the *NEC* standard for all kitchen countertop outlets, it was common practice to split-wire receptacles with half the receptacle being fed from one circuit (line 1, typically black) and the other half fed from another (line 2, typically red). The ungrounded conductors sharing the same neutral would result in a line 1-to-line 2 voltage of 240 and a line-to-ground or neutral of 120 volts. The change to the required GFCI receptacles for kitchen countertops started to occur in the 1970s. With the GFCI technology, the need to split-wire receptacles went out of favor. However, if this is done for any reason, *Section 210.4(B)* requires both ungrounded conductors to be able to be disconnected simultaneously at the "*point where the branch circuit originates.*" For all practical purposes, the branch circuit originates at the panelboard, and a two-pole circuit breaker provides the simultaneous disconnect.

A switched receptacle most likely has some form of lighting outlet plugged into it, and using the bottom receptacle keeps the cord from obstructing the constant power section.

26.3.2 Ground-Fault Circuit Interrupter (GFCI) Receptacle Terminations

A ground-fault circuit interrupter (GFCI) receptacle is a safety device that shuts off the power when it senses a ground-fault condition. The GFCI receptacle can also protect additional outlets wired downstream through a connection known as a *line/load configuration*.

The terms *line* and *load* as they apply to a GFCI receptacle must be understood to make the proper connections based on intended use. A GFCI receptacle has four terminal screws and the hexagonal green grounding screw. Two of the terminal screws are labeled *line*, and two are labeled *load*. The **line terminations** are for the incoming power to the receptacle, and the **load terminations** are for the downstream outlets protected by the GFCI. An incorrectly wired GFCI receptacle trips when power is applied, and it cannot be reset until the wiring is corrected. The GFCI receptacle is sometimes wired in a **line/line configuration** where only the GFCI device shuts off under a ground-fault condition. In this scenario, other connected devices remain unaffected.

Most GFCI receptacles are connected to the conductors differently than a standard duplex receptacle, where the conductor is wrapped around the terminal screw. A GFCI receptacle uses pressure plates that pinch the conductors

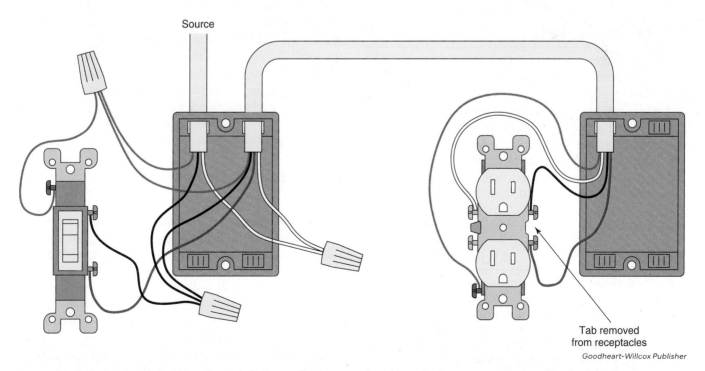

Tab removed from receptacles

Goodheart-Willcox Publisher

Figure 26-9. Removing the connecting tab between the brass-colored terminal screws creates a split-wired duplex receptacle. Half of the receptacle will have constant power while the other half is switch-controlled.

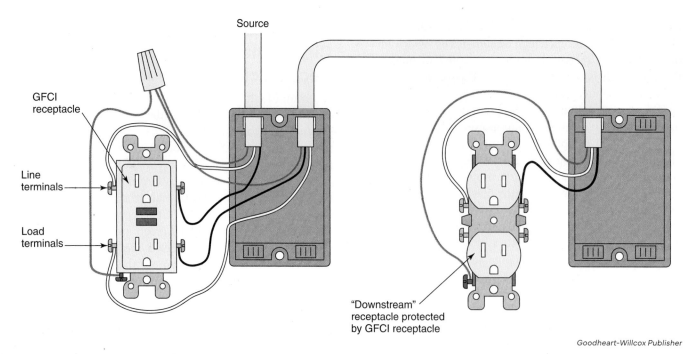

Source

GFCI receptacle

Line terminals

Load terminals

"Downstream" receptacle protected by GFCI receptacle

Figure 26-10. Connecting the GFCI receptacle.

when the terminal screw is tightened, **Figure 26-10**. There are general provisions for up to two conductors per terminal screw. It is necessary to check the manufacturer's requirements for the proper torquing of these screws and to make sure it is safe to land more than one wire under a terminal.

26.4 **Device and Cover Plate Installation**

Once the electrical terminations have been made, the devices are ready to be installed. Be sure that conductors are pushed to the rear of the junction box. If using NM cable with a bare equipment grounding conductor (EGC), it must not be in contact with any component that will become energized, such as a terminal screw for the hot conductor. The junction box should not be packed with conductors, and there should be ample air circulation to allow the device to breathe. Follow the box fill requirements referenced in *Section 314.16(B)*.

PRO TIP   **Allowable Box Fill**

The *NEC* allowable box fill requirements are only the minimum. It is good practice to allow extra space in a box when installing dimmers, USB charger and receptacle combinations, and GFCI or arc-fault circuit interrupter (AFCI) receptacles. Each of these will occupy more space than the standard size switches and receptacles.

Ensure that all single-pole switches are installed in the correct position so that when the switch is down, it is in the OFF position. The *NEC* does not specify an orientation for receptacles. They may be installed either ground up or ground down, but the orientation should be consistent throughout the house.

After installation, the yoke of the device should be snug against the wall. This will allow the cover plate to be flush with the wall. When installing a cover plate with multiple screws, align the the slots on the head of the screws in the same direction for a neat and professional appearance instead of leaving the screw slots oriented in random positions. Be sure that the cover plate is level following installation. Boxes level and installed at the same height, plus trim screws oriented in the same direction, indicate a quality job installed by a craftsperson.

26.4.1 **Installation in Damp or Wet Locations**

Section 406.9 addresses enclosures and covers for receptacles installed in damp and wet locations. In damp locations, *Section 406.9(A)* requires the enclosure for the receptacle shall be weatherproof when the receptacle is not in use, but this requirement does not extend to a situation when the attachment plug is inserted. *Section 406.9(B)* requires a receptacle installed in a wet location to have an enclosure that is weatherproof when the attachment plug is not inserted or while in use, **Figure 26-11**. In addition, weatherproof cover shall be listed and identified as extra duty. This type of a weatherproof cover

Scenario: An electrician needs to install two readily accessible outdoor receptacles in a single-family dwelling—one in the front and one in the back. Consider the relevant code requirements and determine the location and rating requirements for these receptacles:

Section 210.52: This section states these receptacles will be rated 15 or 20 amps at 125 volts. For a single-family dwelling and each unit of a two-family dwelling that is at grade level, at least one receptacle outlet readily accessible from grade and not more than 6.5′ above grade level shall be installed at the front and the back of the dwelling.

Section 406.9(A)–(B): Outdoor receptacles must be rated for wet locations. Receptacles of 15 and 20 amps,

125 and 250 volts installed in a wet location shall have an enclosure that is weatherproof whether or not the attachment plug cap is inserted. An outlet box hood installed for this purpose shall be listed and shall be identified as "extra-duty." Other listed products, enclosures, or assemblies providing weatherproof protection that does not utilize an outlet box hood need not be marked "extra-duty."

Solution: As required by the *NEC*, the required outdoor receptacles shall be listed and identified as the weather-resistant type and be installed in an enclosure that is listed and identified as "extra-duty." The term *extra-duty* refers to the durability of the outlet box hood.

Figure 26-11. A weatherproof receptacle cover.

is acceptable to meet the requirements listed in *Section 406.9(A)* for a damp location.

26.5 Major Appliances

Major appliances are those that are served by an individual branch circuit that supplies no other loads. Some major appliances, such as the cooking range and clothes dryer, are

cord-and-plug connected, while others, such as the water heater and the furnace, are hard wired.

Recall from Chapter 10, *Electrical Devices*, that for new construction, a 120/240-volt four-wire circuit is required for these electric ranges and dryers. A typical four-wire range or dryer receptacle must have four termination points marked as the following:

- *W*—white grounded, neutral conductor
- *X* and *Y*—two ungrounded, hot conductors
- *G*—the bare or green equipment grounding conductor (EGC)

Common residential receptacle configurations are shown in **Figure 26-12**.

26.5.1 Disconnecting Means

The disconnecting means must be within sight of the appliance and is intended to disconnect the appliance from its power source to be safely serviced when repair or maintenance is needed. A disconnecting means is required for those hard-wired appliances not within sight of its circuit breaker. *Section 110.25* addresses locking the circuit breaker in the open position if the circuit breaker is not within sight of the hardwired appliance. Doing this will satisfy the disconnect requirement if the locking provisions remain in place whether the lock is installed or not. Specific appliance disconnect requirements are covered in *Article 422, Part III*.

A disconnect is generally used for 240-volt appliances that do not require a grounded, neutral conductor. A disconnect has four terminal lugs for the power connection, two for the line connection, and two for the load connection. There will also be two terminal lugs for the equipment grounding conductors, **Figure 26-13**. If needed, an option to buy a neutral/grounded conductor kit is available for most disconnects.

Common Residential Receptacle Configurations

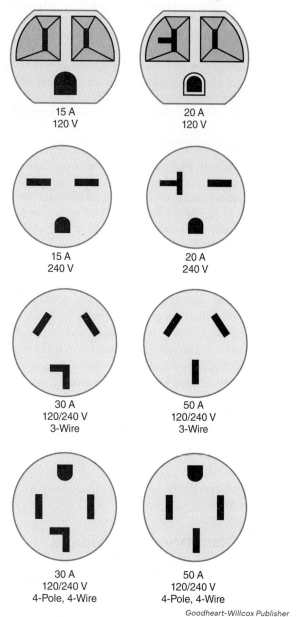

15 A
120 V

20 A
120 V

15 A
240 V

20 A
240 V

30 A
120/240 V
3-Wire

50 A
120/240 V
3-Wire

30 A
120/240 V
4-Pole, 4-Wire

50 A
120/240 V
4-Pole, 4-Wire

Goodheart-Willcox Publisher

Figure 26-12. Common receptacle configurations found in residential electrical systems.

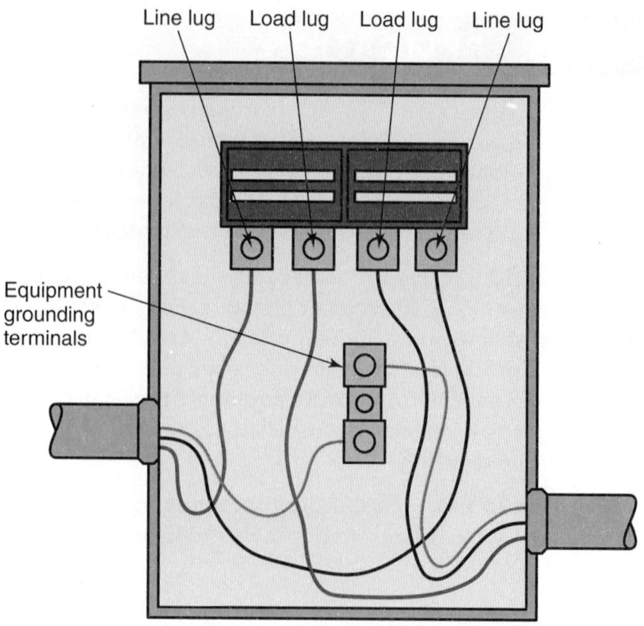

Line lug   Load lug   Load lug   Line lug

Equipment grounding terminals

Goodheart-Willcox Publisher

Figure 26-13. A typical disconnect for a 240-volt appliance. Note that the equipment grounding terminals are attached directly to the enclosure.

Summary

- Trim out refers to installing all devices and cover plates, connecting major appliances, installing the luminaires, and installing the breakers in the distribution panel.
- A terminal loop is a small hook formed at the end of the conductor that wraps around the terminal screw of a device. The terminal loop should be a half-circle with the end of the bare conductor, even with the insulation. Once formed, the terminal loop should be wrapped around the terminal screw in a clockwise direction and squeezed close.
- *Section 404.2(A)* requires all three- and four-way switching circuits to be wired such that all switching is on the ungrounded circuit conductor.

- With a single-pole switch, it does not matter which ungrounded conductor is connected to either of the terminal screws. The switch merely opens and closes the circuit. When installing three-way switches, each switch location will have one common conductor for either the ungrounded, hot conductor or the switch-leg to the luminaire and two traveler conductors connected to each switch. Four-way switches are connected to the traveler conductors only. The common conductor is not connected to a four-way switch.
- A standard duplex receptacle will have a connecting tab between the terminal screws on either side. Removing the connecting tabs allows the duplex receptacle to be split-wired.
- In a line/load configuration, the line terminations are for the incoming power to a GFCI receptacle, and the load terminations are for the downstream GFCI-protected outlets.
- Major appliances are those that are served by an individual branch circuit that supplies no other loads. Some major appliances, such as the cooking range and clothes dryer, are cord-and-plug connected, while others, such as the water heater and the furnace, are hard wired.
- A disconnecting means will be required for those hard-wired appliances if the appliance is not within sight of its circuit breaker. A circuit breaker is acceptable as a disconnecting device if the circuit breaker contains permanent provisions for locking in the open position and is not within sight of the appliance.
- The disconnect is generally used for 240-volt appliances that do not require a grounded, neutral conductor.

Know and Understand

1. The _____ stage of construction involves the installation of all devices, luminaires, major appliances, and circuit breakers.
   - A. rough-in
   - B. trim out
   - C. punch list
   - D. turnover

2. A small hook formed in the conductor for a device termination is called a(n) _____.
   - A. pigtail
   - B. equipment grounding conductor
   - C. switch loop
   - D. terminal loop

3. The conductor should be wrapped _____ around the terminal screw.
   - A. clockwise
   - B. counterclockwise
   - C. two times
   - D. loosely

4. *True or False?* The *NEC* requires all switched receptacles to receive their power at the switch location.

5. *True or False? Section 404.2(B)* requires all switching circuits to disconnect the grounded conductors.

6. A short length of conductor that is used to splice two or more conductors to a device is called a(n) _____.
   - A. pigtail
   - B. equipment grounding conductor
   - C. switch loop
   - D. terminal loop

7. *True or False?* When installing a single-pole switch, it does not matter which ungrounded conductor is connected to either of the two terminal screws.

8. How many conductors are required to be run between two three-way switches?
   - A. Three; two common conductors and one traveler conductor
   - B. Three; one traveler conductor, one common conductor, and one equipment grounding conductor
   - C. Four; one common conductor, two traveler conductors, and one equipment grounding conductor
   - D. Four; two common conductors, one traveler conductor, and one equipment grounding conductor

9. How many conductors are attached to a four-way switch?
   - A. Four; two traveler conductors and two common conductors
   - B. Four; four traveler conductors
   - C. Five; four traveler conductors and one equipment grounding conductors
   - D. Five; two traveler conductors, two common conductors, and one equipment grounding conductor

10. When terminations are made to a standard duplex receptacle, the silver-colored screws are used for the _____ conductors.
    - A. black, ungrounded
    - B. white, grounded
    - C. bare equipment grounding
    - D. None of the above.

11. When terminations are made to a standard duplex receptacle, the brass- or gold-colored screws are used for the _____ conductors.
    - A. black, ungrounded
    - B. white, grounded
    - C. bare equipment grounding
    - D. None of the above.

12. When a GFCI receptacle is wired to protect downstream outlets, the incoming power is connected to the _____ terminals.
    - A. line
    - B. load
    - C. equipment grounding
    - D. None of the above.

13. *True or False?* A GFCI receptacle wired in a line-line configuration does not protect downstream outlets.

14. *True or False?* For a receptacle installed in a damp location, an enclosure that is not weatherproof when a plug is inserted is permitted by the *NEC*.

15. When wiring a receptacle for a range or clothes dryer, the two ungrounded, hot conductors are wired to the terminals marked _____ and _____.
    - A. W; X
    - B. W; Y
    - C. X; Y
    - D. Y; G

Apply and Analyze

1. Why is it essential for the trim out to be performed in a neat, diligent, careful, and professional manner?
2. Explain why the conductors should be wrapped around the terminal screw in a clockwise direction.
3. How many three-way and four-way switches are required to operate a lighting load from four locations?
4. Explain how a standard duplex receptacle is split-wired.
5. When installing a split-wired receptacle, explain why it is a best practice to switch-control the bottom half of the receptacle and have the top half receive constant power.

Critical Thinking

1. What section in the *NEC* references *terminal connection torque*?
2. Discuss the importance of proper mechanical torque. Identify how your local inspectors handle this critical aspect of an electrical installation.

CHAPTER 27

Luminaire Trim Out

VH-studio/Shutterstock.com

CHAPTER OUTLINE

Introduction

Electricians must be familiar with all aspects of luminaires and lighting. This textbook has covered various topics regarding lighting and lighting outlets in a residence. Chapter 12, *Lamps and Lighting* discussed light bulbs and lamps used in residential wiring, and Chapter 18, *Branch Circuit/Receptacle Requirements* reviewed the required locations for lighting outlets in a dwelling. Refer to these chapters for the necessary foundation to cover lighting requirements outlined by the *NEC*. This chapter outlines the installation requirements for luminaires, including recessed lighting, in bathrooms and closets.

27.1 Types of Luminaires

A luminaire is not just the light source but includes everything in the lighting unit. The light source in a luminaire is a key item that has evolved over time. From the original incandescent to fluorescent lamps and further evolving to halogen, mercury vapor, and high- and low-pressure sodium sources of light. The latest evolution is the advent and refinement of the light emitting diode (LED). Most light sources now being used incorporate LED technology. The operating costs are less, the lamp life is long, maintenance costs are reduced in commercial and industrial settings due to fewer lamp replacements, and LEDs generate less wasted or unneeded heat. They are also aesthetically more desirable since they can output millions of colors. Refer to *Article 100* for an extensive, in-depth description of a luminaire.

Some of the common types of luminaires used in dwelling units include recessed lighting, surface-mounted lighting, pendant lighting, and track lighting. Each of these may have different types within a general description.

27.2 Recessed Luminaires

Recessed luminaires have seen an enormous transformation in recent years. Canless LED luminaires have replaced traditional recessed luminaires, often referred to as can lights or pot lights. Unlike the traditional types, *canless luminaires* are very thin, some as thin as the sheetrock, and can be installed just about anywhere. Although traditional recessed can luminaires are still available, they are losing popularity with the expanded use of canless recessed LED luminaires.

Recessed can luminaires are composed of two parts: the housing and the trim, **Figure 27-1.** The housing includes the light bulb socket, a junction box to enclose the needed wires, a frame or means of attachment to the structure, and a metal *top*

Recessed Can Luminaire

Housing

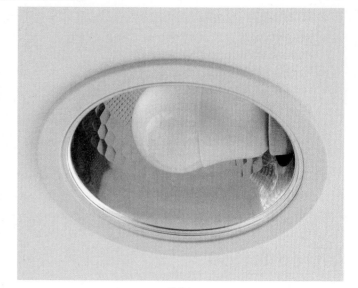

Trim

Chad Robertson Media/Shutterstock.com; Sira Anamwong/Shutterstock.com

Figure 27-1. A recessed can light housing and trim.

hat or can to enclose the bulb. The housing of the luminaire is installed during the rough-in stage, and the trim, which completes the luminaire, is installed during the finish or trim-out phase. The trim of a recessed can luminaire is the portion that is visible after installation. There are a wide variety of styles of trims available. Some are called *baffles* that disperse and diffuse the light for general lighting, and others serve to provide accent lighting to a specific item, such as a piece of art or a fireplace.

27.2.1 Type IC and Non-Type IC Rated Luminaires

Installed incandescent lamps will generate a high amount of heat. When installing recessed luminaires, an electrician must choose the most appropriate and safe materials to include. *Section 410.116(A)* differentiates between two types of recessed luminaires based on whether they can be in contact with thermal insulation: those marked as Non-Type IC and those marked as Type IC. In both cases, *IC* stands for **Insulation Contact**. Most of the newer style canless LED recessed luminaires are Type IC.

The bulb's wattage rating and the type of trim affect whether the luminaire is Non-Type IC or Type IC. For example, higher wattage incandescent bulbs, 75 watts and above, produce a high amount of heat, and some types of trims, such as those with a lens or an "eyeball" style, may not allow for air circulation. Without air circulation, the heat will be trapped in the can housing and transferred to the nearest object. This causes significant dangers if placed near or in contact with combustible materials such as

insulation. *Section 410.16(A)(1)* requires that a minimum of 1/2″ clearance shall be maintained from combustibles, and *Section 410.16(B)* does not allow any insulation within 3″ of the Non-Type IC rated luminaire, including the enclosure, wiring compartment, ballast, transformer, LED driver, or power supply.

Type IC recessed luminaires are listed by the manufacturer to be in contact with the thermal insulation and combustible materials. A Type IC rated luminaire does not have the strict requirements of the Non-Type IC rated types. The use of Non-Type IC recessed can enclosure requires greater care and more installation time to make sure the installation is *NEC* compliant. Oftentimes a listed housing is used to totally enclose the fixture to make sure no insulation or combustible materials make contact with the luminaire. Refer to *Article 410 Part X* for all provisions for installing the different types of recessed luminaires.

27.3 Surface-Mounted Luminaires

Unlike a recessed luminaire, a surface-mounted luminaire is mounted to the surface of the ceiling or wall. A surface-mounted luminaire installed on a wall is called a **sconce**. There are two main categories of surface-mounted luminaires with multiple variations: the flush-mounted and the pendant. In addition, there are many different styles of surface-mounted luminaires, and the luminaire style is a leading factor in the room's décor.

27.3.1 Flush-Mounted Luminaires

Flush-mounted luminaires are mounted directly to the surface of the ceiling or wall, **Figure 27-2**. A *semi-flush luminaire*, also called a *close to ceiling*, has a few inches of space between the ceiling and the luminaire, **Figure 27-3**.

These luminaires are hung from an outlet box installed during the rough-in stage using hardware supplied with the luminaire. A mounting bracket attaches to the outlet box with 8-32 screws, and the luminaire attaches directly to the mounting bracket. For luminaires weighing less than six pounds, *Section 314.27(A)(1) Exception* allows no fewer than two 6-32 screws to provide support.

Although not common, occasionally an electrician needs to install a luminaire weighing more than 50 lb. In this case, the luminaire must be supported independently of the outlet box unless the outlet box is designed to support the weight of the fixture. Check the interior stamping in the box to see the maximum weight it is allowed to support. If the box does not have an interior weight stamped inside, the rating of the box is 50 lb maximum.

27.3.2 Pendant Luminaires

Pendant luminaires are luminaires hung from the ceiling from several inches to several feet, depending on the installation. There are three major types of pendant luminaires: cord-hung, stem-hung, and chain-hung, **Figure 27-4**. Cord-hung luminaires can be adjusted to any height. For a stem-hung luminaire, the length of the stem determines the installed height. Cord-hung and stem-hung luminaires generally support one or two lightweight lamp fixtures. Chain-hung luminaires are usually heavier and can support more light fixtures. A chandelier, for example, is a common multi-lamp pendant luminaire, usually chain hung, used for ambient lighting, typically over a dining room table.

Close attention to the weight of a pendant luminaire is important. If this weight exceeds 50-pounds, a separate support or a box rated for a greater weight needs to be installed. Refer to *Section 314.27(A)(2)* for this guidance.

27.4 Track Lighting

Track lighting is a type of luminaire comprised of two parts: an electrified track and removable lighting heads that can be positioned anywhere along the track. Many different lighting heads are available, including different styles of spotlights that direct light where desired. Track lighting is generally used for task lighting and accent lighting.

The *NEC* rules for track lighting can be found in *Article 410 Part XIV*. Because of the possibility of overloading the track with too many lighting heads, *Section 410.150(B)* requires that the connected load on the lighting track shall not exceed the track's rating. When using LEDs with lower ampacity, overloading a track is more difficult. In addition, *Section 410.150(C)* outlines locations where track lighting shall not be installed, including wet or damp locations, and where extended through walls or partitions.

Section 410.154 requires the lighting track to be securely mounted and suitable for the maximum weight of the lighting heads installed. A single section of track 4′ or shorter shall have two supports unless identified for supports at greater intervals. Where sections of track are installed in a continuous row, each individual section of no more than 4′ shall have one additional support.

27.5 Installing Luminaires

The thermal insulation and weight of a luminaire are not the only concerns an electrician must account for during

Flush-Mounted Luminaire

Anant Jadhav/Shutterstock.com

Figure 27-2. A flush-mounted luminaire is mounted directly to the ceiling.

Semi-Flush Luminaire

Mariana Serdynska/Shutterstock.com

Figure 27-3. A semi-flush mounted luminaire has a few inches of space between the ceiling and the luminaire.

**Cord-Hung
Pendant Luminaires**

**Stem-Hung
Pendant Luminaires**

**Chain-Hung
Pendant Luminaires**

Volodymyr_Shtun/Shutterstock.com; pics721/Shutterstock.com; pics721/Shutterstock.com

Figure 27-4. A comparison of the different types of pendant luminaires typically installed in a residence.

installation. Another key issue is the grounding of all luminaires. Specifically, if the luminaire is metal, a method to ground the fixture must be provided. Exceptions apply to alternate methods, typically found in older construction, where a luminaire is connected to a system that does not have an equipment grounding conductor (EGC) installed with the branch circuit. In these instances, GFCI protection of the circuit is acceptable or, per *Section 250.130(C)*, a separate EGC can be routed back to an accessible point of the grounding electrode system, an equipment grounding terminal bar in a panel, or the grounded service conductor in the service equipment. Although this is not common, many old dwellings will have newer light fixtures installed on branch circuits that have no EGC installed. For specific requirements, refer to sections in *Article 410*.

27.5.1 Installing Luminaires in Bathtub and Shower Areas

Section 410.10(D) addresses the installation of luminaires in the bathtub or shower area of a bathroom. This includes information regarding restricted zones in these locations. A restricted zone extends 8′ vertically and 3′ horizontally from the top of a bathtub rim or shower stall threshold, **Figure 27-5**. No parts of any pendant luminaire, track lighting, or paddle fans shall be located within a restricted zone. Surface-mounted luminaires are allowed within the restricted zone if marked for damp or wet locations in compliance with *Sections 410.10(D)(1)–(2)*.

Additional requirements must be considered for wet locations and GFCI protection. Where luminaires are subject to

shower spray, they shall be marked as suitable for wet locations. GFCI protection is provided only if the luminaire instructions require it.

During the rough-in, the required junction box should be mounted at a height appropriate for the type of luminaire and centered over the space intended for the vanity or pedestal sink. While not required by the *NEC*, luminaires above

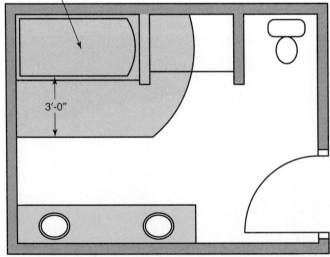

No pendant luminaire, track lighting, or paddle fans allowed in this region, up to 8′ above the tub

3′-0″

Goodheart-Willcox Publisher

Figure 27-5. Pendant luminaires, track lighting, and paddle fans are not permitted to be installed within the restricted zone of a bathtub or shower area.

the bathroom sink and vanity are very popular. The most common location is on the wall above the mirror, resulting in the best light for the mirror. If possible, avoid installing the vanity luminaire on the ceiling; shadowing will likely occur with this type of installation, **Figure 27-6**.

Although it may seem like a simple task to install a luminaire above the bathroom sink, accurate measuring and careful planning are required. If placement is off or in an undesirable location, reworking the location of the junction box after the wall is finished could be required by the homeowner or general contractor. It is cheaper to do the job correctly the first time than to be forced to fix an incorrect installation.

The first step in determining the junction box location is knowing the final choice of vanity cabinet or pedestal sink. The specific details of the vanity cabinet or sink, including dimensions, style, and placement preferences, will aid in determining the installation location of the lighting. High-quality blueprints show the final locations and sizes.

Once the center location for the junction box is determined, it is important to make sure one of the wall studs is not located at the needed center for the luminaire junction box. The wall studs can be modified for the final locations of wall sconces and bathroom lighting fixtures. The final

Artazum/Shutterstock.com

Figure 27-6. Wall-mounted luminaires provide good lighting for a person using a bathroom vanity.

location of the junction box will take priority over the initial placement of a stud.

Another possibility is to allow a larger empty opening behind the drywall, both horizontally and vertically, near the proposed location of the box. During the rough-in portion, keep an adequate length of wire (perhaps 18″) to stick out from a small hole in the drywall at the proposed junction box location in the wall, or temporarily bury a loop of wire (around 18″ to 24″) in the wall to be retrieved when the luminaire is installed. At this point, an *old-work* round cut-in box can be installed at the best location.

Next, consider the desired height to the center of the junction box. This will be determined by three factors: the top height of the mirror, its frame, and the type of luminaire to be installed. It is necessary to know if the luminaire has lampshades extending downward a few inches below the bottom of the junction box.

Some luminaires do not require a junction box. Working with the general contractor or homeowner to determine the exact luminaires to be installed is essential. A few simple actions and questions can save the electrician from a potential rework dilemma.

27.5.2 Installing Luminaires in Clothes Closets

In *Article 100* of the *NEC*, a clothes closet is defined as "*a non habitable room or space intended primarily for storage of garments and apparel.*" Because of the presence of potentially combustible materials, the *NEC* has specific rules concerning the types of luminaires allowed in clothes closets and their permissible locations in *Section 410.16*.

Section 410.16(B) permits only three types of luminaires in a closet:

- Surface-mounted or recessed incandescent or LED luminaires with a light source that is completely enclosed
- Surface-mounted or recessed fluorescent luminaires
- Surface-mounted fluorescent or LED luminaires identified as suitable for installation in a closet space

Note that any pendant luminaires or incandescent luminaires with open or partially enclosed lamps are prohibited by *Section 410.16(C)*. This has not always been the case. Open or partially enclosed fixtures were commonly installed with an application specific minimum of 12″ or 18″ clearance from combustible materials. With the development of luminaires designed specifically for closets, the *NEC* rules have adapted to safer installations. This is primarily a result of the cooler operating temperatures of LED and fluorescent fixtures as opposed to the heat generated by the traditional incandescent lamps.

Clothes closet storage space is defined in *Article 100* and refers to the areas in a closet where combustible materials can be kept. Per *Section 410.16(D)*, luminaires in closets may be installed on the ceiling or the wall above the door if there is sufficient clearance between the luminaire and closet storage space. The minimum required clearance varies based on the type of luminaire. See **Figure 27-7**.

Additionally, surface-mounted fluorescent or LED luminaires are permitted to be installed within the closet storage space where identified for such use.

It should be noted that the *NEC* does not require luminaires to be installed in clothes closets or storage areas. While it may suffice to have no light in a reach-in closet, a good practice is to install adequate lighting in larger closets and storage areas.

Lighting Clearance for Closets

Type of Lighting	Minimum Clearance between Luminaire and Closet Storage Space
Surface-mounted incandescent or LED luminaires with a completely enclosed light source	12″
Surface-mounted fluorescent luminaires	6″
Recessed incandescent or LED luminaires with a completely enclosed light source	6″
Recessed fluorescent luminaires	6″

Goodheart-Willcox Publisher

Figure 27-7. Minimum clearances for luminaires in clothes closets. Note the largest distance is for surface-mounted incandescent or LED lamps.

Summary

- Some of the common types of luminaires used in dwelling units include recessed lighting, surface-mounted lighting, pendant lighting, and track lighting.
- Recessed can luminaires are composed of two parts: the housing and the trim. The housing is installed during the rough-in stage, and the trim is installed during the trim-out phase. *Article 410 Part X* contains the provisions for installing the different types of recessed luminaires.
- *Section 410.116(A)* refers to two types of recessed luminaires based on whether they can be in contact with thermal insulation, those marked as Non-Type IC and those marked as Type IC, where *IC* stands for *Insulation Contact*. The bulb's wattage rating and the type of trim affect whether the luminaire is Non-Type IC or Type IC.
- A surface-mounted luminaire installed on a wall is called a sconce. There are two main categories of surface-mounted luminaires: the flush-mounted and the pendant. Flush-mounted luminaires are mounted directly to the surface of the ceiling. They are hung from an outlet box installed during the rough-in stage using hardware supplied with the luminaire. Pendant luminaires are hung from the ceiling from several inches to several feet. The three major types of pendant luminaires are cord-hung, stem-hung, and chain-hung.
- Track lighting is generally used for task or accent lighting and is comprised of two parts: an electrified "track" and removable lighting "heads" that can be positioned anywhere along the track. The *NEC* rules for track lighting can be found in *Article 410 Part XIV*.
- *Section 410.10(D)* addresses the installation of luminaires in the bathtub or shower area of a bathroom. Where subject to shower spray, luminaires shall be marked as suitable for wet locations.
- Because of potentially combustible materials, the *NEC* has specific rules concerning the types of luminaires allowed in clothes closets and their permissible locations in *Section 410.16*.
- Clothes closet storage space is defined in *Article 100* and refers to the areas in a closet where combustible materials can be kept. The location of a luminaire in a closet limits the distance between the luminaire and the nearest point of closet storage space.

Know and Understand

1. *True or False?* When installing recessed can lights, both the housing and the trim are installed during rough-in.
2. When referring to recessed can lights marked as Non-Type IC, thermal insulation must be kept at least _____ from the luminaire housing and wiring compartment.
   A. 1/4″   C. 6″
   B. 3″   D. 12″
3. The two main categories of surface-mounted luminaires are the flush-mount and the _____.
   A. canless   C. chandelier
   B. sconce   D. pendant
4. Track lighting is a type of luminaire comprised of two parts, an electrified "track" and _____.
   A. the power supply   C. lighting heads
   B. the housing   D. trim
5. *True or False?* Track lighting is ideal for general illumination.
6. *Section 410.150(B)* requires that the connected load on the lighting track shall not exceed the track's _____.
   A. rating   C. voltage
   B. length   D. weight
7. For luminaires installed in bathtub or shower areas of a dwelling, a restricted zone is established that extends _____ vertically and _____ horizontally from the top of the bathtub rim or the shower stall threshold.
   A. 3′; 8′   C. 9′; 4′
   B. 8′; 3′   D. 10′; 5′
8. *True or False?* Surface-mounted luminaires are permitted to be installed in the restricted zone of a bathtub or shower area.
9. Where subject to shower spray, luminaires in the restricted zone shall be marked as _____.
   A. suitable for wet locations
   B. suitable for damp locations
   C. suitable for the restricted zone
   D. Luminaires are not allowed where subject to shower spray.
10. *True or False?* GFCI protection is required for luminaires installed in a bathroom.
11. According to the *NEC*, a non-habitable room or space intended primarily for storage of garments and apparel is the definition of a(n) _____.
    A. attic   C. pantry
    B. crawl space   D. closet
12. *True or False?* Pendant luminaires are allowed in a walk-in closet but not a reach-in closet.
13. A surface-mounted incandescent luminaire with a completely enclosed light source installed on the ceiling or above the door in a closet must be at least _____ from any storage space.
    A. 4″   C. 10″
    B. 6″   D. 12″

Apply and Analyze

1. What is the main benefit of the canless recessed LED luminaire?
2. A _____ is a luminaire installed on a vertical wall.
3. What size of screw is typically used to support a luminaire from the outlet box?
4. What are the three main types of pendant luminaires?
5. What typically determines whether a pendant luminaire is cord-hung or chain-hung?
6. A multi-lamp pendant luminaire, usually chain hung, used for ambient lighting, typically over a dining room table, is known as a _____.
7. An electrician is installing 12′ of track lighting in a continuous row. How many supports does the *NEC* require?
8. What type of luminaire is permitted by the *NEC* to be installed in the restricted zone of a bathtub or shower area?
9. What consideration should be made when installing a luminaire on the ceiling at a bathroom vanity location?
10. What is the minimum distance allowed between a recessed fluorescent luminaire and the nearest point of closet storage space?

Critical Thinking

1. Why are the distance requirements in a closet greater for incandescent luminaires than for fluorescent luminaires?
2. Considering the technology available today with LEDs, what are the best and most cost-effective choices for illumination in closets and storage areas?

Chad Robertson Media/Shutterstock.com

CHAPTER OUTLINE

- Identify the proper termination points for conductors in a distribution panel.
- Identify the correct breaker size based on the conductor's ampacity.
- Determine final ampacity of NM and UF cable installed in all panelboards using the 60°C (140°F) column of *Table 310.16*.
- Identify the overcurrent protection limitations required by *Section 240.4(D)* on conductors sized 14 AWG, 12 AWG, and 10 AWG.
- Identify the proper overcurrent protection for branch circuits wired with 8 AWG and larger NM and UF cables.
- Summarize the installation requirements of single-pole and double-pole breakers.
- Explain how to install GFCI and AFCI circuit breakers.
- Describe the benefits of installing a dual-purpose circuit breaker.
- Explain how to identify branch circuit locations in a distribution panel directory.

bus bar
double-pole branch circuit
double-pole circuit breaker
dual-purpose circuit breaker
single-pole circuit breaker
time current curves

Introduction

A distribution panel, also called a *panelboard*, is the heart of the electrical system and houses the final overcurrent devices that protect the branch circuit conductors. It serves as the termination point for the feeder or service conductors and the origin of all branch circuits. In the trim-out, or *finish*, phase, the distribution panel must be "trimmed out," which involves terminating the equipment grounding conductors (EGCs) and the grounded neutral conductors to their proper bus bar and installing circuit breakers.

If possible, the distribution panel should be trimmed before permanent power is established by the serving utility. Check with the utility provider to determine if permanent power can be brought to the main service before the installation of circuit breakers and the branch circuits. If permanent power is available at the dwelling unit, panel trim should always occur with the main circuit breaker in the de-energized position. Care should be taken even if the main breaker is off. The conductors feeding the main circuit breaker, the line side, and the associated lugs are still energized. Refer to Chapter 11, *Panelboards, Overcurrent Protective Devices, and Disconnects* for additional information on overcurrent protective devices and their operation.

28.1 Conductor Installation

Recall that during the rough-in stage, feeder or service conductors are installed to the main distribution panel, all branch circuit home run cables are brought to the distribution panel, and they are labeled on the outer sheathing as to their purpose. The cables enter the panel through knockouts in the top and bottom, or occasionally the side, and are secured with cable clamps.

During the trim-out phase, the outer sheathing of each cable is removed, and the branch circuit conductors are landed at the proper termination point in the panel trim-out. Some utilities or local codes may require the service disconnecting means to be located outside the house, but most do not. *Section 230.70(A)(1)* requires the service disconnecting means to be installed "*at a readily accessible location either outside of a building or structure or inside nearest the point of entrance of the service conductors.*"

If the service disconnect is located outside, the bare equipment grounding conductors (EGCs) and the white grounded conductors are kept isolated from one another when feeding the main panelboard in the dwelling. In this instance, the panel in the dwelling unit is a subpanel, and the outside disconnect is considered the main panel or disconnect. The bare equipment grounding conductors (EGCs) are terminated to the equipment grounding bar, which is bonded to the enclosure. The white grounded conductors are connected to the neutral bar, which is isolated from the enclosure.

Section 408.41 allows only one neutral conductor per terminal, but some panel brands allow for more than one EGC per terminal. Per *Section 110.14(A)*, some circuit breakers allow more than one ungrounded conductor to be installed. Typically in these cases, the ungrounded conductors must be the same size. Always follow the manufacturer's requirements during installation.

28.2 Circuit Breaker Selection

Panel trim-out requires selecting the properly sized circuit breakers. Recall that the circuit breaker acts as a safety valve that opens the circuit if the current level increases beyond its rating. The higher the overcurrent, the quicker the breaker will open or trip. The smaller the overcurrent is in relation to the size of the circuit breaker, the longer the time to open the circuit. Manufacturers published *time current curves*, which indicate the length of time it will take to open the mechanism on a circuit breaker in relation to the amount of overcurrent present.

SAFETY NOTE

Circuit Breakers and Distribution Panels

The type of circuit breaker must be compatible with the distribution panel, made by the same manufacturer, or approved for such use. Some circuit breakers can be used, if listed, in different brands of panels, but most cannot. It may appear as though a circuit breaker produced by one manufacturer will fit in a panelboard of another manufacturer, but if the fit is not precise, overheating, breaker tripping, or a short circuit can occur and damage the equipment.

The ampacity rating of a circuit breaker establishes the branch circuit's rating, and the circuit breaker is sized to the conductor's ampacity. *Section 334.80*, covering NM cable, and *Section 340.80*, covering UF cable, require using the 60°C (140°F) column of ampacities of *Table 310.16* when selecting the proper breaker size, **Figure 28-1**. Both NM and UF cable types are used in dwelling units. NM cable type is used for the interior wiring, and UF cable type is used for any outdoor, wet location wiring. It is not uncommon to see a larger size conductor terminated on a circuit breaker than what is required. This is generally done to accommodate voltage drops on long conductor runs.

In *Table 310.16*, note the asterisk beside conductors sized 18 AWG through 10 AWG. The footnote to the table refers to *Section 240.4(D)* for sizing overcurrent protection for these sized conductors. (The smallest copper conductor size allowed for residential branch circuit wiring, as referenced in *Section 310.3*, is 14 AWG, so sizes 16 AWG and 18 AWG can be ignored.)

Section 240.4(D) limits the overcurrent protection for 14 AWG, 12 AWG, and 10 AWG to 15 amps, 20 amps, and 30 amps, respectively. These limitations apply to all wiring types, not just NM cables. Similar rules govern the installation of smaller aluminum conductors. However, newer aluminum installations utilizing 10 or 12 AWG are rare.

NM and UF cable conductors sized 8 AWG, 6 AWG, and larger are protected at their 60°C (140°F) ampacities. This overcurrent protection is a 40-amp circuit breaker for 8 AWG and a 60-amp circuit breaker for 6 AWG. The 60°C (140°F) ampacity for 6 AWG conductors is 55 amps, but 55 amps is not a standard overcurrent size listed in *Table 240.6(A)*. *Section 240.4(B)* allows using the next higher standard overcurrent device rating provided three conditions are met:

- The conductors being protected are not part of a branch circuit supplying more than one receptacle for cord-and-plug-connected portable loads.
- The conductor's ampacity does not correspond with the standard amp rating of a fuse or circuit breaker.
- The next higher standard rating does not exceed 800 amps.

28.3 Circuit Breaker Installation

Recall that 120-volt branch circuits are fed from single-pole circuit breakers, and 240-volt branch circuits are fed from double-pole circuit breakers. One ungrounded, hot conductor, usually the black conductor, is connected to a single-pole breaker, while two ungrounded conductors are connected to a double-pole breaker.

Table 310.16 Ampacities of Insulated Conductors with Not More Than Three Current-Carrying Conductors in Raceway, Cable, or Earth (Directly Buried)

Size AWG or kcmil	Temperature Rating of Conductor [See Table 310.4(1)]						Size AWG or kcmil
	60°C (140°F)	75°C (167°F)	90°C (194°F)	60°C (140°F)	75°C (167°F)	90°C (194°F)	
	Types TW, UF	Types RHW, THHW, THW, THWN, XHHW, XHWN, USE, ZW	Types TBS, SA, SIS, FEP, FEPB, MI, PFA, RHH, RHW-2, THHN, THHW, THW-2, THWN-2, USE-2, XHH, XHHW, XHHW-2, XHWN, XHWN-2, XHHN, Z, ZW-2	Types TW, UF	Types RHW, THHW, THW, THWN, XHHW, XHWN, USE	Types TBS, SA, SIS, THHN, THHW, THW-2, THWN-2, RHH, RHW-2, USE-2, XHH, XHHW, XHHW-2, XHWN, XHWN-2, XHHN	
	COPPER			ALUMINUM OR COPPER-CLAD ALUMINUM			
18*	—	—	14	—	—	—	—
16*	—	—	18	—	—	—	—
14*	15	20	25	—	—	—	—
12*	20	25	30	15	20	25	12*
10*	30	35	40	25	30	35	10*
8	40	50	55	35	40	45	8
6	55	65	75	40	50	55	6
4	70	85	95	55	65	75	4
3	85	100	115	65	75	85	3
2	95	115	130	75	90	100	2
1	110	130	145	85	100	115	1
1/0	125	150	170	100	120	135	1/0
2/0	145	175	195	115	135	150	2/0
3/0	165	200	225	130	155	175	3/0
4/0	195	230	260	150	180	205	4/0

Notes:
1. Section 310.15(B) shall be referenced for ampacity correction factors where the ambient temperature is other than 30°C (86°F).
2. Section 310.15(C)(1) shall be referenced for more than three current-carrying conductors.
3. Section 310.16 shall be referenced for conditions of use.
*Section 240.4(D) shall be referenced for conductor overcurrent protection limitations, except as modified elsewhere in the *Code*.

Reproduced with permission of NFPA from NFPA 70, National Electrical Code, 2023 edition. Copyright © 2022, National Fire Protection Association. For a full copy of the NFPA 70, please go to www.nfpa.org

Figure 28-1. To determine the ampacities of NM cable, use the 60°C (140°F) column of *Table 310.16*.

A ***double-pole branch circuit*** can be either a 240-volt three-wire circuit to supply an appliance, such as a water heater, electric furnace, or a baseboard heater, or a 120/240-volt four-wire circuit, such as a clothes dryer or cooking range. A 240-volt three-wire circuit has two ungrounded conductors, one white and one black, and a bare equipment grounding conductor. The white conductor must be re-identified as an ungrounded, current-carrying conductor at the breaker and the appliance using black or red tape or a black or red permanent marker.

For example, a cable identified as a 14-2 will contain three-conductors. However, only two are considered as current carrying under normal conditions. The third or bare conductor is the EGC, which carries current only when experiencing ground-fault conditions. A 14-3 cable will contain four conductors. Three conductors will normally carry a current under normal conditions and the fourth is the EGC in this type of a cable. The same is true for cables containing conductors 12 AWG and larger.

In residential wiring, ***single-pole circuit breakers*** are rated at 15 or 20 amps and take up one space in the panel. ***Double-pole circuit breakers*** are rated at 15 amps and larger, but typically no larger than 100 amps, and take up two spaces in the panel. All circuit breakers receive power from the ***bus bar***, an ungrounded copper or aluminum bar located vertically in the center of the panel. A single-pole breaker stabs or connects onto one of two legs of a bus bar, while a double-pole breaker stabs onto the line-1 and line-2 legs of the two vertical bus bars. Fulcrum points on either side of the bus bars hold the breakers in place so they can be rocked into position and attached to the bus bar, **Figure 28-2.**

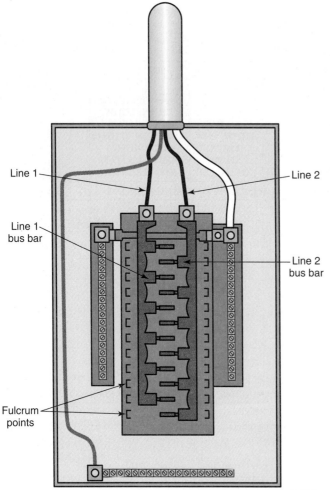

Line 1

Line 2

Line 1
bus bar

Line 2
bus bar

Fulcrum
points

Goodheart-Willcox Publisher

Figure 28-2. In a residential panel, the circuit breakers attach to the bus bar to receive power.

28.4 GFCI, AFCI, and Dual-Purpose Circuit Breakers

Section 210.8(A) requires GFCI protection for all 120-volt 15- and 20-amp receptacles installed in the following locations:

- Bathrooms, laundry areas
- Bathtubs or shower stalls where receptacles are within 6′ of the outside edge of the tub or shower stall
- Kitchens, areas with sinks, spaces for food preparation or cooking
- Basements, garages, crawl spaces, and accessory buildings
- Outdoors, boathouses

GFCI receptacles can protect these outlets, or GFCI circuit breakers can be used to protect the entire branch circuit.

Recall that GFCI devices operate by monitoring the current flowing into and out of the device. *Article 100* states, "*Class A ground-fault circuit interrupters trip when*

the ground-fault current is 6 mA or higher and do not trip when the ground-fault current is less than 4 mA." This trip is accomplished by the electronics within the circuit breaker designed to sense the current flowing to ground and not flowing in the intended return path, the neutral or grounded conductor. If this portion is 6 mA or greater, the GFCI will open the circuit.

With a GFCI circuit breaker, both the branch circuit's grounded conductor and ungrounded conductor are attached to the circuit breaker. This allows the internal current sensing to take place. In non-GFCI-protected breakers, only the ungrounded conductor of the branch circuit connects to the circuit breaker. For a GFCI to work correctly, the circuit breaker may have a white pigtail to connect to the neutral bar. Some manufacturers provide a clip-on feature that allows the GFCI breaker to connect directly to the neutral bar, much the same way as it attaches to the power bus, **Figure 28-3**. The neutral conductor from the branch circuit will connect directly to a silver screw provided in the circuit breaker.

For a dwelling unit, *Section 210.12(B)* requires arc-fault circuit-interrupter protection for all 120-volt 15- and 20-amp branch circuits supplying outlets or devices in the following locations:

- Kitchens and bedrooms
- Family rooms, dining rooms, and living rooms
- Parlors, libraries, and dens
- Sunrooms and recreation rooms
- Closets, hallways, and laundry areas

The AFCI circuit breaker is similar in appearance and installation to the GFCI circuit breaker. The grounded and ungrounded branch circuit conductors are attached to the AFCI circuit breaker, which is connected to the neutral bus bar via a factory-installed pigtail or the clip-on neutral feature available as an option in some panelboards. According to *Article 100*, an AFCI is "*a device intended to provide protection from the effects of arc faults by recognizing characteristics unique to arcing and by functioning to de-energize the circuit when an arc fault is detected.*" This recognition of the unique signature of an electric arc is done electronically by the AFCI circuit breaker or receptacle.

Arc-fault protection requirements are broader than ground-fault protection requirements. It should be noted that the required GFCI protection is specifically for *receptacles*, while the required AFCI protection is for *outlets* and *devices*. However, there is some overlap in GFCI and AFCI protection requirements, notably in kitchens.

There are two ways to protect the kitchen branch circuits to comply with the *NEC*. The first is to use an AFCI circuit breaker and GFCI receptacles. The second, which has become the increasingly common method, is to use *dual-purpose circuit breakers* that provide both AFCI and GFCI protection. Many electricians also install dual-purpose breakers for all 120-volt branch circuits as an added safety measure for

AFCI breaker with clip-on neutral connection

Neutral bus bar clip

Breaker installed in panel

Goodheart-Willcox Publisher

Figure 28-3. Some AFCI and GFCI circuit breakers attach directly to the neutral bus bar in the panel. This eliminates the need for the circuit breaker to have a neutral pigtail to attach to the bus bar. Circuit breakers must be approved for use in the specific panel by the manufacturer.

their clients. This technology, if installed properly, saves lives and is effective in preventing fires caused by arcs.

28.5 Circuit Identification and Panel Labeling

Section 408.4(A) requires every circuit in the distribution panel to be legibly identified to its use with enough detail to differentiate it from all other circuits. Therefore, a circuit directory must be located on the face of, or inside, the panel door or at each circuit breaker. In addition, do not identify circuits based on temporary conditions of occupancy, such as "Sam's Room."

Branch circuits used for the same purpose must be identified as to their locations. Therefore, multiple listings for "Lights" or "Receptacles" do not sufficiently identify the circuits. Acceptable methods of labeling branch circuits include "Front Bedroom" if there is only one front bedroom or "Kitchen Lights." A clear and complete listing of all circuits is a key safety consideration for anyone who needs to access the circuit breakers in an emergency.

Summary

- During the trim-out phase, the outer sheathing of each cable is removed, and the branch circuit conductors are "landed" at the proper termination point in the panel trim-out to start the trim-out phase.
- The bare equipment grounding conductors (EGCs) are terminated to the equipment grounding bar, purchased separately, bonded to the enclosure, and the white grounded conductors are connected to the neutral bar isolated from the enclosure.
- The ampacity rating of a circuit breaker establishes the branch circuit's rating, and the circuit breaker is sized to the conductor's ampacity. Use the 60°C (140°F) column of *Table 310.16* when selecting the proper breaker size.
- *Section 240.4(D)* limits the overcurrent protection for copper 14 AWG, 12 AWG, and 10 AWG to 15 amps, 20 amps, and 30 amps, respectively.
- NM and UF cable conductors sized 8 AWG and larger are protected at their 60°C (140°F) ampacities per the ampacities listed in *Table 310.16*.
- 120-volt branch circuits will be fed from single-pole circuit breakers. Single-pole breakers are rated at 15 or 20 amps and take up one space in the panel. In the panel, a single-pole breaker will stab or connect onto one of two legs of a bus bar.
- 240-volt branch circuits will be fed from double-pole circuit breakers. Double-pole breakers will be rated 15-amps and larger, but typically no larger than 100 amps, and take up two spaces in the panel. In the panel, a double-pole breaker will stab onto the line-1 and line-2 legs of the two vertical bus bars.

- *Section 210.8(A)* outlines the requirements for GFCI protection for 120-volt 15- and 20-amp receptacles. With a GFCI circuit breaker, both the branch circuit's grounded conductor and ungrounded conductor are attached to the circuit breaker.
- The AFCI circuit breaker is similar in appearance and installation to the GFCI circuit breaker. The grounded and ungrounded branch circuit conductors are attached to the AFCI circuit breaker. The circuit breaker is connected to the neutral bus bar via a factory-installed pigtail or the clip-on neutral feature available with some panelboards.
- To protect the kitchen receptacles and comply with the *NEC*, many electricians install dual-purpose circuit breakers that provide both AFCI and GFCI protection.
- *Section 408.4(A)* requires every circuit in the distribution panel to be legibly identified as to its use with enough detail to differentiate it from all other circuits. No circuit shall be identified such that they depend on the temporary conditions of occupancy.

Know and Understand

1. *True or False?* The feeder conductors are installed in the distribution panel during the rough-in stage.
2. *True or False?* When the service disconnect is located at the service equipment outside the house, the bare equipment grounding conductors and the white grounded conductors are isolated from one another in every dwelling unit subpanel.
3. *True or False?* In general, circuit breakers are interchangeable among brands.
4. The rating of a branch circuit is established by the

   _____.
   A. the load
   B. conductor size
   C. rating of the circuit breaker
   D. service size
5. When the wiring method is NM cable, the *NEC* limits the ampacity of the conductors to the _____ column of *Table 310.16*.
   A. 60°C (140°F)   C. 90°C (194°F)
   B. 75°C (167°F)   D. None of these.
6. The smallest conductor size allowed in residential wiring branch circuits is _____.
   A. 18 AWG   C. 14 AWG
   B. 16 AWG   D. 12 AWG
7. *True or False?* If the ampacity listed in *Table 310.16* does not match a standard size fuse or circuit breaker listed in *Table 240.6(A)*, the next lower size fuse or circuit breaker must be used to protect the circuit.
8. *True or False?* In a distribution panel, the ungrounded conductors are always connected to the circuit breaker.
9. The largest branch circuit in a residence is typically rated at _____.
   A. 40 amps   C. 60 amps
   B. 50 amps   D. 100 amps

10. What is the benefit of using a GFCI circuit breaker rather than a GFCI receptacle in branch circuits where GFCI protection is required?
    A. GFCI circuit breakers are less expensive than GFCI receptacles.
    B. GFCI circuit breakers respond faster in clearing a ground fault.
    C. GFCI circuit breakers are more reliable than GFCI receptacles.
    D. GFCI breakers protect the entire circuit from ground faults capable of causing electric shock or electrocution.
11. *True or False?* Lighting outlets are not required to have GFCI protection.
12. *True or False?* The *NEC* requires AFCI protection for branch circuits serving bathrooms, basements, and garages.
13. *True or False?* A circuit directory of branch circuits is required to be affixed to the distribution panel.

Apply and Analyze

1. What must be done to the white conductor at the circuit breaker when installing a 3-wire, 240-volt circuit?
2. Explain the operation of the GFCI circuit breaker.
3. Discuss the two methods of providing both GFCI protection and AFCI protection to kitchen receptacles.

Critical Thinking

1. What is the common feature of areas where the *NEC* requires GFCI protection?
2. Explain how AFCI protection requirements for a dwelling are broader than GFCI requirements per the *NEC*.

CHAPTER **29**

Branch Circuit Testing and Troubleshooting

Grandbrothers/Shutterstock.com

SPARKING DISCUSSION

How does an incorrectly wired receptacle or switch impact the time required to complete an electrical installation?

LEARNING OBJECTIVES

- Summarize the basic rule for troubleshooting an electrical circuit.
- Explain how electrical power is established for the first time.
- Discuss how a standard receptacle is tested using a plug-in tester.
- Identify the manufacturer-approved method of testing GFCI receptacles.
- Compare how GFCI and AFCI breakers are tested.
- Explain how to test single-pole switches.
- Summarize how three-way and four-way switching circuits are tested.
- Explain the purpose for testing a 240-volt circuit.
- Discuss how a 120/240-volt appliance receptacle is tested.

TECHNICAL TERMS

electrical troubleshooting
luminaire testing
receptacle testing

Introduction

When the trim-out phase is completed, all the feeders and branch circuits can be energized, allowing you to test the electrical system. Testing the system is key to ensure proper operation and that there are no short circuits or ground faults present. There are many issues that can be detected during troubleshooting, including a nonworking lighting outlet, receptacle, or a circuit breaker tripping when a light switch is closed. Even the most experienced electrician will encounter electrical faults or miswired circuits from time to time, so you must know how to identify and resolve these situations.

Electrical troubleshooting involves systematically examining a nonworking electrical circuit to determine the root problem. Since problems manifest when the circuit is energized, troubleshooting is the rare instance where electricians might work on an energized circuit. This chapter will provide you with the foundational knowledge and application to begin understanding how to best approach troubleshooting for panelboards, branch circuits, receptacles, luminaires, and major appliances.

SAFETY NOTE   **NFPA 70E**

NFPA 70E, titled the *Standard for Electrical Safety in the Workplace*, permits testing, troubleshooting, and voltage measuring on an energized circuit. This requires that the appropriate safe work practices are followed, and the proper PPE is worn. For circuits 240 volts and less, PPE required is voltage-rated rubber gloves with leather protectors, hard hat, safety glasses, and an arc–rated PPE Category 1 rated long sleeve shirt. Care must always be taken to avoid injury by a fall or shock.

29.1 Establishing Electrical Power for the First Time

Depending on a municipality's rules and regulations, either the AHJ or an electrician calls the serving utility to establish permanent power to a dwelling. Before the arrival of the utility's service truck, the electrician should prepare a few things. Because an electrical arc can occur when the electrical meter is inserted into the meter socket, utilities require the service disconnect and all circuit breakers in the distribution panel to be in the OFF position. An electrician can perform this as well as put all light switches in the OFF position to help later troubleshooting.

When the utility worker has completed their work and the electric meter has been installed, power should be on the line side of the main breaker in the panelboard inside the dwelling unit, provided that there is no exterior disconnect. If there is an exterior disconnect on the dwelling unit and the outdoor service disconnecting means is put in the closed position, a distribution panel without a main breaker can then be energized.

Most main interior panelboards contain a main breaker. In these cases, power is on the line side, and when the main breaker is energized, the entire panel becomes energized. Once this is complete, turn the breakers on at the distribution panel. It is good practice to start with the larger two-pole breakers. Next, turn on each of the remaining single-pole breakers. If the breaker trips when it is first closed, do not close it again; instead, make a note to troubleshoot the circuit.

SAFETY NOTE   **Tripped Breakers**

According to NFPA 70E, a breaker that opens under a short-circuit or ground-fault condition cannot be re-energized until the cause of the fault is determined and corrected. A very important unwritten rule is to always stand off to the side of the panelboard and turn your head away from the breakers every time you turn on any electrical breaker or switch. If a dangerous short circuit does occur, a serious arc fault could result. Sometimes this arc can cause serious injury or death. Although not likely to occur in dwelling units, the possibility does exist in certain municipalities that utilize large amounts of electricity in a small area.

Once electrical power is established, an electrician must go through various tests to troubleshoot the electrical system. This includes branch circuits, receptacles, luminaires, and testing of major appliances that are hardwired (directly connected without a cord and plug). Basic testing of the system also requires voltage testing at the main distribution panel and all subpanels.

29.2 Main Panelboard and Subpanel Testing

Before turning on power to any panelboards, it is best practice to ensure all the terminations are properly torqued. To accomplish this, you must follow the manufacturer's torque requirements during the trim-out portion. Loose connections on electrical wires are always a problem and will lead to overheating and possible shorting out of the connections.

You must test the voltage at the panel before energizing the individual circuit breakers. The voltage between line 1 and line 2 should be approximately 240 volts. Depending on the utility, actual voltage can vary. If it is a long distance from the utility's transformer, the delivered voltage can be lower, especially when measuring voltage under a load. Remember, the greater the amperage draw, the higher the voltage drop.

Voltage from line 1 or line 2 to the grounded, neutral conductor should be exactly half of the line-to-line voltage. Thus, if the line-to-line voltage is 240 volts, the line-to-neutral voltage should be 120 volts. The line-to-neutral voltage should be the same as the line-to-ground voltage because the grounding electrode conductors and the neutral (grounded) conductor are all terminated at the same place for the main electrical service. If the voltage readings are not near these values, a problem may exist.

If the voltage testing indicates a problem, the local utility needs to be contacted. The utility can be experiencing the same issues as the customer, including loose connections, malfunctioning transformers, or weather-related issues. Oftentimes, problems with the incoming electrical power are on the utility side.

29.3 Branch Circuit Troubleshooting

When troubleshooting a circuit breaker for a branch circuit that trips or experiences a short circuit or a ground fault, the number one rule is to fix the problem before attempting to re-energize the circuit breaker. This rule applies to not only branch circuits but also to service or feeder conductors.

If the short is not detected, it is possible (and very undesirable) that a drywall screw has penetrated the NM cable. If this occurs, you must determine which two boxes the system's cable are located on either side of the short. Ensuring both ends of the cable are disconnected, use an electrical meter to indicate if there is a short between the conductors. If it is badly damaged, the entire cable may have to be replaced, which requires opening up parts of the wall.

PROCEDURE

Troubleshooting a Circuit Breaker

Follow the process below to check a circuit breaker after it trips or shorts when it becomes energized.

1. Ensure the power is OFF.
2. Set your electrical meter to test resistance. Go to any receptacle or the main panel and test the resistance between the ungrounded and the grounded (neutral) conductor or the grounding electrode conductor.
3. If you read zero resistance (measured in ohms), this indicates a short circuit or ground fault occurred.
4. Test all light switches in the OFF position, or all the light bulbs, ballasts, or LED drivers while disconnected. Every light source, ballast, or driver has an internal resistance that exists under normal conditions, but this resistance is not low enough to allow enough current to trip the circuit breaker.
5. Work now to determine the location of the short circuit. Examine every switch and receptacle until

the problem is found. Note: if a problem is not detected, the issue may be buried in the wall.

There are a few methods that can be used to try to find the source of the short circuit. First, try the *halfway point*, or *50/50* method:

1. Find the halfway point of a branch circuit.
2. Take the joints apart to isolate and test which part of the wiring contains a fault.
3. Move to the next halfway point on a branch circuit, and repeat the process until the problem is found.
4. If no problem is found at the receptacles or switches, it could be in a light fixture. The problem may be a "pinched" wire or an ungrounded wire that has popped out of its termination device and shorted to ground.

PRO TIP   **Repairing Damaged Cable**

A repair possibility allowed by *Section 334.40(A)* is to use a nonmetallic-sheathed cable interconnector device. This allows the cable to be spliced and concealed without a box and buried behind drywall, but even then, part of the wall must still be opened to repair the cable. Make sure a listed interconnector device is used. Depending on how much contact the screw has made with the cable, it may be the case a standard circuit breaker will not open, but an AFCI breaker will. This demonstrates the AFCI is acting as designed.

Sometimes, a portion of a circuit operates while another portion of the same circuit does not. This condition is usually caused by a box that has been roughed in but has been covered by sheetrock and not trimmed out. The covered box must be found and trimmed out for the circuit to function. There is often a slight bulge in the sheetrock where a box has been covered. It is the responsibility of the electrician to find the covered box, cut the sheetrock, and install the switch or receptacle to get the circuit operating.

29.4 Receptacle Testing

Receptacle testing is used to determine proper wiring and detect circuit conditions, such as open ground, open neutral, open hot, hot/ground reversed, hot/neutral reversed,

and correct wiring. This test uses a three-prong plug-in tester, also simply called a receptacle tester. See **Figure 29-1**.

A receptacle tester combines the necessary steps to determine all the possible operating conditions of an energized

Klein Tools

Figure 29-1. A three-prong receptacle tester with an LCD display. This particular model displays the voltage level and the correctness of the wiring.

Figure 29-2. This receptacle tester tests the wiring condition and the operation of AFCI-protective devices and GFCI-protective devices by simulating an arc-fault or a ground-fault condition in the circuit.

circuit feeding a receptacle. If this type of tester is not available, the electrician can use an electrical meter to determine how the receptacle is wired—but this is not as quick. Any miswired receptacle should be investigated and rewired as necessary. Note that a receptacle wired incorrectly, typically with the ungrounded conductor reversed with the neutral or missing an equipment grounding conductor, still operates but not correctly.

Many receptacle testers also have a button for checking the operation of GFCI devices, whether a receptacle or a circuit breaker. Although the GFCI test button is a convenient method of checking the operation of a GFCI device, the only manufacturer-approved method of testing GFCIs is to use the test button on the receptacle or the breaker.

Arc-fault breakers are tested using the button on the breaker or by using an arc-fault tester, **Figure 29-2.** Each receptacle in the house should be tested for proper wiring, and the functionality of the GFCI and the AFCI protective devices should be tested for those areas where required.

29.5 **Luminaire Testing and Troubleshooting**

Luminaire testing is completed to test for proper operation of all luminaires and their switching circuits. This is required for all single-pole, three-way, and four-way switched circuits. Testing single-pole switched circuits is a simple process. For example, when the switch to a luminaire is first turned on, one of three actions occur:

- The luminaire turns on.
- The luminaire does not turn on.
- The circuit breaker protecting the circuit trips.

If the luminaire turns on with the switch is ON or in the up position for a single-pole switch, the circuit is wired properly.

PROCEDURE | **Troubleshooting Luminaires**

Further troubleshooting is required if the luminaire does not illuminate or the breaker trips when the switch is placed in the ON position. Follow the steps below for each scenario to try to isolate the issue.

Troubleshooting a nonworking luminaire:

1. First ensure the luminaire is functioning. Try a known working one before disassembling switches and the luminaire.

PRO TIP | **LED Technology**

With the new LED technology, the problem may be the driver for the LED lights. If possible, return the entire LED unit to the lighting supplier and receive a replacement.

2. Open the box containing the switch (rather than opening up or dropping the luminaire or fan/light combination).
3. Check to see if the proper voltage is at the switch.
4. If the neutral is present, check the voltage from the line-to-neutral and then line-to-grounding electrode conductor.
5. Turn on the switch to see if voltage is going to the luminaire. If not, this indicates a bad switch. If voltage is on the load side, there is an issue with the neutral or switch leg at the luminaire.
6. Another option to test the switch is to turn the circuit off and test continuity in the ON and OFF position.

Troubleshooting a circuit breaker ground fault or short circuit:

1. Check in the switch box to see if an ungrounded, hot conductor is contacting an equipment grounding conductor or a neutral conductor not properly terminated.
2. If this is the case, resolve the issue and turn on the luminaire.
3. If this is not the case, disconnect the switch leg, equipment grounding conductor, and neutral (if in the switch box).
4. Test continuity between the switch leg, equipment grounding conductor, and neutral. The meter reading should indicate no continuity between any of the conductors. Reading a very low resistance (continuity) will indicate a short.
5. If further investigating for a short, look for a pinched or incorrectly terminated wire. It is possible you discover a wire not properly installed in a wire nut. If either of these are observed, be sure to resolve the issue.

Testing and troubleshooting are more complicated with three-way and four-way switching circuits. In addition to the three simple actions that can occur with single-pole switches, three-way and four-way switched circuits might only work correctly if the switches are in certain positions. A mistake often made in wiring three-way switching circuits is terminating a traveler conductor to the common terminal on one or both three-way switches or incorrectly terminating the wires on a four-way switch, **Figure 29-3**. The common terminal of one of the two switches in a three-way switched circuit is intended for the ungrounded or "hot" feed. The common terminal on the other switch has the switch leg terminated to it.

Note that every three-way switch has two terminals that are the same color designed for the travelers (usually gold or brass) and a third terminal that is a different color (usually brown) for the common conductor, which is either the ungrounded conductor or the switch leg. One of the two three-way switches has an ungrounded or hot conductor, while the other 3-way switch is the landing spot for the switch leg.

The four-way switch must also be wired correctly for the circuit to function as intended. A four-way switch has two sets of two screws. Each set is a different color, such as brown for one set and brass or gold for the other. Each set also has wires from the same cable terminated as travelers from one direction.

The correct method for testing three-way switching circuits is to turn the luminaire ON with Switch A and OFF with Switch B, then ON again with A and OFF again with B. This is called ABAB switching. This method tests both of the traveler conductors and ensures that the circuit works properly with the switches in any position.

Testing a four-way switching circuit is performed in the same manner as with three-way circuits. First, the ABAB switching is performed on the three-way switches. Then, the position of the four-way switch is changed, and the three-way switches are again tested ABAB. Sometimes it helps to temporarily remove the four-way switch from the circuit and get the circuit functioning as a three-way circuit first, then reinstall the four-way switch. The luminaire should be able to be turned ON or OFF with any switch.

29.6 Major Appliance Testing

Recall that major appliances are those served by an individual branch circuit that supplies no other loads. These appliances are typically powered by 240-volt or 120/240-volt circuits. In general, 240-volt appliances are hard-wired and are required to have a disconnecting means if they are not located within sight of the overcurrent device that protects the circuit. Typical 240-volt appliances include water heaters and heating and air conditioning systems. On the other hand, 120/240-volt circuits are generally cord-and-plug connected and are commonly the range and the clothes dryer.

Testing a 240-volt appliance involves using a multimeter to test for the proper voltage at the appliance or the disconnecting means. A 120/240-volt receptacle is tested using a multimeter, **Figure 29-4**. The meter should have the following readings:

- 240 volts between the two "hot" terminals: X and Y
- 120 volts between either "hot" terminal (X or Y) and the neutral terminal W
- 120 volts between either "hot" terminal (X or Y) and the grounding terminal G
- 0 volts between the neutral terminal W and the grounding terminal G

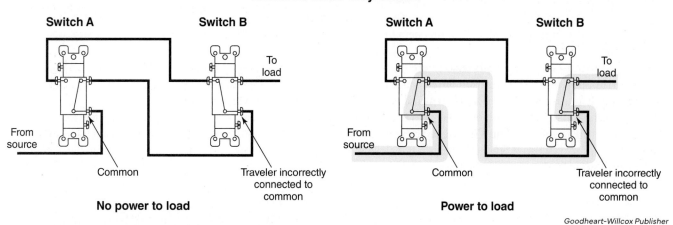

Miswired Three-Way Switch

Goodheart-Willcox Publisher

Figure 29-3. A miswired 3-way switching circuit. Switch B has a traveler incorrectly connected to the common terminal. Switch A is wired correctly. With Switch A in the position shown, toggling Switch B controls the flow of power to the load. However, when Switch A is in the other position, no power can reach the load.

Eaton

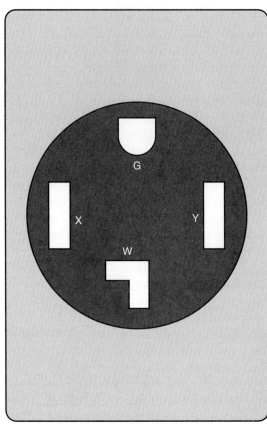

Voltage Testing

X to Y: 240 V
X or Y to G: 120 V
X or Y to W: 120 V
G to W: 0 V

Goodheart-Willcox Publisher

Figure 29-4. Expected voltages when testing a 120/240-volt dryer receptacle.

Summary

- Troubleshooting involves systematically examining a nonworking electrical circuit to ascertain the root problem.
- Many problems only manifest when the circuit is energized, and troubleshooting is one of the rare instances when electricians might work on an energized circuit.
- Depending on the municipality's rules and regulations, either the AHJ or electrician calls the serving utility to establish permanent power to the dwelling.
- Because of the possibility of establishing an electrical arc when the meter is inserted into the socket, the utility requires the service disconnect and all circuit breakers in the distribution panel to be in the OFF position to ensure no electrical load is being drawn.
- One of the most common faults encountered when testing an electrical system for the first time is a circuit breaker that trips upon initial energization, often caused by a ground fault, where a bare equipment grounding conductor is contacting an ungrounded or "hot" terminal screw on a switch or receptacle.
- One method of testing circuit breakers is to turn on the 2-pole circuit breakers at the distribution panel from the largest rating to the smallest. This helps determine any unwanted line-to-line faults in the larger branch circuits.
- If a breaker trips when it is first closed, do not close it again but make a note to troubleshoot the circuit. The cause of the fault must be determined and corrected first.
- The three-prong plug-in tester is used to test receptacles for proper wiring.
- Many receptacle testers also have a button for checking the operation of GFCI devices.
- Each receptacle in the house should be tested for proper wiring, and the functionality of the GFCI and the AFCI protective devices should be tested for those areas where required.
- When testing luminaires, if the breaker holds but the luminaire does not illuminate, the first thing to check is that the bulb or LED is in working order.

- When the switch to a luminaire is first turned on, one of three results will occur: the luminaire will turn one, the luminaire will not turn on, or the circuit breaker protecting the circuit will trip.
- A mistake often made in wiring three-way and four-way switching circuits is to terminate a traveler conductor to the common terminal on one or both three-way switches.
- The correct method for testing three-way switching circuits is to turn the luminaire ON with Switch A and OFF with Switch B, then ON again with Switch A and OFF again with Switch B.
- Major appliances are typically powered by 240-volt or 120/240-volt circuits.
- 240-volt appliances are generally hardwired, and 120/240-volt circuits are generally cord-and-plug connected.

Know and Understand

1. Receptacles are tested for proper connection using a(n) _____.

   A. three-prong plug-in tester
   B. ohmmeter
   C. ammeter
   D. continuity tester

2. *True or False?* The GFCI test button on some three-prong receptacle testers is the only way to test a GFCI receptacle or a GFCI circuit breaker's functionality.

3. *True or False?* Arc-fault circuit breakers can be tested using a GFCI tester.

4. The most straightforward luminaire switching circuit to troubleshoot is the _____.

   A. four-way circuit   C. split circuit
   B. three-way circuit   D. single-pole circuit

5. *True or False?* Three-way and four-way switching circuits are tested in a similar manner.

6. *True or False?* All major appliances are hardwired.

7. *True or False?* All major appliances are cord-and-plug connected.

8. A 120/240-volt receptacle should read _____ volts between the *X* and *Y* terminals.

   A. 120 volts   C. 0 volts
   B. 240 volts   D. None of the above.

9. A 120/240-volt receptacle should read _____ volts between the *W* and *G* terminals.

   A. 120 volts   C. 0 volts
   B. 240 volts   D. None of the above.

10. A 120/240-volt receptacle should read _____ volts between the *X* and *G* terminals.

    A. 120 volts   C. 0 volts
    B. 240 volts   D. None of the above.

Apply and Analyze

1. Why must the electrician use caution when troubleshooting an electrical circuit?

2. Describe the proper sequence for testing a three-way switching circuit.

3. Name three potential outcomes when a luminaire is energized for the first time.

Critical Thinking

1. Explain the difference between wiring a 240-volt appliance and a 120/240-volt appliance.

2. When testing a 120/240-volt receptacle for the clothes dryer, an electrician encounters 240 volts between terminals *X* and *W*, 120 volts between terminals *X* and *G*, and 120 volts between terminals *W* and *G*. There are 0 volts between *Y* and *G*. What is the likely cause of these readings?

3. An AFCI circuit breaker is occasionally tripping out and opening the circuit when a load is plugged into the circuit. A standard breaker is installed to replace the AFCI breaker. The traditional standard breaker is not experiencing the same problems. Is it okay to remedy the problem in this fashion?

SECTION 10

Outdoor Circuits

Alexey Stiop/Shutterstock.com

CHAPTER 30
Outdoor Lights and Receptacles

CHAPTER 31
Swimming Pools and Hot Tubs

In addition to the indoor circuits, most residences also have outdoor electrical circuits for lights and receptacles. Some houses will also have a swimming pool, hot tub, or both. It should be noted that outdoor wiring has additional considerations than indoor wiring. This section sets forth the requirements for outdoor residential circuits.

Chapter 30, *Outdoor Lights and Receptacles*, covers the installation of luminaires and receptacles outside the dwelling. Thus, an emphasis is placed on overhead conductor spans and burial depths for underground conductors for these circuits.

Chapter 31, *Swimming Pools and Hot Tubs*, highlights the installation of permanent and storable swimming pools, spas and hot tubs, and indoor hydromassage bathtubs. Emphasis is given to the *NEC* requirements for these installations.

CHAPTER **30**

Outdoor Lights and Receptacles

Irina Mos/Shutterstock.com

- Describe the different wiring methods for outdoor wiring.
- Identify the minimum burial depths for outdoor residential wiring.
- Calculate the requirements of expansion fittings for PVC conduit.
- Describe the process for installing a post light for a dwelling.
- Explain the requirements for installing outdoor receptacles.

TECHNICAL TERMS

conduit-supported
messenger wire
overhead conductor span
post light
post-supported
raintight
sunlight resistant
weatherhead
weatherproof

Introduction

Residential electrical wiring can extend beyond the walls of a dwelling and includes outdoor circuits, feeders, lights, and receptacles. Most of these outdoor circuits are preferred to be installed underground, although there are a few exceptions. For example, many homes have post lights to illuminate sidewalks or driveways. In addition, there may be a need for a receptacle outlet to be installed for a decorative fountain, seasonal lighting, or general use for yard equipment.

Regardless of the situation, the *NEC* has specific outdoor and underground wiring rules pertaining to burial depths and other installation requirements. This chapter will discuss the methods and requirements for installing outdoor lighting and receptacle circuits.

30.1 Outdoor Wiring Methods

For outdoor wiring, conductors are installed either overhead via exposed wiring methods on the surface of the building or buried underground. *Article 225* covers the rules for installing outdoor branch circuits and feeders. The types of wiring methods used for outdoor installations need to be rated for certain conditions, such as sunlight resistant or suitable for wet locations. Additional items to consider are movement caused by settling and seasonal weather changes.

30.1.1 Overhead Spans

Where circuits or feeders are installed as an overhead conductor span, there are specific requirements associated with these installations. An **overhead conductor span** is typically installed with an insulator wire holder anchor point on each building or structure to support the span of wires. A cable rated for outdoor use is strung between the insulator wire holders. The cable may contain or be attached to a messenger wire to allow it to be pulled tight and increase the ground clearance.

The distance between the insulated wire holders and any intermediate insulating supports will be a deciding factor for the wire size required. The wires used for overhead conductor spans primarily are required to be sized for the rating of the branch circuit overcurrent protective device. The distance between supports may require a larger-sized wire based on the distance between span supports. *Section 225.6* requires the minimum conductor sizes for overhead spans to be 10 AWG copper or 8 AWG aluminum for spans up to 50′. If this distance is longer, a messenger wire is required to support the cable span. A **messenger wire** is a steel cable or a wire that is run along with or integral with a cable or conductor to provide mechanical support.

A triplex cable is a type of cable that can be used for overhead spans. Triplex cable is constructed with outdoor sunlight-resistant insulated aluminum conductors twisted together with a bare conductor. A triplex cable has a hardened messenger wire within the bare conductor. Where the cable attaches to the insulating wire holder, a preformed dead-end grip or wedge clamp is used to grip the bare conductor and connected to the insulated wire holder attachment.

The height above the ground for an overhead span will isolate the circuit from possible contact by persons or vehicles. *Section 225.18* lists the clearance for overhead conductors and cables applicable for residential outdoor wiring. Where the voltage to ground does not exceed 150 volts and the area is accessible to pedestrians only, a clearance height of 10′ is required to the lowest point of the span above the finished grade, sidewalks, or any platform that will allow personal contact. Where the overhead span crosses over residential driveways, a clearance of 12′ is required.

30.1.2 Underground Wiring

The principles of underground wiring are to install conduits or cables below grade to protect them from physical damage and prevent contact with electrical conductors. The types of wiring methods used in many residential applications are UF cable, PVC conduit, and in some cases, intermediate metal conduit or galvanized rigid conduit.

The requirements for underground wiring can be found in *Section 300.5*, with the minimum burial depths covered in *Table 300.5(A)*, **Figure 30-1**. This table has five columns of burial depths based on the wiring method and location of the wiring method or circuit. There are seven specified locations of wiring methods contingent on whether the installation is under concrete, buildings, roadways, parking lots, residential driveways, airport runways, or a location not specified. Column 4 is for residential branch circuits rated 120 volts or less with GFCI protection and for circuits not rated more than 20 amps. A branch circuit that conforms to these conditions need to only be buried at a depth of 12″ in a non-specified area or under a one- or two-family driveway or dwelling-related parking space. The burial depth can be reduced to 6″ if under 2″–4″ of concrete, or 4″ if under 6″ of concrete. A minimum burial depth prevents physical damage to the conductors.

Type UF cable is the most common wiring method for underground residential branch circuits. Recall that Type UF cable has an outer sheathing suitable for direct burial without a protective conduit. However, *Section 300.5(D)(1)* requires that enclosures or raceways shall protect buried cables emerging from grade from the minimum cover distance below grade to a height of 8′ above finished grade. The trench, or *excavation,* where UF cable is installed needs to have backfill that does not contain large rocks, paving materials, cinders, or other sharp or corrosive materials that can damage the cables. Where the UF cable enters a raceway, a bushing is required to be installed on the end of the conduit, and an S-shape or U-shape is bent in the cable to allow for movement that can occur with settling or frost.

If an underground UF cable becomes damaged, it can be repaired with a splice. *Section 300.5(E)* permits direct-buried conductors or cables to be spliced or tapped without a splice or junction box if the splicing means is listed for underground use. Underground splice kits use a set screw splicing mechanism sealed in a water-tight heat-shrink tube, **Figure 30-2**. These splice kits maintain the integrity of the jacket material because the inside conductors from the cable are not suitable for direct burial.

Direct burial twist-on connectors are also available, **Figure 30-3**. These connectors are filled with a silicone-based sealant to protect against moisture and corrosion. Twist-on wire connectors are suitable for use mainly in underground splicing enclosures. The conductors within the cable are required to be protected in the same manner as the original cable sheath, such as the shrink tube from a splice kit. The underground twist-on wire connectors require some type of covering that will keep the integrity of the jacket of the cable over the splices and conductor insulation. This can be done by making the splice in a junction box or by adding a layer of cable jacket repair tape at a thickness equal to the original cable jacket.

If the method for underground wiring is in a conduit, *Section 300.5(B)* states that the interior of raceways installed underground is considered a wet location. This means that any conductors installed in an underground conduit shall be of a type listed for wet locations.

Typically, the conductors will have a *W* in their insulation designation, such as *RHW, THW,* or *XHHW.* These letter designations are required by *Section 310.10(C)*. The letters associated with the conductor insulation types can be viewed in *Table 310.4(A)*.

- The letter *T* indicates a thermoplastic insulation.
- The letter *R* indicates a thermoset rubber insulation, and the *X* indicates a heat-cured, cross-linked polymer insulation.
- The letter *H* designations indicate a high temperature rating of 75°C or *HH* indicates a high temperature rating of 90°C.
- The letter *N* indicates the conductor having a nylon or Teflon jacket to protect the conductor from abrasion during installation.

The most popular wire type for residential underground conduit applications is THWN insulation. This wet location, nylon-jacketed, thermoplastic-insulated wire has an ampacity temperature rating of 75°C or 90°C if a "-2" suffix is added.

Conduits used for underground wiring that attach to a building after they emerge from underground are required

Table 300.5(A) Minimum Cover Requirements, 0 to 1000 Volts ac, 1500 Volts dc, Nominal, Burial in Millimeters (Inches)

Location of Wiring Method or Circuit	Column 1 Direct Burial Cables or Conductors		Column 2 Rigid Metal Conduit or Intermediate Metal Conduit		Column 3 Electrical Metallic Tubing, Nonmetallic Raceways Listed for Direct Burial Without Concrete Encasement, or Other Approved Raceways		Column 4 Residential Branch Circuits Rated 120 Volts or Less with GFCI Protection and Maximum Overcurrent Protection of 20 Amperes		Column 5 Circuits for Control of Irrigation and Landscape Lighting Limited to Not More Than 30 Volts and Installed with Type UF or in Other Identified Cable or Raceway	
	mm	in.	mm	in.	mm	in.	mm	in.	mm	in.
All locations not specified below	600	24	150	6	450	18	300	12	150[1,2]	6[1,2]
In trench below 50 mm (2 in.) thick concrete or equivalent	450	18	150	6	300	12	150	6	150	6
Under a building	0 (in raceway or Type MC or Type MI cable identified for direct burial)	0	0	0	0	0	0 (in raceway or Type MC or Type MI cable identified for direct burial)	0	0 (in raceway or Type MC or Type MI cable identified for direct burial)	0
Under minimum of 102 mm (4 in.) thick concrete exterior slab with no vehicular traffic and the slab extending not less than 152 mm (6 in.) beyond the underground installation	450	18	100	4	100	4	150 (direct burial) 100 (in raceway)	6 4	150 (direct burial) 100 (in raceway)	6 4
Under streets, highways, roads, alleys, driveways, and parking lots	600	24	600	24	600	24	600	24	600	24
One- and two-family dwelling driveways and outdoor parking areas, and used only for dwelling-related purposes	450	18	450	18	450	18	300	12	450	18
In or under airport runways, including adjacent areas where trespassing is prohibited	450	18	450	18	450	18	450	18	450	18

[1]A lesser depth shall be permitted where specified in the installation instructions of a listed low-voltage lighting system.
[2]A depth of 150 mm (6 in.) shall be permitted for pool, spa, and fountain lighting, installed in a nonmetallic raceway, limited to not more than 30 volts where part of a listed low-voltage lighting system.
Notes:
1. Cover shall be defined as the shortest distance in mm (in.) measured between a point on the top surface of any direct-buried conductor, cable, conduit, or other raceway and the top surface of finished grade, concrete, or similar cover.
2. Raceways approved for burial only where concrete encased shall require a concrete envelope not less than 50 mm (2 in.) thick.
3. Lesser depths shall be permitted where cables and conductors rise for terminations or splices or where access is otherwise required.
4. Where one of the wiring method types listed in Columns 1 through 3 is used for one of the circuit types in Columns 4 and 5, the shallowest depth of burial shall be permitted.
5. Where solid rock prevents compliance with the cover depths specified in this table, the wiring shall be installed in a metal raceway, or a nonmetallic raceway permitted for direct burial. The raceways shall be covered by a minimum of 50 mm (2 in.) of concrete extending down to rock.
6. Directly buried electrical metallic tubing (EMT) shall comply with 358.10.

Figure 30-1. Detail of *Table 300.5(A)*, minimum cover requirements for buried electrical conductors.

to have expansion fittings or be flexible where they are subject to movement by settlement or frost. PVC conduits can use a standard expansion fitting to perform this function. Metal conduits or raceways may transition to a liquidtight flexible conduit or use metal expansion fittings.

Where underground conduits enter or leave a building from outside, they are required to be sealed. The seal prevents water or moisture from entering a box or panel and contacting live parts. The seal also prevents condensation moisture from causing corrosion. Typical sealing is done within a junction box or a fitting with a removable cover. The sealing material, called *Duct Seal*, is a noncorrosive, nontoxic, dough-like material that can be pressed into the conduit opening. It remains soft and pliable for a long time. This allows it to seal against water and still be removed if the wiring needs to be repaired or replaced.

Goodheart-Willcox Publisher

Figure 30-2. An example of an underground splice kit. The tubing is a heat-shrinking material to seal the splice from moisture intrusion.

Ideal Industries, Inc.

Figure 30-3. Underground twist-on wire connectors listed for direct burial, below-grade installation.

No matter what type of outdoor wiring method is used for branch circuits or feeders, it is important to consider the following key concepts:

- Keep any energized conductors away from possible contact or damage.
- Keep water or moisture from entering electrical wiring or equipment.
- Using wiring methods suitable for outdoors where sunlight, weather, and movement can cause damage.

30.2 **Exterior Conduits and Cables**

The circuits run as overhead conductor spans are typically transitioned to a different wiring method to enter each building. The outdoor wiring methods can be conduit or cable where not subject to physical damage. The installation of raceways requires a fitting, such as a *weatherhead*,

to prevent water from entering the raceway. The raceways on the exterior of a building need to comply with *Section 225.22*.

All boxes, conduit bodies, and fittings are required to be listed for use in wet locations. The conductors exiting the conduit from the weatherhead, and connecting to the overhead conductor cables, need to be listed as sunlight-resistant. Many thermoplastic insulated conductors with a nylon or Teflon covering 8 AWG and larger can be listed sunlight resistant, but not all manufacturers make them. Smaller 14 AWG through 10 AWG conductors are not listed as sunlight resistant.

When the exterior conduit is electrical metallic tubing (EMT), the fittings need to be *weatherproof* or constructed or protected so that exposure to the weather will not interfere with successful operation. When EMT enters or exits enclosures or fittings installed outdoors, a weatherproof connection is required. The methods to make these connections may depend on if the opening of the box is threaded, uses a weatherproof fitting, uses an O-ring gasket, or enters the box through the side or bottom. Fittings can also be *raintight*, meaning they can withstand exposure to a beating rain. Raintight is the designation given to conduit fittings used with EMT and have a distinctive feature or marking that will indicate them as weatherproof fittings.

Some electricians use fittings known as a Meyers Hub to provide a weatherproof connection when entering a knockout in the top of a box without a threaded entry. These hubs are only listed for use with National Pipe Thread (NPT) taper of 3/4″ per foot. Intermediate metal conduit (IMC) and galvanized rigid metal conduit (RMC) have this type of thread. EMT connectors have a standard straight thread to be used with a locknut and are not technically supposed to be used with fittings designed for tapered threads. Manufactures have identified this issue and make fittings that transition between EMT and the hub or box that uses tapered threads.

When an O-ring, or *sealing ring*, is provided with an EMT connector, it is part of the listing of the fitting for use in wet locations. The entry into the top of a box is approved with this fitting but not advised since these sealing O-rings can become misaligned during installation, deteriorate with age, and crack allowing water to enter. The better electrician may choose to use a different wiring method, use a hub with an adapter fitting or enter the box in the side or the bottom.

When the fitting is installed in the side or bottom of a box, the chance of water entry is minimized. Drip loops should be placed in conductors entering outdoor boxes or conduits, **Figure 30-4**. These will cause water to drip off or run down the side and not onto possible energized parts. By entering the side or bottom of a box or enclosure, water will not usually be transferred along the wire into an adjacent conduit. Careful attention to the wiring methods installed outside can make a big difference when trying to keep water and electricity from causing issues.

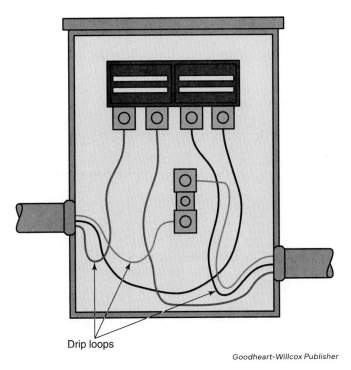

Drip loops

Goodheart-Willcox Publisher

Figure 30-4. Interior of an air conditioner disconnect with drip loops.

Cable wiring methods that can be installed on the exterior of a building were not subject to physical damage are SE cable and UF cable. An interpretation of *Article 300.5(D)(1)* will dictate that any cable installed outdoors below 8′ is required to be protected. In addition, all the cable wiring methods will need to be listed for wet locations and be sunlight resistant.

Where junction or device boxes are used for outdoor wiring methods, they are required to be rated for the environment. *Table 110.28* lists the ratings of these types of boxes. Typical boxes for outdoor residential applications are rated NEMA 3R. These boxes are designed to prevent water from entering and may have provisions to allow water to drain if it gets inside. Small metal and PVC device boxes used for receptacles are typically rated NEMA 3R but do not have any water drain provisions.

Where PVC conduit is installed on the exterior of a building, it is required to be sunlight resistant. PVC plastics, when exposed to ultraviolet radiation, will be degraded and have less impact strength along with discoloration. The *sunlight-resistant* conduit has ultraviolet-absorbing materials which delay the degrading effects of the sun. The installed PVC is required to be type Schedule 80. Schedule 80 conduit has a thicker wall and will maintain stiffness in higher temperatures. Schedule 80 can take more impact abuse than Schedule 40, which has a thinner wall and typically is used for underground conduit installations.

PVC conduit terminations at boxes and fittings will also need to ensure a weatherproof connection. Glued applications into sockets on PVC plastic boxes will inherently be weatherproof when the proper primer PVC cement is applied. The transition between a PVC conduit into a box or fitting with a threaded adapter can be a concern. The female threaded boxes and fittings are listed for use with threaded IMC and galvanized RMC conduit types and the straight thread of a PVC fitting may not work to keep water out. *Section 300.15* states that "*fittings and connectors shall be used only with that specific wiring methods for which they are designed and listed.*" PVC fitting manufacturers make transition fittings for attachment to threaded boxes or fittings which are for the correct tapered thread. When PVC conduit is used with metal boxes or fittings, additional grounding or bonding of the boxes is required.

30.2.1 Calculating Expansion Fitting Requirements

PVC conduit may require expansion fittings where movement of the conduit exceeds 1/4″ between terminations. When PVC conduit is installed between items such as boxes, fittings, luminaires, or other conduit terminations that restrict the movement of the conduit during temperature changes, *Section 352.44(A)* requires that an expansion fitting or coupling be installed to accommodate the movement of the conduit, **Figure 30-5.** The straps used to support the PVC conduit should be of a type to allow the conduit to move when temperature changes cause movement.

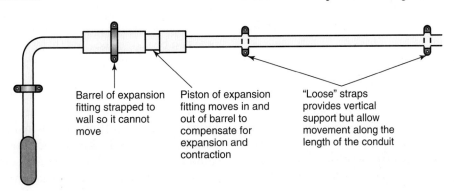

Barrel of expansion fitting strapped to wall so it cannot move

Piston of expansion fitting moves in and out of barrel to compensate for expansion and contraction

"Loose" straps provides vertical support but allow movement along the length of the conduit

Goodheart-Willcox Publisher

Figure 30-5. Expansion fitting in a run of PVC conduit.

To calculate the amount of expansion or contraction of a PVC conduit run, use the following formula developed from data in *Table 352.44(A)*:

$$E = 4.056 \times 10^{-4} \times l \times \Delta t$$

where

E = expansion in inches
l = length in feet
Δt = change in temperature degrees Fahrenheit

Example 1: How much will a 25-foot run of PVC change in length where the expected temperature change is 90°F annually? Will this need an expansion fitting?

$$E = 4.056 \times 10^{-4} \times 25 \times 90$$
$$= 0.9126''$$

Yes, this will need an expansion fitting since the change in length exceeds 1/4″.

To determine the maximum length of conduit allowed before an expansion fitting, use the following formula:

$$\frac{l = 0.25''}{4.056 \times 10^{-4} \times \Delta t)}$$

where

l = length in feet
Δt = change in temperature degrees Fahrenheit

Example 2: How far can a run of PVC conduit be before an expansion fitting is needed when the temperature change between seasons is 80°F annually?

$$\frac{l = 0.25'')}{4.056 \times 10^{-4} \times 80}$$
$$= 7.7' \text{ or } 7'8''$$

The PVC conduit can be 7.7′ or 7′8.5″ before an expansion fitting is needed.

30.3 Outdoor Lighting

The installation and wiring of outdoor lighting is an element of residential wiring that can impact the curb appeal of a home. Decorative lighting on the home, post lights, landscape lights, or architectural lighting can make a difference on how a home looks.

When planning and installing outdoor lighting, consider *Section 410.10* that states, "*luminaires installed in wet or damp locations shall be installed such that water cannot enter or accumulate in wiring compartments, lampholders, or other electrical parts.*" The luminaires selected for outdoor lighting should be rated for wet locations where they are subject to rain, snow, or dew. The industry standard definition for determining a damp versus a wet location is to view an invisible line at 45 degrees from the outer edge of an overhang

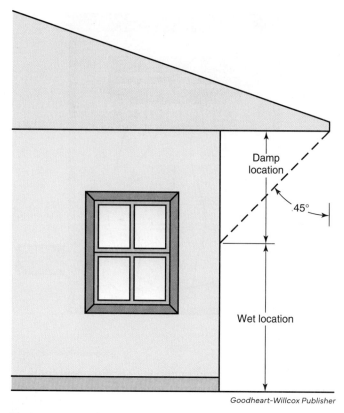

Figure 30-6. Diagram of the distinction between a damp and wet location.

toward the home. Where this line intersects with the house or wall, the area above the 45-degree line is considered a damp location, and the area below is a wet location. See **Figure 30-6**.

30.3.1 Post Lights

Post lights are a popular outdoor luminaire and are typically installed in the front yard of a residence. Post lights are defined by the *NEC* as a luminaire with a support pole. These are wired based on the diagram shown in **Figure 30-7**. Conduit is installed on the side of the house to protect the conductors where they emerge from the ground per *Sections 300.5(D)(1)–(2)*.

UF cable is not permitted to be installed directly in a concrete post base. A short conduit with bushings is installed as a sleeve to allow the cable to enter the post. The base of the pole can be supported directly by casting it in the concrete or as directed by the manufacturer. When casting a pole light in the concrete base, supports or straps may need to be used to hold the post plumb while waiting for the concrete to cure. The UF cable is pulled through the conduit sleeve to the top of the light pole and wired to the luminaire. It is permissible to splice the conductors in the pole below the luminaire provided the splices are accessible by removing the luminaire. Often, an optional photocell control is provided at the top of the pole or in the luminaire to turn the

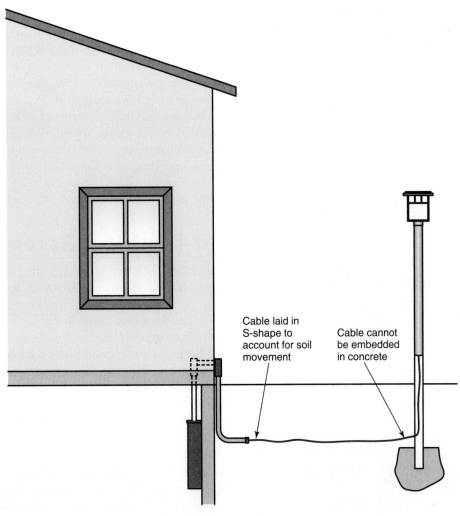

Figure 30-7. Typical installation of a post light in the yard of a residence.

Cable laid in S-shape to account for soil movement

Cable cannot be embedded in concrete

Goodheart-Willcox Publisher

light on at night and off during the day. The standard connections for these are shown in **Figure 30-8**.

30.3.2 Tree Lights

Another popular form of outdoor lighting is installing outdoor luminaires in trees. The luminaires can be pointed up to illuminate the foliage and provide a bit of reflected mood lighting, or they may point down to provide direct lighting below. This practice is permitted by *Section 410.36(G)*. Trees or other vegetation, however, cannot be used to support horizontal conductor spans per *Section 225.26*.

Over time, the tree's growth can cause physical damage to the conductors, so periodic maintenance is required to prevent the tree from growing around the conduit or fixture. A common practice when installing wiring methods on a tree or vegetation is to use liquidtight flexible conduit with an S-shape up the tree. As the tree grows the S-shape

will straighten out, and the flexible conduit will move with the tree as it grows wider.

30.3.3 Pathway and Landscape Lights

Where homes have raised stairs or decks, some step lights may be desired. These step lights can be low-voltage units that are wired as part of the landscape lighting system, or they can be 120-volt powered lights. When the lights are 120 volt they will require a weatherproof box or a listed lighting back box. The wiring methods used to wire the lights are dependent on if it is subject to physical damage.

Outdoor lighting for residential properties can also be decorative landscape lighting. These lights are typically low-voltage lights that use a low-voltage transformer. The landscape lighting uses low-voltage direct burial cables to wire between lighting units. The cable used has individual conductors or twin cables rated for direct burial.

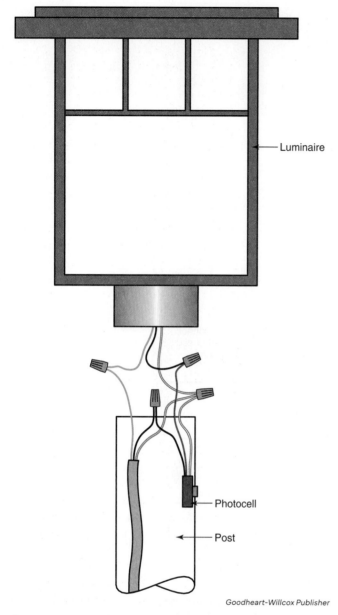

Luminaire

Photocell

Post

Goodheart-Willcox Publisher

Figure 30-8. Standard connections for a photocell.

Underground splicing can be done with rated splice kits or silicone-filled underground rated twist-on wire connectors. Many issues related to landscape lighting is a failure in the underground splices. Using properly rated splice kits or adding jacket repair tape to splices can extend the life of these splices.

The requirements of *Section 300.5* for burial depth of the cables may be applicable. For most installations in garden and lawn edge applications, the low-voltage cables are required to be installed 6″ below grade. There is a note in *Table 300.5(A)* that allows for a lesser depth if the wiring is part of a listed low-voltage lighting system. Many times, to control the landscape lighting, a photocell and time clock

are included with the transformer. For most applications, a switched or unswitched 15 A weatherproof receptacle is installed on the exterior of the home, and the transformer is mounted next to the receptacle and plugged in. Alternatively, the transformer may be directly wired to a junction box that has a switched circuit.

In outdoor lighting installations, the use of properly rated luminaires and proper wiring methods are important. The proper placement of lighting can add beauty to a home and provide safety and security. Outdoor lighting is added to a home as required by *Section 210.70(A)(2)* or just to enhance architectural features.

30.4 **Outdoor Receptacles**

Receptacle outlets in a home are often used to provide power for portable appliances, electronics, tools, or general lighting. *Section 210.52(E)* requires outdoor receptacle outlet locations for a home, one in front, and one in back accessible from grade. Other receptacle outlets may be required for decks, balconies, and where equipment is installed that requires servicing. Special consideration must be given to outdoor receptacles located in the yard rather than on the house's exterior wall.

Because the receptacle is outdoors, it must be installed in a weatherproof box and be GFCI protected per *Section 210.8(A)(3)*. In addition, *Section 406.9(B)(1)* requires the receptacle to be listed and identified as weather-resistant and protected by an enclosure that remains weatherproof when a plug is inserted, and the receptacle is in use. If an outlet box hood is used for this purpose, it must be marked as *extra duty*. Extra-duty covers are more durable than standard in-use covers and are not easily removable. Prior to this requirement in the *NEC*, the covers that were rated for in-use were easily removed and often lost. Removal of these covers would leave receptacles exposed to the weather and create hazards.

Different types of receptacles can be installed in a yard: free-standing receptacle outlets or receptacle outlets with listed assemblies. The free-standing receptacles use standard weatherproof outlet device boxes and weatherproof covers. The listed assemblies are typically designed with the weatherproof boxes as part of the assembly, and they use methods to allow for the installation of weatherproof covers or have designs to make them weatherproof.

There are acceptable methods for supporting free-standing receptacle outlets: conduit-supported and post-supported. *Conduit-supported* receptacle outlets have metal device boxes installed on two or more threaded steel conduits that provide support for the box. *Post-supported* receptacles use another structure, such as a wood post or metal stake, to support the boxes with receptacle outlets.

30.4.1 Conduit-Supported Receptacle Outlets

Section 314.23(F) applies to conduit-supported enclosures with devices. The *NEC* defines an enclosure as "*the case or housing of apparatus...to prevent personnel from accidentally contacting energized parts or to protect the equipment from physical damage.*" A device box for a receptacle outlet qualifies as an enclosure because the receptacle device is an apparatus, and the device box is the enclosure provided to house this device. It states that the enclosure shall be no larger than 100 in³ and must have threaded entry hubs for the supporting conduit. In addition, the enclosure shall be supported by two or more conduits threaded wrench tight into the hubs, and each conduit shall be secured within 18″ of the enclosure. The requirement for securing the conduit within 18″ of the enclosure limits the height of the enclosure to 18″ above grade level.

Receptacle outlets that are conduit-supported work well in small garden areas where they are not subject to damage from lawnmowers or power equipment.

30.4.2 Post-Supported Receptacles

Post-supported free-standing receptacles are not explicitly addressed in the *NEC*. However, standard practice is to use a pressure-treated 4×4 wood post buried no less than 2′ in the ground or by driving a piece of galvanized strut channel into the ground 3′–6′ to provide the necessary stability. A single conduit can protect the conductors where they enter the receptacle outlet box from underground wiring methods. Because the receptacle enclosure is supported by the post or strut and not the conduit, it can be higher than 18″ above grade level. The wiring methods on post-supported receptacles can also include PVC boxes and conduit.

Listed assemblies can provide a receptacle outlet in an enclosure that is flush with the finished grade or is designed to be self-supporting when installed. These listed assemblies can provide a low-profile option for the installation of outdoor receptacle outlets.

30.4.3 Ground Receptacle Outlets

Low-profile receptacle outlets can be provided with listed assemblies that look like a small post, **Figure 30-09**. These provide a clean appearance and require less work than a conduit or post-supported receptacle outlet. The design of these

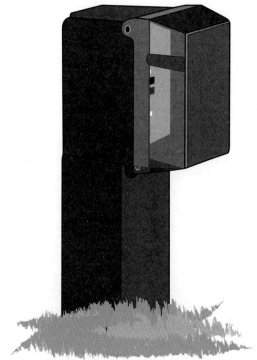

Goodheart-Willcox Publisher

Figure 30-9. A low-profile receptacle outlet that holds the assembly into the ground.

small, listed assemblies provides an anchoring structure to hold the assembly into the ground and integrate a weatherproof cover into the design. Some of these even have provisions to install small landscape lights in threaded entries that are part of the listed assembly. Because they provide the necessary protection for the wiring methods, a UF cable can be installed directly into the bottom of the assembly with a small S-bend, and no additional conduit is required for the protection of the cable. It is recommended to install these in areas where they are not subject to contact with lawn maintenance equipment. They are made from PVC materials and do not stand up long to abuse from lawn trimmers or impact from lawn mowers.

Listed receptacle outlet assemblies can require less materials for assembly and take less time to install. The wiring methods used to supply these assemblies can be conduit or cables. The manufacturers provide methods to transition from the wiring methods to the listed assemblies therefore no additional conduits or fitting are required.

Summary

- Outdoor wiring can be accomplished by installing the conductors overhead or burying them underground. Wiring methods for outdoor wiring are required to be weatherproof and rated for wet locations.
- *Section 225.18* lists the clearance for overhead conductors and cables.
- Most outdoor residential wiring is installed underground at a minimum burial depth to prevent physical damage to the conductors. The requirements for underground wiring can be found in *Section 300.5*, with the minimum burial depths covered in *Table 300.5(A)*.
- Wiring methods used outdoors that are subject to movement require expansion fittings or flexible wiring methods to allow movement and maintain a weatherproof installation.
- When a UF cable enters a conduit for protection, a S-shape or U-shape bend should allow for movement.
- Outdoor luminaires are required to be rated for wet locations where they are subject to weather conditions.
- When casting a pole light in the concrete base, supports or straps may need to be used to hold the post plumb while waiting for the concrete to set. The UF cable is pulled through the conduit sleeve to the top of the light pole and wired to the luminaire.
- Outdoor luminaires for landscape lighting are typically supplied by a low-voltage transformer. Step or stair lighting require a weatherproof backbox or device box for 120-volt installations.
- Outdoor receptacles must be GFCI protected and weather rated.
- There are two acceptable methods for supporting free-standing receptacle outlets, conduit-supported and post-supported. Conduit-supported receptacle outlets have metal device boxes installed on two or more threaded steel conduits that provide support for the box. Post-supported receptacles use another structure, such as a wood post or metal stake, to support the boxes with receptacle outlets. Flush, outdoor in-ground receptacle outlets use design methods like a diving bell to keep water from entering the device compartment if the area were to get wet or flood.

Know and Understand

1. An electrician is installing a branch circuit overhead to supply lighting and receptacles in a workshop in the back yard of a residence accessible only to pedestrians. What is the minimum allowable vertical clearance for the conductors?

   A. 10'
   B. 12'
   C. 15'
   D. 18'

2. *True or False?* Most outdoor residential wiring is installed overhead.

3. Which Table of the *NEC* pertains to burial depths of conductors?

   A. *Table 310.15*
   B. *Table 250.66*
   C. *Table 300.5(A)*
   D. *Table 230.51*

4. What is the most common wiring method for underground residential branch circuits?

   A. Type NM cable
   B. Type MC cable
   C. Type AC Cable
   D. Type UF Cable

5. *True or False?* The *NEC* permits underground conductors to be spliced without a junction box.

6. The *NEC* classifies the interior on an underground raceway as a _____ location.

   A. dry
   B. wet
   C. damp
   D. secure

7. *True or False?* The *NEC* permits luminaires to be installed in trees.

8. The maximum height for a conduit-supported free-standing receptacle is _____.

   A. 6"
   B. 12"
   C. 18"
   D. 24"

9. Free-standing receptacle installed outdoors must be _____.

   A. GFCI protected
   B. Installed in a weatherproof enclosure
   C. Identified as weather-resistant
   D. All of these.

10. *True or False?* The enclosure of a free-standing receptacle shall be weatherproof when a plug is inserted and the receptacle is in use.

Apply and Analyze

1. What are two methods of installing outdoor branch circuits?

2. What is the intent of the minimum burial depths for conductors?

3. What is the maximum burial depth for any type of wiring shown in *Table 300.5(A)*?

4. Conductors used in raceways that are installed underground must have the letter _____ in their insulation designation.

5. Buried conductors that emerge from grade shall be protected to a height of _____ above grade level.

6. What are two acceptable methods of installing free-standing receptacles in the yard of a dwelling?

Critical Thinking

1. What is the burial depth if the branch circuit is rated 20 amps, GFCI protected, and runs under a residential driveway? Refer to **Figure 30-7** and *Table 300.5(A)*.

2. Explain why periodic maintenance is important for luminaires installed in trees.

CHAPTER **31** | Swimming Pools and Hot Tubs

Elena Elisseeva/Shutterstock.com

Introduction

A residential electrician will most likely encounter a home with a swimming pool, hot tub, or hydromassage bathtub at some point during their career. Any object that holds water that a person can enter for recreation or therapy can be considered a pool or tub. These can include swimming pools that are either permanent or storable, hot tub spas, and hydromassage bathtubs. How the water and filtration systems are configured, and the physical size, will determine what type of pool or tub it is.

Pool water filtration systems, chemical treatment systems, heaters, and lighting are just a few pool elements that need electricity. Due to the hazards associated with electricity and proximity to water, an electrician must take great care when installing or maintaining these installations and components. Proper protective technologies, wiring methods, grounding, and bonding of the associated components of pool or tub systems will result in a safe installation. The *NEC* rules for swimming pool, hot tub, and hydromassage bathtub installations can be found in *Article 680*.

31.1 Permanently Installed Swimming Pools

The *NEC* differentiates between a pool and a swimming pool. A *pool* is designed to "*contain water on a permanent or semipermanent basis and used for swimming, wading, immersion, or therapeutic purposes.*" A *permanently installed swimming pool* can be constructed in the ground, partially in the ground, or above ground and must be "*capable of holding water in a depth greater than 42".*" A typical permanently installed residential swimming pool is shown in **Figure 31-1**.

When constructing and installing swimming pools, the electrician must consider a variety of elements. These elements include the following:

- The materials and use of electrical equipment near or in contact with the water
- Proper clearances from electrical wires and lighting
- The locations or environments around pools
- The corrosive nature of the water and water treatment chemicals
- The ability to prevent accidental electrical shock or electrocution

pics721/Shutterstock.com

Figure 31-1. A typical permanently installed residential swimming pool.

31.1.1 Electrical Equipment for Swimming Pools

All associated electrical equipment used for swimming pools, spas, and hot tubs must be listed by a nationally recognized testing laboratory as suitable for use with a pool, hot tub, or spa. The listing requirements of this equipment contain elements for added safety. Some of these elements are systems of double insulation, power supply cords with limited lengths and energy, or isolated power supplies. The materials used to construct a pool can determine how

electrical equipment may be used for the pool. Pools can contain metal ladders, reinforcing structures, and other elements that conduct electricity. These may become hazardous if electricity or an energized conductor comes into contact with the metallic items or the pool water.

With this safety hazard in mind, *Section 680.22(A)* requires all receptacles to be GFCI protected and at least one 125-volt, 15- or 20-amp receptacle to be provided for a permanently installed pool. This receptacle can be on any general-purpose branch circuit and shall be located not less than 6′ and no greater than 20′ from the inside wall of the pool, **Figure 31-2**. Additionally, the receptacle shall not be located more than 6-1/2′ above the grade level of the pool.

To ensure safety, pool covers may have a pump that sits on the cover to remove water from rain. This receptacle allows for the safe connection of power to supply these types of auxiliary equipment. Other appliances, such as radios or beverage blenders in the pool area, can be used without needing extension cords. Extension cords should never be an option when operating electrical equipment near a pool.

Additional receptacles may be required for pools or pool equipment areas. Where receptacles are provided for connection of pool circulation pumps, heaters, or sanitation systems, they need to also be of the grounding type, GFCI protected, and more than 6′ from the inside walls of the pool. As with the receptacles and outlets, all circuits associated with equipment that can potentially interact with swimming pool water are required to have GFCI protection per *Section 680.5*.

Receptacle Requirements for a Permanently Installed Pool

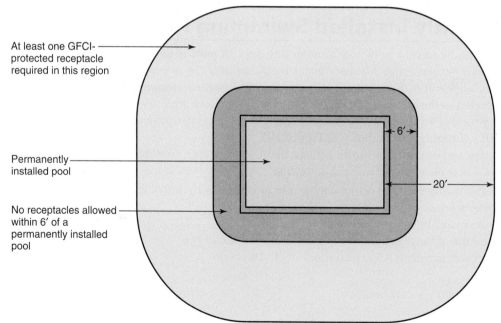

At least one GFCI-protected receptacle required in this region

Permanently installed pool

No receptacles allowed within 6′ of a permanently installed pool

Goodheart-Willcox Publisher

Figure 31-2. Requirements for receptacle locations for a permanently installed swimming pool.

Other electrical equipment required during a swimming pool installation include a maintenance disconnecting means, per *Section 680.13*, to simultaneously disconnect all ungrounded conductors for all pool-related equipment other than lighting. The disconnecting means shall be readily accessible, in sight of the equipment, and located at least 5′ horizontally from the inside walls of the pool. The disconnect will provide service personnel with a method to safely disconnect the power while servicing equipment such as motors, timers, and control panels associated with the pool.

The rules concerning motors for permanently installed swimming pools can be found in *Section 680.21*. Motors that are part of pool filtration or sanitation systems typically need to be removed for annual cleaning or storage. Using a flexible wiring method for the connection of these motors will help the process. Many 120-volt applications for pool pumps include a cord and plug connection. Where cords are used, the minimum size equipment ground of 12 AWG is still needed. The maximum length of a cord is limited to 3′. The pool pumps and motors listed for swimming pools typically have an additional grounding connection for equipotential bonding along with the minimum equipment ground size.

31.1.2 Equipotential Bonding

Equipotential bonding, required by *Section 680.26(B)–(C)*, is the bonding of all metal parts related to the pool and metal objects in the pool vicinity to reduce voltage gradient in the pool water and the pool area. Parts to be bonded together are outlined in *Sections 680.26(B)(1)—(7)*. Items to be bonded together, shown in **Figure 31-3**, include the following:

- Reinforcing steel of the forming shell of the pool
- Perimeter surfaces within 3′ of the inside wall of the pool, metal components of the pool structure, such as metal ladders, underwater lighting, metal fittings, electrical equipment
- Fixed metal parts within 5′ of the inside wall of the pool, such as metal fences, awnings, and door or window frames

The bonding conductor shall be solid copper, not smaller than 8 AWG, insulated or bare. The conductor must be solid to prevent water or moisture from collecting between strands of a stranded conductor, which would accelerate the corrosion of the conductor. This bonding conductor is not required to be extended to the panel, service equipment, or grounding electrode of the residence. The bonding of any pumps or heating equipment will make this connection through the normal equipment grounding conductor paths.

Equipotential bonding of a pool aims to eliminate the voltage gradient in the pool area and is not intended to facilitate the opening of an overcurrent device. A ***voltage gradient*** is the difference in electrical potential or voltage across a distance. It refers to how the voltage spreads or drops between the potential source or the ground itself resulting from stray voltage. See **Figure 31-4**. Notice in this figure that the voltage is highest where the appliance has fallen in the water and lessens the further away from the appliance. Proper bonding helps to eliminate these voltage gradients.

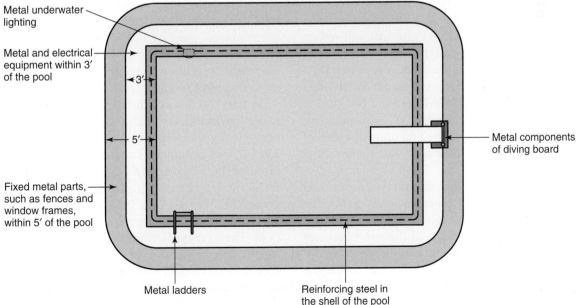

**Equipotential Bonding for a Pool
Examples of Items That Must Be Bonded**

Metal underwater lighting

Metal and electrical equipment within 3′ of the pool

3′

5′

Metal components of diving board

Fixed metal parts, such as fences and window frames, within 5′ of the pool

Metal ladders

Reinforcing steel in the shell of the pool

Figure 31-3. Equipotential bonding of the metal parts of a pool.

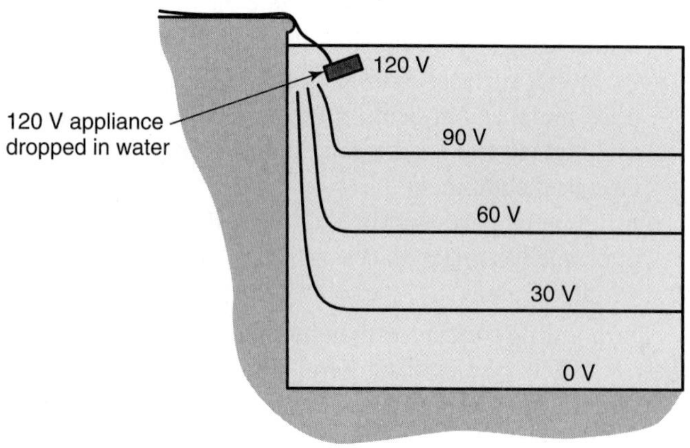

Voltage Gradient in a Pool

Goodheart-Willcox Publisher

Figure 31-4. Voltage gradients resulting from a 120-volt appliance immersed in the pool by accident.

A voltage gradient of one volt per foot can create enough current through a person to cause muscle contraction and prevent the ability to swim or walk. This is hazardous and can create situations where drowning may occur.

The installation of equipotential grounding and bonding is important and needs to be properly installed, inspected, and maintained. The principle of equipotential grounding can best be explained by imagining a bird on a high-voltage wire. The bird is at an equipotential as the wire, and there is not a current path to another potential. In a swimming pool, the water, ladders, surrounding deck, and any other parts of the swimming pool are connected to make sure that all components are the same wire. If something were to energize the pool, the people swimming in the pool are not electrocuted because they are like a bird on a wire.

Where in-ground pools are installed, the concrete reinforcing rods are tied all together with steel tie wires. The wire provides an electrical potential grid that will reduce voltage gradients across the pool. If the concrete reinforcing rods are coated with epoxy or nonconductive materials, a copper grid of 8 AWG solid wire is required to be installed in accordance with *Section 680.26(B)(1)(b)(3)*. If a pool is constructed entirely of fiberglass or PVC nonconductive materials, the equipotential bonding grid under the shell of the pool is not required, but all the additional equipotential grounding elements are required.

The surface of the deck or surrounding pool area is also to be included as part of the equipotential bonding grid. The purpose of this is to keep the voltage potential of the water and surrounding surfaces the same. If the water becomes energized for some reason and a person were to get out of the water, an electrocution hazard would exist where the person becomes a conductive path between the pool water and the surrounding surface.

For deck surfaces, bonding and grounding is done by connecting metallic reinforcing mats of the concrete surface

to the pool shell at four locations surrounding the pool. If the deck is constructed of paver materials, a solid conductor is installed below the surface 18″–24″ from the inside edge of the pool. This conductor surrounds the entire pool and connects to any metallic or conductive objects, such as ladders or stair railings.

The pool water is to be connected to the equipotential grounding grid in one of two ways:

- Connecting the 8 AWG conductor from the bonding grid to a metallic object that has 9″ of surface area in contact with the water, such as a railing or ladder
- Installing a listed fitting into a skimmer overflow, pipe tee, or pipe section that will connect the water to the equipotential bonding grid

All connections for the equipotential bonding are required to be rated for wet locations and direct burial.

31.1.3 Wiring Methods for Corrosive Environments

Proximity to water is not the only safety concern when installing a swimming pool.

To keep algae from growing and pool water sanitized to prevent the spread of sickness or disease, the treatment systems for the water contained in a swimming pool use chlorine, salts, acids, and pH reducers/increasers. These chemicals create *corrosive environments* where they are stored, but also in areas with circulation pumps, automatic chlorinators, filters, open areas under decks adjacent to or abutting the pool structure, and similar locations.

Section 680.14 pertains to the corrosive environments where swimming pool sanitation chemicals are stored. These areas are labeled corrosive environments because the air in such areas is considered to contain acid, chlorine, and bromine vapors. Therefore, wiring methods used in such environments must be listed and identified for such use. Acceptable wiring methods include rigid metal conduit, intermediate metal conduit, rigid PVC, and reinforced thermosetting resin conduit.

Section 680.7 states that wiring methods used in corrosive environments shall contain an insulated copper equipment grounding conductor sized, according to *Table 250.122*, but cannot be smaller than a 12 AWG. Where installed in noncorrosive environments, branch circuits shall comply with the wiring methods described in *Chapter 3* of the *NEC*. Where grounding and bonding conductors connect to terminals, they are required to be listed for wet locations. These types of terminals are typically made of copper, copper alloy, or stainless steel. Aluminum terminals or lugs should never be used near or as part of pool wiring, as the corrosive nature of the moisture will deteriorate aluminum.

The wiring methods used near pool water below ground are restricted due to the nature of the moisture conditions under the deck and adjacent to the pool. *Section 680.11* restricts wiring methods to rigid metal conduit, intermediate metal conduit, rigid polyvinyl chloride conduit, reinforced thermosetting resin conduit, PVC jacketed type MC cable listed for direct burial use, liquidtight flexible metal conduit listed for direct burial, and liquidtight flexible nonmetallic conduit listed for direct burial. Wiring of any kind is not permitted underneath a pool unless it is directly associated with an item that is part of the pool. This means that any pool location cannot be placed over underground service laterals or underground wiring for any site branch circuits or feeders.

The areas above the pool also are a concern for wiring. If an overhead conductor span for a service drop, outdoor branch circuit, or feeder is above the pool or adjacent areas, minimum clearances are required. *Section 680.9* covers the overhead conductor clearance requirements. Typical clearances required are 22-1/2′ above any water and a minimum of 14-1/2′ above any platform, diving board, stair, or tower. The purpose of these higher clearances is for the cleaning tools used with most pools, including long aluminum poles with skimmers or attachments. A person using one of these may encounter an overhead electrical conductor and become electrocuted.

The *Code* does not permit overhead conductors within 10′ of a pool unless the extra high clearances are met. In many areas of the country, there are local public service commission utility regulations that do not permit a pool to be installed below an overhead conductor service drop.

31.1.4 Lighting Requirements for Swimming Pools

For permanently installed swimming pools, *Section 680.22(B)* outlines the requirements for installing luminaires, lighting outlets, and ceiling-suspended paddle fans above a pool area. For a new outdoor pool installation, overhead lighting must have a minimum clearance of at least 12′ above the maximum water level of the pool and extend 5′ horizontally from the inside walls of the pool per *Section 680.22(B)(1)*, **Figure 31-5**.

For installations in indoor pool areas, the same clearances apply. An exception to these clearance requirements is for totally enclosed luminaires or a ceiling-supported paddle fan that is identified for use on porches or patios. In these cases, they are allowed to be closer to the pool so long as they are GFCI protected and installed not less than 7′-6″ above the maximum pool water level per *Section 680.22(B)(2)*, **Figure 31-6**.

SAFETY NOTE Lighting above Pool Areas

While not forbidden by the *NEC*, it is not wise to hang any lighting directly above the water of a pool area. The maintenance needed for changing lamps or cleaning would require someone to put a ladder or scaffolding in the pool.

Lighting Requirements for Permanently Installed Outdoor Pools

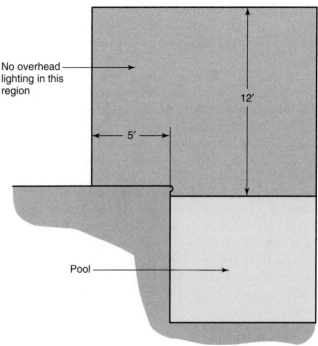

Goodheart-Willcox Publisher

Figure 31-5. Requirements for clearances of luminaries, lighting outlets, and ceiling-supported paddle fans for permanently installed outdoor pools.

Lighting Requirements for Permanently Installed Indoor Pools

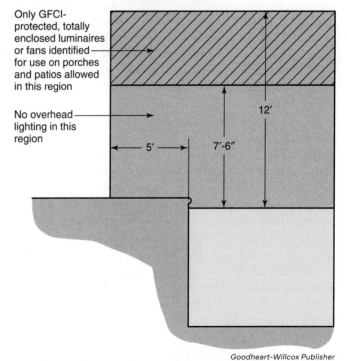

Goodheart-Willcox Publisher

Figure 31-6. Requirements for clearances of luminaries, lighting outlets, and ceiling-supported paddle fans for permanently installed indoor pools.

Swimming pools can also contain **underwater luminaires** within the walls or floor designed to be continuously submerged. These designs incorporate methods to seal the wiring from water entry but provide proper cooling of the lamps and electrical equipment. The requirements of underwater lighting are covered in *Section 680.23*.

Section 680.23(A)(3) requires GFCI protection for all underwater luminaires operating at voltages above the **low-voltage contact limit**, which, according to the *NEC*, is a voltage not exceeding 15 volts AC. The GFCI protection and low-voltage contact limit power supplies help prevent the possibility of electrocution when a failure in equipment or wiring methods happens.

Specialized swimming pool lighting supplied by low voltage is permitted within 5′ of the inside walls of the pool if they operate below the low-voltage contact limits. These lights may be part of architectural features or landscape lighting. If installed within 5′ of the inside walls of the pool, they are required to be supplied by a transformer or power supply marked for use with a fountain, swimming pool, or spa.

Where switches are installed for the control of pool lighting, any area lighting or other pool-related equipment is also required to be at least 5′ from the inside walls of a pool unless they have a barrier that increases the reach distance to 5′. The enclosures containing switches and receptacles are required to be rated for a wet location.

There are three basic types of underwater lighting techniques: the dry-niche luminaire, the wet-niche luminaire, and the no-niche luminaire. A **dry-niche luminaire** is installed in the wall or floor of the pool below the water level in a niche sealed against water intrusion. A **wet-niche luminaire** is installed in the wall or floor of the pool below the water level, and the luminaire is submerged in water. A **no-niche luminaire** is installed above or below the water level without a niche.

When underwater luminaires are installed in a wet niche, they typically have a long cord that is pulled through a conduit up to a deck junction box or transformer that is installed more than 4′ from the inside wall of the pool. The bottom of the box is required to be installed 4″ above the deck or landscaping and a minimum of 8″ above the highest water level, **Figure 31-7**. This allows the connections and splices to the luminaire to be located where they are not subject to immersion.

Wet niche lighting is typically installed with additional cord length stored by wrapping it around the luminaire behind the light to allow for servicing the light without having to be in the water, **Figure 31-8**. Often it is desired to install the wet niche housing where it can be reached without having to enter the pool water. This typically puts the housing within arm's length of the deck and may be a violation.

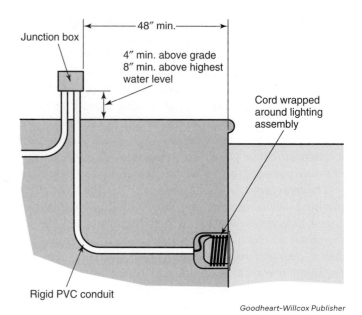

Goodheart-Willcox Publisher

Figure 31-7. A wet niche installation technique for an underwater luminaire.

Goodheart-Willcox Publisher

Figure 31-8. An underwater luminaire.

The location of the niche in the side of the pool is not permitted less than 18″ from the normal water level unless listed for lesser depths. The absolute minimum depth for listed luminaires is 4″. Typically, the heat created by the lighting element in the luminaire needs the water on the surface of the lens to accommodate the cooling. When pool levels are subject to evaporation, the 18″ depth will typically keep the luminaire under the water level. *Section 680.23(A)(7)* requires inherent protection to be included when not submerged.

31.2 Storable Pools

Article 100 defines storable swimming pools as those "*intended to be stored when not in use and are designed for ease of relocation, regardless of water depth.*" These pools are usually sold as a self-contained package and can be assembled and disassembled by the homeowner as they require minimal modification of the pool site. See **Figure 31-9**.

31.2.1 Storable Pool Electrical Requirements

The installation requirements for storable swimming pools can be found in *Article 680 Part III*. Storable pools are not subject to the equipotential bonding requirements of a permanent, in-ground pool. However, *Section 680.31* requires the filter pump to be double insulated with the internal and nonaccessible non-current-carrying metal components grounded via the equipment grounding conductor of the power supply. Cord-connected pool pumps also shall be provided with integral ground-fault protection.

Section 680.32 requires all power equipment associated with the storable pool, including power supply cords, to be protected with ground-fault circuit interrupters. This section further states that all 125-volt 15- and 20-amp receptacles within 20' of the inside walls of the storable pool shall be protected by a ground-fault circuit interrupter.

31.2.2 Luminaires for Storable Pools

Underwater luminaires for storable swimming pools are covered in *Sections 680.33(A)* and *(B)*. *Section 680.33(A)* covers the requirements for luminaires that operate within the low voltage contact limit. It states that the luminaire shall be part of a listed cord-and-plug assembly and have the following features:

- No exposed metal parts
- A lamp that is suitable for the supplied voltage
- An impact-resistant polymeric lens, luminaire body, and transformer enclosure
- A transformer or power supply to step the voltage down from the supply voltage level to a level within the low-voltage contact limit

Section 680.33(B) covers the requirements for luminaires installed in a storable swimming pool over the low-voltage contact limit but not over 150 volts. These luminaires are permitted to be cord-and-plug connected and not use a transformer or power supply and shall have the following features:

- No exposed metal parts
- An impact-resistant polymeric lens and luminaire body
- A ground-fault circuit interrupter with open neutral conductor protection as an integral part of the assembly
- The luminaire lamp is permanently connected to the GFCI with open neutral protection
- Installed in compliance with *Section 680.23(A)*

In addition, per *Section 680.34*, receptacles for lighting or other use shall not be located less than 6' from the inside walls of a storable pool.

31.3 Spas and Hot Tubs

Spas and *hot tubs*, **Figure 31-10**, are defined in *Article 100* as a "*hydromassage pool or tub for recreational or therapeutic use, not located in a health care facilities, designed for immersion of users, and usually having a filter, heater, and motor-driven blower.*" They can be installed indoors or

PixelSquid3d/Shutterstock.com

Figure 31-9. A typical storable residential swimming pool and its water circulating equipment.

Sergio Sergo/Shutterstock.com

Figure 31-10. A typical spa or hot tub.

outdoors, as well as in or on the ground. A spa or hot tub is generally not designed or intended to have its contents drained after each use. The installation requirements for spas and hot tubs are found in *Article 680 Part IV*.

31.3.1 Outdoor Installation Requirements for Spas and Hot Tubs

Hot tubs contain pumps and jets that can create large flows of water and large amounts of suction on the intake ports. Dangerous situations have occurred where children with long hair or baggy swimsuits have been suctioned to the bottom of a spa or hot tub and accidentally drowned. *Section 680.41* requires a clearly labeled emergency shutoff switch for multi-family dwellings. The switch is intended to disconnect the motors that provide power to the circulation and jet systems. It must be installed in a readily accessible location, not less than 5′ away, adjacent to and within sight of the hot tub or spa. Residential hot tubs will typically have multiple suction fittings or fittings designed to prevent these accidental drownings. Many people do not want to see electrical disconnecting switches exposed, but careful planning can provide additional levels of safety during the installation.

Typical installations of packaged hot tubs or spas have all the connections for the pumps and heaters below the skirting covers of the hot tub. Hot tubs that are part of inground pool installations will have the pumps, heaters, and sanitation equipment installed in a room, vault, or separate area. The GFCI protection and disconnecting requirements of the *NEC* apply to this equipment if it is part of a packaged spa or installed as part of a permanent installation.

Section 680.42 covers the electrical requirements for spas or hot tubs installed outdoors. The provisions of *Article 680 Parts I* and *II* also apply to spas and hot tubs in addition to permanently installed pools. The same requirements for maintenance disconnecting means, overhead or underground wiring within the vicinity, ground fault protection, equipotential bonding, underwater luminaires, and wiring in corrosive environments apply.

Some specific wiring items related to hot tubs are stated in *Section 680.42(A)(1)* and *(2)*, which permits listed packaged spa or hot tub assemblies to use flexible connections such as liquidtight flexible metal conduit or liquidtight flexible non-metallic conduit. It also permits the use of a cord-and-plug connection with a cord no longer than 15′ where protected by a GFCI. Some packaged hot tubs are made to be set in place on a patio, deck, or concrete pad. The use of these flexible wiring methods permits easier installation.

The electrocution hazards between the water of a hot tub or spa are the same for swimming pools. If a faulted pump, heater, or electrical items were to cause the water to become energized, the water and the surrounding area need to be kept at the same potential. Bonding requirements for spas or hot tubs installed outdoors are covered in *Section 680.42(B)*.

This section states that equipotential bonding shall not be required where all the following conditions apply:

- The spa or hot tub is listed, labeled, and identified as a self-contained unit for aboveground use
- The spa or hot tub shall not be identified as suitable only for indoor use
- The installation shall follow the manufacturer's instructions and shall be located on or above grade
- The top rim of the spa or hot tub shall be at least 28″ above all perimeter surfaces

Common equipotential bonding situations that arise with hot tub and spa installations are where spas or hot tubs are installed within 5′ of a home or building that has metallic objects that can provide a path for grounding. It becomes difficult to bond railings for decks, downspouts, and metal window frame trims. Careful attention to the layout and location of a hot tub or spa can avoid unnecessary bonding requirements.

31.3.2 Indoor Installation Requirements for Spas and Hot Tubs

Section 680.43 covers the electrical requirements for spas or hot tubs installed indoors and states that the installation shall comply with *Parts I* and *II* of *Article 680* with three exceptions:

- A listed spa or hot tub packaged units rated 20 amps or less shall be permitted to be cord-and-plug connected
- The equipotential bonding requirements for perimeter surfaces shall not apply to a listed, self-contained spa or hot tub installed above a finished floor
- For dwelling units only, where a listed hot tub is installed indoors, the wiring methods of *Section 680.42(C)* shall also apply

A challenging part of indoor installations is where homeowners plan rooms for the hot tub with additional receptacle needs. *Section 680.43(A)* requires at least one 125-volt, 15- or 20-amp receptacle on a general branch circuit shall be located not less than 6′ from and not exceeding 10′ from the inside wall of the spa or hot. Other receptacles shall be located no less than 6′ from the inside wall of the spa or hot tub. In addition, receptacles rated 125 volts, and 30 amps or less shall be GFCI protected if within 10′ of the inside wall of the spa or hot tub. Any receptacles that supply power for a spa or hot tub shall be protected by a ground-fault circuit interrupter per *Sections 680.43(A)(1)–(3)*.

31.3.3 Luminaires, Lighting Outlets, and Paddle Fans for Spas and Hot Tubs

The installation of luminaires, lighting outlets, and ceiling-suspended paddle fans are covered in *Section 680.43(B)*. Unless GFCI protected, the mounting height of a luminaire shall not be less than 12′ above the maximum water level. If GFCI protected, the mounting height can be reduced to not less than 7′6″. Furthermore, recessed luminaires with a glass or plastic lens and a nonmetallic or electrically isolated metal trim and marked as suitable for damp locations, or a surface-mounted luminaire with a glass or plastic globe, a nonmetallic body, or a metallic body that is isolated from contact and marked as suitable for damp locations are permitted to be installed less than 7′6″ over a spa or hot tub. Switches shall be located not less than 5′ from the inside walls of the indoor spa or hot tub, **Figure 31-11.** For indoor spa and hot tub locations, *Section 680.43(D)* requires the following part to be bonded together:

- All metal fittings within or attached to the spa or hot tub structure
- Metal parts of electrical equipment associated with the water circulating system unless part of a listed, labeled, and identified self-contained spa or hot tub
- Metal raceway and piping that is within 5′ of the inside walls of the spa or hot tub
- All metal surfaces within 5′ of the inside walls of the spa or hot tub (an exception is for small conductive surfaces such as drain fittings, towel bars, and other nonelectrical equipment)
- Electrical devices and controls that are not associated with the spa or hot tub but are located less than 5′ away

According to *Section 680.43(E)*, the method of bonding the metal parts associated with the indoor spa or hot tub can be one of three methods:

- The interconnection of metal piping and fittings
- Metal-to-metal mounting on a common frame or base
- The use of a solid copper bonding jumper, insulated, covered, or bare, not smaller than 8 AWG

In addition, *Section 680.43(F)* requires all electrical equipment within 5′ of the inside wall of the spa or hot tub and all electrical equipment associated with the circulating system of the spa or hot tub to be grounded.

As with permanently installed pools, spas and hot tubs require wiring methods suitable for corrosive and wet locations. The GFCI requirements protect against possible electrocution and equipotential bonding.

31.4 Hydromassage Bathtubs

A *hydromassage bathtub* is defined in *Article 100* as a *"permanently installed bathtub equipped with a recirculating piping system, pump, and associated equipment. It is designed so it can accept, circulate, and discharge water upon each use."* They are generally installed in the bathroom in place of a traditional bathtub. See **Figure 31-12.** The rules governing the electrical installation of hydromassage tubs can be found in *Article 680 Part VII*.

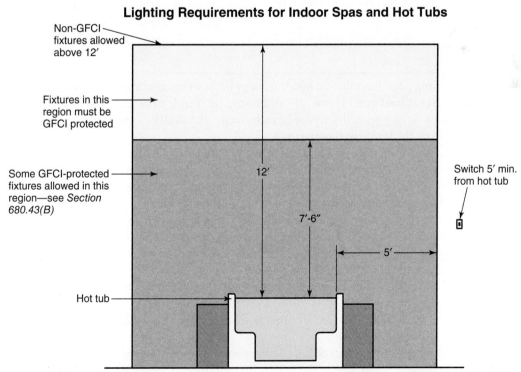

Lighting Requirements for Indoor Spas and Hot Tubs

Non-GFCI fixtures allowed above 12′

Fixtures in this region must be GFCI protected

Some GFCI-protected fixtures allowed in this region—see *Section 680.43(B)*

12′

7′-6″

5′

Switch 5′ min. from hot tub

Hot tub

Goodheart-Willcox Publisher

Figure 31-11. The clearances for electrical equipment installed about an indoor spa or hot tub.

Marko Poplasen/Shutterstock.com

Figure 31-12. A typical hydromassage bathtub.

31.4.1 Hydromassage Bathtub Installation Requirements

Section 680.71 requires a dedicated branch circuit protected by a readily accessible ground-fault circuit interrupter for hydromassage bathtubs and their associated electrical components. In addition, all 125-volt receptacles rated 20 amps or less and located within 6′ of the inside walls of the hydromassage tub shall have GFCI protection.

Section 680.73 requires the electrical equipment associated with the hydromassage tub to be accessible without damaging the building structure. Cord-and-plug-connected hydromassage tubs where the supply receptacle is accessible

through a service access opening, the receptacle shall be installed not more than 1′ from the opening, and its face shall be in direct view. If the access panel requires tools to remove, the receptacle is not readily accessible according to the *NEC* definition, and a GFCI circuit breaker in the distribution panel will be required to protect the circuit.

Section 680.74(A) and *(B)* require the following components of a hydromassage bathtub to be bonded together with a solid copper bonding jumper no smaller than 8 AWG:

- All metal fittings that are in contact with the circulating water
- Metal parts of electrical equipment associated with the water circulation system
- Metal-sheathed cables, raceways, and metal piping within 5′ of the inside walls of the tub and not separated by a permanent barrier
- All exposed metal surfaces that are within 5′ of the inside walls of the tub
- Electrical devices and controls that are not associated with the hydromassage tub and located within 5′ of the tub

In many hydromassage bathtub installations, the plumbing supply lines are plastic, and the piping associated with the hydromassage bathtub is plastic. Additional equipotential bonding is not required if there are no conductive paths to ground from small metal fittings or other bathroom accessories. Careful evaluation of the conditions of an installation will help determine if bonding is required during the rough-in stage rather than trying to fish or make connections after final hydromassage tub installations.

Summary

- The *NEC* rules for swimming pool, hot tub, and hydromassage bathtub installations can be found in *Article 680*.
- *Section 680.22(A)(1)* requires at least one 125-volt, 15- or 20-amp receptacle to be provided for a permanently installed pool. This receptacle can be on any general-purpose branch circuit and shall be located not less than six feet and no greater than twenty feet from the inside wall of the pool.
- A maintenance disconnecting means is required by *Section 680.13* to simultaneously disconnect all ungrounded conductors for all pool-related equipment other than lighting.
- The rules concerning motors for permanently installed swimming pools can be found in *Section 680.21*.
- *Section 680.14(A)* pertains to the corrosive environment of areas where swimming pool sanitation chemicals are stored. In addition, areas with circulation pumps, automatic chlorinators, filters, open areas under decks adjacent to or abutting the pool structure, and similar locations shall be considered a corrosive environment.
- Equipotential bonding is the bonding of all metal parts related to the pool and metal objects in the pool vicinity to reduce the voltage gradient in the pool water and the pool area.
- A voltage gradient is defined as the difference in electrical potential, or voltage, across a distance. Proper bonding will help eliminate these voltage gradients.
- *Section 680.22(B)* outlines the requirements for installing luminaires, lighting outlets, and ceiling-suspended paddle fans above a pool area.
- A dry-niche luminaire is installed in the wall or floor of the pool below the water level in a niche sealed against water intrusion. A wet-niche luminaire is installed in the wall or floor of the pool below the water level, where the luminaire is entirely surrounded by water. A no-niche luminaire is intended to be installed above or below the water level without a niche.

- The low-voltage contact limit is defined as a voltage not exceeding 15 volts AC.
- *Article 100* defines storable swimming pools as those *"intended to be stored when not in use and are designed for ease of relocation, regardless of water depth."*
- The installation requirements for storable swimming pools can be found in *Article 680 Part III. Section 680.32* requires all power equipment associated with the storable pool, including power supply cords, to be protected with ground-fault circuit interrupters.
- The installation requirements for spas and hot tubs are found in *Article 680 Part IV. Section 680.42* covers the electrical requirements for spas or hot tubs installed outdoors. *Section 680.43* covers the electrical requirements for spas or hot tubs installed indoors.
- A hydromassage bathtub is defined in *Article 100* as *"a permanently installed bathtub equipped with a recirculating piping system, pump, and associated equipment. It is designed so it can accept, circulate, and discharge water upon each use."*
- *Section 680.71* requires a dedicated branch circuit protected by a readily accessible ground-fault circuit interrupter for hydromassage bathtubs and their associated electrical components.

CHAPTER 31 REVIEW

Know and Understand

1. A permanently installed swimming pool is capable of holding water in a depth greater than _____.
   A. 12″
   B. 24″
   C. 36″
   D. 42″

2. The rules for swimming pools, spas and hot tubs, and hydromassage bathtubs can be found in *Article* _____ of the *NEC*.
   A. *310*
   B. *410*
   C. *680*
   D. *682*

3. A disconnecting means for a permanently installed swimming pool, as required by *Section 680.13*, shall be readily accessible, in sight of the equipment, and located at least _____ from the inside wall of the pool.
   A. 2′
   B. 5′
   C. 10′
   D. 20′

4. *True or False?* The requirement that the pump motor of a permanently installed swimming pool be GFCI protected applies only to cord-and-plug connected motors.

5. *True or False?* Wiring methods in a corrosive environment must contain an insulated equipment grounding conductor.

6. The clearance for luminaires, lighting outlets, and ceiling-supported paddle fans for a new outdoor, permanently installed swimming pool shall be no less than _____ above the maximum water level of the pool.
   A. 6′
   B. 8′
   C. 10′
   D. 12′

7. For an indoor pool, the clearances for totally enclosed luminaires or ceiling-supported paddle fans that are on a GFCI-protected branch circuit shall not be less than _____ above the maximum water level of the pool.
   A. 3′6″
   B. 5′6″
   C. 6′6″
   D. 7′6″

8. For both indoor and outdoor permanently installed swimming pools, the restricted area for luminaires, lighting outlets, and ceiling supported paddle fans shall extend _____ horizontally from the inside walls of the pool.
   A. 3′
   B. 5′
   C. 10′
   D. 20′

9. The type of underwater pool luminaire that is sealed against water intrusion is the _____ luminaire.
   A. dry-niche
   B. wet-niche
   C. no-niche
   D. waterproof-niche

10. The type of pool luminaire that can be installed above or below the water level is the _____ luminaire.
    A. dry-niche
    B. wet-niche
    C. no-niche
    D. waterproof-niche

11. *True or False?* Equipotential bonding conductors of a permanently installed swimming pool shall be extended to the panel, service equipment, or the grounding electrode of the premise wiring.

12. *True or False?* The purpose of equipotential bonding of a permanently installed swimming pool is to facilitate the operation of an overcurrent protective device under a fault condition.

13. The difference in voltage across a space or distance is defined as _____.
    A. potential difference
    B. voltage gradient
    C. Ohm's law
    D. ampacity

14. A storable pool is one designed for a maximum depth of _____.
    A. 42″
    B. 48″
    C. 60″
    D. 72″

15. *Section 680.31* requires the filter pump of a storable pool to be _____.
    A. dual-voltage rated
    B. hard-wired
    C. waterproof
    D. double insulated

16. *True or False?* Underwater luminaires are not permitted in a storable pool.

17. *True or False?* For spas and hot tubs in a single-family dwelling, a clearly labeled emergency shut-off switch is required to be installed.

18. An exception to *Section 680.43* permits an indoor spa or hot tub package rated _____ or less to be cord-and-plug connected.
    A. 20 amps
    B. 30 amps
    C. 40 amps
    D. 50 amps

19. Unless GFCI protected, the mounting height of a luminaire over an indoor spa or hot tub shall be not less than _____ above the maximum water level.
    A. 6′
    B. 8′
    C. 10′
    D. 12′

20. *True or False?* A hydromassage bathtub requires a dedicated branch circuit.

Apply and Analyze

1. What is the purpose of the disconnecting means required by *Section 680.13* for a permanently installed swimming pool?
2. *Section 680.21(C)* requires outlets supplying single-phase 120-volt through 240-volt motors be provided with _____.
3. Name four acceptable wiring methods for areas around swimming pools considered to be a corrosive environment.
4. GFCI protection is required for all underwater luminaires that operate at voltages above the _____.
5. What is the purpose of equipotential bonding of the metal parts associated with a permanently installed swimming pool?
6. The equipotential bonding conductor shall be a solid copper, insulated or bare, not smaller than _____.
7. List five features required of underwater luminaires installed in a storable swimming pool if the luminaires operate above the low-voltage contact limit.

8. Equipotential bonding of a spa or hot tub installed outdoors shall not be required if what four conditions are met?
9. List three methods of bonding the metal parts associated with an indoor spa or hot tub in accordance with *Section 680.43(E)*.
10. List five items that must be bonded together when installing a hydromassage bathtub.

Critical Thinking

1. What is the primary difference between a spa or hot tub and a hydromassage bathtub?
2. Explain why a GFCI receptacle cannot be used to protect a hydromassage bathtub if the receptacle is located behind an access panel that requires screws to secure.

Elena Elisseeva/Shutterstock.com

In previous chapters, you've learned the fundamental skills required for a residential electrician and how those skills are applied throughout the installation of an electrical system. This section focuses on some specific, specialized topics.

Chapter 32, *Voice, Data, and Signaling*, presents an introduction to residential communication wiring. Chapter 33, *Solar Photovoltaic Systems*, provides a basic overview of PV system types, components, and installation methods.

Chapter 34, *Garage Wiring and Electric Vehicle Charging Power*, covers electrical requirements and load calculations for attached and detached garages, along with an introduction to electrical vehicle charging installations. Chapter 35, *Standby Power Systems*, introduces standby power system types, sizing, connections, and installation.

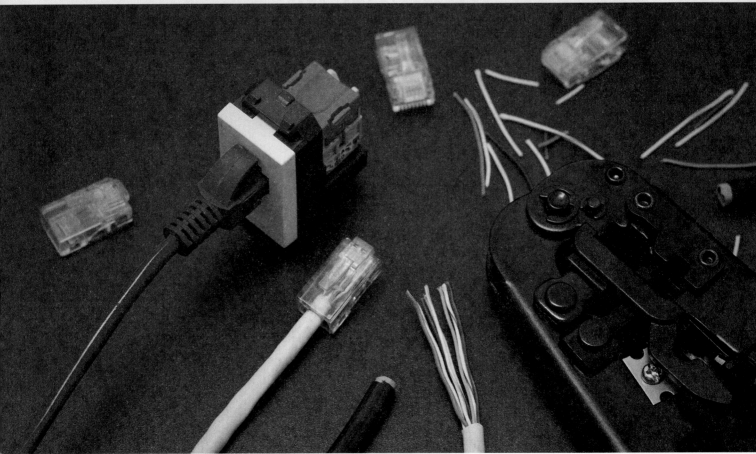

Worklike/Shutterstock.com

SPARKING DISCUSSION ⚡

How do the roles and responsibilities of an electrician apply to communications wiring?

(Continued)

LEARNING OBJECTIVES

After completing this chapter, you will be able to:

- Identify communication and signal requirements for residential construction.
- Compare types of cable required for specific applications.
- Explain installation methods for communication and signal cables and components.
- Analyze testing and certification requirements for communication and signal installations.

TECHNICAL TERMS

bandwidth
coaxial cable
crosstalk
ethernet cables
fiber optic cable
intersystem bonding termination (IBT)
keystone jack
punch down tool
twisted pair cable

Introduction

The advancement of technology has brought us a world where a multitude of devices and appliances can communicate with the homeowner and with each other. The addition of these devices and connection methods must now be considered during the construction of dwellings. This includes devices for safety, security, entertainment, and communication.

Low-voltage and signal systems found in homes can include internet connections, security and alarm systems, and life safety signals such as smoke, heat, flame, and carbon monoxide detection. Audio and video, home entertainment, theater systems, and home automation are found with increasing frequency in new homes.

Low-voltage communication and signal cables and outlets are installed by specialty contractors, but depending on the contract and the situation, an electrical contractor or electrician can be required to install or troubleshoot these systems. Therefore, it is necessary for an electrician to become familiar with the equipment and terminology and the requirements outlined in the *National Electrical Code*.

32.1 Introduction to Residential Communication Systems

Similar to electrical service, communication systems have a service entrance location to provide a means for communications service providers to make connections from the utility side to the customer side. Telephone and cable TV service can be brought to the customer through underground conduit or from a conduit riser and overhead service. The service provider is usually responsible for bringing the conductors to a premises and often install a termination enclosure or necessary equipment, such as a telephone network interface device (NID), **Figure 32-1**. The NID is the transition point between the communication utility company's wiring and the premises wiring.

In some cases, an electrical contractor may be required to install the interface.

Communication service equipment that carries electrical signals on copper conductors, such as twisted pair and coaxial cable (coax), or optical fiber conductors, must be bonded to the service grounding equipment through an intersystem bonding termination device. See **Figure 32-2.** According to the *NEC,* an ***intersystem bonding termination (IBT)*** provides a means for a connecting communication systems grounding conductors at the service equipment or at the disconnecting means for buildings or structures supplied by a feeder or branch circuit. See **Figure 32-3.**

LSqrd42/Shutterstock.com

Figure 32-1. A telephone network interface box is the transition point from the utility's service to the customer's system.

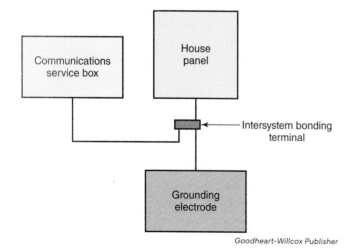

Goodheart-Willcox Publisher

Figure 32-2. Communication systems must be bonded to the electrical grounding system.

**Grounding electrode
conductor connected to IBT**

**Grounding conductors bonded to
grounding electrode conductor**

Arlington Industries, Inc.

Figure 32-3. An intersystem bonding termination device is often used to bond communication systems to the electrical grounding system.

In new construction and most existing structures, there is a grounding means. *Section 800.100(B)* states the requirements for grounding and directs us to *Section 250.94(A)*, which provides the following requirements for an IBT:

- Be accessible for connection and inspection.
- Consist of a set of terminals with the capacity for connection of not less than three intersystem bonding conductors.
- Not interfere with opening the enclosure for service, building or structure disconnecting means, or metering equipment.
- At the service equipment, be securely mounted and electrically connected to an enclosure for the service equipment, to the meter enclosure, to an exposed nonflexible metallic service raceway, or be mounted at one of these enclosures and be connected to the enclosure or to the grounding electrode conductor with a minimum 6 AWG copper conductor.
- At the disconnecting means for a building or structure, be securely mounted and electrically connected to the metallic enclosure for the building or structure disconnecting means or be mounted at the disconnecting means and be connected to the metallic enclosure or to the grounding electrode conductor with a minimum 6 AWG copper conductor.
- Be listed as grounding and bonding equipment.

There are several *NEC* requirements based on the type of cable used. *Section 820.100* states that the outer shield of coax cable must be grounded. While optical fiber does not carry an electrical signal, if the fiber optic cable from the service provider is armored, that metal armor must be bonded to the IBT as well.

The *NEC* requirements for the bonding conductor are as follows (these apply most notably to *Articles 800, 820,* and *770*):

- **Insulation.** The bonding conductor or grounding electrode conductor shall be listed and shall be permitted to be insulated, covered, or bare.
- **Material.** The bonding conductor or grounding electrode conductor shall be copper or other corrosion-resistant conductive material, stranded or solid.
- **Size.** The bonding conductor or grounding electrode conductor shall not be smaller than 14 AWG. It shall have a current-carrying capacity not less than that of the grounded metallic member(s). The bonding conductor or grounding conductor shall not be required to exceed 6 AWG.
- **Length.** The bonding conductor or grounding electrode conductor shall be as short as practical, not to exceed 20′ in length. If the bonding conductor must exceed a length of 20 feet, an additional ground rod can be added to the grounding electrode system to make the installation acceptable.

- **Run in straight line.** The bonding conductor or grounding electrode conductor shall be run in as straight a line as practical.
- **Physical protection.** Bonding conductors and grounding electrode conductors shall be protected where exposed to physical damage. Where the bonding conductor or grounding electrode conductor is installed in a metal raceway, both ends of the raceway shall be bonded to the contained conductor or to the same terminal or electrode to which the bonding conductor or grounding electrode conductor is connected.

32.2 Cables, Connections, and Tools

Several types of cable can be found for communications and signaling circuits. The most common found in dwellings are twisted pair, coaxial cable, and optical fiber. Each of these cable types have many variations that make them suitable for specific applications.

32.2.1 Twisted Pair

The most common type of cable used for telephone and networks is twisted pair. *Twisted pair cable* is made of pairs of separate twisted insulated wires that are ran parallel to each other.

Current flowing through a conductor creates a magnetic field around the conductor. The strength of the magnetic field is proportional to the amount of current. Thus, if the current in the conductor is changing, the strength of the magnetic field is also changing. A changing magnetic field induces or causes current to flow in a conductor. Thus, the changing magnetic field generated by one conductor can affect the current being carried by a second nearby conductor. The undesirable electrical interference caused in the second conductor is called *crosstalk*.

Fortunately, crosstalk can be reduced by twisting pairs of wires together. The opposing directions of two signals on each conductor of a pair will have opposite polarity and cancel much of the effects of induction, reducing crosstalk.

Many types of twisted pair cables have developed over the decades and are distinguished from each other by several category designations. The physical construction of each category has progressively improved their operating characteristics, resulting in increases in the capacity of data that can be transmitted, the speed at which the data is transmitted, the maximum length of the cable without a major loss of signal strength, and reduction in crosstalk and external interference.

Homes built prior to 1983 have an old unshielded twisted pair (UTP) cable with an unofficial designation of Level 1 (sometimes called CAT1) cable. A CAT1 cable is limited to voice signals and not suitable for data transmission. The early 1990s saw the introduction of CAT3 cable, which can carry voice and data. More recent homes might have CAT5 or CAT5e (e for *enhanced*) cable. See **Figure 32-4**.

CAT6A is available in a multitude of varieties depending on the application and the manufacturer. Some of the variations found in the construction of CAT6A cable include insulation type, conductor material, unshielded or shielded, shielding on entire cable, shielding on pairs, drain wire, arraignment of the conductors in the cable, and the type of interior support spline or separator (if included). Certain applications might require a specific type of cable. Not all manufacturers of the cable will provide every option.

32.2.2 Coaxial Cable

Community Antenna Television, or CATV, is more commonly known as *cable TV*. CATV is covered in *Article 820*. Coaxial cable is not limited to television signals but is now also used to provide broadband internet service to homes.

Coaxial cable, often referred to as *coax cable*, consists of concentric layers of conductor and insulation. The most basic construction includes an inner conductor covered by a dielectric (insulator), surrounded by a conductive shield, and covered with an outer insulating jacket, **Figure 32-5**. This construction reduces interference from other electrical

sources. Variations of coax construction exist for many applications. Differences include conductor size, conductor material, dielectric thickness, dielectric material, shielding method, and insulation type.

Various coax types are differentiated by an identifier that begins with the letters *RG*, followed by a number, and sometimes followed by forward-slash and *U* (RG-59/U, for example). *RG* is an abbreviation for *radio guide*, and the *U* is for *universal*, indicating general use.

There are several common types of coax cables, such as RG-59, RG-11, and RG-6. The type used most frequently for installation for residential CATV and internet is RG-6.

32.2.3 Fiber Optic Cable

Fiber optic cable, also called *optical fiber cable*, transmits light signals in the form of high-speed pulses through a very small-diameter glass or high-density plastic fiber. The small diameter and low angle of reflection allow the light signal to travel long distances.

The following are the basic components of a fiber optic cable (**Figure 32-6**):

- **Core.** Glass or plastic fiber that transmits the light signal.
- **Cladding.** Reduces signal loss and assists with the internal reflection, serving a similar purpose as insulation on a traditional electrical conductor.

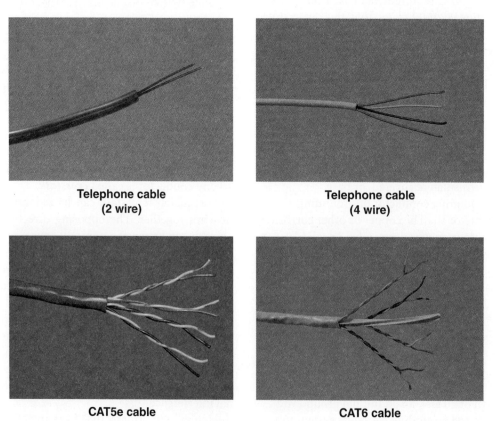

Telephone cable
(2 wire)

Telephone cable
(4 wire)

CAT5e cable

CAT6 cable

Goodheart-Willcox Publisher

Figure 32-4. Many types of communications cable are found in homes.

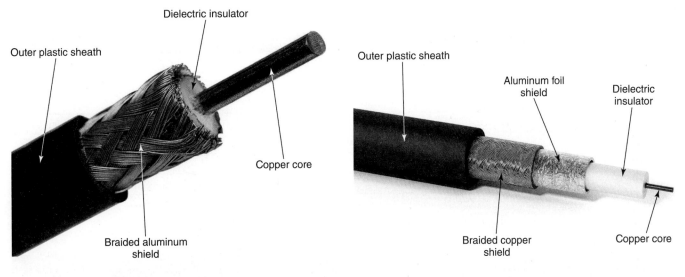

Standard coaxial cable

Double-shielded coaxial cable

Flegere/Shutterstock.com; cigdem/Shutterstock.com

Figure 32-5. Coaxial cable is used for CATV systems and internet systems.

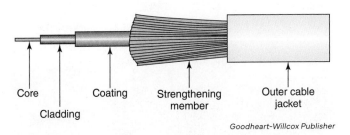

Goodheart-Willcox Publisher

Figure 32-6. Basic components of fiber optic cable.

- **Coating.** Provides a layer of protection for the fiber core.
- **Strengthening fibers.** Provides structural support to the cable and prevents damage to the core during installation.
- **Outer cable jacket.** Outer protection for the entire cable assembly.

Fiber optic cable is used for high-speed data transmission and is available in many areas for residential broadband internet service. It is common for the fiber optic cable to be brought to the service provider's equipment and then converted to electrical signals delivered via coax cable throughout the home. Some equipment uses fiber optic cable to connect individual components. In these cases, cables with appropriate connectors preinstalled can be purchased to make these connections. See **Figure 32-7.**

32.2.4 Bandwidth

Bandwidth is the rate of data transmission and is measured in bits per second (bps). This can be modified with a metric prefix, such as k, M, G, or T, representing kilo, mega, giga, and tera, respectively.

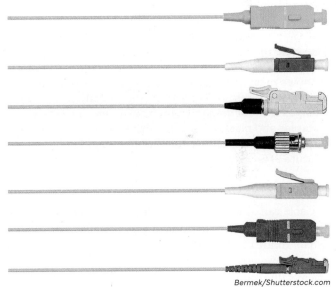

Bermek/Shutterstock.com

Figure 32-7. Fiber optic cables have many types of connectors.

The type of technology used in a system determines its bandwidth, and the cable installed must be capable of transmitting data at that rate. For a system to have the ability to carry a certain bandwidth, it also must support a corresponding system operating frequency.

Bandwidth, frequency, and cable type also determine the distance that signals can be transmitted. **Figure 32-8** lists characteristics of cable and ethernet standards. Actual system values vary, especially with fiber optic cables, where there are many standards and applications.

Characteristics and Cables for Ethernet Standards

Ethernet Standard	Max Bandwidth	Frequency	Cable Type
10BASE-T	10 Mbps	16 MHz	CAT3
100BASE-TX	100 Mbps	100 MHz	CAT5
1000BASE-T	1 Gbps	100 MHz	CAT5e
10GBASE-T	10 Gbps	250 MHz	CAT6
10GBASE-T	10 Gbps	500 MHz	CAT6A
1000BASE-LX	100 Gbps	500 MHz to 1.5 GHz	Optical fiber
RG-6, RG-11	100 Mbps	750 MHz	Coaxial

Goodheart-Willcox Publisher

Figure 32-8. Characteristics of cable and ethernet standards.

32.3 Rough-in Installation

For new construction, cable for communication and signaling is installed during the rough-in phase. Some cables can be installed originating at the service to each individual location or to an initial location and then daisy-chained to each subsequent location. Other cable types or network configurations require dedicated lines from the source to each outlet location.

> **PRO TIP**   **Fastening Cables**
>
> Cables must be fastened using staples or cable ties. Many products are made to secure and support cables without damaging them, including insulated staples and plastic or metal cable supports. Unused cables must be removed unless identified for future use.

Cable can also be installed through a conduit. Conduits can be installed to the location of use or to a location identified for future use. Conduit types that are commonly used for this purpose are electrical nonmetallic tubing (ENT), flexible metallic conduit (FMC), flexible metallic conduit (FMC) and electrical metallic tubing (EMT). Any conduit used for installation should be a minimum size 21 (trade size 3/4).

Electrical nonmetallic tubing (ENT) is the preferred conduit because it is flexible, lightweight PVC. It is common to use ENT that is orange for communication and signaling circuits, **Figure 32-9**. ENT must be secured not more than 3′ from where it enters a box or enclosure at intervals not exceeding 3′. Bends between pull points cannot exceed 360 degrees.

NEC requirements for ENT are found in *Article 362*. Because ENT is a flexible conduit, you must ensure that any bends made in the tubing are not less than the minimum radius for the size of tubing being used. This information is found in *Table 2* of *NEC Chapter 9*. See **Figure 32-10**.

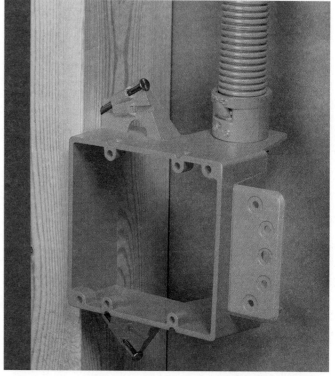

Carlon, Lamson & Sessions

Figure 32-9. Orange electrical nonmetal tubing (ENT) is a common wiring method for communications wiring.

There are times when other conduit types might be more appropriate. This depends on the location and potential exposure to damage or weather. In cases such as this, EMT might be the better option.

Whether the cable is run with or without a conduit, at the cable termination point a method for attaching a connection method must be installed in the wall. This can be a metal or plastic device ring with holes tapped for 6-32 screws to secure a wall plate, **Figure 32-11**.

When selecting the mounting ring, it is important to know the thickness of the wall and the width of the devices. The depth of the ring should not be more than the thickness of the wall material.

Bend Radii for Electrical Nonmetallic Tubing (ENT)

Tubing Size (nominal inch)	Minimum Bend Radius (inches)
1/2	4
3/4	5
1	6
1 1/4	8
1 1/2	10
2	12
2 1/2	15
3	18

Goodheart-Willcox Publisher

Figure 32-10. Bending radii for common sizes of ENT.

32.3.1 Ethernet Connections

Ethernet cables are network cables used for high-speed wired connections between devices with ethernet ports. These cables can be purchased with connectors installed, or connectors can be field installed. In this case, the correct connectors and tools are needed.

Connectors for ethernet cable are called RJ-45 connectors, installed on the cable using a special crimp tool. See **Figure 32-12.** RJ-45 connectors are not the same for every type of ethernet cable, so if connectors need to be installed on cables, ensure that the connector matches the type of cable. It will be necessary to know if it is CAT5e or CAT6 and shielded or unshielded.

Industry standards for telecommunications are determined by Telecommunications Industry Association (TIA) and Electronic Industries Alliance (EIA). Two common standards for installing ethernet cable in plugs or jacks are T568A and T568B, **Figure 32-13.** T568B is common for commercial applications, while T568A is seen more in residential installations. All connectors and jacks must match, so if you select T568A, keep all connections the same. There are exceptions to this, such as crossover cables which will not be covered in this text.

A *keystone jack* is an RJ-45 female connector generally installed in wall plates. Keystone jacks have color codes for the two standards printed on the jack itself, **Figure 32-14.** The connection points on the keystone jack are a type used for telephone and ethernet terminations and require a special tool called a *punch down tool*. A punch down tool uses a pressure-actuated spring to push the wires into the connector while simultaneously cutting off excess

Double-gang device ring for low voltage

Double-gang box with one gang for electrical and one for communications

Arlington Industries, Inc.

Figure 32-11. Nonmetallic device rings for new construction.

RJ-45 pass-through
connector

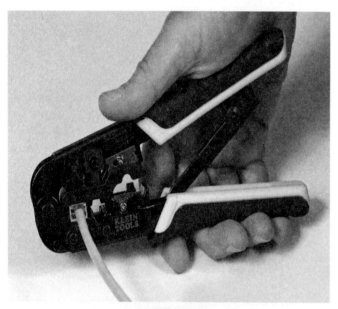

Crimping tool

Klein Tools

Figure 32-12. RJ-45 connectors are used on CAT5, CAT5e, CAT6, AND CAT6A cables.

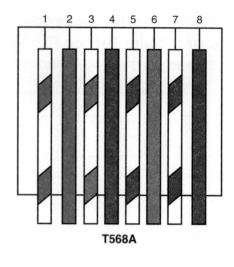

T568A

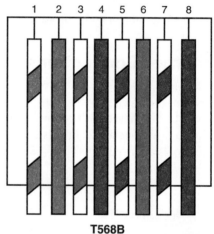

T568B

Goodheart-Willcox Publisher

Figure 32-13. Wire sequence for RJ-45 connectors.

wire, **Figure 32-15**. The wire is placed across the connector, and the tool is placed over the wire. The tool is pushed into the connector and when the correct amount of pressure is applied, the spring is actuated, forcing the wire into the connection. It is important to place the tool in the right direction so that the blade cuts the correct part of the wire. The jack can be inserted into a wall plate designed to hold one or more keystone jacks, **Figure 32-16**. Some wall plates are manufactured with the jack as part of the plate.

32.3.2 Coaxial Cable Connections

There are many different types of connections available for coax cable, and each is used for a specific application. The type used for CATV and internet is the F connector. Even within each class of connector, there are different installation types: twist-on, crimp, and compression. Preparing the cable for the installation of the connectors is common for each type of connector. The inner core extends 5/16″

Ideal Industries, Inc.

Figure 32-14. A keystone jack provides a female connection for an RJ-45 connector. A color-code on the keystone jack identifies to wire connection locations.

Ideal Industries, Inc.

Figure 32-15. A punch down tool presses a wire into a slot to make a secure connection.

beyond the dielectric, and the shield is exposed for 1/4″. See **Figure 32-17**. The cable can be cut and stripped using cable cutters and a knife, but tools are available to strip the cable in the appropriate dimensions and at the correct depth in one shot, **Figure 32-18**.

Coaxial cable can be terminated with twist-on, crimped, or compression connectors. See **Figure 32-19**. Twist-on connectors are the least expensive and easiest to install. They do not require any special tools except for those needed to strip

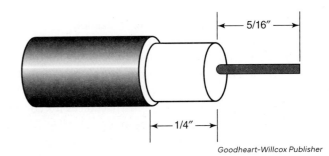

Goodheart-Willcox Publisher

Figure 32-17. Coaxial cable stripping dimensions.

Eaton

Figure 32-16. Wall cover plates for RJ-45 and other communications system connectors.

Ideal Industries, Inc.

Figure 32-18. Three-blade coaxial cable stripping tool.

Coaxial Cable F Connectors

Twist-on connectors

Crimped connectors

Compression connectors

Ideal Industries, Inc.

Figure 32-19. Common types of F connectors for coaxial cable.

the cable. A disadvantage of this type of connector is that the integrity of the connection is not always ensured since it uses a method of attaching the device to the cable.

More secure than twist-on connectors, a crimp connection is a type of compression connection where pressure is applied to a connector with a crimping tool to make a solid electrical connection. For coaxial cable connectors, the crimp tool presses the connector into a hexagonal shape over the cable. This creates a good solid connection, but the hexagonal shape of the crimp die distorts the shape of the cable at the connecting point. This can have a negative effect on the signal quality at the point of utilization.

Compression connections are the best choice for connection integrity and signal quality. The connectors are more expensive than most other options, and the connector installation requires a special compression tool.

At the signal outlet, there are many choices for connection points. Two common options are a wall plate with an F connector, **Figure 32-20**, and a modular F connector that can snap into a wall plate.

Splitters can be used to share the signal. Note the signal direction when connecting cables to the splitter: a splitter divides the signal, so the branches receive 25-50% of the incoming signal. The manufacturer data sheet for each type of splitter can provide you the details about how the signal is divided between the outputs because they are not all the same.

32.3.3 Fiber Optic Connections

Installing fiber optic connectors onto fiber optic cable is not something that electricians typically do. Fiber optic cable installation is an area of specialization. Connection methods may be specific to the type of fiber optic cable being used.

32.4 Testing Cables and Connections

There are many manufacturers of testing equipment. Each have a wide variety of test equipment options, ranging from very simple to extremely complex. Be sure to select the equipment that best suits your needs and budget.

Testing the installed cables is a critical part of installation. See **Figure 32-21**. It ensures conductor integrity between the source and termination points. A tester might have a transmit terminal and receive remote terminal. It may test for opens, shorts, signal quality, the installed connector and correct order of pairs for twisted pair cable. Some testers have connections for coaxial cable, RJ-11 (phone), and RJ-45 (ethernet).

Testers for fiber optic cable use a light signal to test integrity of the cable and the quality of the connections.

32.5 Specialty Signaling Devices

Today's homes can contain many wired and wireless devices related to areas such as entertainment, security, safety, home management, and energy efficiency. The following sections focus on two important devices often required to be installed in residences: smoke detectors and carbon monoxide detectors. Many of these devices can be installed

Eaton

Figure 32-20. Wall plate–mounted F connector for coaxial cable.

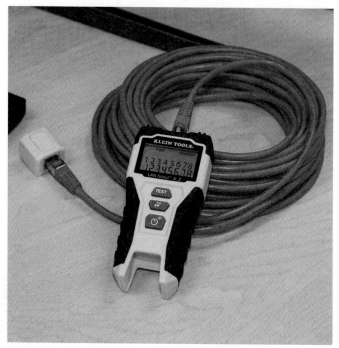

Klein Tools

Figure 32-21. A cable tester checks the cable and terminations at the connectors for opens, shorts, miswiring, and split pairs.

by the homeowner, but some require an electrician to perform the work.

32.5.1 Smoke Detectors

Requirements for smoke detectors and alarms are not in the *NEC* (NFPA 70), but NFPA 72, National Fire Alarm Code. In the United States, while specific requirements can vary, laws regarding the installation of smoke detectors exist in every state. Check with your AHJ for the requirements regarding the type of detector, the acceptable installation locations, and any other important information.

There are two main categories of smoke alarms used for dwellings: photoelectric and ionization. Photoelectric smoke detectors contain a light sensor and a light source in an internal chamber where smoke can enter. The smoke obstructs the light sensor and triggers the alarm. Ionization sensors use a small amount of a radioactive material placed between two electrically charged plates. The charges transfer to the air molecules in the chamber, causing current to flow through the ionized air between the plates. Smoke that enters the chamber causes the ions to disperse, disrupting the flow of current and triggering the alarm. The two methods of detection offer different protection: ionization types are better for detecting smoke caused by flaming fires, while photoelectric types are more responsive to smoke caused by smoldering fires. The NFPA recommends using both types of smoke detectors.

Depending on local regulations, it may be required, or only recommended, that a smoke detector be installed in each bedroom, hallway, living room, family/recreation room, and basement area. Avoid areas where fumes, smoke, or steam interfere with the operation of the detector, such as garages, kitchens and cooking areas, attics, and bathrooms. In addition, do not install smoke detectors near ceiling fans.

If mounting on a wall, the detector should be no more than 12″ from the ceiling. If mounted on a ceiling, it should not be closer than 4″ to a wall. If installed in a pitched ceiling, the detector should be installed 4–12″ from the peak. These recommendations ensure the detector operates correctly and avoids false alarms.

Some other variations in smoke detectors include the type of power source, combination smoke detector/carbon monoxide detector, and dual sensor that contains both an ionization sensor and a photoelectric sensor. Smoke detector power sources are generally one of three types: 9 V battery, 10-year lithium battery, and hardwired to 120 V, usually with a battery backup. See **Figure 32-22**.

Alarms are available with strobes necessary for those who are deaf or hard of hearing, but they also are a useful notification for when an alarm is not heard. Auxiliary strobes can be used in interconnected systems and can be activated by smoke, heat, or carbon monoxide detectors. Vibration notification devices activated by a smoke alarm are available to wake an individual in the event of an alarm. The NFPA covers more information on their website.

Heat detectors are required in some jurisdictions. They can be installed inside the home and in areas where smoke detectors cannot be used, such as attics and attached garages. The alarm is initiated after the temperature exceeds a preset level, such as 135° F (57°C).

PRO TIP **Ionization Smoke Detectors**

Ionization smoke detectors and any detectors that contain lithium batteries must be disposed of according to local regulations.

Smoke detector installed on ceiling

Hard-wired smoke detector with battery backup

Smoke detector powered by 12-volt battery

Zigmar Stein/Shutterstock.com; Michael O'Keene/Shutterstock.com; SpeedKingz/Shutterstock.com

Figure 32-22. Smoke detectors vary by detection method and power source.

32.5.2 **Carbon Monoxide Detectors**

Carbon monoxide (CO) is a colorless, odorless, tasteless, and poisonous gas produced by the combustion of fuels. Carbon monoxide is present in low concentrations in homes. However, if the CO concentration in a home rises, people may experience physical symptoms such as headache, fatigue, dizziness, and nausea. High CO concentrations can cause death.

The danger from carbon monoxide is based on the concentration (measured in parts per million, or ppm) and length of exposure. A person may need to be exposed to a slightly elevated concentration for many hours before experiencing mild symptoms. Exposure to a relatively high concentration, however, may result in severe symptoms in less than an hour. See **Figure 32-23**.

Carbon monoxide is produced in the home by fuel-burning appliances such as furnaces, cooktops, and heaters. Homes with an attached garage can be affected by carbon monoxide released in vehicle exhaust.

Due to the dangers of carbon monoxide poisoning, most municipalities require a carbon monoxide detector to be installed in homes. See **Figure 32-24**. Always be familiar with the building requirements for the home's location.

In a multistory house, installing a CO detector on each floor of a house is the best practice. The best location is near sleeping areas, so an alarm will wake anyone who is sleeping. CO detectors are commonly installed on a wall at a height of approximately 5′ above the floor. Avoid installing CO detectors in excessively humid areas such as bathrooms, in direct sunlight, near fuel-burning appliances (min. of 15′

away), or near a source of blowing air such as a fan, vent, or open window.

32.5.3 **Interconnected Systems**

Safety devices can be interconnected so that an alarm that is initiated in one part of the house activates the alarms throughout the house. This helps alert anyone who did not hear the initial alarm. Interconnected devices are available as wired or wireless systems.

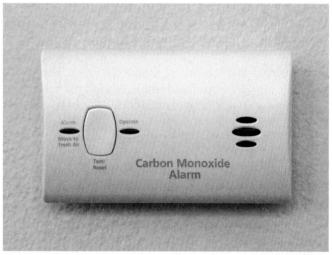

Andy Dean Photography/Shutterstock.com

Figure 32-24. Carbon monoxide detectors are generally best installed mounted to a wall and approximately 5′ above the floor.

Impacts of Carbon Monoxide Concentration

Carbon Monoxide (CO) Concentration (in ppm)	Possible Effects and Notes
0.5–5	Typical concentration in homes without gas cooktops
5–15	Typical level in homes with properly adjusted gas cooktops
up to 30 or higher	Typical level in homes with poorly adjusted gas cooktops
30–69	In this range, UL 2034 CO detectors must not alarm in the first 8 hours, but may alarm after 8 hours
50	OSHA permissable exposure limit for an 8-hour exposure
70–149	In this range, UL 2034 CO detectors must not alarm in the first 60 minutes and must alarm within 4 hours
100	Headache, fatigue, dizziness within 2 hours of exposure
150–399	In this range, UL 2034 CO detectors must not alarm in the first 10 minutes and must alarm within 50 minutes
200	Fatigue and headache within 1 to 2 hours of exposure
400	Headache and nausea after 1 to 2 hours, life-threatening after 3 hours. At concentrations at 400 ppm and above, UL 2034 CO detectors must not alarm in the first 4 minutes and must alarm within 15 minutes
800	Headache, dizziness, and nausea in 45 minutes, life-threatening within 2 hours

Goodheart-Willcox Publisher

Figure 32-23. Carbon monoxide concentrations and corresponding possible effects and alarm limits.

A wired interconnection requires a conductor to connect to the interconnection lead in each detection device. Depending on the manufacturer, this could be a wire with red or yellow insulation. If it is planned to interconnect the safety devices during construction, this requires NM cable to include one extra wire. For example, if the devices need 120 V to operate and are all on the same circuit, the first device will have a 14-2 NM cable from the circuit breaker, but each subsequent device needs a 14-3 NM cable to connect the interconnection wire.

A wireless system eliminates the need for an extra interconnection wire. Each manufacturer has specific instructions for connecting compatible devices.

32.6 Old-Work Installations

The installation of communication cables and outlets is not much different from an electrical old-work installation, as covered in other chapters. Cables can be run in attic spaces and crawl spaces and also fished in walls.

With old work, you may encounter old, obsolete cables, such as older telephone coaxial cables. If there are plans to use this old cable in the future, it must be tagged to identify that it will be used. If this cable will no longer be used, the *NEC* categorizes this cable as *abandoned*. Abandoned cable must be removed, or as much of the abandoned cable that is accessible, to reduce its contribution to fire loading (becoming available fuel for a structure fire).

When fished into an existing wall, cable access and wall plate mounting can be accomplished with cut-in low-voltage rings. Drywall is cut to the size of the back part of the ring, and the ring is held in place between the clamps and the wider front part of the ring. See **Figure 32-25**. Surface-mounted raceways and boxes are another option for old work additions.

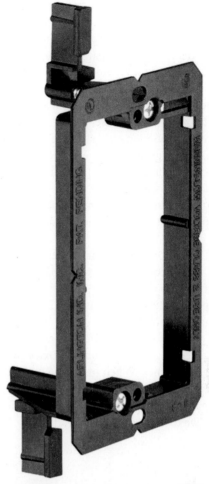

Arlington Industries, Inc.

Figure 32-25. Old-work nonmetallic rings have clamps that attach to the drywall for easy installation.

Summary

- Low-voltage and signaling systems include telephone, cable TV, ethernet, security systems, fire and smoke detection systems, and home entertainment.
- Low-voltage and signaling systems are often installed by specialty contractors, but an electrical contractor could be responsible for all or part of the installation.
- Communication systems have a service entrance location to provide a means for communications service providers to make connections from the utility side to the customer side.
- Telephone and cable TV service can be brought to the customer through underground conduit or from a conduit riser and overhead service.
- The service provider will usually bring the conductors to the premises and install a termination enclosure or necessary equipment, but the electrical contractor might be required to install the interface.
- Communication service equipment and conductors must be connected to an intersystem bonding termination device (IBT) at the service.
- An intersystem bonding termination (IBT) provides a means for a connecting communication systems grounding conductors at the service equipment or at the disconnecting means for buildings or structures supplied by a feeder or branch circuit.

- Several types of cable can be found for communications and signaling circuits. The most common that are found in dwellings are twisted pair, coaxial cable, and optical fiber. Each of these cable types has many variations that make them suitable for specific applications.
- In twisted pair cable, the pairs of conductors are twisted together to prevent crosstalk between pairs. CAT6A is the current standard for twisted pair.
- Fiber optic cable transmits light signals, in the form of high-speed pulses, through a very small diameter glass or high-density plastic fiber. The light signal transmitted through fiber optic cable must be converted to an electrical signal for equipment to use.
- For new construction, cable for communication and signaling is installed during the rough-in phase. Some cables can be installed originating at the service to each individual location or to an initial location and then daisy-chained to each subsequent location. Other cable types or network configurations require dedicated lines from the source to each outlet location.
- Rough-in of communication systems uses similar methods and materials as electrical rough-in.
- RJ-45 plugs and jacks specifically for use with CAT6A must be used and require the right crimping and punch down tools.
- T568A and T568B are the standards for connecting twisted pair cable for ethernet.
- CATV F connectors for coaxial cable are twist-on, crimp, and compression. Compression connectors retain the best signal integrity.
- Testing the installed cables is a critical part of installation. It ensures conductor integrity between the source and termination points.
- Requirements for smoke detectors and alarms are not in the *NEC* (NFPA 70) but NFPA 72, National Fire Alarm Code.
- Due to the dangers of carbon monoxide poisoning, most municipalities require a carbon monoxide detector to be installed in homes. When installing connectors for communications systems, it is important to test the cables and connections.
- Surface metal raceways and surface non-metallic raceways can be used for installations in existing buildings as an alternative to fishing cables in walls.

Know and Understand

1. *True or False?* Depending on location, cables for telephone and cable TV service can be underground or overhead.
2. *True or False?* It is not necessary to bond communication equipment to the premises grounding system.
3. A(n) _____ is required to bond the communications equipment to the grounding electrode.
   A. grounding electrode conductor
   B. intersystem bonding termination device
   C. equipment grounding conductor
   D. supply-side bonding jumper
4. According to the *NEC*, the bonding conductor for an IBT shall not be smaller than _____ and shall not be required to exceed _____.
   A. 16 AWG; 4 AWG   C. 14 AWG; 6 AWG
   B. 6 AWG; 1/0   D. 12 AWG; 10 AWG
5. The type of coaxial cable most frequently used for residential installation is _____.
   A. RG-6   C. RG-59
   B. RG-11   D. RG-450
6. *True or False?* ENT must be secured not more than 5′ from where it enters a box?
7. For ENT, bends between pull points cannot exceed
   _____.
   A. 90°   C. 270°
   B. 180°   D. 360°
8. Industry standards for telecommunications are determined by who?
   A. Telecommunications Industry Association (TIA)
   B. Electronic Industries Alliance (EIA)
   C. Both the TIA and EIA
   D. Neither of these.
9. *True or False?* Photoelectric smoke detectors contain a tiny amount of radioactive material.

Apply and Analyze

1. What are some of the challenges to installing communication wiring in an existing home? What "old work" methods did you learn in previous chapters that you could use for the installation of communication systems?

Critical Thinking

1. *NEC Chapter 9 Table 2, Radius of Conduit and Tubing Bends* lists the minimum radius of bends depending on the method used to make the bends. The bending radius is permitted to be larger than what is listed but not smaller. Why is this important, and what would be the result of violating these rules?
2. Some studies have found that standard smoke alarms might not wake sleeping children, and older adults might not respond to the alarm at the frequency that most smoke detectors operate. Standalone smoke detectors that are activated behind a closed door or on a different level of a home might not be heard by the occupants of the house. What changes to current regulations would you recommend to ensure that alarms are more effective at notifying people in the case of a fire or other emergency?

CHAPTER 33

Solar Photovoltaic Systems

Jason Finn/Shutterstock.com

SPARKING DISCUSSION

How can the energy of the sun be transformed to be used for our daily electrical needs?

Introduction

The converting of light energy into electrical energy is known as *photovoltaics (PV)*. The advancements in technology and costs of photovoltaic materials have made it possible to install photovoltaic power systems on homes to produce needed electricity. In this chapter, the topics of photovoltaic power systems will be covered in a general overview.

33.1 Solar Photovoltaic (PV) Power Systems

Designers of PV systems use solar data sources such as the National Renewable Energy Laboratory to design PV systems. Other factors that are considered include roof angle, orientation to the sun's path, and sunlight availability.

Many PV system designers use computer software to input the latitude, longitude, tilt angle, azimuth, and elevation to predict the amount of power a system will produce over time. Prior to designing any PV system, a site assessment is done to ensure proper design. One of the critical elements of a site assessment is a shading analysis. If a PV system is installed near large trees or where portions of a roof are shaded from the sun, the electrical production of the system will be reduced when in the shade. Items such as plumbing vents, satellite dishes, rooftop antennas, and chimneys are often overlooked and cause reduced power outputs.

33.1.1 PV System Types

The site assessment helps determine how much energy the customer needs. The goal in sizing any PV system installation is dependent on the type of system installed. Photovoltaic systems are sized for either off-grid or grid-tied applications. An *off-grid PV system* provides for all the electrical needs of the residence, eliminating the need for a connection to the electric utility. Off-grid PV systems use batteries to store power for night use or when it is cloudy.

A *grid-tied PV system* is connected with the service from the electric utility. Grid-tied PV systems can provide excess power back to the utility grid and draw power from the utility when the PV system cannot meet the load demands. These PV systems rely on a source of voltage from the utility to operate and provide power for house loads.

33.1.2 PV System Components

PV systems are an assembly of components that produce electricity, provide mechanical or structural support, and control the transfer of electricity to the electrical system.

PV Cells

At its core, a PV system is composed of semiconductor *solar cells*. When sunlight strikes a solar cell, the energy from the photons in the sunlight create electrical power in the cell. The voltage of a single solar cell is approximately 0.5 volts. The amount of electrical current produced by the solar cell is dependent on the intensity of the sunlight—the greater the intensity, the greater the electrical current.

PV Modules

PV manufacturers assemble solar cells into *PV modules*, often called *solar panels*. Modules are made by connecting many cells in series. Common PV modules have 60 or 72 cells. The output voltages are 30–38 volts dc for a 60-cell module and 36–48 volts dc for a 72-cell module. Layers of silver paste and conductive strips connect each cell to one another. If viewed closely, the fine lines of conductive silver can be seen on the surface of the cell, **Figure 33-1**.

PV Module Construction

A PV module contains several layers of materials. A sheet of polymer film is placed on both sides of the solar cells. A pane of glass that has an anti-reflective coating is placed on the polymer film on the front side of the module, and a backsheet is attached to the polymer film on the back side of the solar cells. This assembly is sent through a vacuum chamber and oven to heat-seal the layers together. The cells are wired to an electrical junction box on the back of the panel. Once the assembly is complete, the metal frame is placed around the assembly and the module is tested. Refer to **Figure 33-2**.

The junction box on the back of the modules contains bypass and reverse current blocking diodes and the lead connections. Reverse blocking diodes are manufactured into a photovoltaic module to prevent higher voltage modules from causing current to flow backward in lower voltage parallel connected modules. Bypass diodes allow for current to continue to flow if part of the module becomes shaded. The leads are typically photovoltaic wire. This wire has a cross-linked polymer insulation type XLPE with an overall sunlight-resistant jacket. On the ends of the PV module leads are typically MC-4 electrical connectors.

The use of manufactured PV modules is the most popular method of creating a solar PV system. Other types of solar electric modules can be incorporated into the construction of a building. Solar PV shingles and solar PV awnings are examples of integrated building systems.

PV Arrays

Photovoltaic modules are connected to create a *PV array*. If PV modules are connected in series, the voltage of the PV array will be the total of the voltages of the modules. If the modules are connected in parallel, the current of the PV array will be the total of currents of the modules, **Figure 33-3**.

Many PV arrays installed on rooftops are dependent on the available support structure. The added weight of a PV array can add 5–10 lbs per ft^2 to a structure. If the home is older or the roof design does not have any additional strength, the roof-mounted array may be limited in size.

eenevski/Shutterstock.com

Figure 33-1. Solar cells contain fine lines of silver conductive strips to control the current produced when sunlight strikes the cell.

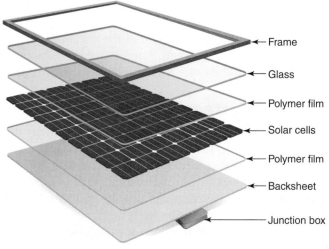

Frame
Glass
Polymer film
Solar cells
Polymer film
Backsheet
Junction box

Alejo Miranda/Shutterstock.com

Figure 33-2. The components of a PV module provide support and protection for the solar cells.

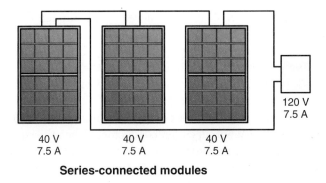

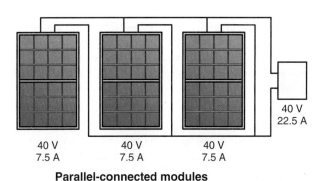

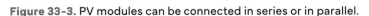

Series-connected modules

Parallel-connected modules

Goodheart-Willcox Publisher

Figure 33-3. PV modules can be connected in series or in parallel.

33.1.3 PV Inverters and Power Conversion Equipment

All PV modules produce dc power, and most homes operate on ac power. An electrical conversion device called an *inverter* converts the dc power into ac power. The methods used to make these conversions are microinverters, string inverters, and battery charge controllers with battery inverters.

A *microinverter* is a small electronic inverter connected to a single PV module. Each microinverter will have ac power run to them. Some PV modules have been manufactured with the microinverters as part of the module assembly. These are called ac photovoltaic modules.

The *NEC* defines microinverters installed on or as part of the module assembly as an ac module system. The ac circuit that is connected to the output of the microinverters is known as the *inverter output circuit*. *Section 690.8(B)(1)* requires that PV output circuits be sized based on 125% of the output rating of the parallel connected inverters.

String inverters are used when the PV modules' dc circuits are connected in series or parallel, **Figure 33-4**. The dc PV source circuit is connected to a single-string inverter. These residential 240-volt inverters can be sized from 1 kW to 20 kW. The PV module circuits need to be configured so that the input voltages are kept in the working range of the inverters.

The microinverters and string inverters are typically used for grid-tied systems. This does not mean that they are not capable of being used in an off-grid system. *Section 705.6*

requires that systems connected to the utility have anti-island technology to turn the inverter system off when ac power is lost.

Anti-island design provides safety for electricians and utility workers. For example, if a powerline is knocked down, the utility power system has protective relays and switches that disconnect the power coming from the utility generation plant. The utility workers safeguard the generation circuits before working on the downed power line. The utility worker expects the power to be coming from the utility source, and that is where it will be safeguarded. If a person has a PV system that remains operational on the customer (house) side of the power line, the power produced by the PV system will back-feed through the transformers and energize the downed power line. This energized line could then electrocute a utility worker.

If a grid-tied system is configured for operation when utility power is lost, a transfer switch or microgrid interconnect device disconnects the utility power from the backup system when used in backup power mode, **Figure 33-5**. Some automatic systems can sense when utility power is lost or restored and will switch to the desired mode.

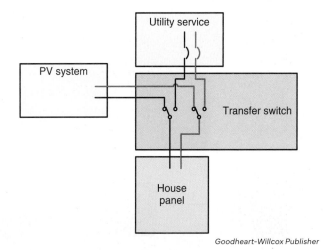

Goodheart-Willcox Publisher

Figure 33-5. A transfer switch isolates the PV system from the utility service to prevent back-fed current when the utility service is de-energized.

Figure 33-4. A string inverter converts the dc power supplied by the PV array to ac power that can be used to supply household loads.

The final type of PV system configuration is the charge controller and island-type (off-grid) inverter. These are mainly used for off-grid applications and in small applications such as bus shelter lighting. The PV array dc output circuit is connected to a dc battery charge controller. The charge controller takes the power generated from the PV systems and regulates it for charging batteries. An inverter is connected to the batteries to invert the dc battery voltage to ac to supply house loads, **Figure 33-6**.

The battery bank allows the power supplied by the PV system to be used at times when the PV system is not producing power, such as cloudy days or nights. Most off-grid PV systems are typically sized to provide a minimum amount of power for up to three days.

PV inverter configurations can be grid-tied but also provide battery storage. Microinverters or string inverters can be used with storage battery systems. Different methods can be used to charge a battery when the solar PV system is overproducing power during the day. During low production periods, the power stored in the battery is discharged to provide power for the house.

PV System Rapid Shutdown

With all the arrangements of systems available and the inability to turn off the sun to remove power, *Section 690.12* requires that a system be put into place to shut down the power from all circuits leaving the PV array if the utility power is lost or an emergency shutdown switch is initiated. The term associated with the emergency power-off condition is known as **rapid shutdown**. These are required on installations on buildings when other than microinverters are used. The *NEC* has requirements to provide rapid shutdown devices (RSD) to all PV output circuits. The location of the devices that provide rapid shutdown is at the array location or within 3′ of the array boundary where the output circuit leaves the array location.

Rapid shutdown can be achieved using module-level power electronics (MLPE) dc-to-dc converters that provide rapid shutdown. Alternatively, RSDs can be incorporated into the PV output circuit for each string inverter. Some PV shingle modules require rapid shutdown devices for each set of a specific number of shingles.

Rapid shutdown devices are designed to provide a small dc voltage on the PV output circuit conductors. When the rapid shutdown device receives a "keep alive" signal from the connected inverter or rapid shutdown switch, the rapid shutdown device will provide the output power developed from the module. If the "keep alive" signal is received from the inverter on the dc power conductors, the system will operate normally. If ac power is lost, the inverter will shut down, the "keep alive" circuit signal will be lost, and the RSDs will reduce the module output back to the lower dc voltage.

33.1.4 PV Array Mounting Systems

The most popular mechanical mounting system for PV modules are rooftop racking systems. Rooftop racking systems use a system of flashing, attachment anchor points, extruded aluminum support rails, and module hold-down clips. In addition to the major components of these systems, manufacturers offer wire management clips and hooks, junction boxes, and grounding elements.

Other PV array mounting systems are pole-mounted and ground-mounted arrays. These systems are typically designed with all the necessary components to complete the installation. Larger systems require a structural engineer to design proper support structures.

One of the key elements of mounting to roofs is making sure that the flashing systems will keep out water or ice melt from the structure, **Figure 33-7**. Long-term moisture in wood-framed structures can lead to rot and weakened attachment points. Always follow the flashing installation instructions provided by the racking manufacturer.

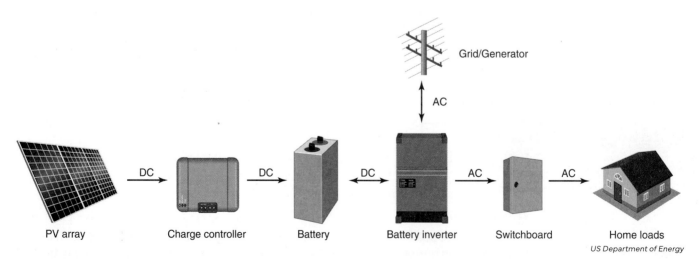

US Department of Energy

Figure 33-6. A charge controller uses the dc power supplied by the PV array to charge the battery.

US Department of Energy

Figure 33-7. Proper flashing installation is critical to prevent roof leaks.

33.2 **Residential Power and PV System Integration**

The connections between the PV system and a home electrical power system are known as the *system integration*. The connections made for utility interactive (grid-tied) inverters typically require a disconnecting means on the exterior of the home to allow for fire department and utility personnel to de-energize the systems without having to enter the structure.

The connections between the PV systems and the home electrical system are designed for power to be transferred in a bidirectional relationship. *Bidirectional power* production sources are regulated under *Article 705*. All components and devices used in a bidirectional system must be listed for bidirectional use. If the terminals on devices, breakers, or switches are marked "Line" and "Load," they are not able to be used in a bidirectional manner.

System integration using energy storage is becoming popular with PV systems and electric vehicles. Manufacturers of inverters are designing units with the capability of connecting electric vehicle batteries to the dc side of the inverter systems. The electric vehicle connection allows the vehicle to charge directly from the solar PV system and also allows the vehicle battery to supply power to the household loads.

Other battery storage systems use batteries to store the excess energy produced by the PV array and draw from the energy storage when the demand is required from the home. Some homes with battery storage have the capability to provide power independent of the utility. This arrangement is commonly known as a *microgrid*, which is a local, self-sufficient electrical grid separate from the utility grid. These microgrid systems use the battery energy storage as a holding tank for the PV power created and draw from them to power the house. Automatic transfer switches disconnect the utility power from the microgrid when it is in island mode. Island mode with a transfer switch prevents power from the PV system from back-feeding into the utility system.

33.3 **Installing Rooftop PV Systems**

The roof of homes is the most popular location to install PV systems. The types of roofing materials use will dictate the type of mounting and flashing systems that are used. Typical roof types of residential homes are composition (asphalt) shingles, metal standing seam roofs, metal tile shingles, clay and concrete tile roofs, and flat roofs, **Figure 33-8.**

Composition shingles

Metal standing seam roof

Metal tile shingles

Clay roof

Concrete roof

Flat roof

Evannovostro/Shutterstock.com; U. J. Alexander/Shutterstock.com; Bilanol/Shutterstock.com; tamara321/Shutterstock.com; Emagnetic/Shutterstock.com; Photomann7/Shutterstock.com

Figure 33-8. The type of roof will determine the type of anchors and flashing required for a PV array installation.

PROCEDURE

Preparation for PV Installation

1. Determine a proper and safe means to access the roof—typically by either ladder or scaffolding.
2. Establish the type of fall protection that will be used for workers. Most residential roofs are pitched at an angle, so typically personal fall arrest systems with vertical lifelines, and shock absorbers are used.
3. Install a fall arrest system anchor point either under or above the shingles. Screws or nails are used to attach directly into framing member truss or rafter.
4. Attach a vertical lifeline to the anchor point.
5. Ensure the lifeline or the lanyard attached to the lifeline contains a shock absorber. This shock absorber is designed to unravel during a fall to decelerate a person rather than being jerked to a stop.
6. Donning a personal fall protection harness, connect it to the vertical lifeline with a lanyard. Refer to **Figure 33-9**.
7. Set up roof jacks and planking by first nailing the roof jacks (metal brackets) into the roof structure to support planking and provide a walking surface. Refer to **Figure 33-10**.
8. Coordinate the location of the roof jacks and planks with the layout, so they do not interfere with one another.
9. Once all is set up, determine a location to store tools and material needed to assemble the PV array.

US Department of Energy

Figure 33-9. Proper fall protection is needed when installing PV systems on rooftops.

Goodheart-Willcox Publisher

Figure 33-10. Installing roof jacks and planking provides a stable work surface for installing rooftop PV systems.

33.3.1 Lay Out the Array

Racking systems are designed based on the structure of the home. The placement of the anchor points for the racking are laid out on the roof. The flashing that comes with the roof attachment points have tolerance for shingle overlap or tile flashing overhang. If these are not followed, there may be a chance for a water leak.

PRO TIP   **Installing Roof Tiles**

Many installations on composition tile roofs will use the shingles as a guideline for installation. Shingles may not be installed in perfectly straight lines or be square with the house. It is good practice to measure between the peak of the roof and the facia or eave and find a center line. A chalk line can be snapped across the shingles on this center line. This reference line can now be used to lay out and install the anchors for the racking systems.

Also, when marking roof surfaces, pencil or permanent marker does not show up well. Sidewalk chalk is great for marking on roof surfaces as it is easily seen and washes away with rain when the job is done.

When possible, stagger racking anchor points on different trusses or rafters. Refer to **Figure 33-11**. The racking will provide better support if alternate framing members are used. The wind loading and weight distribution is better when anchors are not all installed in the same framing member.

Installing Anchor Points

An anchor point is installed into a truss or rafter. If the truss or rafter does not work with the racking attachment point, additional roof blocking may need to be added. Blocking should be attached to the rafters or trusses using lumber and hardware that meets structural requirements.

PRO TIP   **Flashings**

Installing flashings typically requires the use of a roofing shingle bar or flat pry bar. Careful attention should be paid when removing composition shingles or roof tiles so that minimal repairs will need to be made after the installation of flashing and anchor points.

Racking anchors require a specific length of lag bolt embedded into the wood for attachment points. If the roof truss is not of sufficient thickness to provide the number of embedded threads required for the anchor point, sistering of additional lumber to the side of the truss will be required, **Figure 33-12**.

Installing Racking

When installing the racking for the modules, the location and direction will typically be predetermined. The support rails are the long supports the modules are attached to. These rails will have hardware, brackets, or clips attached to the anchor points, **Figure 33-13**. The brackets and anchor points have some adjustment to allow for waviness in the roof. A string can be used to keep the support rails in line while attaching the hardware. More than one structural framing member should be used to support the rail. Mounting a support rail vertically with all anchor points on one framing member can overstress a structural truss or rafter.

Support rails may need to be coupled together to create longer lengths. Careful attention should be given to where the couplings are placed in relation to anchor points and PV module attachments. Some racking manufacturers require couplers to be installed within a specific distance of an anchor point.

PRO TIP   **Racking Rails**

In many installations, the rails for the racking are purchased longer than necessary. Do not cut them to length before installation. Instead, overhang the extra length on each end of the array. This allows for alignment where couplings need to be installed at specific distances from supports. It is often easier to come back and cut off the excess rails then to try and get a perfect alignment with the modules from the start.

Prepare for Module Installation

Once the racking rails are installed, microinverters or MLPE devices may be mounted to the rails, depending on the type of installation. These will typically be installed directly below or adjacent to the module that they serve.

The interconnecting cables between MLPEs or microinverters are installed. If a separate rapid shutdown device is installed for each circuit, these also are mounted and wired with the cabling. Depending on the arrangement of the MLPE or microinverter circuits, some manufacturers provide prewired cables for these interconnections, **Figure 33-14**. Other cabling may require that the series and parallel configurations be made. Before any modules are installed, all these connections should be verified.

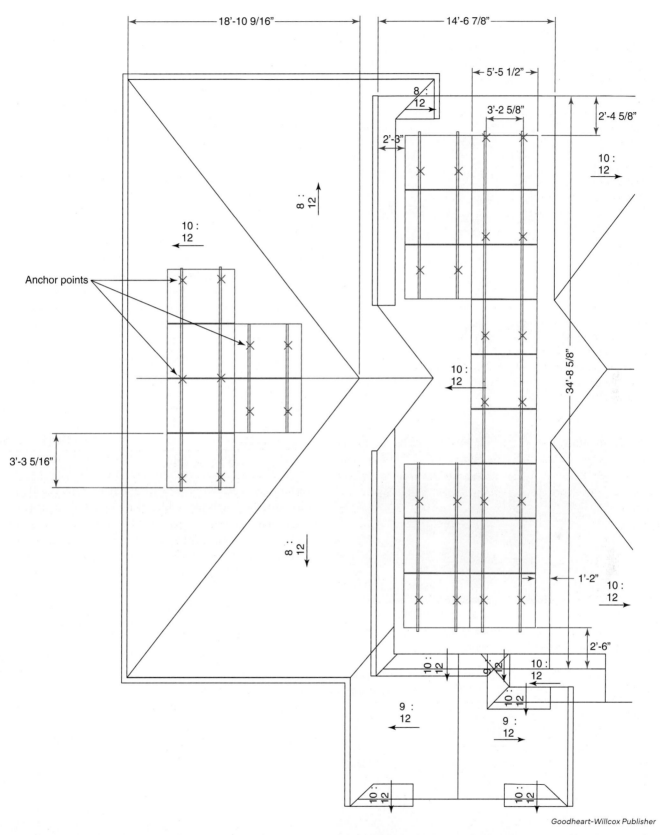

Goodheart-Willcox Publisher

Figure 33-11. This plan drawing of a rooftop PV system installation shows racking anchor points positioned on alternate rafters.

Goodheart-Willcox Publisher

Figure 33-12. If a roofing truss or rafter is not sufficient to support an anchor point, sistering the rafter or truss may be needed.

Support rails

Goodheart-Willcox Publisher

Figure 33-13. After the anchor points are installed, support rails are connected to the anchor points.

Goodheart-Willcox Publisher

Figure 33-14. Microinverters are attached to the support rails before the PV arrays are installed. These microinverters have prewired cables for easy installation.

Section 690.43 requires that all exposed noncurrent-carrying metal parts of PV module frames, electrical equipment, and conductor enclosures of PV systems are connected to an equipment grounding conductor. If the racking system is listed by the manufacturer for grounding, the connections between the modules and the racking will be made by the racking module hold-down clips. Equipment grounding conductors will need to be installed between the rails of the racking.

The circuits that transmit power from the PV system to the main electrical equipment can be run inside or outside of the house. Per *Section 690.31(D)(1)*, dc PV source output circuits operating over 30 volts must be installed in type MC cable or a raceway. These circuits are also required to be marked every 10′ and at junction and splice points with a label that identifies them as a "Photovoltaic Power Source."

The dc conductors that connect the PV array and the inverter are also required to be color coded to identify polarity. Black is used to indicate the negative and red to indicate the positive. If a conductor for a PV system is grounded through a ground detection system, this grounded conductor will be required to be identified as white. If a white grounded conductor is used, an additional label or tag is required to indicate if it is a positive or negative conductor.

Section 310.10(D) and *Section 300.6(C)(1)* require that any conductors or nonmetallic raceways installed exposed to direct sunlight are required to be sunlight resistant. The cables that interconnect modules and MLPEs or microinverters are typically made with PV wire. This is a 1000-volt insulated cable with an additional overall sunlight-resistant jacket. These wires are permitted to be run exposed without a raceway so long as they are under the array, not readily accessible, and not subject to physical damage.

The connections between the PV array wiring and the building wiring methods that contain the PV output circuits are typically done on the roof with a weatherproof junction box. The junction boxes are required to be accessible. The removal of a PV module to access the junction box is permitted per *Section 690.34*. The junction box may be weather resistant, but *Section 300.9* considers all raceways and junction boxes on the exterior of a building to be wet locations. When making connections in these junction boxes, proper wet location splices are required. A typical wet location twist-on wire connector is suitable for this, **Figure 33-15**.

Installing Modules

One of the most difficult parts of installing a PV system is installing the PV modules. A 60-cell module typically weighs about 40–50 lbs. The pitched roof walking surfaces and PV racking make this stage of installation challenging.

To move PV modules from the ground to the roof, many installers use a ladder hoist. Scaffolding becomes the next best method to use when raising PV modules to a roof area. Using a swivel hoist arm with a pulley and rope allows PV

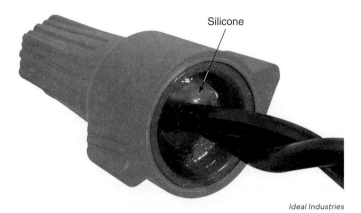

Silicone

Ideal Industries

Figure 33-15. This weatherproof twist-on wire connector contains silicone gel to protect the connection from the elements.

Wichien Tepsuttinun/Shutterstock.com

Figure 33-17. MC-4 connectors with positive and negative leads.

modules to be hoisted safely to the roof. On lower single-story home roofs, it may be possible to hand PV modules to a person on the roof or scaffold.

When installing the modules, begin by getting the first module straight and level, located in the correct position, connected to the wiring, and securely fastened, **Figure 33-16.** Many installers position the module near its location and rest the module on its side or on the roof jacks and planks to properly make all the needed electrical connections. The MC-4 specially designed positive lead and negative lead connectors keep connection polarity easy. Refer to **Figure 33-17.**

After the connections are made, the module is aligned with the racking. The hold-down clips are then inserted and properly tightened. Many PV racking manufacturer hold-down clips are selected specific to the module frame

thickness. Installers need to verify the module to clip match prior to this point. Some hold-down clips or fasteners are designed to hold two modules with one bolt. These mid-clips are loosened when placing the next module.

Typically, before each additional module is placed, the wiring below the modules is cable tied or clipped to the frame using PV wire clips. *Section 690.31(C)(1)* requires that conductors are secured and supported at intervals not exceeding 24″. Cable ties are required to be sunlight resistant. Once the modules are placed, and all the hardware is torqued, one final stage after start-up is installing screening around the array. The screening prevents animals from making homes under the PV array and damaging the wiring. Nests and other materials under the array will also prevent airflow and proper cooling.

33.4 Installing Pole-Mounted and Ground-Mounted PV Systems

The differences between roof-mounted solar PV systems and pole-mounted or ground-mounted PV systems are foundations, support structure, underground wiring methods, and access for assembly. Refer to **Figure 33-18.** The use of ground-mounted or pole-mounted systems may be because the house desiring to use the PV power is in a wooded area, and shading is an issue. There may be an area on the property that is not shaded and is a better location for a PV system. Pole-mounted arrays also allow for angle adjustment and tracking type systems that change the angle of the array with the sun's position. Keeping the PV modules aligned with the sun increases the power output.

US Department of Energy

Figure 33-16. Preparing to install the first PV module. Ensuring that the first module is installed in the correct position and alignment makes the installation of the remaining modules easier.

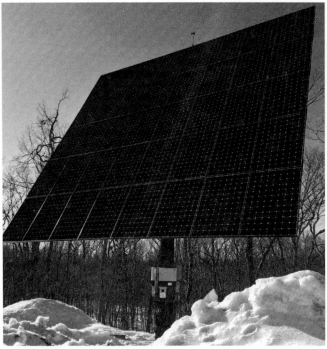

US Department of Energy

Figure 33-18. Pole-mounted PV systems provide greater options for locating the PV array when compared with roof-mounted systems.

33.4.1 Pole-Mounted Arrays

Installing a pole-mounted array starts with a proper foundation. The base of a pole-mounted array is typically concrete. A hole is dug large enough to fit a concrete form tube. The size of the tube will be dependent on the size of the pole and the number of solar modules mounted on the assembly.

Poles can be as small as 2″ pipe for small arrays, or a large array may require multiple 8″ poles. The pole support or an anchor bolt assembly is inserted into the concrete form tube and leveled. Reinforcing steel rebar are placed into the form. Before the concrete is poured into the form, the underground conduits are placed to allow the electrical to run up the side of the pole to the array. The *Code* requires that if conduits experience deflection during changing seasons, expansion fittings are required. If conduits are subject to movement when not installed in the concrete, expansion fittings are required.

Grounding is another item that needs to be completed as part of a pole-mounted array. This is typically done while the ground is open from digging the concrete base. The *Code* permits additional auxiliary grounding electrodes be installed at the array location. Typically, ground rods are driven and connected to the pole or to the equipment grounding conductors of the system.

After the concrete is poured, the pole is either already in place or is set on the anchor bolts. Installed at the top of the pole is the pole mount and framing assembly. These are manufactured kits designed for specific numbers of

modules. Depending on the pole top mount, access to mount the arrays becomes a challenge. Some pole-top mounting kits are designed to allow assembly of the entire array at ground level and then use a chain hoist to lift the array into place. Other pole-top array assemblies may require scaffolds or stepladders to complete assembly.

Once the pole top structure is assembled, the racking, MLPEs, wiring, and grounding are nearly the same as that of rooftop arrays. The nice aspect of a pole top array is that the modules can be placed on the racking, and the connections for the wiring can be completed from below. This also is nice for troubleshooting as modules do not need to be removed to access the wiring.

One final step may be required if the pole top array modules are not installed above 8 feet. The wiring and cables interconnecting the arrays are required to be protected by installing in a raceway or effectively guarded. In most areas, inspectors will accept a height of 8 feet above grade as an acceptable method for making the conductors not readily accessible, but that is not an option for all pole-mounted arrays. The wiring below 8 feet will need some sort of barrier, such as a fence, guarding screens, or solid barrier underneath the arrays to restrict access to the wiring.

33.4.2 Ground-Mounted Arrays

Ground-mounted arrays are very similar to pole-mounted arrays, but instead of a single concrete foundation, there are multiple foundations that support the racking structure. The assembly is similar to pole-mounted arrays but may have multiple ground supports for racking. The requirements to protect the exposed conductors on the underside become more prevalent. Many ground arrays will have the underside screened or covered to prevent unintended contact with the wiring and to keep animals from making nests or chewing on the wiring.

33.5 Wiring Connections into Residential Power Systems

Depending on the area of the country or the utility that provides power to a home, specific interconnection requirements may differ. The electrical wiring, devices, equipment, conduits, and cables are called the **balance of system** components. These components and their configuration will balance the installation of the PV systems with the electrical and utility needs of the home.

The primary concerns for the utility connections are to monitor the amount of power that flows to and from the home and how to provide safe de-energizing procedures for both the utility and the homeowner. In some areas of the country, municipalities or states may require that all new homes be built solar ready. If a PV system is to be installed in a solar-ready home, the basic infrastructure for the metering, disconnects, and interconnection points are put in place with the construction of the home.

Section 690.13 requires a photovoltaic system disconnect be provided to disconnect the photovoltaic production equipment, energy storage equipment, and utilization equipment from its associated premises wiring. The location of the disconnect is required in a readily accessible location.

Some utilities and municipal codes require this disconnect to be located on the exterior of the home. Manufacturers of electrical distribution equipment for homes make meter panel combinations that meet all the criteria for the *NEC* regulations and most municipal or utility requirements, **Figure 33-19**.

33.5.1 **Grid-Tied Microinverters**

Typical wiring system layouts for grid-tied interactive systems using microinverters with no energy storage include the dc wiring between each module and the microinverter and the 240-volt ac power circuits for the different microinverter circuits. If a single circuit is used, it can run directly through the PV system disconnect and then into a back-fed circuit breaker in the main panelboard, into a back-fed circuit breaker in a solar-ready exterior meter main panelboard, or in some cases, utilities provide a separate connection at the meter ahead of the main circuit breaker for the house, **Figure 33-20**.

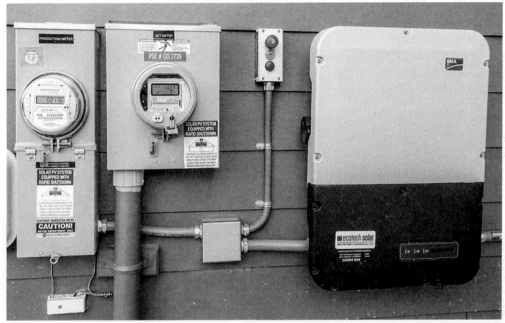

US Department of Energy

Figure 33-19. This installation includes a dedicated PV meter and disconnect.

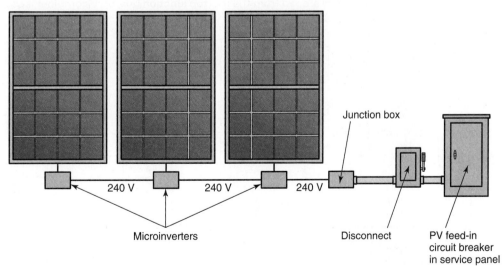

Goodheart-Willcox Publisher

Figure 33-20. PV modules with microinverters connected to a PV feed-in circuit breaker in the main service panel.

Multiple 240-volt circuits with microinverters will need to utilize an ac combiner panel. This will add up the individual circuits of the multiple microinverter circuits to a larger feed circuit to the main service panelboard. If the main service panelboard is a meter mains panel, there may be no additional PV system disconnect requirements. If the combiner panel is going to back-feed a breaker in the main panel, an additional disconnect may be required between the PV combiner panel and the house main panel on the exterior of the home, **Figure 33-21.**

When any circuit breaker is back-fed as an additional source to a home main panelboard, *Section 705.12* has specific requirements for these connections. The back-fed additional source circuit breaker is required to be installed at opposite ends of the bus bar from the main overcurrent device or main feed connections. The rating of the bus bar is required to be 120% of the total output current of all sources that are supplying the bus, plus the rating of the main breaker.

For example, consider a home has a 200-amp main breaker panel with a bus bar rating of 225 amps. The total rating of the overcurrent devices protecting the bus bar cannot exceed 120% of the rating of the bus bar. The 225 amps multiplied by 120% gives a total amp rating of 270 amps. Subtracting the 200-amp main circuit breaker leaves only 70 amps available for a PV system back-feed breaker.

The overcurrent devices that supply power from a PV system need to be rated for a continuous load since PV systems produce power for more than three hours. The maximum load permitted to be connected to the above example would be 56 amps (56 × 125% = 70 A). The total wattage for the PV system would be around 13,440 watts. This may be a large system of approximately 38–40 PV modules.

33.5.2 DC String Inverters with MLPE or RSDs

In PV systems that use MLPE dc-dc maximizers and a string inverter, the dc power conductors from the module configurations are run from the maximizers to the dc disconnect location near the inverter, **Figure 33-22.** The dc disconnect

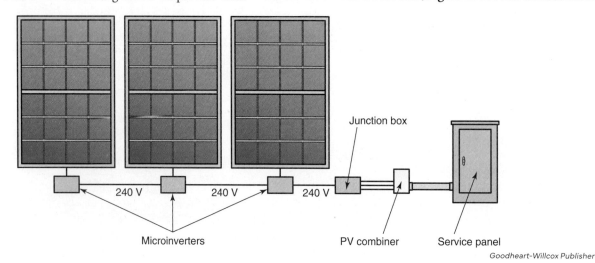

Goodheart-Willcox Publisher

Figure 33-21. When multiple microinverter circuits are fed from the rooftop junction box, a PV combiner is used to combine the microinverter output circuits and feed the service panel.

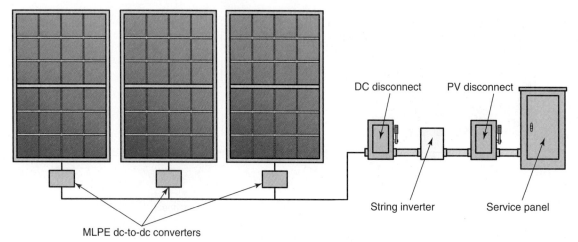

Goodheart-Willcox Publisher

Figure 33-22. A system using MLPE dc-to-dc converters may need a dc disconnect ahead of the string inverter.

may be an integral part of the inverter assembly or may be a separate disconnect. Depending on the manufacturer, there may be additional rapid shutdown devices as part of this wiring between the MLPE and the disconnect.

The string inverter typically provides the "keep alive" signal on the circuit, so the MLPE devices change the output from 1 Vdc to the maximum voltage and current capacity for the solar conditions. Some inverters may have a separate rapid shutdown switch where multiple strings are connected through a combiner.

The string inverter locations depend on the system configuration or the utility requirements. If a roof-mounted array is used, the inverter may be located outside the home on the exterior near the electrical service that has a PV-ready meter main. If the string inverter is added to an existing home, the location of the inverter may be adjacent to the house electrical panel with a separate disconnect for the ac side at a readily accessible location outside of the dwelling.

Many PV system designers will gather the information during the site assessment and will plan the disconnect location prior to the PV system installation. Most installations require some type of utility or local AHJ plan review prior to installations. The PV installations that provide backup power or energy storage typically have two configurations. The first configuration uses the dc power from the PV array to directly power dc loads. This is popular in off-grid applications, mobile homes, and RVs.

The second type of PV system with energy storage is the use of an interactive inverter with a battery storage system. The inverters in these types of systems typically monitor the energy produced by the PV array and store the energy in the batteries when the PV system is producing more than a home needs and gives it back from the battery when the home demand is higher than what the PV system can produce.

These energy storage systems can also provide backup power for the home if utility power is lost. Additional transfer switches or microgrid interconnecting devices are installed to make sure that a back feed on the utility is not possible.

PV microgrid systems are going to be evolving quickly as more energy storage options become available. The use of electric vehicle batteries as energy storage and bidirectional charging systems are among the newest incorporated designs. For these types of installations, additional requirements of *Article 702* and *Article 705* must be considered.

Summary

- Designers of photovoltaic (PV) systems will use or complete site assessments prior to the installation to determine the conditions to which a design is based.
- Photovoltaic systems are sized for either grid-tied or off-grid applications. When a system is grid-tied, the utility power is interactive with the systems, and any additional energy produced by the PV system will be exported to the utility grid. When energy is needed that is not produced by the PV system, it will be imported.
- Photovoltaic cells are assembled into modules. Modules can be in the form of glass panels with frames or can be part of a building integrated system such as solar shingles.
- PV modules will have an integrated junction box as part of the assembly that contains the by-pass and reverse blocking diodes.
- Photovoltaic modules are assembled onto locations, and the assembled PV modules are called *arrays*. The configuration of the wiring within the array will be dependent on the types of power conversion equipment used.
- Sizing the circuits in the PV arrays will be typically based on manufacturer-specific requirements of the electronic equipment.
- There are three typical system integrations used with photovoltaics: the grid-tied utility interactive system, off-grid battery-based system, or grid-tied system with energy storage.
- Microinverter systems convert the dc power to ac power right at the module location.
- PV modules are typically mounted on racking structures installed. The racking installation consists of roof flashing, attachment points, mounting rails, hold-down clips, and other module electronics mounting hardware.
- The racking, module hold-down clips and other associated hardware for the racking need to be tightened to a torque value specified by the manufacturer. The failure to properly torque fasteners can lead to system failures. The racking, microinverters, MPLEs, module frames, and other exposed non-current carrying parts of a PV systems are required to be grounded.
- Module installations require that proper connections are made, the modules are installed straight, and the cable management systems properly support wiring on the underside. Part of the proper module installation is safely getting them to the installation location. Proper safe hoisting or material management systems are needed.

- Pole-mounted and ground-mounted arrays are similar to roof-mounted arrays with the addition of the support structure. Pole-mounted arrays will need to have foundations installed and proper underground wiring to the array location. The ground-mounted arrays will also have foundation elements, but they may be designed differently due to the configurations.
- The interconnections of PV systems into the home power systems can be at a meter mains device that has the capability to install a back fed breaker. This balance of system components and connections will need to be designed to meet *NEC* requirements and utility interconnection requirements.
- All PV systems will need a PV system disconnect located at a readily accessible location. Additional disconnects may be required.

CHAPTER 33 REVIEW

Know and Understand

1. Prior to designing any photovoltaic (PV) system, what needs to be done to ensure proper design?
   A. Site assessment
   B. Purchase materials
   C. Measure power
   D. None of these are the first step.
2. *True or False?* The smallest portion of a solar PV system that is field assembled is the cell.
3. *True or False?* To increase the voltage of a PV system, the modules must be connected in parallel.
4. *True or False?* To increase the current of a PV system, the modules must be connected in parallel.
5. Which of the following are manufactured into a photovoltaic module to prevent higher voltage modules from causing current to flow backward in lower voltage parallel connected modules?
   A. Reverse blocking diodes
   B. Bypass diodes
   C. PN junction cells
   D. None are correct.
6. The connectors used on module leads are _____.
   A. NEMA
   B. MC-4
   C. dc-ac
   D. PV-4
7. *True or False?* Roof attachments are attached to the truss or rafter on all rooftop installations.
8. A PV system that is grid-tied or utility interactive is required to be listed with _____ technology.
   A. anti-islanding
   B. off grid
   C. micro-grid
   D. back feed
9. Which of the following are considered module-level power electronics?
   A. ac power circuit breakers
   B. dc-dc power converters
   C. String inverters
   D. Bypass diodes
10. Rapid shut down devices will limit the output of the dc module to _____ dc unless a "keep alive" signal is received.
    A. 1
    B. 2
    C. 4
    D. 12
11. Circuit breakers and other electrical equipment used for interconnected power production ac inverters of a photovoltaic systems are rated for _____ use.
    A. dc power use only
    B. ac power only
    C. bidirectional
    D. noncontinuous
12. *True or False?* All conductors and electrical equipment used in solar PV systems are required to be sunlight resistant.
13. PV wiring under modules and along racking are required to be supported at intervals not to exceed _____.
    A. 12"
    B. 24"
    C. 36"
    D. 48"
14. *True or False?* Ground-mounted arrays require foundations to be installed to support the structures.
15. The photovoltaic system disconnect is required to be installed _____.
    A. as a circuit breaker in a panel
    B. in a readily accessible location
    C. adjacent to the array on the roof
    D. within 50' of the array location

Apply and Analyze

1. The conversion of solar energy into electrical energy is called _____.
2. If a photovoltaic output circuit has a total current of 12.8 amps, what is the minimum amp rating that would be used to size the conductors and overcurrent devices?
3. If a back-fed circuit breaker is installed in a panelboard with a 125 A bus rating and a 100 A main circuit breaker. What is the largest size breaker that can be installed?

Critical Thinking

1. When installing the electrical wiring associated with photovoltaic systems, considering the inability to de-energize the modules, what safety items can be included in the design to minimize dangers?
2. What items are needed for safe roof installation of photovoltaic systems?
3. Describe the three basic photovoltaic system integrations used in residential solar power systems.

CHAPTER 34 Garage Wiring and Electric Vehicle Charging Power

Konstantin L/Shutterstock.com

SPARKING DISCUSSION

How often have you been in a garage with adequate lighting, enough receptacles where extension cords are not needed, and can provide for all the needs of the homeowner?

CHAPTER OUTLINE

34.1 Garage Environments

34.2 Detached Garages and Outbuildings

34.2.1 Required Receptacle Outlets for Detached Garages

34.2.2 GFCI Protection for Detached Garages

34.2.3 Supplying Power for Detached Garages, Accessory Structures, or Outbuildings

34.2.4 Disconnect and Grounding Requirements

34.2.5 Feeder and Subpanel Requirements

34.3 Attached Garages

34.3.1 Fire-Rated Construction

34.3.2 Required Receptacle Outlets for Attached Garages

34.4 Power Circuits for Workshops or Vehicle Maintenance

(Continued)

Copyright Goodheart-Willcox Co., Inc.

34.5 Electric Vehicle (EV) Charging Outlets

    34.5.1 Electric Vehicle Supply Equipment (EVSE)

34.6 Calculating Loads for Garage Power

LEARNING OBJECTIVES

- Identify lighting and receptacle requirements for detached garages with power.
- Describe proper power installations for detached garages and outbuildings.
- Identify lighting and receptacle requirements for attached garages.
- Describe safe wiring practices for attached garages.
- Describe the basic circuit requirements for electric vehicle charging equipment located in garages.
- Identify methods of load management for EV charging equipment installed in garages.
- Determine additional load calculations required for garages.

TECHNICAL TERMS

appliance branch circuit

attached garage

continuous load

detached garage

electric vehicle supply equipment (EVSE)

multiwire branch circuit

Introduction

When wiring residential garages, accessory structures, and outbuildings, electrical power installations can be overlooked by electricians. Power needs for a garage are minimal by the *Code*, but homeowners often find that the wiring in garages can be substandard after moving into a home. This can lead some to attempt do-it-yourself wiring, which can be extremely unsafe. When planning wiring of garages, the power requirements need careful consideration for overhead garage door openers, task lighting, receptacle outlets for hand tools or vacuums, electric vehicle charging power, and other items.

The power and wiring requirements for a garage, attached or detached from the home, can be different. Detached garages often require separate panels installed, whereas attached garages can have power requirements met by the main panel in the home. This chapter will help determine the correct needs for electrical receptacles, power, and lighting outlets in garages.

34.1 Garage Environments

A garage's interior can be either finished or unfinished. Many garages have open studs and exposed roof structures. The types of wiring methods used in garages may depend on if the inside will have drywall, plaster, or plywood finishes installed.

Type NM cable typically is permitted by *Section 334.10* for wiring in garages. When run exposed, it is required by *Section 334.15* to closely follow the surface of the building finish or of running boards installed to protect cables. In areas where the cables cannot be installed in this manner, protection from physical damage is required using an approved raceway.

The exposed attic or roof structure areas of a garage are often used for storage of personal items. If type NM cables are run in these areas, care should be taken to avoid running across the top of framing members or the face of studs and rafters. Building codes may not permit drilling of the framing members which may require guard strips at least as high as the cable to be installed for protection. In many installations, electrical metallic tubing (EMT) is installed to protect exposed NM cable.

34.2 Detached Garages and Outbuildings

A *detached garage* is a freestanding structure, much like an accessory structure or outbuilding, that does not require the main building's structural elements for support. For garages that are separate from the home, the power supply, grounding, and disconnect requirements are additional elements of concern. Note that the requirements in the *NEC* for garages are not in just one part of the *Code*, thus you must have a good working knowledge of the requirements to ensure proper installations and know where to locate them.

34.2.1 Required Receptacle Outlets for Detached Garages

The minimum required receptacle outlets for a garage, as stated in *Section 210.52(G)(6)*, is at least one 120-volt 15-amp receptacle outlet is to be installed for each vehicle stall. These receptacles are not permitted to be installed more than 5-1/2′ above the floor. These receptacles are installed and are required by *Section 210.11(C)(4)* to be connected

to an individual 20-amp branch circuit with no other outlets allowed. If a garage is a single-car garage, the receptacle installed must be a duplex receptacle because the *Section 210.21(B)(1)* requirements does not permit a single 15-amp receptacle on a 20-amp branch circuit.

Required Lighting Outlets for Detached Garages

Section 210.70 (A)(1) outlines the minimum installation for lighting a detached garage. This section requires a lighting outlet controlled by a wall-mounted switch or control device for general lighting of the interior of the garage area. Refer to **Figure 34-1**.

In addition, *Section 210.70(A)(2)* requires a minimum of one exterior lighting outlet controlled by a wall-mounted switch or control device for the personnel door to light the exterior entrance. A vehicle door is not a personnel entrance or exit. Therefore, lighting on the exterior of a vehicle door is not a requirement, but installing it may be good practice.

Garage Door Opener Appliance Requirements

Typically, most garages have electric garage door openers installed. To power a garage door opener for a typical installation, you need to provide a receptacle in the ceiling above the location of the garage door opener. A garage door opener receptacle is not permitted to be connected to the 20-amp circuit that supplies the receptacles for the garage for each vehicle stall.

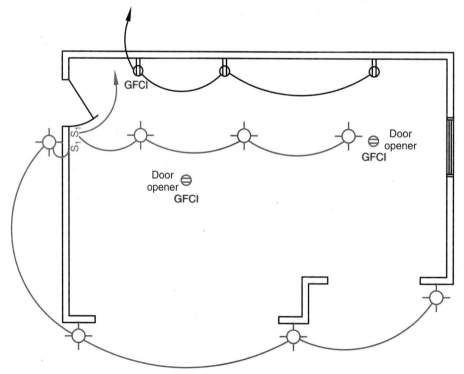

Goodheart-Willcox Publisher

Figure 34-1. Typical garage wiring plan.

Receptacles for garage door openers or other items may be connected to the lighting circuit run to supply the garage lighting. *Section 210.23(B)(2)* limits the power for permanently installed utilization equipment fastened in place and connected to a circuit with lighting or receptacles to 50% of the branch circuit rating. Garage door opener ratings cannot be greater than 50% of the circuit rating if lighting or receptacles are also connected to the circuit. A standard 15-amp branch circuit has a total capacity of 1800 VA. If the garage door rating is greater than 900 VA, an additional individual branch circuit is required. A 20-amp branch circuit allows for a 1200 VA garage door opener.

Some garage door openers must be installed on an individual branch circuit. If the installation manual states it, *Section 110.3(B)* requires it. If the manufacturer does not require an individual branch circuit, then it is permissible to connect the receptacle for the garage door opener on a lighting circuit provided it complies with the 50% requirement outlined in *Section 210.23(B)(2)*. Often the brand of garage door opener is unknown at the time of rough-in, so many

electricians install the garage door opener receptacle on an individual branch circuit.

34.2.2 GFCI Protection for Detached Garages

All receptacles and receptacle outlets installed in garages are required to have ground-fault circuit-interrupter protection.

Section 210.8 requires GFCI protection to be in a readily accessible location. If a receptacle supplies appliances or garage door openers, the GFCI protection needs to be readily accessible. A receptacle mounted above a garage door opener in a ceiling is not readily accessible. Using a GFCI circuit breaker or adding an additional dead front or blank GFCI protective device in a readily accessible location in the garage to supply these receptacles is necessary. Refer to **Figure 34-2**.

34.2.3 Supplying Power for Detached Garages, Accessory Structures, or Outbuildings

Section 210.11(C)(4) requires at least two 120-volt circuits to run to the detached garage. One to supply the required receptacle outlets for each vehicle parking space required by *Section 210.52(G)(1)* and one to supply the lighting outlets required by *Section 210.70(A)(2)*. In many cases these two 120-volt circuits can be supplied by a single *multiwire branch circuit* from the main panel of the home. This circuit can be installed underground or overhead to the separate building or structure. There are minimum conductor sizes and vertical clearances needed for overhead wiring run

---

PRO TIP

Garage Door Opener Installation

The wiring requirements for garage door openers also include low-voltage controls. Typically, a manual operation button controller is installed inside the garage near the personnel door to the garage, and sensors may need to be installed near the bottom of the door frame. Always follow the specific installation instructions provided with the garage door opener.

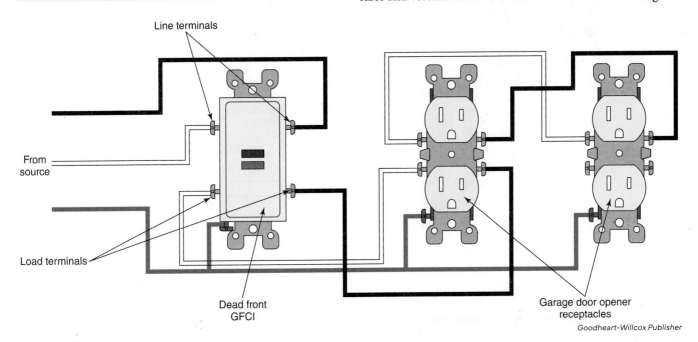

Goodheart-Willcox Publisher

Figure 34-2. Using a dead front GFCI to provide protection to garage door opener receptacle. The power source is connected to the line terminals on the GFCI, and the downstream receptacles are connected to the load terminals.

to a garage or outbuilding. Approved wiring methods and burial depths must be considered for underground power to a detached garage or outbuilding. Refer to Chapter 30, *Outdoor Lights and Receptacles* for details on outdoor wiring.

Running power to a separate building or structure requires a separate disconnecting means and additional grounding requirements. Part of the installation is to set up a means to turn off power to a building in the event of an emergency. Proper grounding must also provide safe touch potentials on exposed non-current-carrying metal equipment and fault paths that facilitate the operation of overcurrent devices.

Per *Section 225.30*, only one branch circuit or feeder is permitted to supply a separate building or structure. A multiwire branch circuit is a single branch circuit for this purpose. Therefore, two 120-volt 20-amp branch circuits installed as a multiwire branch circuit to supply a garage complies with the *Code*. One application uses two single-pole breakers, one on line 1 (L1) of the panel and the other on line 2 (L2) of the panel with a shared neutral wire. *Section 210.4(B)* requires the breakers to have a handle tie or use a two-pole breaker. Refer to **Figure 34-3**.

34.2.4 Disconnect and Grounding Requirements

When running power to a separate building or structure, a method to disconnect the power is required at the structure by the NEC. The disconnect is required by *Section 225.31(B)* to be readily accessible and as near as possible to where the conductors enter the structure. This can be done with a separate disconnecting switch installed on the exterior or interior of the structure. In the case of a single circuit containing two ungrounded legs of a multiwire branch circuit, two switches in the same box or a two-pole switch listed for 240-volts and two circuits can be suitable for this installation.

If the power run to a garage is a service supplied by the utility, a service disconnect, or an emergency disconnect is required by the *NEC* on the outside of the building in a readily accessible location. This is because firefighters or utility workers need access to shut off power to a building in an emergency. You can put in a disconnect and lock it in the on position if necessary. If the power to the detached garage comes from the home, the house emergency disconnect switch serves this purpose.

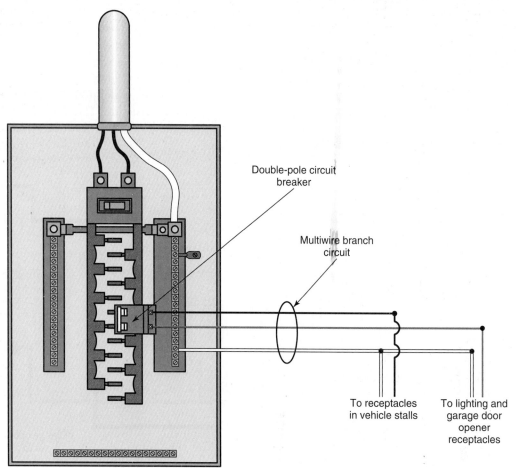

Goodheart-Willcox Publisher

Figure 34-3. Garage loads can be supplied by a multiwire branch circuit from the home's main service panel.

CODE APPLICATION

Sizing Equipment Grounding Conductors and Grounding Electrode Conductors for Detached Garages

Scenario: A 100-amp feeder is installed from the home main electrical panel to supply power to a detached garage. The installation is done using rigid PVC nonmetallic conduit with three 4 AWG copper THWN insulated conductors. An equipment grounding conductor is required to be run with the circuit conductors, and a grounding electrode conductor is also to be installed from the subpanel in the garage to a ground rod. Consider the relevant code articles to determine the installation requirements and sizes of these grounding conductors.

Section 250.32(A): A building(s) or structure(s) supplied by a feeder(s) or branch circuit(s) shall have a grounding electrode system and grounding electrode conductor installed in accordance with *Article 250, Part III*.

Section 250.66: The size of the grounding electrode conductor and bonding jumper(s) for connection of grounding electrodes shall not be smaller than given in *Table 250.66*, except as permitted in *Section 250.66(A)–(C)*.

Table 250.66 Grounding Electrode Conductor for Alternating-Current Systems

Size of Largest Ungrounded Conductor or Equivalent Area for Parallel Conductors (AWG/kcmil)		Size of Grounding Electrode Conductor (AWG/kcmil)	
Copper	Aluminum or Copper-Clad Aluminum	Copper	Aluminum or Copper-Clad Aluminum
2 or smaller	1/0 or smaller	8	6
1 or 1/0	2/0 or 3/0	6	4
2/0 or 3/0	4/0 or 250	4	2
Over 3/0 through 350	Over 250 through 500	2	1/0
Over 350 through 600	Over 500 through 900	1/0	3/0
Over 600 through 1100	Over 900 through 1750	2/0	4/0
Over 1100	Over 1750	3/0	250

Notes:
1. If multiple sets of service-entrance conductors connect directly to a service drop, set of overhead service conductors, set of underground service conductors, or service lateral, the equivalent size of the largest service-entrance conductor shall be determined by the largest sum of the areas of the corresponding conductors of each set.
2. If there are no service-entrance conductors, the grounding electrode conductor size shall be determined by the equivalent size of the largest service-entrance conductor required for the load to be served.
3. See installation restrictions in 250.64.

Tables reproduced with permission of NFPA from NFPA 70, National Electrical Code, 2023 edition. Copyright © 2022, National Fire Protection Association. For a full copy of the NFPA 70, please go to www.nfpa.org

Section 250.32(B): An equipment grounding conductor shall be run with the supply conductors and be connected to the building or structure disconnecting means and to the grounding electrode(s). The equipment grounding conductor shall be sized in accordance with *Section 250.122*. Any installed grounded conductor shall not be connected to the equipment grounding conductor or grounding electrode(s).

Section 250.118(A): Each equipment grounding conductor run with or enclosing the circuit conductors shall be a copper, aluminum, or copper-clad aluminum conductor. This conductor shall be solid or stranded; insulated, covered, or bare; and in the form of a wire or a busbar.

Section 250.122(A): Copper, aluminum, or copper-clad aluminum equipment grounding conductors of the wire type shall not be smaller than shown in *Table 250.122*.

Table 250.122 Minimum Size Equipment Grounding Conductors for Grounding Raceway and Equipment

Rating or Setting of Automatic Overcurrent Device in Circuit Ahead of Equipment, Conduit, etc., Not Exceeding (Amperes)	Size (AWG or kcmil)	
	Copper	Aluminum or Copper-Clad Aluminum*
15	14	12
20	12	10
60	10	8
100	8	6
200	6	4
300	4	2

Notes: Where necessary to comply with 250.4(A)(5) or (B)(4), the equipment grounding conductor shall be sized larger than given in this table.
*See installation restrictions in 250.120.

Tables reproduced with permission of NFPA from NFPA 70, National Electrical Code, 2023 edition. Copyright © 2022, National Fire Protection Association. For a full copy of the NFPA 70, please go to www.nfpa.org

Solution: The *Code* requires that a separate equipment grounding conductor be installed with the feeder conductors. The equipment grounding conductor is sized and installed based on the overcurrent protective device protecting the feeder or branch circuit. The feeder breaker that is to supply the detached garage is 100 amps. Referring to *Table 250.122*, 100 amps indicates that an 8 AWG copper equipment grounding conductor is required to be run with the 4 AWG feeder conductors. The grounding electrode conductor is to be installed between the equipment grounding conductor in the subpanel to the grounding electrode. The grounding electrode conductors are sized based on the feeder supply conductors. Referring to *Table 250.66*, a minimum of a size 8 AWG conductor is required to be installed to connect the equipment ground of the subpanel to the grounding electrodes.

Equipment Grounding and Grounding Electrodes

An equipment grounding conductor must be installed with the supply conductors when supplying a separate building or structure by a branch circuit or feeder. This conductor provides a fault current path back to the source in the event of a ground fault.

A general rule of *Section 250.32(A)* is to also install grounding electrodes at the separate building or structure. The main reasons for this are to reduce the effects of lightning or line surges and keep everything at the same potential. In the event of a lightning strike or line surge, the grounding electrodes provide a point of connection to the earth for this to dissipate. The ground, concrete, metal structural elements, and any exposed non-current-carrying metal parts are all connected to the same ground potential as the supply circuit. The connection between the grounding electrode and equipment grounding keeps the buildings at the same potential, and circulating currents will be minimal.

Grounding electrodes of the types described in the *Section 250.52(A)(1)* through *(A)(3)* are required to be connected to the equipment grounding conductor run to the detached garage. These can be metal underground water pipes, in-ground metal support structure, and concrete encased electrodes. If these electrodes are not present, *Section 250.50* requires grounding electrodes be created by installing ground rings, ground rods, or ground plates.

The grounding electrode conductor is connected between the grounding electrodes and the equipment grounding conductor run with the circuit conductors. This connection is typically made at the disconnect enclosure location or where conductors enter the structure. The size of this conductor is based on the size of the conductors that supply the separate building or structure. This information can be found in *Section 250.66* and *Table 250.66*. For instance, in the case of the 20-amp multiwire branch circuit supplying the two 120-volt circuits, an 8 AWG copper conductor is required to be installed as the grounding electrode conductor.

An exception in *Section 250.32(A)* permits a multiwire branch circuit to be installed without the grounding electrode system. Since the conductors serving a multiwire branch circuit are smaller, the chance of large currents from lightning strikes getting back into the home are not likely.

If the conductors are installed overhead to the detached garage, it is highly recommended that the grounding electrode be installed. Overhead conductors are subject to lightning strikes, and the absence of a grounding electrode can allow the surge to be transmitted back to the house. If a grounding electrode is also installed at the garage, this provides an additional point of earth grounding.

34.2.5 Feeder and Subpanel Requirements

If additional outlets or circuits are desired for garage door openers, power tools, electric vehicle charging equipment, or additional power outlets, a feeder needs to be installed to the detached garage. A separate feeder is typically run as a 240/120-volt three-wire power circuit that terminates in a subpanel with a main circuit breaker that serves as the disconnecting means. A 120-volt 20-amp circuit to supply receptacles for vehicle spaces and the required lighting outlets are connected to the same subpanel. Other outlets can then be added. Note that multiple branch circuits are not permitted to be run to detached garages.

When a feeder is installed to a subpanel in the detached garage, the size of the feeder is dependent on the power requirements of the detached garage. Grounding electrodes must be installed and connected to the equipment grounding conductor run with the feeder circuit. The grounded (neutral) conductor that is run with the feeder remains isolated. The bonding jumper between the grounded (white) conductor and the equipment grounding (green/bare) conductor is not allowed in the garage or outbuilding. Refer to **Figure 34-4**.

> **PRO TIP**
>
> ### Solar Ready or EV Ready Structures
>
> This is not a requirement of the *National Electrical Code*, but areas of the country may have state or municipal laws that require homes to be built solar ready or electric vehicle ready. If these requirements are present, there may be a need for additional conduits or feeder capacity provisions. Consult local and state rules regarding requirements when installing power to detached garages, accessory structures, or outbuildings.

34.3 Attached Garages

An **attached garage** is a portion of the house designed for vehicle parking and has access to the home. This area is typically separated from the main home by fire-rated walls and doors, but it can rely on the main building for structural support. When supplying power to items within an attached garage, these fire ratings require attention. Many garages can contain appliances, utility equipment, central vacuum appliances, and laundry areas in addition to required receptacle and lighting outlets. The power requirements for these areas need to be determined when garage power is planned.

34.3.1 Fire-Rated Construction

Typically, the walls between the home and the garage are fire-rated construction. Fire-rated construction is

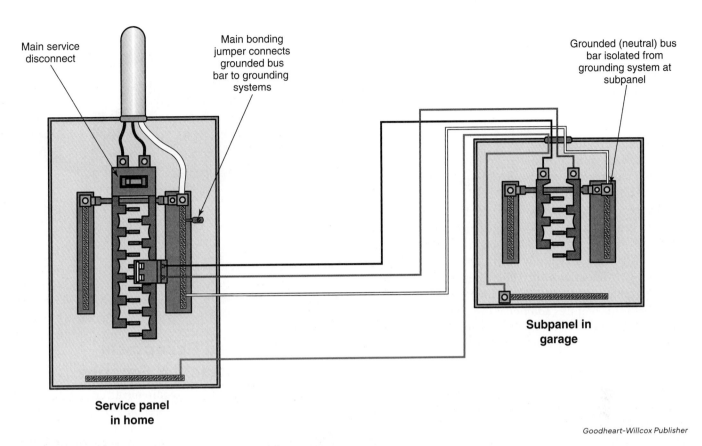

Goodheart-Willcox Publisher

Figure 34-4. Feeder and subpanel wiring for a garage panel.

addressed in *Section 300.21*. It is required to install fire-stopping materials around electrical penetrations into or through fire-resistant-rated walls, partitions, floors, or ceilings. Building codes also contain restrictions on fire-rated membrane penetrations on opposite sides of a fire-resistance-rated wall assembly.

If a home is a single-story and the fire wall between the home and garage extends up to the underside of the roof, this is the only wall considered for fire-rated construction. If the wall between a home and its garage does not extend to the underside of the roof, the garage has a common attic with the home. This requires the entire garage ceiling to be fire-rated along with common walls.

In homes with multiple floors, fire-rated construction consists of upper floors that share structural elements with the garage, walls with structural elements, the ceiling of the garage, and any walls and floors that share a common attachment with the garage.

The type of fire-rated construction used is determined by UL fire-rated assemblies or the home design professional. Methods of fire stopping and fire caulking need to be done in strict adherence to the firestop manufacturer's instructions for through penetrations and membrane penetrations. Refer to **Figure 34-5.**

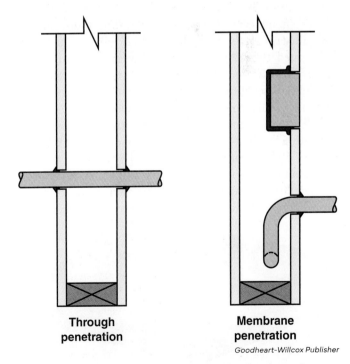

Through penetration   **Membrane penetration**

Goodheart-Willcox Publisher

Figure 34-5. Details of firestop application for through penetrations and membrane penetrations in fire-rated walls.

Drilled holes in framing studs, joists, and rafters may need to be fire-stopped to prevent the spread of fire. Any penetrations in drywall or plaster need to be evaluated for the need to provide fire stopping. When a hole is drilled and wiring is run through a fire-rated wall, the opening surrounding the cable or conduit needs to be filled with a fire-stopping material. This is typically done using a fire-rated caulk or putty, **Figure 34-6**.

The boxes and penetrations that are put inside fire-rated-constructed walls and ceilings cannot be standard plastic boxes. The boxes installed are typically steel or fiberglass fire-rated boxes. There are maximum opening areas permitted in fire-rated construction walls and ceilings. The maximum penetration opening allowed for fire-rated construction is typically 16 in². This is about the area of a two-gang outlet or switch box. Similarly, the area of a round lighting outlet box is less than 16 in². Any opening larger than this requires additional fire-stopping materials. Electrical panels are typically not permitted to be installed flush inside the fire-rated walls.

Fire-rated construction often limits boxes from being installed within the same vertical stud space with penetrations on each side of the wall. If this is needed, additional fire protection methods must be applied. The minimum distance between boxes that penetrate the fire-rated construction boxes is 24″ horizontally. Additional fire-stopping methods can be done for wall openings using protective materials, known as *fire-rated putty pads*, around the back and sides of the box to increase the fire resistance of the outlet box or use fire-rated gaskets on the underside of the device cover plate.

Some homeowners prefer to install recessed downlights, also called *can lights*, in the garage. If the ceiling of the garage is fire rated, the opening size of most recessed downlights is too large. This requires additional fire-rated assemblies to be made on top of the lights or special fire-rated luminaires for the fire-rated construction. Refer to **Figure 34-7**.

LED lighting configurations can provide similar lighting with fewer fire-rating issues. Many downlight installations in a garage may be accomplished by using a fire-rated round outlet box with an LED surface-mounted wafer light. Consideration should be taken when calculating box fill as these lights typically mount with a driver device inside the

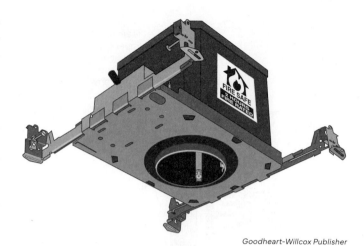

Goodheart-Willcox Publisher

Figure 34-7. Fire-rated down lights.

box and a two-conductor volume allowance will need to be considered toward box fill.

34.3.2 Required Receptacle Outlets for Attached Garages

Receptacle outlets and GFCI requirements do not differ for an attached or detached garage. *Section 210.52(G)* requires a minimum of one 15-amp 120-volt receptacle for each vehicle parking area in a garage. These receptacles must be supplied by a 20-amp branch circuit, and no other outlets are permitted to be connected to this circuit. Depending on how a home is constructed, additional required receptacle outlets associated with other parts of home construction may be required.

If an attached garage contains appliances or utility equipment that serves the home, additional receptacles may be required for this equipment. If the equipment requires servicing, such as a separate heating or cooling system, *Section 210.63* requires a receptacle installed on the same level and within 25′.

Appliances such as central vacuum cleaner systems and power ventilation fans for water heaters, refrigerators, and freezers need to be supplied by appliance branch circuits. According to the *NEC*, an **appliance branch circuit** supplies energy to at least one outlet to which appliances are connected and has no permanently connected luminaires that are not a part of an appliance. These appliance branch circuits are to be installed separately from the required garage receptacles and luminaire circuits. The circuit size will need to be calculated in accordance with the applicable parts of *Section 220.14*. Lighting and receptacles installed for garage door openers can be installed on the same general branch circuit, provided the garage door opener power requirements do not exceed 50% of the branch circuit.

Required Lighting Outlets for Attached Garages

Attached garages have similar lighting outlet requirements to detached garages. A switched lighting outlet controlled

Eaton

Figure 34-6. Sheets of moldable firestop putty.

by a wall switch or control device is required for the interior of a garage. When a garage has at least one stair between the home and the garage, an additional lighting outlet and a three-way switch may be required if the stairs have more than six risers. If the garage has an exterior exit at grade level, a lighting outlet controlled by a wall switch or control device is required to be installed at each exit.

According to *Section 210.70(A)(2)*, additional lighting outlets may be required for servicing these appliances in instances where utility rooms are part of the garage or equipment or appliances are present. These lighting outlets are to be controlled by a wall switch or control device near the utility room or area entrance.

For any electric service equipment or subpanel installed in a garage, a lighting outlet is required to provide illumination for the electric panel location. The lighting outlet for the garage is permitted to be the source of illumination for the electric panel, but if the garage lights are controlled by an automatic means such as an occupancy sensor, then it is required to have a manual override. *Section 110.26(D)* does not permit lighting for general illumination of electrical equipment to be controlled by an automatic means.

34.4 Power Circuits for Workshops or Vehicle Maintenance

There are many instances where residents may convert an area of their garage into a workshop or have provisions for vehicle maintenance. Power requirements for these areas are in addition to those outlined in the *NEC*. These areas require additional branch circuits, and the loads for these additional circuits need to be calculated and added to the service load calculations. Refer to Chapter 20, *Service Calculations*, for more on this topic.

The types of power often needed for workshop areas of a home can include 240-volt receptacles for compressors, welders, or portable electric heaters. These items can be hardwired with disconnects, or receptacle outlets can be installed. If receptacles are installed, there are different configurations and GFCI protection requirements that need to be considered. Typical configurations for 240-volt receptacles are shown in **Figure 34-8**.

34.5 Electric Vehicle (EV) Charging Outlets

There are additional power requirements for garages that store electric vehicles. Charging these vehicles in a garage depends on the type and charging method desired. Most electric vehicles have the means to control the power to charge the batteries as part of the vehicle. Power connections

to building electrical systems from EV charging equipment communicate with the vehicle to determine how much power the vehicle needs. It also helps to provide protection in the event of a ground fault or missing equipment grounding.

Electric vehicles have standardized charging plugs and receptacle configurations. The standardization of the connections determines the types of charging that can occur. Most residential electric vehicle charging equipment operates on 120 or 240 volts. Some equipment can interconnect with the vehicle battery DC conductors to interface with energy storage or photovoltaic systems.

34.5.1 Electric Vehicle Supply Equipment (EVSE)

Electric vehicle supply equipment (EVSE) is used to transfer energy between the premises wiring and the electric vehicle. The type of circuits and connections needed are based on EVSE requirements, **Figure 34-9**.

Level I equipment is intended to use a standard NEMA 5-15R 125-volt 15-amp receptacle. The *NEC* requirements to provide a receptacle for each vehicle stall may serve these types of chargers when portable means of charging are used.

Section 625.42 requires the circuits used to supply Level II equipment to be large enough to supply the capable load. These 240-volt circuits can run at the full-rated current for the total time to charge the vehicle. The time required to charge a vehicle depends on the depth of discharge of the vehicle battery, the size of the battery in the vehicle, and the rating of the EVSE.

When sizing the branch circuit to supply EVSE, including bidirectional EVSE, they are required by *Section 625.41* to be sized for continuous duty and shall have a rating of not less than 125% of the maximum load of the equipment. In most cases, the circuits used to charge electric vehicles operate as **continuous loads**, or electrical loads where the maximum current is expected to continue for three hours or more. The circuit breaker and conductor ampacity ratings are required to be 125% higher than the equipment rating. The size circuit breaker used determines the conductor size needed, and these conductors must also be protected for their rated ampacity.

Outlets for supplying power for EVSE equipment can be cord-and-plug connected or hardwired. Refer to **Figure 34-10**. The Level II type EVSE typically has supply cords and vehicle cords integrated into the assembly. The maximum length of cord allowed to connect to the supply receptacle is 12″ for portable cord and plug equipment. For permanently mounted and connected equipment, the supply cord is permitted to be up to 6′ long. The EVSE is required to be mounted at a height where the cord does not touch the floor.

The output cable is a cord used to connect the EVSE and the vehicle. It has a maximum length of 25′. The EVSE must

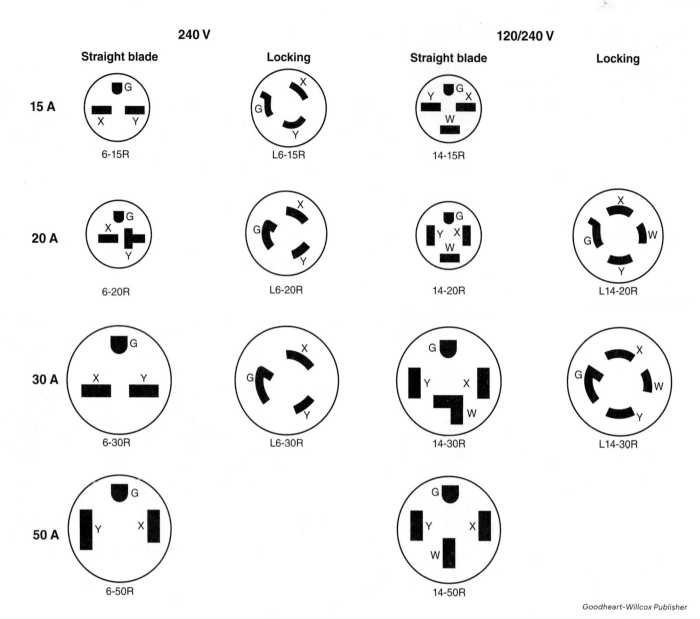

Figure 34-8. NEMA standard 240-volt outlet configurations.

Goodheart-Willcox Publisher

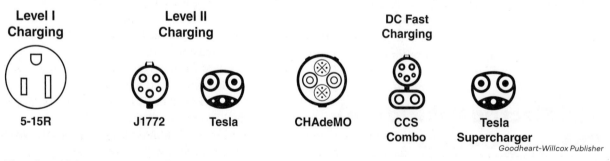

Figure 34-9. Electric vehicle connector configurations.

Goodheart-Willcox Publisher

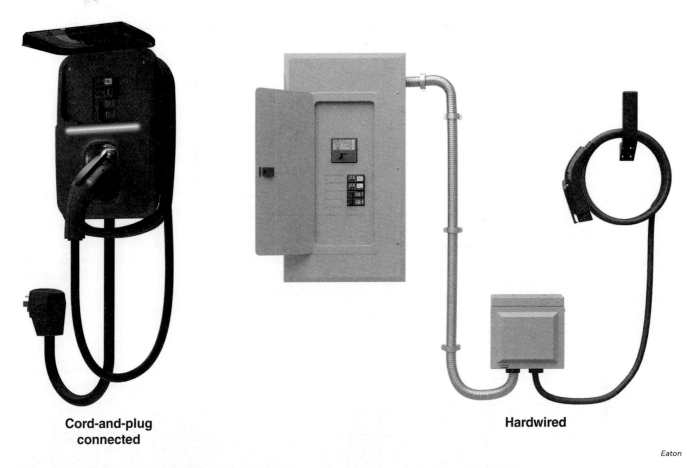

**Cord-and-plug
connected**

Hardwired

Eaton

Figure 34-10. Electric vehicle supply equipment can be cord-and-plug connected or hardwired.

have technology built into it to de-energize the output to the vehicle cord if the cord experiences cord strain or a ground fault. This is done with a circuit within the cord where its resistance values change when the connector is properly inserted in the vehicle.

For permanently mounted EVSE equipment that is not cord-and-plug connected or rated more than 60-amps, a disconnecting means is required. The disconnect must be in a readily accessible location and have permanent provisions for locking it in the open position.

The Level II EVSE equipment using a cord-and-plug connection is found in multiple configurations. The receptacle that needs to be installed should match the configuration of the EVSE. The most popular residential receptacle configurations are the NEMA 6-50, NEMA 14-30, and NEMA 14-50, **Figure 34-11**.

The NEMA 6-50 cord-and-plug configuration supplies single-phase, 240-volt power without a neutral connection. This requires a two-pole circuit breaker, two correctly sized ungrounded conductors in accordance with the breaker

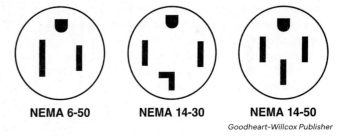

NEMA 6-50   **NEMA 14-30**   **NEMA 14-50**

Goodheart-Willcox Publisher

Figure 34-11. Common EV 240-volt outlet receptacles.

amp rating, and an equipment ground sized in accordance with *Section 250.122* based on the breaker amp rating.

The NEMA 14-30 and NEMA 14-50 cord-and-plug configurations are the same for those used in most electric dryers and ranges. These require a grounded neutral conductor to be run with the ungrounded conductors. This circuit is sized the same for the amp rating of the breaker. The receptacle amp rating must be the rating of the branch circuit, and the branch circuit rating is the amp rating of the circuit breaker.

Example: A 9.6kW 240-volt Level II EVSE with a NEMA 6-50P plug connection is to be installed in a residential garage. The amp rating for the EVSE is 40 amps. The wiring method to be used is type NM-B cable installed concealed in an attached garage.

In this scenario, we know that the minimum overcurrent device required to supply this EVSE is 125% of the amp rating.

$$40 \text{ A} \times 125\% = 50 \text{ A}$$

Thus, according to *Section 310.16* and *Table 310.16*, the conductors to supply the branch circuit need to have an ampacity large enough to be protected by a 50 A branch circuit. Next, select a conductor from *Table 310.16* using the 60°C ampacity column. A 6 AWG copper conductor's ampacity is 55 amps. Thus, a 6/2 AWG ground cable can be installed for this application. Overall, this circuit would require a NEMA 6-50R receptacle to be installed in a minimum device box with 25 in³ of volume.

PRO TIP   **DC Fast Charging**

DC fast charging methods are large and quite expensive for installation in a home. As the electric vehicle market is evolving, manufacturers are also creating additional methods to charge and store energy. The DC power connections of electric vehicles allow for bi-directional power to store excess power from renewable energy sources and charge electric vehicle batteries. If the demand for power in a residence is greater than the renewable energy source supply, the vehicle battery can provide power backward in these times of need. This can reduce the demand on the power grid, and utilities may offer incentives for these capabilities.

34.6 Calculating Loads for Garage Power

The power needed for garages can be as small as the minimum required branch circuit, but when a subpanel is installed, *Section 215.5* requires an electrician to produce calculations for the load that is supplied. The *Article 220* outlines what size is needed for calculating a feeder and a separate subpanel to power a garage. *Section 220.41* states that receptacles installed for general-use receptacle outlets for wall spaces and areas required for general lighting do not need additional load calculations added to the service. These loads are included in the 3 VA per square foot for the dwelling general lighting load calculations. Garage wiring

for lighting and the required receptacles are permitted to be included in these values.

When these loads are supplied by a subpanel installed in an attached or detached garage, they are to be included in the calculation for the feeder. The load calculations for dwelling units do not include any garage additional power circuits. The additional circuits installed in a garage for appliances, overhead door openers, electric vehicle charging equipment, and other individual circuits are required to be added to the service calculations. These same circuit loads need to be calculated to determine the minimum feeder size for a subpanel installed in a garage.

Receptacles supplied from a feeder and subpanel in a garage are to be calculated at 180 VA per device or as required in *Section 220.14(I)*. For example, consider a two-car garage has two-duplex receptacles installed as part of the required 20-amp circuit to supply receptacles. The load calculation for this circuit is as follows:

$$180 \text{ VA} \times 2 = 360 \text{ VA}$$

This meets the minimum requirements of the *NEC*. Another suggested method is to treat the 20-amp circuit as an appliance circuit in service calculations at 1500 VA per circuit.

Section 220.14(D) requires lighting circuits to be calculated based on the rating of the luminaires. If a luminaire has a rating of maximum lamp size 60-watts incandescent, it needs to be calculated at this value even if an 8-watt LED lamp is installed. If an 8-watt integrated LED wafer light is installed, then the load can be calculated on the fixed value of the luminaire. For example, consider four porcelain socket lighting outlet devices are installed in a garage with a rating of 300 watts, but there are only 8-watt LED lamps installed. The total feeder calculation for this circuit is as follows:

$$300 \text{ W} \times 4 = 1200 \text{ W}$$

Additional circuits for power tools, compressors, welders, or vacuums need to be included in the feeder calculation based on the appliance branch circuit rating. Motors installed must be in accordance with *Section 220.14(C)* and application of *Article 430* or *Article 440*.

Any electric heating equipment installed in a garage that is supplied by a subpanel will need to be included in the feeder calculation at 100% of the rating of the heating equipment. Electric vehicle charging equipment is considered a continuous load, and the amp rating of the EVSE will need to be multiplied by 125% when included in the feeder calculation. The summation of all the loads supplied by the subpanel will need to be included in the feeder calculation. The feeder calculation will determine the minimum conductor ampacity and overcurrent devices required.

Example: A feeder is to be installed to supply a detached two-car garage with the following loads:

- Four duplex receptacles installed for the vehicle parking spaces
- One exterior LED luminaire rated at 25 watts
- The interior lighting consists of four LED shop lights rated at 30 watts
- One garage door opener rated at 845 VA with maximum rated lamps installed
- Four additional receptacles are to be included for a workbench
- A 9.6 kW 40-amp rated EVSE is to be installed
- An additional 240-volt 500-watt heater is to be included

First, we should determine the receptacles for required 20-amp circuit:

$$180 \text{ VA} \times 4 = 720 \text{ VA}$$

Next, we know our exterior lighting is 25 VA. To calculate interior lighting, we must multiply the four LED lights by their wattage.

$$30 \text{ W} \times 4 = 120 \text{ VA}$$

The garage door opener is 875 VA, but we must then calculate the workbench receptacles:

$$180 \text{ VA} \times 4 = 720 \text{ VA}$$

Next, we calculate the EVSE:

$$40 \text{ A} \times 125\% = 50 \text{ A}$$
$$50 \text{ A} \times 24 \text{ V} = 12,000 \text{ VA}$$

Finally, as we know the heater is 500 VA, we must add all these values to receive our total load.

$$720 \text{ VA} + 25 \text{ VA} + 120 \text{ VA} + 875 \text{ VA} + 720 \text{ VA} + 12,000 \text{ VA} + 500 \text{ VA} = 14,960 \text{ VA}$$

Per *Section 215.3*, calculate the minimum OCPD for the feeder.

$$14,960 \text{ VA} \div 240 \text{ V} = 62.3 \text{ A}$$

Per *Section 240.6*, the next standard size OCPD is a 70-amp circuit breaker. For this example, the conductor size for this would be 4 AWG copper or 2 AWG aluminum. The minimum equipment grounding conductor will be an 8 AWG copper or 6 AWG aluminum.

Summary

- Receptacles and lighting outlets required for garages provide the basic power needs in a garage.
- The required receptacles for each vehicle stall will provide basic electrical service for any vehicle maintenance or power needs. The requirements for an individual branch circuit to supply these receptacles will accommodate any garage activity using handheld power tools and portable vehicle power needs.
- The lighting requirements for vehicle garages are to provide wall-switched controlled lighting for both inside and outside a personnel door for safe entry, use, and exit.
- Garage door opener circuits may need to be individual branch circuits or part of a lighting or additional receptacle circuit. This will depend on the manufacturer or circuit conditions.
- GFCI protection is required for all receptacles installed in a garage and must be installed in a readily accessible location.
- Power to detached garages is required to have disconnecting means at the point where the wiring enters the structure.
- Grounding electrodes are required for detached garages. These are to be connected to the equipment grounding conductor run with the circuit conductors with a properly sized grounding electrode conductor.
- If a feeder and subpanel are installed for a detached garage, the equipment ground and neutral conductors are to be kept separate to prevent parallel neutral current paths. The grounding electrode conductor is attached to the equipment grounding conductor in the subpanel.
- Attached garages typically have fire-rated walls or assemblies to separate the dwelling from the garage. Proper identification of fire ratings, penetrations of fire-rated assemblies, and needed fire stopping are important in homes with attached garages.
- Additional appliances, HVAC equipment, electrical service equipment, electrical vehicle charging power, or other power outlets are all part of garage wiring.
- Electric vehicle supply equipment (EVSE) in garages is typically 120- or 240-volt for Level I or Level II charging. The circuit supplying EVSE is considered a continuous load.

- Receptacles installed for electric vehicle charging are typically 30- or 50-amp rated and are standard NEMA configurations.
- Load calculations for feeders and subpanels in garages are based on the requirements of *Section 220.14*. Dwelling unit loads are ignored in this application.

Know and Understand

1. *True or False?* Detached garages require additional disconnect and grounding electrodes.
2. What is the minimum number of circuits required for a garage?
   - A. 1
   - B. 2
   - C. 3
   - D. 4
3. *True or False?* GFCI receptacles are required to be provided for all garage outlet locations.
4. *True or False?* An individual appliance branch circuit can have lighting outlets connected to the circuit if the appliance is not rated more than 80% of the circuit rating.
5. What is the minimum number of lighting outlets required to installed on the inside of a detached garage?
   - A. 1
   - B. 2
   - C. 3
   - D. 4
6. What size copper equipment grounding conductor would be required to be run with a 60-amp feeder used to supply power to a subpanel in a garage?
   - A. 10 AWG
   - B. 8 AWG
   - C. 6 AWG
   - D. 4 AWG
7. What is the maximum size opening for a device box in a fire-rated wall between a house and garage?
   - A. 8 in^2
   - B. 12 in^2
   - C. 16 in^2
   - D. 24 in^2
8. *True or False?* Attached garages required receptacles are the same as detached.
9. What is the horizontal spacing required between outlet boxes in a fire-rated wall between a house and attached garage?
   - A. 12″
   - B. 16″
   - C. 24″
   - D. 32″
10. If a person wants a 240-volt 20-amp receptacle for a compressor in their garage what NEMA configuration would be best?
    - A. NEMA 5-20R
    - B. NEMA 6-20R
    - C. NEMA 14-20R
    - D. NEMA 6-60P

11. *True or False?* EVSE output cables are never permitted to hang so that they touch the floor.
12. A Level I EVSE is rated for _____ and a maximum of _____.
    - A. 120 V; 8 A
    - B. 120 V; 12 A
    - C. 240 V; 16 A
    - D. 240 V; 80 A
13. What is the maximum length of cord permitted for the supply cord to a permanently mounted EVSE in a garage?
    - A. 1′
    - B. 2′
    - C. 4′
    - D. 6′
14. If a detached garage has a feeder that supplies power that is rated 100-amps and is supplied with 2 AWG copper conductors, what is the minimum size copper grounding electrode conductor required to connect to the ground rods installed?
    - A. 10 AWG
    - B. 8 AWG
    - C. 6 AWG
    - D. 4 AWG

Apply and Analyze

1. What is the maximum rating of a garage door opener that can be installed on a 120-volt 15-amp lighting circuit that has two lighting outlets connected to the same circuit in a detached garage?
2. What type of receptacle would provide 240-volt 50-amp power for an EVSE?
3. What would the service calculation be for two lamp holder sockets rated for 150 watts with 12-watt lamps installed in an attached garage?

Critical Thinking

1. When installing overhead conductors to supply power to a detached garage, what is the main reason why an equipment grounding conductor and grounding electrodes are required?
2. What are two methods of providing fire-rated device boxes in rated walls of a garage?
3. How can parallel paths be created when installing subpanels?

CHAPTER **35** Standby Power Systems

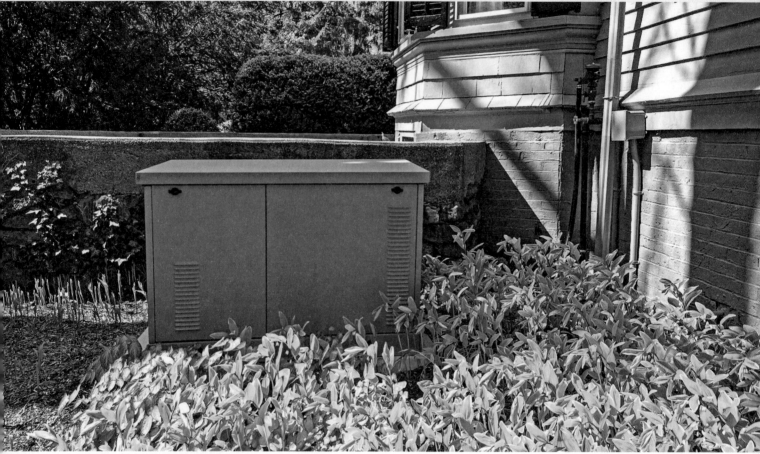

Steven Phraner/Shutterstock.com

SPARKING DISCUSSION

What are the potential benefits and detriments of a residential standby power system?

CHAPTER OUTLINE

(Continued)

LEARNING OBJECTIVES

After completing this chapter, you will be able to:

- Identify types of standby power systems.
- Interpret residential load data to determine the size standby power system desired.
- Describe differences between automatic and manual transfer systems.
- Explain the steps and methods for installing standby power generators.
- Identify types of battery inverter home standby power systems.

TECHNICAL TERMS

automatic transfer system

interconnection

load diversity

manual transfer system

power factor

reactive load

resistive load

standby power system

transfer panel

transfer switch

Introduction

Everyone relies on electricity in their daily life. However, due to failed equipment or natural disasters, there is always the chance of power outages. In situations where utility power is not available, a home standby power system is useful. Residential *standby power systems* are intended to supply power to a property. These systems are not intended for life-saving power options.

When tornados, ice storms, and hurricanes damage large areas of a utility grid, homes with standby power systems can remain habitable. This chapter covers the basics of standby power, including the *National Electrical Code* installation requirements.

35.1 **Types of Standby Power Systems**

Electricity for standby power is provided by sources such as generators, renewable solar systems, or even batteries. The following list are some of the standby power systems that can be installed in a home:

- Permanently installed generators
- Portable generators
- Battery-based inverters
- Renewable energy standby power systems
- Electric vehicle battery bidirectional standby power systems
- Electric vehicle inverter standby power systems

35.2 **Sizing Standby Power Systems**

Most home standby power sources are rated in kilowatts (kW) and are capable of supplying power up to the kW rating. The rating for generators is sized for resistive loads. *Resistive loads* are those that consume power by converting electricity to heat and light as a direct result of electrical current. Not all loads are purely resistive loads. Some loads, such as motors and coils, are considered reactive loads. *Reactive loads* use magnetic fields that can store energy and return some of the stored energy back to the source. If motors or other reactive loads are added to the standby power system, the power factor of these loads needs to be considered. *Power factor* is the ratio of power consumed by the resistive loads to the electrical energy required by the generator to operate the circuit.

Home standby power systems are typically rated for loads without a power factor below 100%. These loads consume all the power that is made by the standby source. If motors or appliances with power factors below 1.0 are going to be supplied by a standby power source, additional kW capacity or generator de-rating may be required.

When determining the size of a generator, there is an additional surge or standby rating provided by the manufacturer. This rating is the amount of power that can be taken for short intervals without damaging the generator, typically in the 2–5 second range. If surge currents last longer than the allowed time, additional higher kW ratings are required.

Motors can draw four to six times the amount of current during the starting and acceleration periods than needed at running. This is known as inrush current, and it is needed when calculating the size of a standby power system. The surging overloads can trip overcurrent devices, stall generators, or damage inverters if large motors, such as air conditioning compressors, are not considered during sizing.

35.2.1 Determining Load Type for Standby Power

In most design applications, it is important to identify the type of loads that are connected to a power system. When sizing a standby power source, a load wattage worksheet is completed to determine the size of standby power system needed, **Figure 35-1**. The loads connected to the standby power system are classified as the following:

- **Cycling loads.** Loads that run at intervals but not the same interval all the time. Examples include refrigerators, garage door openers, and microwaves. These loads are desired but not always on or off.
- **Static loads.** Loads that are on at the same wattage for long periods of time and do not change in wattage value. Examples include televisions, lights, and computers.
- **Phantom loads.** Loads that draw small amounts of power all the time. Examples include internet modems, Wi-Fi routers, clocks, and electronics that use remote controls.
- **Surge loads.** Loads that use motors that start with full voltage applied and have starting inrush current. Examples include air conditioners, sump pumps, and well pumps.

In many situations, the loads for the home are not exactly known for new construction. The appliances may not be determined, or other loads may be missing before the standby system is designed. This requires the use of load charts to help determine the wattages for the load estimation worksheets.

35.2.2 Estimating Load Size for Standby Power

The total connected loads for a standby system can be calculated using a load estimation worksheet, **Figure 35-2**. Some residential applications connect just selective essential loads that are needed to maintain basic operation, such as furnaces, well pumps, sump pumps, refrigerators, and minimal lighting. Other applications for standby power systems provide enough power for the entire home.

When a total home standby source is needed, all loads may be considered using load diversity calculations. *Load diversity* is the total expected power, or load, to be drawn during a peak period. Many standby generator manufacturers can provide software-based calculators to estimate whole house loads using diversity factors. You can also use load management devices to turn off nonessential loads when a standby source is nearing capacity. Careful discussion with the homeowner before the design of a standby power system will determine if the entire home or selected loads are to be powered.

35.3 Standby Power Connections

To select the proper standby power connection, an electrician must consider the type of standby power system and how it integrates with the home electrical system. There are three basic configurations for connecting a home standby power system:

- Whole house transfer switch and standby power source
- Essential loads panel with transfer switch on the load side of the main breaker
- A microgrid interconnect device that integrates batteries, solar, or generator power into the home ahead of the main breaker or standby subpanel

Standby power systems require these *transfer switches*, also called *microgrid interconnect devices*, to disconnect the home power system from the utility wiring. This device is necessary to prevent a back feed to the power grid from the standby power source that can hurt a utility worker. Connecting a standby power system without separating the utility connections can cause the standby power source to overload.

When standby power sources are connected into electrical systems, the methods used to link the normal utility

Standard Wattage Load Chart—Reference

Load or Appliance	Rated Power (Running)	Rated Power (Surge / Starting)
Light Bulb (Incandescent) 75-watt	75 W	N/A
Light Bulb (LED) 75-watt equivalent	12 W	N/A
Downlight LED 4"	4 W	N/A
Downlight LED 6"	8 W	N/A
Well Pump (1/3 HP)	1000 W	3000 W
Sump Pump—Septic Pump	1000 W	3000 W
Security System	180 W	N/A
Ceiling Fans w/ LED light	300 W	400 W
Bath Exhaust Fan (no heater)	100 W	300 W
Bath Exhaust Fan (with heater)	1400 W	1600 W
Stereo / TV Sound System	200 W	N/A
Television (40" LED)	75 W	N/A
Television (60" Plasma)	300 W	N/A
Cable Box / Streaming Device / Blu-ray-DVD player	75 W	N/A
Personal Computer (Desktop)	300 W	N/A
Personal Computer (Gaming Desktop)	800 W	N/A
Printer (Laser)	500 W	N/A
Printer (Inkjet)	75 W	N/A
Refrigerator / Freezer	800 W	1600 W
Microwave	1100 W	N/A
Dishwasher	1500 W	N/A
Garbage Disposer	200 W	450 W
Gas Range w/ Convection Oven	180 W	N/A
Electric Range (4 burner)	8000 W	N/A
Coffee Maker	1000 W	N/A
Toaster Oven	1200 W	N/A
Garage Door Opener	480 W	1000 W
Water Heater (Natural Gas Power Vent)	360 W	600 W
Water Heater (Electric Tank Style)	4500 W	N/A
Water Heater (Electric Tankless)	13–36 kW	N/A
Clothes Washer	700 W	1300 W
Clothes Dryer (Electric)	5000 W	6000 W
Clothes Dryer (Gas)	500 W	1000 W
Furnace Fan / Heat Pump Air Handler	750 W	1500 W
Space Heater	1500 W	N/A
Air Conditioner (Central Air, 3 Ton)	10,800 W	19,500 W
Air Conditioner (Room Through Wall or Window)	1200 W	2400 W
Humidifier	100 W	150 W
Dehumidifier	1000 W	1200 W
Electric Baseboard Heater	1500 W	1500 W

Goodheart-Willcox Publisher

Figure 35-1. Typical power loads for common household appliances and equipment.

Load Estimation Worksheet (Example—Selective Loads)

Load Description	Quantity	Watts per Item	Load Type	Total Connected Watts	Total Surge Watts
Refrigerator	1	800 W	Cycling	800 W	1600 W
Lights (LED) 75 W Equal	10	12 W	Static	120 W	N/A
Downlights—LED 4"	5	4 W	Static	20 W	N/A
Sump Pump	1	1000 W	Surge	1000 W	3000 W
Well Pump	1	1000 W	Surge	1000 W	3000 W
Furnace (Natural Gas)	1	750 W	Cycling	750 W	1500 W
Microwave	1	1100 W	Cycling	1100 W	N/A
Water Heater (Natural Gas)	1	360 W	Cycling	360 W	600 W

Total Connected Watts = 5150 W

$$\text{Generator Rating (kW)} = \frac{\text{Total Connected Watts}}{1000}$$

$$= \frac{5150 \text{ W}}{1000}$$

$$= 5.15 \text{ kW}$$

$$\text{Generator Surge Rating (kW)} = \frac{\text{Highest Single Surge Rating} + \text{Total Connected Watts}}{1000}$$

$$= \frac{3000 \text{ W} + 5150 \text{ W}}{1000}$$

$$= 8.15 \text{ kW}$$

Goodheart-Willcox Publisher

Figure 35-2. Example of sizing a standby power system.

source and the standby source is known as *interconnection*. *Section 702.5* requires that interconnection or transfer equipment "*shall be listed, designed, and installed so as to prevent the inadvertent interconnection of all sources of supply in any operation of the equipment.*"

Transfer switches are the most common method of providing interconnection of a standby power system to a home power system. There are typically two types of switch configurations that transfer the supply from the utility to the home standby power system: manually operated or automatically operated. *Manual transfer systems* require someone to switch the power source from the utility to the standby system. *Automatic transfer systems* are instead powered by a standby power source and use magnetic solenoids to automatically switch the home from the utility to the standby source.

The location of the standby power connections and transfer equipment determines some of the requirements for installation. Home standby power sources can power only essential loads selected, while other systems can be sized large enough to power most of the home. There are additional load management contactors that can be added to an electrical installation to control the load to prevent overloading of the standby source. Refer to **Figure 35-3**.

The proper design before installation can address many of the requirements for standby power systems. There are circumstances where generators or other standby power systems are retrofitted into existing homes and codes are overlooked, and as a result, additional hazards are created. The close attention to grounding configurations, switching configurations, and connection of the loads provide for a reliable system when the homeowner is in a time of need.

35.3.1 Whole House Transfer Switch

Standby systems that power the entire home require a larger transfer switch with a main breaker ahead of the main electrical panel. The whole house transfer switch has *NEC* requirements. Any connections on the utility side of a main breaker have minimum short circuit ratings that need to comply with *Section 110.9* and requirements for service ratings that follow *Section 230.66(A)*. This section requires equipment used for electrical services to be identified as suitable for use as service equipment. The standby power transfer switch with a main breaker and grounding provisions will need proper service and short circuit interrupting ratings.

Any switch or circuit breaker integral to the transfer switch that may be considered the service emergency

Standby Power Connection Configurations

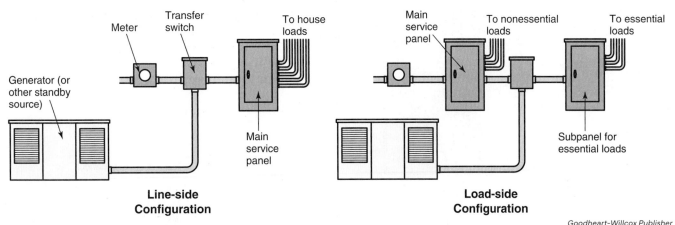

Figure 35-3. Standby power connection configurations.

Goodheart-Willcox Publisher

disconnect required by *Section 230.85* will need to be readily accessible and not behind a cover that would require a tool to remove. The utility side conductors terminating in the transfer switch enclosure need to have barriers installed in accordance with *Section 230.62* to prevent inadvertent contact with these energized conductors.

If a home contains a main breaker, the service grounding requirements must be part of the transfer switch equipment for the standby power source that supplies the entire home. The connection for the main bonding jumper, grounding electrodes, and equipment grounding conductors need to be part of the service-rated standby transfer switch enclosure. All the panels on the load side of the transfer switch must have the neutral and equipment grounds isolated from each other. Refer to **Figure 35-4**.

35.3.2 Essential Loads Transfer Switch

In some cases, a standby power system is not large enough to power all the desired loads. Many homes have electrical equipment that can be considered essential when normal power is lost. Examples of essential loads can be sump pumps, furnaces, refrigerators, freezers, and minimal lighting. A *transfer panel* or *subpanel* is typically used for essential loads powered by a standby power system. The selected loads are connected to the subpanel, and the standby power system uses an automatic or manual transfer switch located between the main electrical service panel or meter main enclosure.

Subpanel installations can contain a main breaker with a back-fed generator breaker and interlock kit. These breaker kits turn a regular main breaker panel into a transfer switch panel. These are found on standard panelboards with listed kits and can add a generator backup to an existing main breaker panel. If this generator interlock kit is added to a main service panelboard, the installing electrician should provide written instructions for the homeowner to remove

nonessential loads by turning breakers off to avoid overloading a standby power source.

For some essential load installations, some generators offer a prewired manual transfer panel that makes a retrofit easier on existing homes. These panels require splices to existing wiring in the main panelboard. *Section 312.8(A)(2)* requires that splices made in the wiring space in these enclosures do not exceed 75% of the cross-sectional area.

35.3.3 Microgrid Interconnect Devices

Microgrid interconnect devices enable standby power systems to use renewable energy sources, such as solar photovoltaic power, wind energy, or battery storage. The smart controllers associated with these technologies can provide power for home loads using a solar photovoltaic system inverter, generator, or battery storage with inverters. Smart technologies used as microgrid interconnect devices can allow the utility to safely disconnect from the standby power system. Refer to **Figure 35-5**. These systems are listed by a nationally recognized testing laboratory to comply with *Section 705.70* and need to be installed based on manufacturer instructions.

PRO TIP **Electric Vehicle Batteries**

Some of the newest technology in home standby power is the connection of electric vehicle batteries to provide standby power for the home. There are different methods to make these connections. One method is to use the vehicle-mounted ac inverter with a transfer switch to supply power to the home like as a portable generator. Another method is to use a bidirectional smart electric vehicle charger that integrates the dc battery of the vehicle and an inverter to supply standby power to the home.

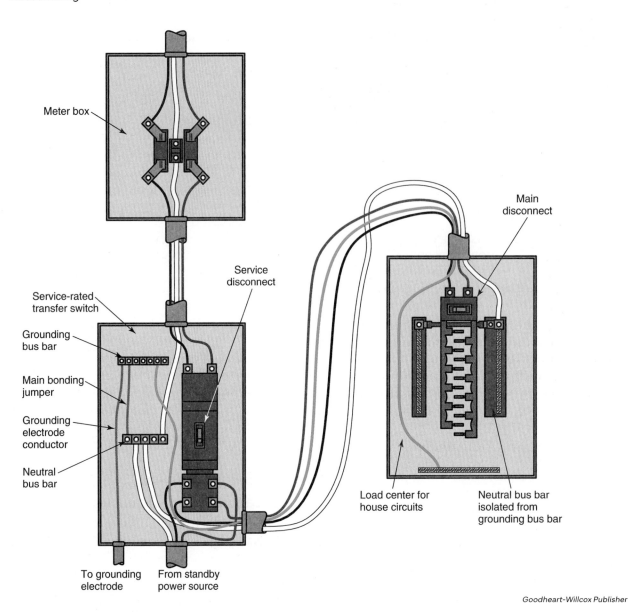

Figure 35-4. Connection of transfer switch used as service disconnect.

Goodheart-Willcox Publisher

Figure 35-5. Smart energy controller.

Goodheart-Willcox Publisher

35.4 Installation of Permanently Installed Generators

The most popular choice for a standby power system is a permanently installed home standby generator. These generators are made of two main components: engines and alternators. The battery that is part of the generator assembly is a key component for automatic transfer systems. This is the only power that can control the switch when the utility power is out. Other electronics and controls included in the generator are also used to set up the control configurations. These electronics can notify the homeowner if there are problems with the unit using an annunciator panel or Wi-Fi connection and software application. These units are typically exercised on a fixed schedule to ensure they are ready in the event of a power failure.

35.4.1 Generator Location

A best practice is to install a generator near the electrical power service to keep wire runs short. A location near the electric service may not always be allowed due to building codes. Other considerations may include the noisiness of a generator and landscaping.

Section 110.26(A)(1) requires that all serviceable equipment be installed to allow 36″ of working space around the front and sides of the generator. The back of the generator can be installed with just 18″ of space. Typically, there are no electrical serviceable parts on the rear of the generator, but connections for the fuel sources and power require clearance.

If a generator is installed near the home's existing electrical equipment, the clearances of *Section 110.26* also need to be maintained for the existing electrical equipment. The generator cannot be installed in front or underneath any electric meter, disconnect switch, transfer switch, or panelboard. Most standby generators have a hinged top that also require clearance to open, and if it were installed under a transfer switch or panel, the hinged top would interfere with working clearances.

The *NFPA 37, Standard for Stationary Combustion Engines and Gas Turbines* requires a generator to be installed a minimum 5′ from combustible structures or shrubs. This can be minimized to 18″ if the structure is determined to be noncombustible, such as brick or stone. Refer to the manufacturer's instructions of the generator to determine if the building clearances may be reduced when weatherproof housing is included.

If a home has a fresh air intake vent for the HVAC system, the International Residential Code requires a minimum clearance of 10′ from any fresh air intake. The distance of 10′ between the exhaust of the generator and the intake vent allows for rising warm exhaust gases to mix with the surrounding air. This dilutes any carbon monoxide or other harmful pollutants. If possible, do not install a generator on the same side of the home as any fresh air intake system.

35.4.2 Underground Preparations and Foundations

After determining the location for a home standby generator, the next step is to prepare the ground where the generator is to be installed. It is critical to have proper preparation of the underground and concrete pads for a generator installation. Generator manufacturers have warranty guidelines that also include proper placement and foundations. Follow all codes and instructions when preparing generator locations and foundations.

The ground underneath a generator should always be firm and free from organic material. Any topsoil, mulch,

or other material that can decompose needs to be removed. A base layer of crushed rock or pea gravel should be added below the generator pad to allow for drainage and provide a firm level area to start.

Some generators are installed on a poured concrete pad, while others can utilize prefabricated concrete, composite, or fiberglass pads. Regardless of base type, the foundation is a key step during the generator installation process. Moving or shifting of the generator pad when it is running can cause gas leaks in the fuel supply line, damage to electrical conduits, or even damage to the generator itself.

When forming a poured-in-place concrete pad, make the concrete pad 4–6″ larger than the generator. The height of the pad should be 2–3″ higher than the surrounding ground or landscape. The thickness of a poured-in-place concrete pad should never be less than 3″. Refer to the generator manufacturer's installation instructions for minimum pad thickness.

When a precast concrete or fiberglass pad is used, the pad should be installed with a crushed gravel base that is level and has been compacted thoroughly. During the pad installation, it is good practice to place a thin 1″ layer of sand below the pad to allow for easier leveling of the pad. The sand will self-compact when the pad is set. Do not use too much sand because it can wash away during a flood or heavy rain.

For composite pads, there needs to be enough mass for the generator installation. These pads may need to be filled with a polymer or water after they have been set. In colder climates, the chance of freezing exists, and they may need to be filled to 80% maximum or have a glycol additive to keep them from freezing. Some fiberglass generator pads have reinforcing fibers run throughout and are not subject to cracking when the generator is running.

A generator must be anchored to a concrete pad to help prevent the generator from moving during operation. Some manufacturers allow access to anchor mounting holes in the base plate after the generator has been placed on the foundation. Depending on the installation, holes may be predrilled prior to setting the generator in place.

35.4.3 Generator Wiring Methods

Wiring between a generator and transfer switch can be installed underground or above ground. Since most generators are installed outside the home, underground wiring methods are common. Underground wiring typically uses schedule 80 rigid PVC conduit, galvanized RMC, or IMC. The *NEC* requirements for burial depths for conduit can differ. Rigid and IMC are typically installed 6″ below grade while Schedule 80 rigid PVC conduit requires a depth of 18″.

Each generator needs at least one electrical conduit run between the transfer switch and the generator for power. Most generators allow the utility sense wires, the 12-volt battery power, and the start and transfer signal wires for the automatic transfer switch to be installed using 600-volt wire with the generator main output power wires. Other accessory wiring may also be permitted to be installed with the power wiring if the insulation meets the requirements of *Section 300.3(C)*. If a generator has communication wiring for internet connections or an alarm annunciator panel, an additional conduit may be required.

Use flexible wiring methods for the connection between the underground conduits and the generator. For conduit that exits underground installations, *Section 300.5(J)* requires an expansion fitting or flexible wiring method to be installed. This accommodates earth movement during freeze-thaw cycles or settling. Transitions between the conduit wiring methods and the flexible wiring methods can be done using fittings identified for the purpose.

Any exposed or concealed conduit installations run between the home and the generator should also contain a portion that utilizes a flexible conduit connection. A generator moves with seasonal temperature changes or experience settling; thus, *Section 300.7(B)* requires wiring methods to accommodate the expansion and deflection of the conduit due to this movement.

35.4.4 Power Wiring

Power wiring between a generator and transfer switch typically contains a single-phase, three-wire circuit. This circuit is composed of two ungrounded conductors and a neutral (grounded) conductor. Since a nonmetallic PVC or a flexible nonmetallic wiring method is used, an additional separate

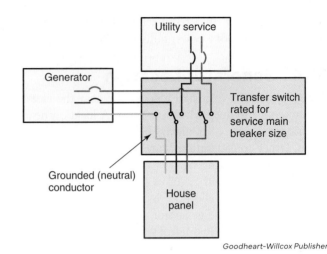

Goodheart-Willcox Publisher

Figure 35-6. Generator power wiring to the transfer switch.

wire-type equipment grounding conductor is required to be installed. Refer to **Figure 35-6**.

A generator typically has an output circuit breaker based on 115% of the kilowatt capacity of the generator. This circuit breaker is used to determine the size of the conductors between the generator and the transfer switch. Common conductor sizes are listed in **Figure 35-7**.

35.4.5 Control and Communications

For automatic transfer switch wiring, control wires run between the generator and automatic transfer switch. Some generator manufacturers sell a seven-conductor, 600-volt

Circuit Breaker and Conductor Sizing for Generators

Generator Rating (in kW)	Max. Circuit Breaker Size (in amps)	Min. Ungrounded Copper Conductor Size (in AWG)	Min. Copper Equipment Grounding Conductor Size (in AWG)
8	40	8	10
10	45	8	10
12	50	6	10
14	60	4	8
16	70	4	8
18	80	3	8
20	90	3	8
22	100	3	8
24	100	3	8
26	110	2	6
30	150	1/0	6
36	175	2/0	6
45	200	3/0	6

Goodheart-Willcox Publisher

Figure 35-7. Generator main breaker ratings and main power conductor sizes.

rated cable that can be installed with the power conductors for this wiring. If a cable is not provided, separate 600-volt conductors can be installed with the power conductors for this wiring.

A typical control wiring connection at the transfer switch will include the following:

- Two conductors to sense the 240-volt ac utility power
- A conductor for the battery charger power
- A dc positive and negative conductor for transfer switch control power
- A conductor for signaling the transfer switch to switch from normal utility power to standby power

Not all manufacturers use the same terminal or wire identifications, but the number of wires and functions are similar. Some transfer switches provide an additional engine start signal, whereas some generator controls can start automatically when there is a loss of ac utility power. Some transfer switches automatically transfer when power is present on the wires from the generator's 240-volt output, and additional control wires are not needed.

If a generator has load-shedding capabilities, additional control wiring may be needed between the generator controller or transfer switch to the load management contactors. These control wires disconnect loads as the generator is loaded. Load shedding can be given priority in the programming of the generator as to which loads are shed first.

Other controls or accessory wiring may be battery blanket heaters, oil heaters, or coolant heaters. These are typically 120 volt or 240 volt, ac-operated, and draw power from the ac utility sense terminals, or need an additional circuit run between the transfer switch and the generator. When an additional 120-volt circuit is provided for the battery heater and engine heater, a separate disconnect switch may be installed so that maintenance can be done in a safe manner.

PRO TIP   Connecting Control Wires

Connect the control wires in the transfer switch with care. Many manufacturers recommend installing the battery and connecting the positive terminal last. Also, be sure to remove the control fuses until all wiring is complete to prevent fuses from blowing during installation.

35.4.6 Grounding Connections

The equipment grounding conductor runs with the power wiring between a generator and transfer switch. Some generators require an additional auxiliary grounding electrode installed at the generator and bonded to the generator frame and equipment grounding conductor. This grounding electrode is typically a ground rod driven at the generator location.

For most home standby generators, the neutral (grounded) circuit conductor is not switched by the transfer switch, and the generator is not considered a separately derived system. The grounding electrodes never connect to the neutral unless the transfer switch has a switching action in the neutral. This requires the generator to have additional system bonding jumpers between the neutral and the equipment ground, along with a grounding electrode connection.

In cases where a generator uses an automatic transfer switch before the main panel to the home, the equipment grounds and neutrals need to be separated in the main panel like a subpanel. When retrofitting generators into an existing home, the location of the main bonding jumper may need to be moved, along with adding additional equipment grounding bars, off the grounded (neutral) connection bus and onto the new equipment grounding bus.

35.4.7 Types of Fuel

Engines for standby power systems need a fuel source to provide power. Most permanently installed standby generators in residential homes use methane (natural gas) or propane as the source fuel.

Natural gas is available in many cities as a fuel source for heating homes. It is supplied by a utility system for a home, and it does not require storage tanks, so it can be available for extended periods of time. In some earthquake regions, however, the supply can be turned off until damage assessments can be made.

Propane is used in areas where natural gas is not available. Propane is clean burning like methane and can be stored as liquified propane (LP) in large or small tanks. LP has a higher energy density than natural gas, and some generators have a higher power rating for LP rather than natural gas. It is also stored onsite, so it is available when utility sources of natural gas are shut off. However, fuel tanks for LP must be sized based on the estimated time that standby power is needed. Large LP tanks take up space, and small tanks can only provide power for limited time periods.

In areas where LP and natural gas are not available, diesel is the fuel of choice. Diesel generators typically have built-in storage tanks that require refilling. Diesel generators can often be noisier and have a smell when operating. In colder weather, diesel can be difficult to start, and additional cold-weather accessories for the engine are needed.

PRO TIP   Fuel Supply Selection

When home standby generators are designed, the type of fuel is determined before installation to best suit the homeowner's choices of standby power needs. The conditions of the fuel supply are communicated at the time of purchase so that proper regulators and storage tanks can be purchased for the installation.

35.4.8 **Fuel Connections**

Most generators have fuel connections or fuel tanks that are separate from the generator enclosures. A few diesel models do have a diesel tank incorporated into the skid frame assembly, but the fuel is typically in a separate connection. All generators are required to have a fuel shut-off valve external to the weatherproof housing readily accessible and near the generator location.

Similar to gas appliances, most natural gas generators run on a fuel pressure of 4–14″ of water column. If the generator is connected to the gas meter on the utility side of the home, an additional natural gas regulator is required to accommodate fuel needs.

Liquid propane-supplied generators have a fuel supply line run from the liquid propane tank location and the generator. This supply line also contains a fuel pressure regulator. Prior to performing startup of the generator, a fuel pressure gauge should be used to adjust the regulator to the required LP pressure based on the manufacturer's instructions. These pressures are also typically in the range of 7–14″ of water column.

The connection between the hard piping of the fuel supply and the generator is done by a flexible pipe connection. These should be kept at the minimal practical length and not pass through a wall or enclosure. Any time a gas pipe is installed, it should be capped until final connections are made. This helps to ensure that no accidental valve opening creates an explosion hazard.

If gas piping is installed underground to the generator location, the piping is typically installed as plastic HDPE or is provided with corrosion protection or cathodic protection with sacrificial anodes. Check with local codes to make sure they permit underground natural gas lines to be installed.

35.4.9 **Other Installation Considerations**

In areas where generators start in freezing temperatures, there may be cold start programming functions that need to be enabled when commissioning a generator. These cold start features will delay the transfer switch from operating until the generator engine has a chance to get up to speed and with the oil flowing. This is typically a delay of 10–30 seconds between the initial outage and the transfer.

Many generators can run weekly on an exercise schedule, with or without a power transfer. This schedule is necessary so the monitoring systems for the generator engine and alternator can evaluate the condition of the generator, ensuring it is ready for an actual outage. The weekly exercise will find out if problems exist and they can be repaired before an outage requires the generator.

35.5 **Installation of Portable Generators**

Portable generators are also a popular source of providing standby power. Using manual transfer switches, a power inlet device can be installed on the exterior of the home for the connection of a portable generator, **Figure 35-8**.

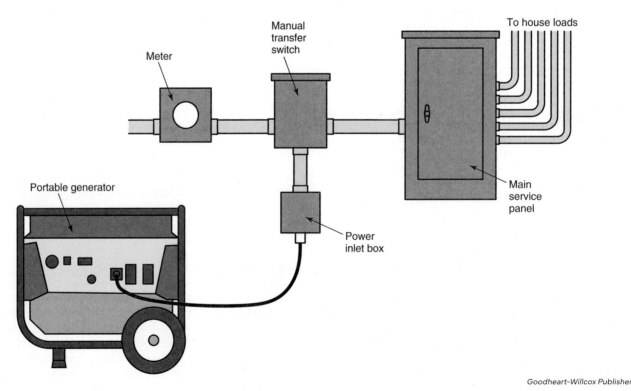

Goodheart-Willcox Publisher

Figure 35-8. Portable generator configuration.

The loads that can be supplied by a portable generator will depend on the rating of the generator. Most portable generators with a 240-volt output produce either 20 or 30 amps.

All installations of wiring for standby power using a portable generator incorporate a transfer switch or means to isolate the utility connection. Use of homemade cords or non-listed cords with two male plugs to back feed power to a home is a violation of the requirements of the *NEC* and can create hazards.

Never put a gasoline generator indoors or in an area near windows or doors. When refueling, always shut off the generator and allow it to cool first. Make sure that the generator output cord is dry and out of standing water.

35.6 Installation of Battery-Based Inverters

A battery-based inverter operates silently and has no moving parts. It consists of a storage battery and inverter. These installations may be included as part of renewable energy systems or are installed to be recharged when the normal utility power is present. The manufacturers can provide batteries that can supply power to a home for weeks, depending on the consumption rate. These originally were developed for off-grid solar applications but have become popular with home standby power needs.

A transfer switch or microgrid interconnect device is used to integrate the backup power into the home loads. The microgrid interconnect devices may be called a MID, Gateway, or Smart Switch, depending on the manufacturer. Using a microgrid interconnect device allows the battery to charge when a renewable energy system is overproducing for the home loads. The microgrid interconnect device remains closed, and any additional energy produced after the battery is charged is exported to the utility.

When the utility power is lost, the microgrid interconnect device opens the connections between the home and the utility, and the battery inverter then powers the home in island mode until the utility power is resumed. These utility interactive systems have additional requirements in the *Article 705*. These require the systems to be listed for the specific purpose and tested upon installation. Many utilities require a commissioning session or documentation to verify that the system is not capable of a back feed to the power grid.

Part of the installation of the battery inverter power systems is the location of the batteries. Many of the new systems use lithium type storage batteries. These battery enclosures can be quite heavy. The proper assessment of the location and the method to fasten or anchor the battery cabinet is a concern. Many battery cabinets require structural blocking or anchoring into a framing member of the wall.

SAFETY NOTE   **Lithium Batteries**

When the batteries are charged and discharged, they will produce heat. The location of the battery cabinet will need to be selected to allow for proper cooling. Lithium batteries can be dangerous if they are overheated and get to what is known as thermal runaway temperatures. This causes lithium-oxygen reactions that will burn very hot, and adding water to the fire will just make things worse.

The wiring for the battery inverter system is like a solar inverter system with battery backup. Each system has its own inverter and does not share any dc circuits between them. The microgrid interconnect devices are very similar to a generator's automatic transfer switch, but there is little need to run any control circuits in between the equipment. The battery inverter systems can be configured to provide power for the entire home or essential loads. This decision will determine the placement of equipment and configurations of the wiring.

Summary

- The sources that can serve as home standby power are both portable and permanently installed generators, batteries with inverters, renewable energy standby power systems, and bidirectional electric vehicle battery chargers with inverters.
- Transfer switches or smart switches are required to be used with standby power systems to protect utility workers from possible back feeds.
- Automatic transfer switches work with the power from the standby power source batteries, whereas manual transfer switches require someone to be present to switch between normal power and standby power.
- Whole house transfer switches need to be installed ahead of the main panel. These may be required to be service rated if they contain the main service disconnect for the house.
- Essential loads can be supplied by standby power sources using subpanels or generator transfer panels.
- Smart system energy controllers can integrate renewable energy systems, batteries, and generators to provide home standby power.

- The selection of a generator for home standby power may depend on fuel availability.
- The types of fuel most common for generators are natural gas (methane), LP (liquid propane), and diesel.
- Portable generators typically use gasoline engines.
- Generator locations need to be determined by evaluating distances from combustible building surfaces, building openings, intake vents, and working clearances.
- Generator wiring methods need to be suitable for installation outdoors or underground and have flexible connections where transitions are made where movement is necessary.
- Power wiring run between the generator and the transfer switch is sized according to the rating of the output breaker that is part of the generator.
- Most generators require control wiring between the transfer switch and the generator.
- The use of batteries and inverters is becoming popular as home standby power systems. These units use transfer switches or microgrid interconnect devices to provide standby power and protect any back feed to the utility.
- Electric vehicle standby power sources are batteries with inverters. The battery is part of the vehicle, and the inverter is integrated into the charging and microgrid power system for the home.

Know and Understand

1. *True or False?* Home standby power systems need to be large enough to power the entire home.
2. *True or False?* Home standby power batteries are rated for their capacity in kW hours.
3. Which of the following loads need to be considered when determining a generator standby rating?
   A. Cycling
   B. Static
   C. Phantom
   D. Surge
4. *True or False?* All loads in a home draw the same amount of current all the time.
5. A _____ load will draw small amounts of power all the time.
   A. cycling
   B. static
   C. phantom
   D. surge
6. What can be used to estimate the size of generator needed for a home standby power system?
   A. Standard wattage load chart
   B. Load estimation worksheet
   C. Manufacturers sizing software
   D. Any of these may be utilized.
7. *True or False?* Power factor is the ratio of the amount of power consumed to the reactive power needed to power a load.
8. *True or False?* Automatic transfer switches need power in order to operate the solenoid.
9. *True or False?* The fuel connection to a generator is required to have a shut off valve near the generator in a readily accessible location.
10. *True or False?* Wiring methods for a generator need to be flexible non-metallic conduit run from the generator to the transfer switch.

Apply and Analyze

1. What switch configuration requires a person to be home to connect loads to a standby power system?
2. What is the distance from a door or window required for a generator installation?
3. What size conductors would be required between the generator output breaker and the transfer switch for a 20 kW generator?

Critical Thinking

1. What must an electrician evaluate when retrofitting a whole house transfer switch and generator into an existing main electrical panel? What may need to be changed when completing this task?
2. A customer would like a home standby system to supply essential loads. What size backup generator would be needed to supply the following loads?
   - 1 standard refrigerator
   - 1 small chest freezer
   - 1 well pump
   - 1 sump pump
   - 1 septic pump
   - 7 LED 75 W equivalent lights
   - 1 natural gas furnace
   - 1 natural gas water heater

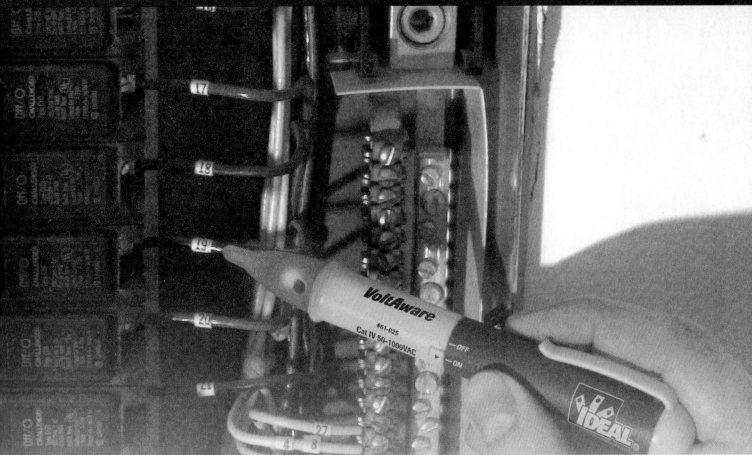

Ideal Industries, Inc.

CHAPTER 36
Troubleshooting Basics

CHAPTER 37
Removing and Adding
Devices and Luminaires

CHAPTER 38
Adding a Circuit

CHAPTER 39
Service Upgrades

This section focuses on the tasks an electrician may be responsible for in an existing dwelling. Chapter 36, *Troubleshooting Basics*, focuses on providing guidance on the types of problems that occur in residential electrical systems and techniques to identify or troubleshoot the problems. Chapter 37, *Removing and Adding Devices and Luminaires*, provides a foundation to determine the right type of device or luminaire to use in the installation, and the *Code* compliance, safety, and functionality of the installed equipment.

Chapter 38, *Adding a Circuit*, covers the steps and considerations of adding a circuit in an existing dwelling. Chapter 39, *Service Upgrades*, covers when and why to perform electrical service upgrades.

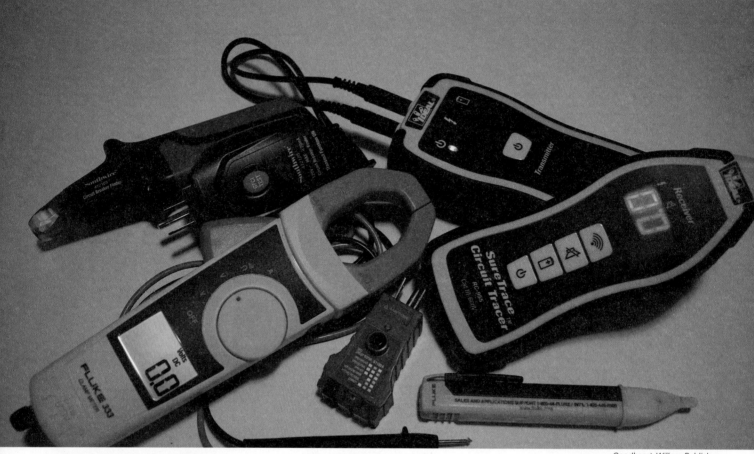

Goodheart-Willcox Publisher

CHAPTER OUTLINE

(Continued)

LEARNING OBJECTIVES

After completing this chapter, you will be able to:

- Describe common causes leading to overload conditions in a circuit.
- Identify tripped or open overcurrent protective devices.
- Indicate the steps of the half-split method of troubleshooting.
- Diagnose circuits containing open grounded conductors.
- Contrast the terms overload, short-circuit, and ground-fault.
- Determine if a fault is related to faulty equipment or faulty wiring.
- Relate voltage variations to open neutral conductors.
- Describe the effects of loose connections.
- Diagnose faults causing arc-fault circuit interrupter tripping.

TECHNICAL TERMS

continuity
half-split method
inrush current
load center
overload
troubleshooting

Introduction

Troubleshooting is a series of steps used to identify and repair the cause of a malfunction in an electrical system. The troubleshooting process typically begins with an overview of the issue. Talking with the homeowner can provide more information about the problem, such as how long the problem has been occurring and what the homeowner has already done to try to fix the issue.

Simple questions and observations can often lead to clues to resolving the problem:

- Has the homeowner replaced devices such as luminaires, receptacles, or switches?
- Are there devices in the circuit that look new or different from the old ones?
- Is there any evidence of water damage or rodent infestation?
- Is the smell of overheated conductor insulation present?
- Are there char marks on the wall near devices indicating possible arcing?
- Are the lights dimming?
- Is the problem occurring in one circuit, several circuits, or all circuits within the dwelling?

Four main problems occur in electrical systems: overloads, open circuits, short circuits or ground faults, and undervoltage or overvoltage conditions. Troubleshooting these problems in residential electrical circuits is a skill that an electrician learns over time. As you develop your troubleshooting skills, you can identify subtle clues that lead to the problem. This chapter will focus on providing guidance on the types of problems that occur in residential electrical systems and techniques to identify or troubleshoot the problems.

36.1 **Overloads**

Overloads can be one of the simplest issues within the electrical system to troubleshoot. An *overload* is an overcurrent in which low to moderate amounts of current flow in excess of the circuit rating. A branch circuit protected by a 20-amp overcurrent device that has 30 amps of current flow is an example of an overload condition. Overload conditions differ from other types of overcurrent because the current stays in the conductive pathway of the circuit. If the current is present long enough, it can cause heating of the conductors that may damage the conductors or equipment. Overcurrent protective devices, like fuses and circuit breakers, open to protect the circuit when overload current becomes high enough for a long enough period.

36.1.1 **Excess Loads**

A common cause of overloads in residential circuits is having too many loads connected to the circuit. If conductors have been subjected to an overload, the excess current causes the conductors to generate more heat than they can dissipate. The heat buildup compounds over time and can lead to conductor insulation damage. A distinct smell of heated insulation may be present and provides one of the first indications of an overload condition.

To troubleshoot the circuit for overloads, place a current clamp meter around the ungrounded conductor of the circuit, **Figure 36-1**. Compare the measured current value to the circuit rating. If the actual current is greater than the circuit rating, an overload condition exists. Removing utilization equipment from the circuit will reduce the current and remove the overload condition. Adding an additional circuit to provide an alternate source for excess loads may be necessary. Visually inspect the overloaded circuit conductors' insulation for damage. Damaged insulation may look discolored, be deformed due to melting, or be charred or blackened. Insulation damage may be hard to visually detect because a great deal of the wiring system is inaccessible within the building structure. A megohmmeter can be used to test insulation resistance to determine if the insulation of the wiring system contains a defect. If the conductor insulation has been compromised, you will need to replace any damaged wiring to reduce potential hazards.

36.1.2 **Motor Overloads**

Another common cause of overloads in residential circuits are motor overloads. Motors are integral to many pieces of utilization equipment and are often subject to inrush current. *Inrush current* is excess current that occurs during motor startup that can be as high as 600% of the circuit's rating. Often, this inrush current does not trip the overcurrent protective device because it quickly drops below the

Goodheart-Willcox Publisher

Figure 36-1. An overload condition exists within this circuit. The conductor is carrying 24.6 amps. The circuit breaker, rated at 20 amps, has not yet tripped. If this condition continues for enough time, the heat generated in the conductor could damage the conductors and equipment.

circuit's rating after the motor comes up to its full running speed, **Figure 36-2**. This overload condition does generate heat in the conductors, but it does not usually damage the conductor insulation since the overload is brief enough that the conductor is able to dissipate the heat before the insulation is damaged.

Motor inrush can be increased or persist longer when there are problems with the motor. Possible problems include friction in the motor bearings, too much load on the motor, or failure in other motor components, such as the starting or running capacitors. Refer to **Figure 36-3**.

Troubleshooting overloads associated with utilization equipment containing motors can be as simple as monitoring current flow with a clamp-on ammeter, both at startup and during normal operation. Replacement of the motor or components associated with the motor may be necessary.

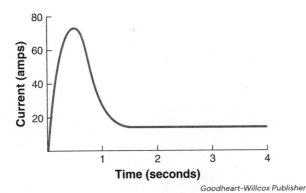

Figure 36-2. Inrush current of a motor is significantly higher than the circuit's rating, but the inrush current declines to normal running current within a few seconds.

Figure 36-3. In-sink waste disposers or garbage disposals are common household motor-operated appliances. Overload conditions can be created in disposal circuits when too much food is placed in them, corrosion builds up due to infrequent use, or if the rotor becomes locked due to wedging of food or kitchen utensils.

36.2 **Open Circuits**

The movement of electrons through the conductive surfaces of a circuit, known as current flow, causes work to be done by the circuit. If the pathway for current flow throughout the circuit is interrupted by a break or opening of the conductive path, there will be no current flow and no ability to power a device, light a lamp, or spin a motor. This is called an open circuit.

Open circuits are one of the most common problems an electrician must troubleshoot. Open circuits are the result of broken wires, poor connections, tripped or open overcurrent protective devices, and equipment failure. The first step in identifying an open circuit typically starts with measuring voltage of the equipment that is not working, **Figure 36-4.** Many electricians use a noncontact

Figure 36-4. Open circuit conditions will result in a loss of voltage at the effected equipment.

voltage tester to identify if there is voltage present on the ungrounded conductor of the system. The following sections outline the necessary steps to troubleshoot various types of open circuit conditions.

36.2.1 **Open Overcurrent Protective Devices**

If a voltage test indicates there is no voltage present, the next logical place to begin the troubleshooting process is at the circuit's source: the overcurrent protective device. Overcurrent protective devices, such as circuit breakers, are contained in the load center. The **load center**, often referred to as a *breaker panel*, is a metal enclosed panelboard containing circuit breakers that supply the home's power and lighting circuits. Some homes are protected by fuses rather than circuit breakers, and these overcurrent protective devices are contained in a fuse box. Before removing the cover of a load center, complete a quick visual scan of the circuit breaker handles to see if any of them are in the tripped position. Most circuit breakers have a middle trip position that is between the on and off position, **Figure 36-5.**

Goodheart-Willcox Publisher

Figure 36-5. Identification of tripped circuit breakers can often be done through visual inspection. These breakers have a visual indicator that helps the identification process. Notice the middle breaker has an orange window adjacent to the breaker handle, indicating that it is in the tripped position. The tripped breaker handle is also in the middle or tripped position, causing it to be out of line with the breaker above and below.

PRO TIP

Identifying a Tripped Circuit Breaker

It is sometimes difficult to identify tripped circuit breakers by visual inspection. Manually pressing the breaker handle toward the on position should allow you to distinguish between a breaker in the on position versus the tripped position. A circuit breaker that has not tripped will feel firm when moved toward the on position. A tripped breaker will move easily when the handle is moved toward the open position.

If the visual inspection does not provide evidence of the problem, the next step is to remove the equipment cover. A noncontact voltage tester can be placed next to the ungrounded conductor connected to the overcurrent protective devices, and it will indicate if any of the breakers or fuses have tripped, **Figure 36-6**.

SAFETY NOTE

Electrical Arc Flash Hazard

Once the cover of the load center has been removed, shock and arc flash hazards exist. The *Electrical Safety Related Work Practices Standard, NFPA 70E*, requires electricians to protect themselves from electrical hazards when working around exposed energized circuit parts and conductors. If you work around an open energized electrical panel, you should don the proper personal protective equipment as recommended by the 70E standard before continuing the troubleshooting process.

Goodheart-Willcox Publisher

Figure 36-6. If there is no visual indication of a tripped circuit breaker, a noncontact voltage indicator can be used to test for the presence of voltage in the conductors connected to the breakers.

If you find a tripped breaker, do *not* immediately reclose the circuit or replace the fuse. OHSA and *NFPA 70E* require verification that it is safe to re-energize the circuit before manually re-energizing. If you know that the overcurrent device opened due to an overload condition and have corrected the issue, you can replace the fuse or reclose the circuit breaker. If you are not sure of the cause, further investigation to determine the cause of the opening of the overcurrent protective device is required. Troubleshooting problems such as these will be covered later in the chapter.

36.2.2 Half-Split Method

If you do not find a tripped overcurrent protective device, the break in the circuit is present somewhere between the source and the utilization equipment. Many electricians use the ***half-split method***, a technique for troubleshooting that separates the circuit at a point halfway between the source and the load. For this method, first measure for voltage at the halfway point. If voltage is present at that point, you know the circuit is operational up to the point you are measuring, and the problem must be in the last half of the circuit. If there is no voltage present at the halfway point, then the problem is in the first half of the circuit. Continue to split the circuit in half until you locate the problem in the circuit, **Figure 36-7**.

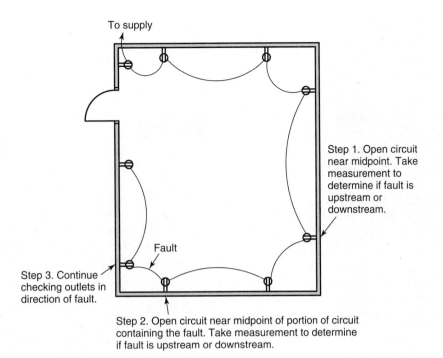

To supply

Step 1. Open circuit near midpoint. Take measurement to determine if fault is upstream or downstream.

Fault

Step 3. Continue checking outlets in direction of fault.

Step 2. Open circuit near midpoint of portion of circuit containing the fault. Take measurement to determine if fault is upstream or downstream.

Goodheart-Willcox Publisher

Figure 36-7. To complete the half-split method, open the circuit somewhere near the middle. Using the appropriate test instrument, determine if the problem is between your location and the source or between your location and the end of the circuit. Repeat the process until the problem is located.

36.2.3 Open Grounded Conductors

If voltage is present on the ungrounded conductor during the initial test of the utilization equipment, but the equipment is not working, the next step is to measure voltage using a voltmeter. Some electricians prefer using solenoid voltage testers, while others choose digital multimeters. Either tester is sufficient for this step. Using a voltmeter, measure across the conductors of the system. The circuit voltage should be 120 volts from an ungrounded conductor to the grounded conductor, 120 volts from an ungrounded conductor and an equipment grounding conductor, and 240 volts or 208 volts between two ungrounded conductors.

Nominal Voltage of Residential Electrical Systems

Most single-family homes are fed by 3-wire, 120/240-volt single-phase systems. This system typically contains two ungrounded conductors (red and black) and one grounded conductor (white). Voltage measured between the two ungrounded conductors is 240 volts.

Multi-family dwelling unit services are sometimes fed by three-phase systems. Each individual dwelling unit is still fed by a 3-wire single-phase system, but the voltage between the ungrounded conductors is 208 volts rather than 240 volts.

A voltage reading of 0 volts from the ungrounded conductor to the grounded conductor and 120 volts from the ungrounded conductor to the equipment grounding conductor indicates an open grounded (neutral) conductor, **Figure 36-8**. Use the half-split troubleshooting method to locate the break in the grounded conductor between the equipment and the source.

36.3 Short Circuits and Ground Faults

A short circuit is considered an overcurrent in which significantly more current flows than the circuit is rated for. Short circuits are created when current flows outside the normal pathway and circumvents the load. This shortened pathway has very low resistance, which results in high levels of current flow. Sometimes tens of thousands of amps can flow in a 20-amp circuit.

Ground faults are short circuits that contain the equipment grounding conductor or the normally noncurrent carrying metal parts of the electrical equipment in the short circuit current pathway.

Short circuits are caused by connecting either two ungrounded conductors of opposite phases together or an ungrounded conductor and a grounded conductor, **Figure 36-9**. Ground faults are caused by connecting an ungrounded conductor and the equipment grounding conductor or grounded equipment together. This issue can happen because of insulation failure, miswiring of devices, poor installation practices, or failure of the equipment connected to the circuit.

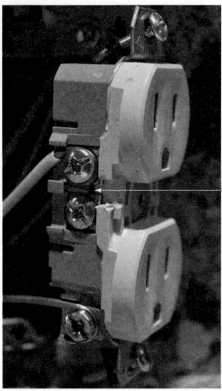

Goodheart-Willcox Publisher

Figure 36-8. The bottom half of this receptacle is in an open neutral condition. Notice that the tab between the grounded conductor screws is removed, breaking the connection to the bottom half of the device. A meter connected between the ungrounded conductor and the equipment grounding conductor would indicate 120 volts, but between the ungrounded and grounded conductor the voltage would be 0 volts.

Tab has been removed, so upper terminal is no longer connected to lower terminal

36.3.1 Faulty Equipment

The process of troubleshooting ground faults and short circuits is similar. One common source for these types of faults is faulty equipment connected to the circuit. The first step

in determining if a short circuit exists is to remove any load connected to the circuit. Equipment plugged into receptacle outlets should be unplugged; luminaires can be disconnected by opening the switches that control them.

PRO TIP
Reading Continuity through Loads

If you choose to use an ohmmeter or continuity tester to troubleshoot short circuit conditions, be sure to remove all loads from the circuit before starting the test. If a continuity tester is applied to a circuit containing a load, such as an incandescent light bulb, the tester will read through the lamp's filament, which can be mistaken for a short circuit.

Once all the loads have been removed and the circuit is de-energized, you can now use an ohmmeter, multimeter, or continuity tester to determine if there is a low-resistance connection between the conductors.

SAFETY NOTE
Ohmmeters and Energized Circuits

Ohmmeters are designed to be used on de-energized circuits. A meter's internal power supply, typically a battery, applies a small voltage to the conductors that can create current flow in a closed circuit. When an ohmmeter is connected to an energized circuit, the high external voltage produces more current than the ohmmeter is designed for. The meter contains fuses designed to protect the circuit, which then opens, causing the meter to no longer work until the fuses are replaced. In extreme situations, an arc can occur while interrupting the current, which may lead to meter damage and potential injury.

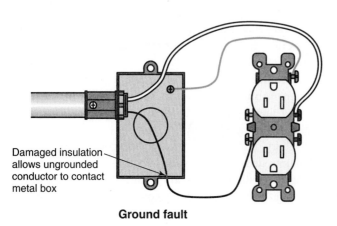

Damaged insulation allows ungrounded conductor to contact metal box

Ground fault

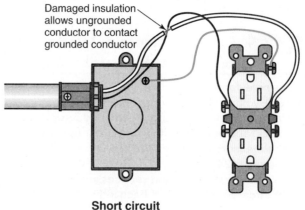

Damaged insulation allows ungrounded conductor to contact grounded conductor

Short circuit

Goodheart-Willcox Publisher

Figure 36-9. A short circuit condition would exist if there were an unintentional connection between the ungrounded conductors or an ungrounded conductor and the grounded conductor or equipment grounding conductor.

Continuity is an indication of an unbroken connection between two conductive points of a circuit. If a continuity tester is applied to opposite ends of an unbroken conductor, the tester indicates continuity. If the conductor is broken somewhere between the points at which the meter is connected, the tester indicates no continuity. Continuity or resistance readings near zero are an indication that a short-circuit condition exists.

If the resistance test completed after disconnecting all loads shows that no fault exists in the wiring system, the problem may have been in a piece of utilization equipment that was connected to the circuit. At this point, the overcurrent protective device can be reset, and the equipment can be placed back into service. Plugging in each piece of equipment individually until the short circuit presents itself again might seem like a logical solution, but doing so is unsafe, as it would likely create an arc fault at the point of reconnection and can result in injury and equipment damage.

A safer method of reconnecting equipment is to do a resistance test on each piece of equipment and compare that resistance to the expected resistance of the equipment. For example, equipment rated 12 amps at 120 volts should have a resistance of 10 ohms, **Figure 36-10**. If the resistance measured by the tester is lower than the expected value, it may be the source of the short-circuit condition. Continue checking each load as it is added back into the system until all loads have been reconnected.

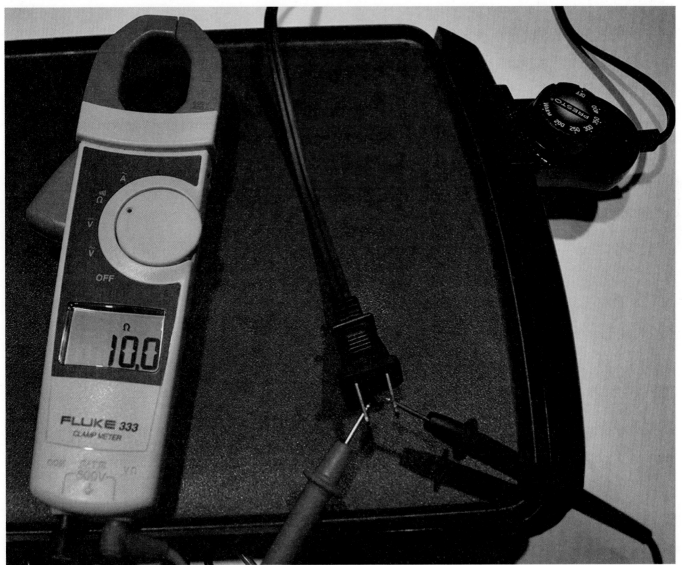

Goodheart-Willcox Publisher

Figure 36-10. This electric griddle has a rated current of 12 amps. With a 120-volt source and a 12-amp draw, the expected resistance is 10 ohms. An ohmmeter can be connected to see if a fault is present in the equipment before it is reconnected to the circuit. This griddle is in good working order, as the predicted resistance closely aligns with the measured resistance.

36.3.2 Faulty Wiring

If the resistance test shows that a fault condition still exists after removing all equipment from the circuit, that is an indication of faulty wiring. Common sources of faulty wiring within the wiring system are typically due to failure of the conductor insulation. This can be caused by damage during installation, fraying, abrasion, penetration from nails or screws, and heating due to overloads. Miswiring and failure of devices, luminaires, and equipment can also lead to short-circuit conditions, **Figure 36-11**. The half-split troubleshooting method can be used to isolate the source of a fault. With the loads removed, split the circuit in half and continue to search for the source of the short circuit or low-resistance connection between the conductors. Once the faulty section of wiring is found, correct the condition before re-energizing the circuit.

36.4 Overvoltage and Undervoltage Conditions

Undervoltage and overvoltage conditions can negatively impact the operation of electrical equipment. The National Electrical Manufacturers Association (NEMA) recommends a tolerance of no more than +10% of the nominal rated voltage. While most sources of voltage variation come from the utility, certain conditions within the residential electrical system can contribute to the problem.

Goodheart-Willcox Publisher

Figure 36-11. An improper connection caused a short-circuit condition, resulting in equipment and conductor damage.

36.4.1 Voltage Drop

The most common source of voltage variation in residential circuits is voltage drop. Voltage drop is the reduction of voltage at the connected load due to losses in the conductor feeding the equipment. Long conductor runs can contribute to voltage drop, **Figure 36-12**. The *NEC* recommends no

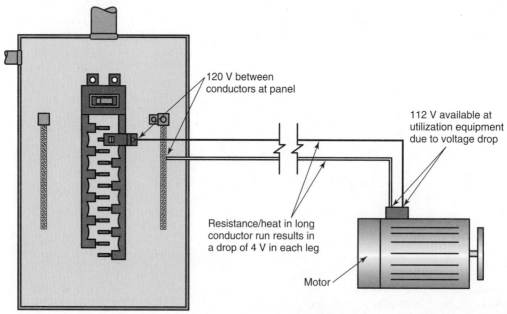

120 V between conductors at panel

112 V available at utilization equipment due to voltage drop

Resistance/heat in long conductor run results in a drop of 4 V in each leg

Motor

Goodheart-Willcox Publisher

Figure 36-12. All conductors inherently contain resistance. Resistance in the conductors of the system is often disregarded, but resistance can contribute to voltage variation issues. The current flowing through the resistance of the wire acts like a resistor in series with the load. Resistance increases as conductor length increases. This series resistance creates a voltage drop, thereby reducing the voltage at the equipment. Voltage drop can be reduced by decreasing the length of the conductor or increasing the size of the conductor.

more than a 3% voltage drop on branch circuits. To troubleshoot voltage drop, measure voltage at the source and at the equipment while the equipment is being used. Subtract the voltage measured at the equipment from the voltage measured at the source to find the voltage drop. Take the voltage drop value and divide by the supply voltage, then multiply by 100 to determine the percentage of voltage drop in the circuit. If a loss greater than 3% is present, you may need to alter the circuit conditions to reduce the voltage variation.

Another source of voltage drop is due to inrush current of large motors or loads. You may have noticed the lights dim when the air conditioning or another large load first turns on in your home. This dimming is due to the voltage drop created in the conductors of the electrical system. As a load becomes energized, the inrush current draws a large amount of current through the conductors. This large current contributes to increased voltage drop in the conductors, which reduces the voltage to the rest of the equipment in the house. The reduced voltage results in the dimming of incandescent lighting, flicker in fluorescent and LED lighting, and potential color change in LED lighting. The dimming does not last long because the inrush current is reduced very quickly after startup. Undervoltage due to these conditions can be limited by increasing the size of the service conductors and reducing the inrush current of the large loads.

36.4.2 Open Neutrals

Another source of undervoltage and overvoltage in residential single-phase, 3-wire circuits is the open neutral. A three-wire circuit can essentially become a series/parallel combination circuit when the neutral conductor is opened under load. The loss of a neutral connection can originate

from inadvertently disconnecting a neutral conductor from a splice or termination bar. It can also come from a break in the circuit conductor or a loose connection. Opening the neutral will create a condition resulting in overvoltage of equipment with high resistances and undervoltage of equipment with low resistances. Refer to **Figure 36-13**. Sensitive electronic equipment, like computers, is typically high-resistance equipment. In the right conditions, the voltage to the equipment can exceed the recommended maximum 10% nominal value and cause damage to the equipment. Lower resistance, higher current drawing equipment, like motors, may not operate properly during undervoltage conditions and can lead to overcurrent conditions and equipment damage.

PRO TIP	Visual Clues to Identify Open Neutrals

Open neutrals contribute to voltage variation in the electrical system. An indicator of an open neutral condition is varying light output from a lamp. The lamp will often get dimmer or brighter as the circuit conditions change. Energize or de-energize loads and watch for the light output to dim or brighten. If light level variability is present, an open neutral condition may exist within the electrical system.

36.4.3 Loose Connections

Loose connections can also cause problems related to undervoltage conditions. When connections between conductive parts of the circuit are not properly tightened, the

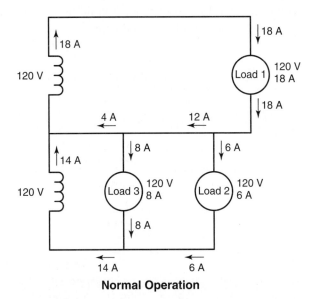

Normal Operation

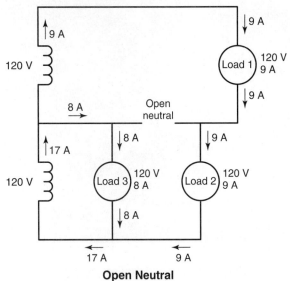

Open Neutral

Goodheart-Willcox Publisher

Figure 36-13. The voltage applied to utilization equipment under open neutral conditions can vary. Opening the neutral conductor can create a series-parallel circuit in which the loads share the system voltage. The amount of voltage each load receives is dependent on the load's resistance.

connection adds resistance to the circuit. This additional resistance creates a voltage drop across the connection and can reduce the voltage available for the connected equipment. The resistance of the loose connection also contributes to the buildup of heat at the connection, which in turn, contributes to another increase in resistance. If the loose connection is not corrected, it can lead to equipment heating, conductor insulation damage, or even fire, **Figure 36-14**.

36.5 **Arc-Fault Circuit Interrupters**

Troubleshooting circuits protected by arc-fault circuit interrupters may take some additional considerations. Arc-fault circuit interrupters protect property and inhabitants from arcing faults that can lead to electrical fires. These arcing faults can come from the circuit itself or

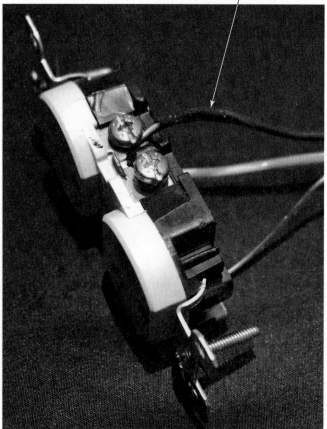

Discolored and damaged insulation due to loose connection

Goodheart-Willcox Publisher

Figure 36-14. This receptacle had a loose connection between the screw and the conductor. The loose connection led to heat buildup causing discoloration and insulation damage to the conductor.

from the utilization equipment connected to the outlets. This device also protects the circuit by monitoring for unwanted connections between the grounded (neutral) and equipment grounding conductors. Arc-fault protection can be provided by the circuit's overcurrent protective device, at the receptacle outlet, or fed through an outlet upstream.

Troubleshooting arc-fault circuits that are tripping usually starts at the device. Most manufacturers of arc-fault protectors provide diagnostic capabilities to determine why the arc-fault device is tripping. Consult the manufacturer's installation instructions for further information on the specific device being investigated. For example, Eaton BR series arc-fault breakers have a diagnostic LED on the face of the breaker. The LED flashes a trip code that can be interpreted by counting the number of flashes and comparing that to a trip code table, **Figure 36-15**.

36.5.1 **Arc-Faults in Series or Parallel Circuits**

Troubleshooting circuits for series or parallel arc-faults is a difficult task. The first step is to remove any loads connected to the circuit. If the arcing fault is stemming from the equipment, the device should not trip after being reset. If the device still trips due to arcing conditions on the wiring system, use the half-split troubleshooting method to identify which section of wiring is causing the problem. The portion of the wiring containing the arcing fault will need to be replaced to remove the fault causing the device to trip.

Another common cause for arc-fault circuit interrupter tripping is an unintended connection between the grounded conductor and the equipment grounding conductor of the circuit. The grounded conductor and the equipment

Example AFCI/GFCI Circuit Breaker Trip Codes

LED Flashes	Meaning
1	Series arc fault
2	Parallel arc fault
3	Overload
4	Overvoltage
5	Ground fault
6	Self-test failure

Goodheart-Willcox Publisher

Figure 36-15. AFCI devices can be equipped with a flashing light to aide in the troubleshooting process. Count the number of flashes to indicate the trip code. Use the trip code table to determine the reason for breaker operation.

grounding conductor are already connected at the electrical service equipment of the dwelling. If an additional connection is made within a circuit due to insulation failure, poor installation, or wire and cable damage, a parallel pathway is established between the grounded conductor and equipment grounding conductor. This creates a dangerous condition because current travels in all paths of a parallel circuit. Current that is normally on the insulated grounded neutral conductor flows on the equipment grounding conductor and the normally non-current carrying metal parts of the circuit. This current can be a source for electrical shock and arcing across loosely joined metal components and is dangerous if not corrected.

An unintentional connection can be found with a continuity tester or ohmmeter. After removing the load wiring from the protective device, test for a connection between the grounded conductor and the equipment grounding conductor. If a connection is present, you will get continuity or a low resistance on the ohmmeter. Disconnect the utilization equipment from the circuit to determine if the faulty condition is within the wiring or the equipment. If the connection still exists after removing the equipment from the circuit, use the half-split troubleshooting method to identify the location of the unwanted connection between the circuit conductors. Replacement of damaged wiring methods may be necessary to remove the hazard from the circuit.

PRO TIP   **The Troubleshooting Process**

Step 1: Gather information.
- Talk with the homeowner about their observations.
- Identify affected components.
- Use sight, smell, and sound clues to identify a possible problem.

Step 2: Predict the likely problem.
- Based on observations, predict if the fault is an overload, open circuit, short-circuit, or voltage variation.

Step 3: Locate the problem.
- Implement testing based on the predicted problem.
- If necessary, use the half-split method to narrow down the circuit until a problem can be isolated.

Step 4: Repair the issue.
- Remove and replace faulty components from the wiring system.
- Replace damaged wiring.
- Repair circuits and devices that have been miswired.

Step 5: Verify problem resolution.
- After making the necessary repairs, re-energize the circuit.
- Test for proper circuit operation.

Summary

- Troubleshooting falls into four main categories: overloads, open circuits, short circuits, and voltage variations.
- Overloads are typically caused by connecting too much utilization equipment to the circuit and motor inrush current.
- Open circuits are the result of broken wires, poor connections, tripped overcurrent protective devices, and equipment failure.
- A non-contact voltage tester can be used to verify if there is an open in the ungrounded conductor but does not indicate an open in the grounded (neutral) conductor.
- The half-split method of troubleshooting can be used to locate problems in a circuit by breaking a circuit in half and identifying which half contains the fault. This method can be repeated until the fault has been located.
- Open grounded conductors can be identified when voltage is present between the ungrounded conductor and the equipment grounding conductor but not between the ungrounded conductor and the grounded conductor.
- Ground faults are short circuits that contain the equipment grounding conductor or normally noncurrent carrying metal parts in the current pathway.
- Ohmmeters and continuity testers read through the connected loads, so the equipment must be disconnected from the circuit before troubleshooting short circuits in the wiring system.
- Faulty wiring is typically a result of damage during installation, fraying, abrasion, penetration from nails or screws, and heating due to overloads.
- Voltage variation should be limited to not more than 10% above or below nominal.
- Voltage drop in conductors should be limited to not more than 3% for branch circuits.

- Open neutral conductors can contribute to voltage variations, increasing voltage in high-resistance loads and decreasing voltage in low-resistance loads.
- Loose connections can result in equipment heating, conductor insulation damage, and fires.
- Most arc-fault circuit interrupter devices have a diagnostic feature for troubleshooting.

Know and Understand

1. Overload conditions are often caused by _____.
   A. environmental damage to conductors
   B. faulty wiring
   C. excess loads
   D. faulty equipment
2. *True or False?* Motor inrush current is sustained for a long period of time.
3. *True or False?* Tripped circuit breakers should not be reset until the problem has been verified.
4. The _____ method is a technique for troubleshooting in which the circuit is broken near the middle to determine the location of a fault.
   A. segment-split
   B. half-split
   C. quick-split
   D. circuit-split
5. When troubleshooting a circuit with an open grounded conductor, the voltage reading between the ungrounded conductor and the grounded conductor is _____.
   A. 120 Volts
   B. 208 Volts
   C. 240 Volts
   D. 0 Volts
6. What is the first step in determining the existance of a short-circuit condition?
   A. Reset the circuit breaker
   B. Remove the connected load
   C. Check for continuity across the conductors
   D. Measure resistance across the conductors
7. Faulty wiring is typically due to which of the following?
   A. Failure of the conductor insulation
   B. Manufacturer defects
   C. Installer error
   D. Failure of the utilization equipment
8. *True or False?* Voltage drop should be limited to less than 3% in branch circuits.
9. Voltage variation due to voltage drop can result from all of the following *except* for _____.
   A. motor inrush current
   B. loose connections
   C. undersized service conductors
   D. open neutral conductors
10. Open neutrals in _____ circuits can create voltage variations across loads of differing resistances.
    A. grounded
    B. single-phase three-wire
    C. single-phase two-wire
    D. ungrounded
11. *True or False?* A common cause for arc-fault circuit interrupter tripping is an unintended connection between the grounded conductor and the equipment grounding conductor of the circuit.

Apply and Analyze

1. If a fuse trips due to overload, why is it dangerous to simply replace the fuse with a larger one?
2. Explain why the use of a noncontact voltage tester is not sufficient for diagnosing an open grounded conductor?
3. List the steps of locating an unintentional connection between a grounded conductor and equipment grounding conductor using the half-split troubleshooting method.
4. What steps would you use to determine if a short-circuit fault was located in the connected equipment or the wiring system?
5. When troubleshooting short-circuit conditions, what must be done before connecting an ohmmeter to the circuit?
6. In a typical single-phase, three-wire electrical system, what is the expected voltage between the two ungrounded conductors, between the ungrounded conductors and the grounded conductor, between the ungrounded conductors and the equipment grounding conductor, and between the grounded conductor and equipment grounding conductor?
7. An Eaton BR series arc-fault circuit interrupter-style circuit breaker has tripped. The trip indicator flashes 5 times. What type of fault is the breaker indicating?

Critical Thinking

1. What clues might indicate an overload condition?
2. What clues might indicate a short-circuit or ground-fault condition?
3. What clues might indicate an undervoltage or overvoltage condition?

CHAPTER 37

Removing and Adding Devices and Luminaires

JR-stock/Shutterstock.com

(Continued)

LEARNING OBJECTIVES

After completing this chapter, you will be able to:

- Determine the minimum allowable voltage and current rating of a receptacle.
- Compare the available options for receptacle replacement of ungrounded circuits.
- Explain split-circuit and switched receptacle applications.
- Identify requirements for receptacle replacements in locations requiring GFCI and AFCI protection.
- Determine if an existing circuit can be extended or if a new circuit should be installed.
- Describe the process for cutting in a box based on building structure, insulation, and finish materials.
- Select proper boxes for additional electrical openings.
- Summarize factors related to selection of new and replacement luminaires.

TECHNICAL TERMS

airtight
current rating
direct lighting
gem box
general diffuse light
general illumination
indirect lighting
insulation contact (IC) rated
knob and tube system
lath and plaster
sheetrock
split-circuit receptacle
switched receptacle
ungrounded receptacle
vapor barrier
vermiculite insulation
voltage rating

Introduction

Electricians regularly install new or replace existing receptacles and luminaires in the course of their work. Determining the right type of device or luminaire to use in the installation affects the *Code* compliance, safety, and functionality of the installed equipment. An electrician must have a working knowledge surrounding device and luminaire selection. This means they must understand special protections like GFCI and AFCI protection, as well as techniques for cutting in boxes and supporting them securely. This chapter will focus on these topics to build your knowledge of residential wiring tasks you'll encounter in your career.

37.1 Replacing Existing Receptacle Devices

Replacing receptacle devices on existing circuits is a routine task for electricians. When doing these types of replacements, there are several factors to consider:

- Receptacle rating, configuration, and grounding scheme
- Special applications, such as split circuit installations and switched receptacles
- Receptacle replacement in locations that require GFCI and AFCI protection

37.1.1 Receptacle Ratings and Configurations

Receptacle devices must be installed in accordance with their ratings. There are two ratings that should be considered: voltage and current. The receptacle configuration must also be considered. Receptacles come in several configurations based on the application in which they are being used. Receptacle configurations have been standardized by the National Electrical Manufacturers Association (NEMA). **Figure 37-1** indicates NEMA configuration and includes a visual representation of the plug and receptacle connections.

Some receptacles, referred to as *single receptacles*, have only one opening to connect utilization equipment. Typical

residential receptacles, referred to as *duplex receptacles*, have two openings for the attachment of equipment, **Figure 37-2**. Receptacles designed for typical equipment are considered straight blade receptacles because the attachment plug has straight blades that are inserted into the receptacle.

The **voltage rating** of a receptacle is the maximum voltage that should be applied to the receptacle. Residential electrical systems are primarily three-wire ac systems. The voltage rating on residential-grade receptacles is typically 125 volts, 250 volts, or 125/250 volts. A 125-volt rated receptacle is routinely used on a 120-volt nominal electrical circuit. Typical receptacles in bedrooms of residential dwellings are rated at 125 volts. The use of a 125-volt receptacle on a 240 volt-circuit is a violation of the receptacle rating because the circuit voltage exceeds the receptacle's rating. Receptacles designed for 250-volt loads should only be used to connect loads that are designed to operate more than 120 volts but less than 250 volts.

An example of a typical 240-volt load is a wall-mounted room air conditioner. Receptacles with a 125/250-volt rating are designed for equipment that utilize a 240-volt and a 120-volt connection, such as residential dryers and ranges, **Figure 37-3**. Dryers and ranges use the 240-volt circuit to supply the resistive heating loads in the appliances, while the 120-volt circuit supplies electronics, controls, and lights.

	125 V	**125/250 V**	**250 V**
15 A	15 A 120 V	14-15R	15 A 240 V
20 A	20 A 120 V	14-20R	20 A 240 V
30 A	30 A 120/240 V 3-Wire	30 A 120/250 V 4-Pole, 4-Wire	6-30R
50 A	50 A 120/240 V 3-Wire	50 A 120/250 V 4-Pole, 4-Wire	6-50R

Goodheart-Willcox Publisher

Figure 37-1. Common straight blade receptacle configurations for residential electrical systems. Extensive charts of receptacle and plug configurations can be accessed with a simple Internet search.

Single receptacle   **Duplex receptacle**

Eaton

Figure 37-2. Standard straight blade receptacles in a duplex and single configuration.

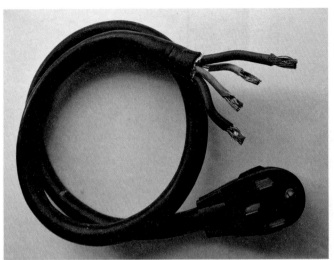

Goodheart-Willcox Publisher

Figure 37-3. This 50-amp, 125/250 volt, four-wire range cord contains both an equipment grounding conductor and grounded neutral conductor.

Selecting a receptacle's current rating should be coordinated with the amount of connected load and the rating of the branch-circuit overcurrent protection device. The *current rating* of the receptacle is the amount of current the device can carry without sustaining damage. A receptacle's current rating must be equal to or greater than the connected load. If the load is supplied by an individual branch circuit, the single receptacle must have a rating greater than the connected load and not less than the rating of the circuit. For example, consider an individual branch circuit that is installed to feed a cord-and-plug-connected microwave. The microwave draws 12 amps and is connected to a 20-amp circuit. A single receptacle installed to supply the microwave is required to have a 20-amp rating because the current rating of the device must not be less than the current rating of the individual branch circuit.

For duplex receptacles and multiple outlet installations, the rating of a branch circuit has an impact on the selection of the receptacle. According to the *Section 210.21(B)*, if a circuit is supplied by a 15-amp overcurrent protection device, the current rating of the receptacle should not exceed 15 amps. For circuits supplied by 20-amp breakers, the current rating of the receptacle is permitted to be either 20 amps or 15 amps. Referring to the earlier example of the microwave, this time consider the microwave is plugged into the 20-amp circuit that feeds the kitchen countertop outlets. The microwave still draws 12 amps and is still connected to a 20-amp circuit. In this installation, the receptacle can be rated at 15 amps because it is not a single receptacle fed by an individual branch circuit.

Utilization equipment, such as microwaves, have an attachment plug that is associated with their current rating. A microwave that draws 12 amps has a 15-amp attachment

plug. If the microwave draws more than 12 amps, such as 16 amps, the equipment will come with a 20-amp attachment plug. Utilization equipment does not typically draw more than it is rated for, so there is limited danger in placing a 15-amp receptacle on a 20-amp circuit because the connected load is unlikely to exceed the rating of the device. The *Code* does not allow placement of 20-amp receptacles on circuits rated for 15 amps, **Figure 37-4.** This is because the potential load exceeds the rating of the circuit and thus can result in an overload condition that may damage the wiring system and devices.

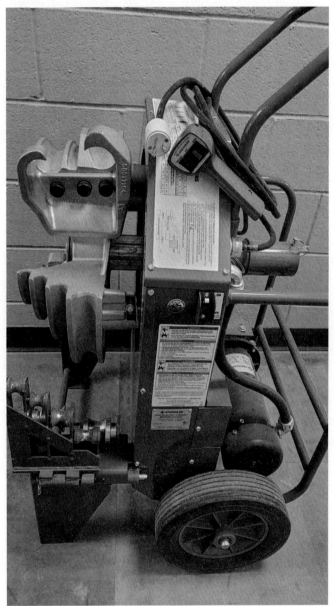

Goodheart-Willcox Publisher

Figure 37-4. The attachment plug of equipment is matched to the expected load. This conduit-bending machine has a 20-amp, 125-volt plug. Notice that one blade of the attachment plug is vertical, and the other is horizontal, which would not allow this plug to be connected to a 15-amp receptacle.

37.1.2 Grounded and Ungrounded Receptacles

When replacing receptacles, consider the grounding pathway and connection. In general, receptacles must be connected to the equipment grounding conductor of the circuit using the grounding terminal on the receptacle. Most newer wiring systems contain an equipment grounding, but some older wiring systems do not contain a grounding pathway. For example, the **knob and tube system**, **Figure 37-5**, is a system of open wires installed on insulating knobs, or two-wire nonmetallic sheathed cable, that does not contain a grounding pathway.

If no grounding pathway exists and an existing receptacle needs to be replaced, there are options for receptacle replacement. The first option is to replace the existing ungrounded receptacle with a new ungrounded receptacle. Note that an **ungrounded receptacle** is a contact device installed at the outlet for the connection of an attachment plug that does not have a means for attachment of an equipment grounding conductor, **Figure 37-6**. According to *Section 406.4(D)(2)*, ungrounded receptacles are only allowed to be installed as replacements. They cannot be utilized for new installations or extensions of existing circuits.

Providing a receptacle with a grounding connection allows for connection of newer electrical appliances that come with attachment plugs containing an equipment ground. Another option for ungrounded receptacle replacement is to provide a ground connection by running an equipment grounding conductor from the premises' grounding system directly to the outlet where a grounded receptacle is used for replacement. If a separate equipment grounding conductor is installed, *Section 250.130(C)* provides installation requirements on conductor sizing, terminations, and physical protection.

Adurable Creations/Shutterstock.com

Figure 37-6. Ungrounded receptacles like this one are permitted to be replaced with new ungrounded receptacles.

It is not always practical to modify the wiring system without damaging building finishes. An option that does not require significant modification is installing a GFCI device. GFCI protection does not provide a ground connection, but it does protect against electrical shock in the absence of a ground connection. A GFCI device can be installed at the outlet where the receptacle is being replaced or an outlet upstream of the device if the wiring feeding receptacles being replaced with a grounded type are protected by the GFCI. *Section 406.4(D)(2)(b)* requires a warning label to be provided at the protected receptacle locations stating the lack of an equipment ground, **Figure 37-7**.

There are special grounding considerations for replacing receptacles for electric ranges and dryers. Existing systems often use three-wire circuits to supply power to these appliances. The three-wire circuit does not contain an equipment grounding conductor. Since no equipment grounding connection is supplied through the wiring system, *Section 250.140* allows the frame of the appliance to be connected to

Alessandro Cancian/Shutterstock.com

Figure 37-5. Knob and tube wiring is a system of open conductors supported by insulators, also known as knobs. If the conductors pass through framing members, they are installed in a porcelain sleeve known as a tube.

Goodheart-Willcox Publisher

Figure 37-7. When GFCI protection is provided at replacement locations of ungrounded receptacles, the cover plate must be clearly marked with the warning statement "no equipment ground."

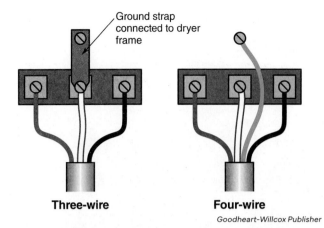

Goodheart-Willcox Publisher

Figure 37-8. Electric dryers and ranges in existing locations are permitted to be supplied by three-wire systems not containing an equipment grounding conductor. The frame of the appliance is connected to the grounded neutral conductor through the use of a bonding jumper or strap.

provide the equipment with a pre-wired attachment cord because there are too many installation variations.

PRO TIP   **Connecting New Appliances**

You may be asked to connect a new appliance for a homeowner. If the branch circuit contains an equipment grounding conductor, install a four-wire cord and receptacle. If the branch-circuit does not contain an equipment grounding conductor, install a three-wire cord and receptacle.

37.1.3 Split-Circuit and Switched Receptacles

Duplex receptacles are used in special installations such as split circuit and switched receptacle applications. A ***split-circuit receptacle*** is a duplex receptacle that has each receptacle outlet connected to a different circuit. In the past, split-circuit wiring was common in residential circuits feeding kitchen countertops. *Section 210.11(C)(1)* requires the installation of at least two circuits to serve the countertop outlets. The split-circuit installation allows for each of the two circuits to be available at every counter outlet. This makes it possible to plug in two appliances to the same receptacle and have them fed from two different circuits. Once GFCI requirements came into place for kitchen countertop outlets, split-circuit receptacle installations became rare.

In split-circuit installations, the two halves of the duplex receptacle are not electrically connected. The halves are separated by removing the small metal tab between the termination screws on the device, **Figure 37-9**. If the device is part of a multi-wire branch circuit, the tab is removed between the terminals for the ungrounded conductor.

the grounded (neutral) conductor. This connection is made by installing a bonding jumper or strap between the frame and grounded conductor terminal inside the appliance, **Figure 37-8**. The bonding jumper ensures a pathway for fault current and reduces the electrical shock hazard for the appliance. This technique is only allowed on existing installations but can still be installed for receptacle replacements if the existing wiring remains in place.

New installations are no longer allowed to be supplied by an ungrounded receptacle. The wiring system for circuits supplying ranges and dryers must contain an equipment grounding conductor. The grounding terminal of the receptacle is connected to the equipment ground from the supply circuit. The attachment cord for the appliance also contains an equipment grounding conductor, which is terminated on the appliance frame. Appliance manufacturers do not

Goodheart-Willcox Publisher

Figure 37-9. Removing the tab between the top and bottom half of this receptacle will separate them electrically allowing them to be connected to different circuits or to control one of the halves with a switch while leaving the other half energized at all times.

If each half of the receptacle is connected to a separate two-wire circuit, both the ungrounded and grounded tabs need to be removed. Failure to remove a tab between the circuits can lead to short-circuit conditions, which may cause fuses to blow or circuit breakers to trip.

A *switched receptacle* is a duplex receptacle in which half of the duplex receptacle is connected to an unswitched circuit, and the other half is connected to a switch. *Section 210.70(A)(1)* requires habitable rooms, kitchens, laundry areas, and bathrooms to be equipped with a wall-controlled switch. If the room does not have a permanently installed light, switching the receptacles will meet the requirements of the *Code*. Lamps can be plugged into the switched half of the receptacle that is controlled by the wall switch. The other half of the receptacle provides the outlet to serve the wall space, as required by *Section 210.52(A)*.

In switched receptacle installations, the two halves of the duplex receptacle are not connected in tandem with one another. Removing the tab between the ungrounded conductor terminals separates the switched half from the unswitched half. Failure to remove the tab between the halves back feeds power to the switched conductor, regardless of the switch's position. Thus, the receptacles no longer can be controlled and are energized at all times. If you or a homeowner replaces a duplex receptacle, and there is a loss of switch control within the circuit, it is possible that someone neglected to remove a tab. The tab for the grounded conductor should stay in place for switched receptacle applications. If the tab is removed, the grounded conductor connection for half of the receptacle will be open, and anything plugged into that half will have 0 volts available.

37.1.4 GFCI and AFCI Replacements

When replacing receptacles in existing installations, consider the physical location to determine if GFCI or AFCI protection may be required. Receptacles installed near water, in kitchens, in bathrooms, or in unfinished basements often require GFCI protection. If the existing receptacle does not have protection, but the location is now required to be protected, a GFCI receptacle must be installed during replacement.

The *NEC* has greatly expanded the locations where AFCI protection has been required over the last few decades. If an existing dwelling unit was built before 2002, there is a high likelihood that AFCI protection was not required at the time of construction, but it may be now. *Section 406.4(D)(4)* requires all receptacles to be replaced in locations that require protection to be AFCI protected. You have the option of providing that protection by installing an AFCI circuit breaker, an AFCI receptacle upstream of the device, or an AFCI receptacle, **Figure 37-10**, as the replacement device.

37.2 Installing a New Receptacle Outlet Location

Installing a new outlet location is a common service provided by electricians in dwelling units. You must decide if you plan to feed the new location by installing a new circuit or extending an existing one. Specialized boxes are used to provide openings for receptacles cut into existing room finishes.

37.2.1 Feeding New Outlets

There are several things to consider when determining if an existing circuit should be extended to a new outlet. The load on the existing circuit should be assessed first before adding additional outlets. Too much load on a circuit causes overload conditions that can lead to dangerous overheating and possibly fire. If the new outlet is supplying a large load or the existing circuit is already heavily loaded, a new circuit should be run to the new outlet.

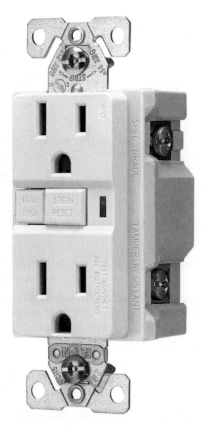

Eaton

Figure 37-10. AFCI receptacles can be used to provide protection when replacing receptacles in locations that now require AFCI protection.

In addition, you must always consider the equipment being connected. Some types of utilization equipment are required to be fed by a dedicated branch circuit based on the manufacturer's installation instructions. Equipment must be installed in accordance with its listing, labeling, and instructions in compliance with *Section 110.3(B)*.

Grounding may factor into the decision to extend an existing circuit or install a new one as well. If there is no equipment grounding conductor in the existing wiring method, it may be best to provide a new wiring method containing one.

Finally, the location of the new outlet must factor into the decision-making process. If the new outlet is installed on the second floor and the load center is in the basement, you must find a pathway through the finished first floor. Basements with finished ceilings, attics with solid floors, and a host of other structure-related issues impact the accessibility of the new location.

37.2.2 Cutting in Boxes

When new receptacle outlets are added to an existing dwelling unit, the opening for the outlet must be cut into finished walls. Cutting in a box seems like an easy task, but it can be complicated by building structure, insulation, and finish materials.

Care must be taken to identify a location for the new box that is within the void space of the wall. The location of a building's structural elements, such as studs, can be found through several different methods. Specialized tools are used to identify the position of the studs behind the surface. Basic stud finders are the most common type of tool used for this job, but digital wall scanners that can detect wood, metal, and wiring are also available. If you are cutting into sheetrock but do not have a stud finder or scanner available, you can use a tapping method to determine the approximate location of the studs.

PROCEDURE

Finding a Stud using the Tapping Method

The tapping method is used to determine the approximate location of studs found in a building's structure. The following steps describe this method:

1. Knock on the wall surface with short quick taps. Listen to the sound produced by the tap.
2. Continue to knock with short quick taps as you move horizontally across the wall surface.
3. Listen to the noise your knocking makes. The sound that is produced by the taps changes as you move closer to and farther away from the studs—the sound will be dull and sound hollow over the void space.
4. Continue to tap as you move across the surface. As you move toward a stud, the sound will be sharper and denser.
5. Once the location of the stud or joist is determined, repeat the steps in the opposite direction to identify the void between the framing members where you will cut in the box.

Insulation can be a challenge when cutting in boxes. Any insulation filling the void can make it more difficult to fish wire to the location. If possible, locate the box in an interior wall that does not contain insulation. When it is necessary to cut into a wall that contains insulation, you must take care to prevent damage to the integrity of the insulation. Some older homes contain *vermiculite insulation*, or a pebble-like mineral product used to fill the wall cavity, **Figure 37-11**. Be careful when cutting into a wall containing vermiculite because the insulation may come pouring out of the opening, leaving the top portion of the wall cavity no longer filled.

SAFETY NOTE   **Asbestos Exposure**

Some types of wall coverings and insulations, such as vermiculite, may contain asbestos fibers. Asbestos is harmful to humans when the fibers are inhaled. Disturbing products like vermiculite may cause the fibers to become airborne, and if the proper protection is not in place, the airborne fibers may be inhaled.

Firn/Shutterstock.com

Figure 37-11. Vermiculite insulation is inserted into wall cavities of some homes. This pebble-like insulation can be difficult to work around, as it will pour out of a hole cut into the wall cavity. Some vermiculite contains asbestos, which can be a health hazard when electricians are exposed to it.

Igor Meshkov/Shutterstock.com

Figure 37-12. Spray foam insulation expands to fill the voids in exterior walls and is very difficult to fish conductors through for new light and outlet locations.

Fiberglass insulation may also be contained in the wall cavity. Fiberglass can be an irritant to the lungs and skin, so take proper precautions when working around the insulation.

Many exterior walls contain a paper or plastic vapor barrier on the backside of the finish material. A *vapor barrier* is an unbroken layer of material that prevents moisture from transferring from one side of the barrier to the other. Damage to the vapor barrier can result in moisture buildup due to condensation. Be sure to avoid damage to the vapor barrier. Special gaskets can be used behind the finish plate of newly installed devices to maintain the vapor barrier across the opening.

Newer homes may have wall cavities filled with spray foam insulation, **Figure 37-12**. Spray foam expands and fills the gaps and voids within the wall cavity, leaving a solidly filled cavity that is very difficult to fish wire into.

The materials used in a wall's construction can also have an effect on the process of cutting in a box. Walls constructed with wood lath covered in plaster are much more difficult to cut into than walls constructed of sheet rock. *Lath and plaster*, a system of wood slats affixed to the wall framing and covered with a layer of plaster, was a common interior wall finish before the use of sheetrock became common.

Plaster can be hard and brittle, **Figure 37-13**, so you may need to be much more careful during the cutting process. Areas of plaster much larger than the intended opening can come loose or crack due to the movement of the wood lath behind the plaster. Many electricians use an oscillating tool

Chris Rokitski/Shutterstock.com

Figure 37-13. Lathe and plaster is a common wall covering in older homes. Care must be taken when cutting new openings in the plaster because it can become brittle, and it can become loose and fall off requiring a lot of repairs.

with a plunge cutting blade to cut openings in plaster. Plaster dulls cutting tools very quickly. To extend the life of your blades, it is recommended to remove the plaster first before cutting the wood lath.

Some plaster installations also use metal mesh near door openings or at the intersection of the ceiling and wall. After removing the plaster, you may need to cut the mesh with diagonal pliers before using a wood cutting blade to remove the lath.

Cutting a Box into Lath and Plaster

Use the following procedure to locate and cut in a box in a lath and plaster wall:

1. Determine the general location of the desired opening, then mark the location of the studs.
2. Leave plenty of clearance around the studs.
3. Hold the box level and plumb at the desired location.
4. Trace the box opening onto the wall surface.
5. Score the lines of the opening with a utility knife.
6. Using a chisel, gently break the paster while holding the chisel about 1/8″ to the inside of the scored line.
7. Once the paster is removed, use an oscillating tool to cut the lath.
8. Fish wire to the new opening.
9. Install the box in accordance with the manufacturer's installation instructions.

The process of cutting in a box in sheetrock is similar to that of lath and plaster. *Sheetrock* is a sturdy product made of gypsum sandwiched between two layers of paper. Compared to lath and plaster, sheetrock is easier to cut. Most electricians use a keyhole saw to do the job. You can use an oscillating tool to cut openings in sheetrock, but you are more likely to damage hidden obstructions, like existing wiring.

Checking for Hidden Obstructions

Before cutting in a box, it is advantageous to check for obstructions behind the sheetrock or plaster. Once the box location has been marked out in pencil, choose a point near the middle of the box, and cut or drill a 1/4″ hole through the surface material. Once the hole has been made, insert an offset screwdriver shaft into the void to a point where the shaft can be held parallel with the surface. Slowly rotate the screwdriver to feel for unseen obstructions in the space, **Figure 37-14**.

37.2.3 Selecting the Proper Box for Installation

You must be able to select the right box for a specific job. When cutting into an existing wall finish, old work boxes are used. Old work boxes are specifically designed to be

Goodheart-Willcox Publisher

Figure 37-14. An offset screwdriver can be inserted at the center of an opening that will be cut-in. Once inserted, the screwdriver can be rotated 360 degrees to determine if there are any obstructions in the wall before cutting the opening.

installed through the wall finish and can be supported by the wall itself. Recall that boxes come in two main categories: nonmetallic and metallic, **Figure 37-15**. Nonmetallic boxes are widely used when the wiring method is nonmetallic sheathed cable. They are easy to install and often contain a set of tabs that secure the box to the wall surface. They also have integral means of retaining the cable to the box that does not require an additional connector. One limitation of nonmetallic boxes is that they are not allowed to be used for metallic cabling methods.

Metallic boxes are typically used with metal sheathed cables but can also be used with nonmetallic sheathed cables. The most common metallic old work box is a gem box. A *gem box* is a metallic box that has a set of adjustable ears to hold the box flush with the surface after installation and removable sides to allow multiple boxes to be combined or ganged for multiple device openings. Gem boxes come in a variety of depths to accommodate wall depth, wire fill, and the device being installed. They can be purchased with internal cable clamps that can accommodate either metallic cables or nonmetallic sheathed cables. Gem boxes without clamps are also available, but a cable connector is required to be added to the box to secure the cabling entering the box.

Most gem boxes are secured in the opening using metal box supports, a pair of thin metal strips shaped like the

Old Work Boxes

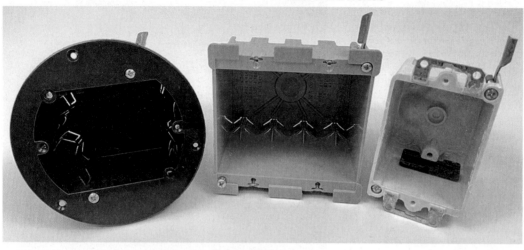

Nonmetallic boxes

Metallic box

Goodheart-Willcox Publisher; Eaton

Figure 37-15. Old work boxes are produced in both metallic and nonmetallic materials. Nonmetallic boxes are often used with nonmetallic cabling methods and metallic boxes can be used with both metallic and nonmetallic cabling. Metallic cut-in boxes are often supported using a box support.

Greek letter π that creates tension between the box and wall surface to hold the box in place. These supports are inserted on each long side of the box in the opening between the wall material and the side of the metallic box. Two metal tabs are then folded over the edge of the box, securing it in place.

37.3 Replacing Existing Luminaires

Replacing a luminaire is not always as easy as a quick swap. There are many aspects of the installation to consider. The lighting technology must also be selected based on factors such as efficacy, color, and lamp life. For instance, the light needs to be properly supported to the box, and the box must be supported to the structure of the building. Refer to Chapter 12, *Lamps and Lighting*, for a review of these lighting factors.

37.3.1 Selecting Light Technology, Color, and Intensity

Selecting lighting technology often comes down to customer preference, but it may make a big difference in the power consumption of a dwelling unit. Lamp efficacy is one factor to consider. Recall that the higher the efficacy of the lamp technology, the less energy it takes to create higher light output. LED technology has very high efficacy and has become a very popular lighting choice. Incandescent lighting has very low efficacy due to the wasted energy converted to heat during the lighting process. Compact fluorescent lighting (CFL) technology has an efficacy near the middle range of LED and incandescent.

Homeowners also must consider the color of light given by a lamp when selecting a light technology. The color of the light can affect the ambiance of the room. Recall that the color temperature of light given off by a lamp is expressed in degrees Kelvin. Typical incandescent lamp color is around 2700 K and gives a warm, yellow tone. A daylight LED might be rated at 6500 K and gives off a white or blue tone, **Figure 37-16**.

The output color for fluorescent technology is based on the lamp selected. Multiple colors are available for standard lamp styles and sizes. Output color for LED technology is specific to the luminaire. Some LED lights come with a specific color rating, while others have a selector switch that allows the customer to choose the color that fits their needs. Incandescent lamps typically have a 2700 K color output. Halogen lamps have a whiter color than standard incandescent lamps with a temperature of 3000 K.

Lamp life is another important consideration when selecting a replacement luminaire. Incandescent lamps have the lowest life span, meaning that they will need to be replaced at the least number of hours of service. CFL technology is the next highest lamp life, with about 10 times the life of an incandescent. LED lamps have the longest lamp life at up to 50,000 hours, or 25 times the life of an incandescent.

The intensity or brightness of a light may also weigh in on the selection of a lamp. Intensity is the total amount of light given off by a lamp measured in lumens. When replacing a fixture, it is important to know if the customer is looking for a similar level of brightness, more light, or less light. If you are comparing the brightness of the existing lamp to a new LED fixture, compare the lumen output of a similar wattage lamp of the same technology to the lumen output of the LED.

Figure 37-16. The color of light given off by a luminaire can vary as the color temperature increases. The light on the left is in the 2700 K range and the light on the right is in the 6500 K range.

37.3.2 Installation and Support

Luminaires can become an electrical hazard if they are not installed properly. Nationally recognized testing laboratories test products and provide them with a listing. This testing process is rigorous and ensures that the equipment performs well enough for it not to be a hazard when installed in accordance with the manufacturer's instructions. Because many homeowners today are looking to replace incandescent lighting with LED, you should always use the listed retrofit kits in the conversion of luminaires from incandescent or fluorescent to LED.

Section 110.3(B) requires equipment to be installed in accordance with the manufacturer's installation instructions. Many luminaires come with mounting brackets, fixture studs, and hardware. The installation instructions will answer questions about using all the provided hardware or if you can simply connect the fixture to the box using a wood screw. If a luminaire is not properly supported, it can fall. This could result in potential property damage, dangerous short-circuit conditions, or an impact injury.

Ceiling-suspended paddle fans installed in dwelling units tend to be the biggest culprit of failed support. The weight and motion of the fans contribute to the failures. Imagine going to sleep with a paddle fan running above your bed. You wake up in the night to realize the fan was not supported properly, and the still spinning fan and box are swinging wildly by the nonmetallic sheathed cable feeding the box. This scenario is the reason why boxes that support paddle fans are required to be listed for the purpose.

When removing a luminaire and replacing it with a fan, you will need to ensure the box is designed for the purpose. Boxes that are listed for fan support will be labeled as rated for fan support. If the existing box does not contain a label, the existing box should be removed and replaced with a rated one, **Figure 37-17**. *Section 422.18(A)* allows for a fan to be directly supported to the structure independently of the box. Structural framing does not typically allow for this type of installation, so one option is to use an expanding bracket that can be installed through the ceiling opening and secured to the joists or rafters on each side of the cavity.

Lisa F. Young/Shutterstock.com

Figure 37-17. A fan-rated box is installed at an existing lighting outlet.

A fan-rated box is then supported by the bracket and can be used to support the fan.

37.4 Adding New Luminaires

There are similar requirements for adding new luminaire locations as replacing luminaires. In addition to the information discussed earlier in this chapter, further examination is needed to determine the right luminaire type to achieve the best light distribution, ensure air sealing of the building envelope, and properly support the luminaire.

37.4.1 Light Distribution

Light distribution from luminaires is dependent on the style and construction of the fixture and the installed lamp. If a homeowner asks about adding a light in their home, you will need to understand the intended purpose of the light. Lighting is classified by the purpose and direction of light emitted. *General illumination* provides a baseline light level across a space. An example of general illumination is a single overhead luminaire in the middle of a bedroom ceiling.

Task and accent lighting, both covered in Chapter 12, *Lamps and Lighting,* must also be considered. An example of task lighting is undercabinet lighting intended to provide light for the kitchen countertop surface, while accent lighting is often used in dwelling units to provide illumination of artwork, fireplaces, and cabinetry.

The light direction from a luminaire can also play into a customer's luminaire preference. A luminaire that focuses 90% or more of the light produced in the downward direction is considered *direct lighting*. Direct lighting is a great choice for task lighting applications. A luminaire that focuses 90% or more of the light produced in the upward direction is considered *indirect lighting*. Wall-mounted sconce fixtures are often used as a form of indirect lighting and contribute to general illumination and accent lighting. A luminaire that allows light to spread in all directions is categorized as a *general diffuse light.* The lights mounted to the bottom of a ceiling-suspended paddle fan are a good example of general diffuse light that contributes to general illumination of a room.

Luminaires are either surface-mounted, suspended, or recessed. Surface-mounted and suspended luminaires are typically used for general illumination and task lighting, while recessed lighting is used for general illumination, task lighting, and accent lighting. Recessed luminaires come with a variety of trim kits that allow light to be directed downward or angled toward a wall or object. Eyeball trims are most often used for accent lighting, and baffle trims are typically used for general illumination and task lighting.

The type of lamp installed into a recessed trim depends on whether general or task lighting is needed. Flood lamps create a widespread pattern and are most often used in general illumination applications. Spot lamps are designed to create a light spread from 10 to 30 degrees so the light can be focused on a specific area. Spot lamps are often used for task and accent lighting. LED lamps and luminaires are also available with a variety of light spread characteristics.

One of the most popular types of recessed downlighting on the market today is the wafer light, **Figure 37-18**. Wafer lights have a similar look and light output to traditional recessed incandescent lighting but have the advantage of a long-life cycle, reduced power consumption, and more flexible installation. A typical recessed incandescent luminaire has a large housing that is recessed within the void space above the ceiling finish. The size of the housing limits the placement of the luminaires due to structural

members, duct work, plumbing pipes or other obstructions. Wafer lights are often less than, or the same depth as, the ceiling finish material. Because of this limited depth, they can be placed with a portion of the luminaire over a framing member or other obstruction. Wafer lights have a termination box that can be placed in the ceiling void. If a fixture is installed through the existing building finish, the termination boxes are not required to be supported in most cases. When these types of lights are placed in exposed framing where the building finish has not yet been applied, some manufacturers require the luminaire to be supported by a metal bracket.

37.4.2 Air Sealing of the Building Envelope

Any time a new lighting opening penetrates the building envelope or vapor barrier, the opening needs to be sealed. The vapor barrier is typically installed between the heated spaces of a home and the home's exterior walls and ceilings. Any break in the barrier allows warm moist air to mix with cold air, resulting in condensation and leakage of conditioned air into unconditioned spaces. Recessed luminaires designed for installation in these environments will be labeled *airtight*. Airtight recessed lighting should be supplied for locations in spaces with unheated attics above. If the luminaire is installed in an area where insulation is closer than 3″, it must be rated for IC. *Insulation contact (IC) rated* recessed fixtures are rated for contact with combustible materials and insulation because the design of the luminaire limits fire hazards associated with heat buildup.

37.4.3 Supporting Luminaires

Proper support of luminaires is critical for their safe installation. In general, boxes are rated to support fixtures up to 50 lb. If a large fixture, such as a chandelier, weighs more than 50 lbs., *Section 314.27(A)(2)* requires the luminaire to be supported independently or with a listed box or support system. *Section 314.27(C)* requires boxes to be fan-rated if they are installed for lighting outlets in ceilings of habitable rooms and the locations can potentially permit the installation of a ceiling suspended paddle fan.

Vadim Shkarbul/Shutterstock.com

Figure 37-18. A wafer style recessed LED downlight is a versatile option for installing new lighting openings in existing ceilings. This wafer light has a depth similar to a typical 5/8″ sheetrock wall covering.

Summary

- Receptacles are available in multiple configurations and are rated based on circuit voltage and current.
- The voltage rating on residential-grade receptacles is typically 125 volts, 250 volts, or 125/250 volts. A 125-volt rated receptacle is routinely used on a 120-volt nominal electrical circuit.
- Ungrounded receptacles may be replaced with ungrounded receptacles, replaced with grounded receptacles protected by GFCI devices, or be served by an individual equipment ground.
- Three-wire 125/250-volt receptacles for ranges and dryers can only be used on existing branch circuits that do not contain an equipment ground.

- Breaking the tabs on a duplex receptacle will allow each half of the receptacle to be controlled or fed separately.
- A split-circuit receptacle is a duplex receptacle that has each receptacle outlet connected to a different circuit. A switched receptacle is a duplex receptacle in which half of the duplex receptacle is connected to an unswitched circuit and the other half is connected to a switch.
- Receptacles installed near water, in kitchens, in bathrooms, or in unfinished basements often require GFCI protection. *Section 406.4(D)(4)* requires all receptacles to be replaced in locations that require protection to be AFCI protected.
- If the new outlet is supplying a large load or the existing circuit is already heavily loaded, a new circuit should be run to the new outlet. Grounding may factor into the decision to extend an existing circuit or install a new one as well. Finally, the location of the new outlet must factor into the decision-making process.
- Cutting in a box seems like an easy task, but it can be complicated by building structure, insulation, and finish materials. The materials used in a wall's construction can also have an effect on the process of cutting in a box.
- Old work boxes should be used when openings are cut into existing finishes.
- Metallic boxes should be used with metallic wiring methods.
- Nonmetallic boxes should be used with nonmetallic wiring methods.
- Fan-rated boxes are required when supporting ceiling-suspended paddle fans.
- Luminaires should be selected based on efficacy, color temperature, intensity, purpose, and desired light direction.

Know and Understand

1. An existing 20-amp branch circuit feeds a single outlet for an air compressor in a residential garage. The compressor draws 9 amps at full load. A _____ 125-volt receptacle is required for the replacement.
   A. 20 amp
   B. 15 amp
   C. 15 or 20 amp
   D. 10 amp

2. *True or False?* A receptacle's current rating must be equal to or greater than the connected load.

3. A cord is being connected to a new electric dryer. The existing branch circuit does not contain an equipment grounding conductor. The frame of the dryer should be connected to the _____.
   A. equipment grounding conductor
   B. service grounding electrode
   C. grounded conductor
   D. ungrounded conductor

4. *True or False?* New installations are allowed to be supplied by an ungrounded receptacle.

5. A(n) _____ condition can occur if there is a failure to remove the tab between the ungrounded terminations of a duplex receptacle being used in a split circuit application.
   A. short-circuit
   B. ground-fault
   C. overload
   D. inrush

6. Which of the following is *not* a factor in determining if a new circuit should be installed for an additional outlet or if the existing circuit can be extended?
   A. Accessibility to the new opening from the load center
   B. Convenience
   C. Load on the existing circuit
   D. Equipment manufacturer's installation instructions

7. The recommended tool for cutting through wood lath for electrical openings is the _____.
   A. sawzall
   B. keyhole saw
   C. jigsaw
   D. oscillating tool

8. Boxes specifically designed to be installed through the wall finish and can be supported by the wall itself are referred to as _____.
   A. old work boxes
   B. remodel boxes
   C. repair boxes
   D. surface boxes

9. Standard incandescent lamps have a color rating of _____.
   A. 6500 K
   B. 5000 K
   C. 2700 K
   D. 2100 K

10. When choosing a light technology, efficacy effects power consumption, _____ has the highest efficacy amongst the common residential lighting technologies.
    A. CFL
    B. incandescent
    C. halogen
    D. LED

Apply and Analyze

1. Explain the hazard created by bonding the frame of an electric range to the grounded conductor if an equipment grounding conductor is also supplied to the range.

2. What can be done to ensure the continuity of the vapor barrier when cutting openings into the building envelope?

3. Why is it recommended to remove the plaster from an opening before cutting through the wood lath?

4. What type of box would you select if you were cutting a receptacle outlet into a sheetrock wall and feeding it with nonmetallic sheathed cable?

5. What factors contribute to the intended purpose of down lighting from recessed luminaires?

Critical Thinking

1. A customer is requesting an additional outlet to be added to serve the top surface of a counter in an exterior wall. Directly below the desired location is an existing ungrounded outlet. The load center is in the basement on the opposite end of the house. The homeowner indicates that the walls are insulated with vermiculite insulation. What considerations would you make as you begin to plan out the installation?

2. A customer would like to add some task lighting above a new kitchen island that measures 3′ by 10′. The homeowner prefers the yellow glow of traditional incandescent light. Make a recommendation for lighting the island. The recommendation should include the number and spacing of lights, the lighting technology, the luminaire type, the lamp type, and the light disbursement.

Ideal Industries

SPARKING DISCUSSION

A client wants to add a dedicated circuit to an existing bathroom on the second floor of their home. When assessing their situation, what should you look for? What potential issues would you discuss with the homeowner before starting the work?

Introduction

There are many considerations to review before the addition of a circuit in an existing dwelling:

- Determine the location for routing the circuit conductors within the building structure.
- Select the proper fishing tools and methods based on the circumstances of the installation.
- Protect the conductors from damage and properly support them.
- Examine the panelboards to determine the proper circuit breakers used for the supply of the new circuit.
- Determine if the panelboards are at capacity and whether adding additional breaker space through special equipment or a subpanel is necessary.
- Assess the service to determine if the additional load can be accommodated.

This chapter will cover these important considerations, as well as walk you through the steps to complete each of these tasks.

38.1 Fishing Conductors

Fishing conductors through concealed spaces is a skill developed over time with proper practice. An experienced electrician is aware of possible routes through an existing structure, such as openings around chimneys and plumbing vents. They can select and utilize the proper fishing tools to reduce the need for creating access openings that will later have to be repaired. They also ensure that added circuit conductors are protected from damage and properly secured to the building structure.

38.1.1 Fishing Tool Best Practices

Selecting the correct fishing tool depends on what type of application you are fishing through. For many years, steel fish tapes were the standard tool for the job. Recall from Chapter 3, *Hand Tools*, that a fish tape is a coil of steel, nylon, or fiberglass tape used as a tool for the installation of electrical conductors in raceways, wall cavities, and concealed spaces. They are stiff enough to be pushed through a wall containing insulation, yet flexible enough to bend and move around objects in the wall. When using fish tape, pay close attention to the head of the fish tape and the bend of the tape.

Fish tapes come with a formed head that includes a loop of material for the attachment of conductors or cables, **Figure 38-1.**

When fishing in walls or other tight spaces, it is recommended to close off the head by applying some electrical tape. Electrical tape can help prevent the fish tape from hanging up within the wall. Sometimes the manufactured head becomes deformed or breaks off and needs to be replaced. You can bend a new loop on the end of the tape using a linesman's pliers.

When fishing an insulated wall from the basement to the attic, leaving an open loop on the end of the fish tape can be advantageous. For example, a hole is drilled at the top and bottom of the wall. Pushing the fish tape directly from the top hole through the bottom hole is an almost impossible task. Instead, pushing a fish tape with an open end into the wall from the top and bottom allows the two fish tapes to

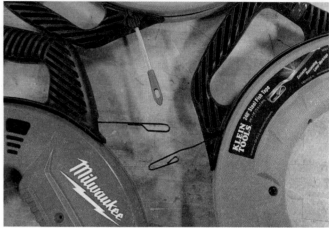
Goodheart-Willcox Publisher

Figure 38-1. The steel fish tapes shown here have a preformed hook, while the fiberglass fish tapes have a nonconductive eyelet.

| PROCEDURE | **Bending a New Loop on a Metal Fish Tape** |

It sometimes becomes necessary to replace the manufactured head on a steel fish tape. Use the following steps to create a head that is more likely to keep its shape and less likely to become brittle and break, **Figure 38-2.**

1. Using linesman's pliers, cut the fish tape to create a clean end.
2. Turn the pliers so they are in the same plane as the tape. Grasp the last inch or so of the tape and create a 15- to 30-degree bend.
3. Heat the last 3″ of tape with a torch. The process of gently heating and slowly cooling allows you to use

less force to create a tighter bend without making the tape brittle.
4. Grab the tape with the pliers just behind the first bend. Bend a loop in the tape until the portion of the tape that was previously bent comes to rest against the straight fish tape.
5. Move the pliers to hold the bent section against the straight section until the tape cools.
6. Wrap the section where the end meets the straight section with electrical tape.

Bending a New Loop on a Metal Fish Tape

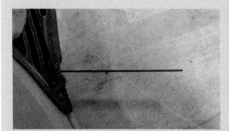
Step 1. Cut fish tape to create a clean end.

Step 2. Bend to 15 to 20 degrees.

Step 3. Heat the last 3″ of the tape.

Step 4. Bend a loop and hold until cool.

Step 5. Wrap with electrical tape.

Completed loop.
Goodheart-Willcox Publisher

Figure 38-2. Forming a new hook on a steel fish tape.

pass one another in the wall. Rotating one or both fish tapes allows you to potentially hook one fish tape with the other, slowly pulling it back out of the hole. The open head allows the two fish tapes to hook together and enables you to pull one tape trough both holes and the wall section.

Fish sticks are used in a similar fashion as fish tapes. Recall from Chapter 3, *Hand Tools*, that fish sticks are flexible fiberglass rods that are threaded together and used to install electrical cables through concealed spaces. Fish sticks are more rigid and less flexible than fish tape. They are much better than fish tape for fishing an attic space. For example, if you cut a series of recessed lights into a ceiling that has an attic above, a fish stick can be directed from light opening to light opening. A steel fish tape has too much flexibility and tends to curl up above the ceiling, making it difficult to fish an open space like an attic. Fish sticks can also be used in walls like fish tapes, but they are much less flexible and often get stuck on obstructions within the wall. Metal fish tapes bend around those objects, while fish sticks will not.

Due to their stiffness, fish sticks can also be pushed in a straighter line. They come in fixed lengths, so as additional sections are added, you can get a better sense of the length of the fish stick within the space. Fish sticks also come with different end attachments, most commonly a bullet nose or hook, **Figure 38-3**. When fishing in a concealed location, such as a wall, it is best to use the bullet nose until you get it to the desired location. Then, you can remove the bullet and insert the hook to pull the cable within the space.

In insulated walls, chain is most often used as a fishing tool. Jack chain can be very effective in the right circumstances and is typically limited to applications where gravity can be applied. A *jack chain* is a metal link chain made of bent loops, often shaped in figure eight pattern, and used for supporting light objects like suspended luminaires.

When fishing from an attic into an uninsulated wall, using a jack chain is easy and effective. You can lower a section of jack chain through the opening in the top of the wall and then simply reach into the stud cavity and grab the chain.

Lightweight string or cord can also be used in a similar application by tying a small weight or hardware nut to the end of the string and allowing gravity to move the string down into the space.

Chain is also a method for fishing spaces along chases or pipes to find pathways between attics and basements of two-story homes. The chain can be lowered from the attic through an inaccessible wall. If the chain gets caught in the wall, an experienced electrician can feel the difference in the weight. If you bounce the chain up and down, that helps it fall through openings or around objects. You can also drop a chain into a wall and insert a fish tape from the bottom, rotating the fish tape to hook the chain.

38.1.2 Cable Routing, Support, and Protection

Adding new circuits to an existing home requires routing cables within the finished structure. Finding routes from the source of electricity, usually in the basement, to the utilization equipment can be a challenge. Routes for installation may include chases, chimneys, and plumbing vents. In cases where no practical route can be found inside the home, you may decide to run a conduit on the exterior surface.

Chases are voids in the building construction, allowing for the installation of pipes, wire, and heating and cooling ducts vertically through the building. These spaces offer a pathway from one area of the home to another without disrupting the building's finishes. Soffits that are above cabinets, around beams, or installed for decorative purposes can also be used as a wire chase.

Chimneys were used for venting fuel-burning appliances, like furnaces and water heaters, in older homes. The fuel-burning appliances were usually installed in the basement of a home, and the chimney extended all the way through the house and out the roof. There are often gaps in the framing around the chimney that may allow for the installation of cables.

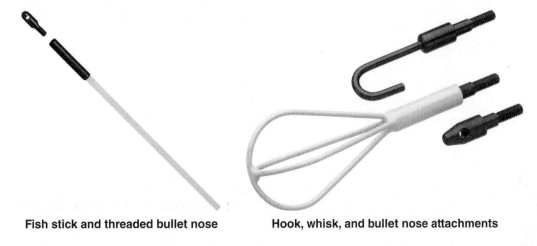

Fish stick and threaded bullet nose   Hook, whisk, and bullet nose attachments

Ideal Industries

Figure 38-3. Fish sticks come in joinable lengths and have interchanging tips.

Plumbing systems also require venting. Vent stacks extend through the framing and out through the roof of a home, **Figure 38-4**. Plumbing vent stacks are often large-diameter pipes. In older homes, plumbers likely cut square holes through the framing materials and inserted round pipes. This leaves a gap in the corner of the opening that can provide a route for cables from a basement to an attic. However, newer homes may not have these gaps since hole saws are often used to drill round holes for the vent stacks.

Always use caution when routing cables through or near return air spaces. Forced air heating and cooling systems use return air pathways to bring environmental, or outside, air back to the heating and cooling equipment, where it can be conditioned and forced through ductwork back into the home. Wall-stud and floor-joist cavities are sometimes used for return air pathways instead of installing a return air duct, **Figure 38-5**. In general, the *NEC* does not permit installing nonmetallic sheathed cable or other nonmetallic wiring methods in spaces used for environmental air. According to *Section 300.22*, cables are allowed to pass through return air cavities but only if they are run perpendicularly to the long dimension of the cavity. When cutting in boxes or finding routes for cables, verify that the wall or floor cavity is not used as a return air. Return air cavities in walls can be identified by visually inspecting for grilles that allow air from the space to enter. Return air cavities in floor joist systems

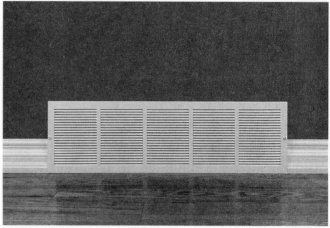

Claude Huot/Shutterstock.com

Figure 38-5. Cables are not permitted to be installed in parallel with floor joists and wall-studs inside spaces used for environmental air. This return air grille provides an indication that the stud spaces above this grille should be avoided for cutting in boxes and routing cables.

can be identified by looking for sheet metal installed across the joist face used to enclose the cavity.

Cable protection and support are other challenges when routing cables to add a new circuit. *Section 334.30* requires nonmetallic sheathed cables to be supported at intervals not exceeding 4 1/2′. Any cable installed in open attic spaces and basements must be supported in accordance with the *NEC*. A listed exception to *Section 334.30* is when cables are fished through concealed spaces. Any wiring within the concealed space can remain unsupported.

Cables also need to be protected from physical damage. *Section 320.23* requires any cables routed within 7′ of an attic opening to be installed parallel to the framing members. If the cables are installed perpendicularly to the framing members and are run across the face of the rafters, they will need to be protected by a guard strip, **Figure 38-6**. The guard strip protects the cables from being stepped or kneeled on when someone is accessing the attic space.

Conductors run in exposed areas of basements should be protected. *Section 334.15(C)* permits cables to be supported by the floor joists, but the cables should run along the side of the joists. If the cables are below the floor joists, or are extending down an open wall, they should be enclosed in a raceway. Nonmetallic sheathed cables that contain 2/6 AWG, 3/8 AWG, or larger conductors are allowed to be secured to the bottom of the framing members without physical protection, **Figure 38-7**.

38.1.3 Boxes and Load Centers

Fishing conductors into boxes and panelboards requires an opening in the enclosure, securing cable, and providing protection from electrical hazards. Removing a preformed

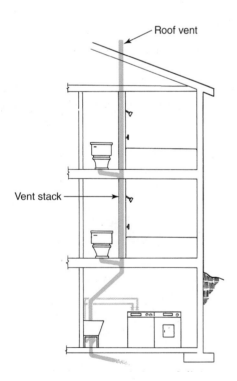

Goodheart-Willcox Publisher

Figure 38-4. Plumbing stacks often provide vertical pathways for routing cables due to the space around the pipes through framing members.

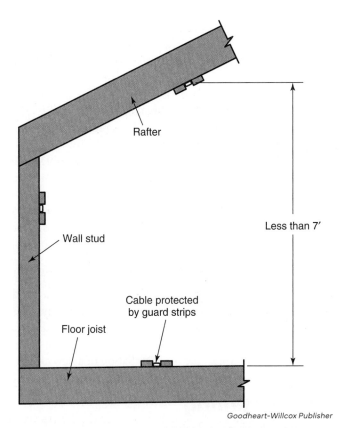

Goodheart-Willcox Publisher

Figure 38-6. Cables installed within 7' of the attic access should be run through bored holes in the joists or be run on the side of the joists parallel to the opening. If the cables are routed on the topside of the joist, they should be secured to a guard strip.

NM Cable in Unfinished Basements

Any size cable can be supported by drilled holes in the joists

Cables no smaller than 6-2 NM or 8-3 NM can be attached to the bottom edges of the joists

Any size cable can be attached to a running board attached to the bottom edges of the joists

Goodheart-Willcox Publisher

Figure 38-7. Cables installed in exposed areas of unfinished basements should be routed through the floor joists or installed in a raceway. If the cables are installed below the floor joist, they must be secured to a running board. Cables containing conductors larger than 8/3 AWG or 6/2 AWG are permitted to be stapled directly to the bottom of the joist.

PRO TIP   Removing a Knockout from Inside an Enclosure

There are methods to remove knockouts from the interior of a box or panelboard. The first method is to pry a gap between the knockout and the interior surface of the box. To do this, apply a chisel, awl, or center punch against the side of the knockout. Gently tap the tool, trying to pry the knockout down or push it sideways to create a gap in the surfaces. In some applications, a self-tapping screw can be driven at the edge of the knockout to force a small gap. Once a gap is created, you can use a screwdriver to pry the knockout open further. Needle-nose pliers can be used to remove the knockout from the opening once enough space has been created.

Another method to remove knockouts is to create a pulling point. Do this by inserting a self-tapping screw into the knockout itself. Once the screw has been driven into the material enough to engage the threads, stop driving it. Using pliers or a screwdriver, pry the knockout away from the enclosure.

knockout from a metal box is a necessary skill when fishing circuits into existing boxes and cabinets. A *knockout* is a plug that is partially punched in an electrical enclosure designed for easy removal to allow for the installation of wiring and raceways. Knockouts are designed to be pushed into the enclosure from the outside. When boxes and panelboards are installed in finished locations, the outside of the enclosure is typically inaccessible. The limited access makes it much more difficult to remove. Forcing knockouts backwards, from the inside to the outside, is impracticable. You must find a method to bring the knockout in rather than push it out.

Nonmetallic boxes often come with preformed cable entries. Most of these boxes contain a means of attaching the cable to the box. Cautiously pry open these integral clamps to ensure that the cable securement method is not compromised.

You must also secure the cable to metallic boxes. This is often done using a cable connector that attaches to the cable and is secured to the box with a locknut, **Figure 38-8.** If this type of cable connector is used, it needs to be attached to the cable before it is fished into the box. The locknut is then threaded over the conductors and attached the connector securely to the box. Nonmetallic connectors are available that can be inserted from the inside of the enclosure that do not require the use of a locknut, making for a much easier installation, **Figure 38-9**.

Arlington Industries

Figure 38-8. A typical nonmetallic sheathed cable connector for a metallic box would require connecting the clamp to the wire before pulling it down the wall and then threading the locknut into place after fishing the conductor into the box.

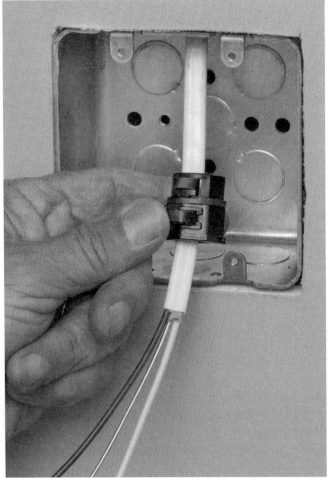

Arlington Industries

Figure 38-9. A nonmetallic sheathed cable connector designed for installation from the inside of a metallic box can make the task of fishing wire much easier.

Fishing cables into panelboards can be especially hazardous if the panels are not de-energized. Using steel fish tapes and uninsulated tools increases the risk of shock and arc flash events. The first overcurrent protection device in most residential load centers is the main breaker. The conductors coming into the main breaker from the utility have limited overcurrent protection, and an arc flash event on the load side of the main can have a significant amount of energy. Removing the utility meter is often the only way to remove power in residential panelboards, making it difficult to eliminate electrical hazards.

Most power utilities regulate access to the meter enclosure with a lock, **Figure 38-10**. Removing the lock violates many utility regulations. Thus, coordination with the utility to disconnect and reconnect power may be necessary to install the new circuit. *Section 230.85(A)(1)* requires an emergency disconnecting means mounted outside the building, **Figure 38-11**. This provides emergency responders with a means of cutting all power. The advantage of the new requirement is that it can be used to de-energize the panelboard for service work, like the installation of a new circuit.

38.2 Installing Circuit Breakers

Installing a circuit breaker is a simple task. Two of the most common violations surrounding the installation of breakers are choosing a breaker that does not match the rating of the equipment and installing it in violation of the manufacturer's instructions. When breaker space is limited in a load center, special equipment can be used to provide more circuits without increasing the size of the panelboard.

Viktorus/Shutterstock.com

Figure 38-10. Most electric utilities place a lock at metering equipment to prevent the entry of unauthorized persons. To de-energize power to a panelboard fed by this meter, the utility company would have to remove the lock and meter.

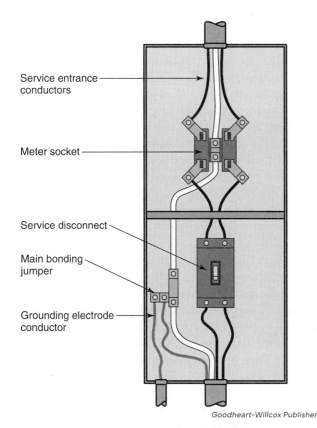

Service entrance conductors

Meter socket

Service disconnect

Main bonding jumper

Grounding electrode conductor

Goodheart-Willcox Publisher

Figure 38-11. The *NEC* now requires the installation of an emergency disconnecting means in an accessible location on the exterior of single and two-family dwellings.

38.2.1 Choosing the Correct Breaker Type

Circuit breakers come in a wide variety of ratings and are produced by multiple manufacturers. For a circuit breaker to perform correctly, it needs to be compatible with the panelboard it is being installed in. Breakers from the same manufacturer are not always compatible, and variations of breakers manufactured by the same company are not always intended to be interchangeable. Some breaker manufacturers produce products with similar physical characteristics like size and shape. If a circuit breaker physically fits into a panelboard, it is not an indication that the breaker will operate properly. You must verify that the breaker and panelboard can work together. Check the label inside the panelboard and the installation instructions provided with the breaker for compatibility.

Adding a circuit to legacy equipment can be a challenge. Even though the equipment is still in service, the manufacturer may no longer be producing new equipment. When new original equipment is no longer available, you can substitute equipment. Several breaker manufacturers are producing substitute equipment that is compatible with legacy panelboards. There is typically a business relationship, like a consolidation, merger, or acquisition, between the current

manufacturer and the original equipment manufacturer. For example, Eaton Corporation offers a BR series circuit breaker that can be used in Bryant panelboards.

38.2.2 Tandem Breakers

Tandem circuit breakers are a special type of circuit breaker that contain two separate switches designed to provide overcurrent protection for two single-pole circuits from a single panelboard opening, **Figure 38-12.** Tandem breakers are also called *twin breakers*, *duplex breakers*, or *double breakers*. The purpose of these breakers is to provide a means for installing more circuits in a panelboard that has no spaces remaining. For example, consider a central air conditioning system added to an existing dwelling. The condensing unit outside requires a two-pole breaker connection. If the panel only has one opening available, you can replace two single breakers with one tandem, creating the extra space needed to add the two-pole breaker for the air conditioner.

Tandem breakers should be installed in existing equipment with caution. Some panelboards are not rated for the use of these devices. Panelboards are constructed with a specific number of openings but sometimes have a rating allowing more circuits than openings. For example, if a 12-space, 12-circuit panelboard contained 11 standard breakers, the addition of a tandem breaker would be in violation of the rating if the addition resulted in 13 circuits. Other panelboards may have a rating of 12 spaces, 24 circuits, allowing for the installation of tandem breakers without violating the rating of the panelboard. Some panelboards are designed to accept a limited number of tandem breakers. When this is the case, there are specific locations that are designed to accept them. These are typically on the opposite end of the bus from the main breaker. In residential panelboards, this position is normally in the bottom few rows of breakers.

Tandem breakers share a single connection point to the electrical busing of a panelboard. Both breakers are

Tandem breaker

Goodheart-Willcox Publisher

Figure 38-12. Tandem circuit breakers allow for the installation of two circuits in the same space as a single standard circuit breaker.

connected to the same phase of the electrical distribution system. This can be a problem when the breakers are installed in systems using multiwire branch circuits. A three-wire, nonmetallic sheathed cable that contains a black, red, white, and bare conductor is used for a multiwire circuit. If one circuit of the tandem goes to the black, and one circuit goes to the red, then the two ungrounded conductors are connected to the same phase. With a shared grounded conductor, the line to neutral load would be additive in the grounded conductor and could easily exceed the ampacity of the grounded conductor. Exceeding the ampacity causing heat-related damage to the conductor and equipment.

Tandem breakers are also limited in application because they do not come equipped with AFCI or GFCI protection. Additional circuits are required to be installed in accordance with the current code requirements, and most residential circuits have GFCI or AFCI protection requirements.

38.3 Subpanels

Electrical distribution systems often contain more than one panelboard. Subpanels are panelboards that are supplied by an overcurrent protection device in another panelboard. They can be installed in various locations:

- In dwelling units to provide additional circuit breaker capacity
- Remotely from the service equipment to provide power distribution in a location closer to the load
- In unfinished spaces near kitchens, because of the large number of circuits required to supply the kitchen loads

Note that sometimes a single larger feeder and subpanel may be cheaper or easier to install than running all the circuits across the residence to the main panel, **Figure 38-13**.

Another common location for subpanels is in a garage or accessory building. *Section 225.30* limits the supply to a separate building to one feeder or branch circuit. If a detached garage associated with a single-family dwelling is fed from the main structure, there is a limit of one feeder or one branch circuit. Providing a feeder to the building and installing a subpanel provides more flexibility for the future needs of the separate structure.

Subpanels are sized in accordance with the connected load. Refer to service and feeder calculations in Chapter 20, *Service Calculations*. Once the load calculation has been completed, the minimum capacity of the subpanel has been established. A feeder circuit breaker is installed in the supply panelboard to provide overcurrent protection for the feeder conductors. Conductors are then sized based on the load being served and the rating of the breaker.

A panelboard is sized from the rating of the feeder breaker. A subpanel rating must be equivalent to or greater than the rating of the feeder circuit breaker supplying it. A 100-amp subpanel can be connected to a 60-amp feeder without issue because the 100-amp subpanel is limited to 60 amps of load by the feeder. A 60-amp subpanel cannot be connected to a 100-amp feeder because the feeder breaker exceeds the subpanel's rating.

Subpanels are not required to have a main breaker. They can be constructed with main lugs only. If a subpanel is installed in a separate building, *Section 225.31(B)* requires the building to have a disconnecting means outside or immediately adjacent to the feeder conductors' entry to the building. Installing a subpanel with a main breaker complies with the requirement. An alternative to providing a main breaker is to provide a separate disconnecting means on the line side of a main lug subpanel.

Many subpanels are located immediately adjacent to the service panelboards. The main reason for this application

Chad Robertson Media/Shutterstock.com

Figure 38-13. A subpanel can be added in a dwelling unit to reduce the length of multiple circuits or to feed loads in areas that are not easily accessible from the main service.

is to provide additional breaker space. A conduit nipple is usually installed between the main panelboard and the subpanel. It may contain additional conductors other than the feeders, but always follow derating requirements to ensure the conductors are installed within their ampacity. If more than one conduit nipple is installed, *Section 300.3(B)* requires all the conductors of a circuit to originate from the same source and be contained within the same raceway. Make sure the ungrounded conductors and grounded neutral conductors of the circuits are fed from the same panelboard and installed in the same nipple when transitioning between the panelboards.

The same working space clearances for other electrical equipment applies to subpanels as well. Ensure there is adequate space in front of the equipment to provide safe working conditions. The working space clearances apply to all subpanel installations, even if they are installed in garages, separate buildings, unfinished basements, or closets. *Section 240.24(D)–(F)* does not allow overcurrent protection devices to be installed over stairs, in bathrooms, or in areas near easily ignitable material like clothes closets. Subpanels contain overcurrent protection devices, so their installation is also limited in those areas.

Subpanels are not contained within the service equipment enclosure. Any panelboard in the distribution system not associated with the main service disconnecting means, where the main bonding jumper is installed, requires separating the equipment grounding conductors and the grounded conductors. Because of this, you must install a separate equipment grounding terminal bar within the subpanel and only connect grounded conductors to the isolated grounded neutral terminal bar supplied with the subpanel, **Figure 38-14**. The connection between the two at the subpanel creates a parallel pathway between the equipment grounding conductor and the grounded conductor. This leads to unwanted current on the equipment ground, causing a potential electrical shock hazard.

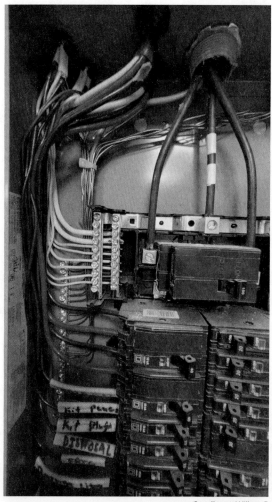

Goodheart-Willcox Publisher

Figure 38-14. In this subpanel, the equipment grounding conductors and the grounded neutral conductors are separated. Notice how the bare equipment grounding conductors are connected to a bus bar directly mounted to the panel enclosure while the grounded conductors are terminated on a bar that is isolated from the enclosure. The subpanel does not contain a main bonding screw or jumper.

Summary

- Steel fish tapes are typically used when fishing walls or other concealed cavities.
- Fish sticks have less flexibility than steel fish tapes, making them ideal for larger spaces like attics.
- Chases and openings around chimneys and plumbing vents are possible cable routes.
- Knockouts in metallic boxes can be removed from the inside using several techniques.
- Circuit breakers must be compatible with the panelboard in which they are installed.
- New circuit breakers can be installed in legacy equipment in limited applications.
- Tandem circuit breakers can be used to save space in some existing panelboards.
- Caution should be used when installing tandem breakers with multiwire branch circuits.
- Subpanels are used to provide breaker capacity and for ease of circuit installation.
- Subpanel installations have similar requirements as other electrical distribution equipment.

Know and Understand

1. Heating a steel fish tape during the process of creating a new pulling head allows for which of the following?
   A. A reduction in the brittleness of the material
   B. The material can bend with less force
   C. A tighter loop to be created
   D. All of these.

2. Which of the following is an advantage of using fish sticks over metal fish tapes?
   A. Fish sticks can bend around obstructions in the wall due to their greater flexibility.
   B. Fish sticks can be pushed in a straight line through open spaces where steel fish tapes tend to curl up.
   C. Hooking two fish sticks inside a wall cavity is easier than hooking two fish tapes.
   D. New ends can easily be formed when it becomes necessary to replace a portion of the fish stick.

3. Openings in the framing around _____ can often provide a pathway for cables through the building structure when installing new circuits.
   A. plumbing vents
   B. recessed luminaires
   C. water supply pipes
   D. bathroom vents

4. In general, nonmetallic-sheathed cable _____ installed in air return cavities being used for environmental air handling.
   A. shall be
   B. shall not be
   C. shall be permitted to be
   D. should be

5. Cables installed perpendicular to the framing and routed within _____ of an attic scuttle hole shall be protected by guard strips
   A. 3′
   B. 5′
   C. 6′
   D. 7′

6. If a circuit breaker fits into a panelboard, it is an indication that the breaker _____.
   A. will operate properly
   B. is intended to be interchangeable
   C. is permitted as a substitution for legacy equipment
   D. physically fits but may not be an indication that the breaker will operate properly

7. _____ breakers are designed to add circuit capacity without additional breaker spaces.
   A. Tandem
   B. Sandwich
   C. Reduction
   D. Slim

8. One limitation of tandem breakers is that they fail to provide _____ protection.
   A. overcurrent
   B. GFCI
   C. overload
   D. short-circuit

9. The *NEC* limits the number of feeders supplying separate buildings, such as a detached garage, to _____ feeder(s).
   A. one
   B. two
   C. four
   D. six

10. Subpanels installed in _____ would constitute a violation of the *NEC*.
    A. accessible attics
    B. garages
    C. bathrooms
    D. unfinished basements

Apply and Analyze

1. When fishing from an attic down an uninsulated interior wall to an opening cut into the wall for the installation of a new receptacle, which type of fishing tool might you choose and why?

2. Describe potential access points for cable routing in an existing home built approximately 50 years ago.

3. What steps would you take to remove a knockout in the top of an existing panelboard that is recessed in a wall?

4. Predict possible hazards that could be created by installing a circuit breaker that is not compatible with the panelboard in which it is installed.

5. Discuss potential solutions for adding a circuit to a panelboard that has no openings for additional circuit breakers.

6. Provide a justification for installing a subpanel on the second floor of a single-family home containing electrically operated laundry equipment on the second floor.

7. In your own words, explain why the equipment grounding conductor terminal and the grounded conductor terminal in a subpanel are isolated from one another.

Critical Thinking

1. Your boss sends you to a century-old two-story house to evaluate the potential for providing a new grounded circuit for the second-story bedrooms. What information would you try to obtain and provide to your boss to create a material list and estimate.

2. An additional circuit for an electric car charger is routed from a panelboard surface mounted to the concrete exterior wall of the basement to the garage. The circuit crosses the basement. It is then fished up through the first-floor wall and into the attic. The cable then traverses the attic and enters the garage, where it is routed through the exposed rafters and down an exterior wall to the charger. Where and when is the cable required to be supported? How would you protect the cable from damage along its route?

kevin brine/Shutterstock.com

SPARKING DISCUSSION

A client has service equipment rated at 60 amps and the insurance company is refusing to offer homeowners insurance unless the service is upgraded. When considering service upgrades, what do you plan to review? How will you prepare for the onsite meeting with the client?

LEARNING OBJECTIVES

After completing this chapter, you will be able to:

- Identify common reasons for an electrical service upgrade.
- Identify outdoor equipment location considerations.
- Discuss factors relating to indoor equipment locations.
- Select service equipment based on the conditions of the installation.
- Describe the advantages of using cable methods or raceways for service entrances.
- Discuss considerations for selecting copper or aluminum conductors.
- Describe condensation issues in service equipment and how it is mitigated.
- Explain compliance issues surrounding system grounding and service upgrades.
- Discuss panelboard labeling requirements.
- Summarize the power transfer process.

TECHNICAL TERMS

affidavit
electrical service upgrade
point of entrance
point of service

Introduction

An *electrical service upgrade* is a process of replacing the existing service equipment with new, typically larger equipment. These upgrades are performed for a variety of reasons:

- To provide higher current capacity with more space for circuit breakers
- To replace fuses with circuit breakers
- To replace poorly performing legacy equipment with higher performing breakers
- To meet industry recommendations or requirements of insurance companies to make a home insurable

There are several considerations to account for before beginning the work of replacing an electrical service. New equipment must be in areas of convenience, code compliant, and updated to meet the customers' needs. Service equipment materials must be selected for proper installation. Careful consideration of conduit and cable entry to panelboards and conduit sealing is critical. You will also be required to inspect existing systems to ensure the service equipment is installed properly and safe for operation.

39.1 Service Equipment Locations

Electrical service upgrades allow equipment to be relocated to comply with today's codes and standards and provide better access for operation and maintenance. It is the electrician's responsibility to evaluate and recommend changes to the location and placement of equipment to the homeowner. Upgrading this equipment provides an opportunity to make the equipment location more convenient, reduce circuit length and provide protection from outdoor hazards by moving from overhead to underground supplies.

Service upgrades are often done because the existing equipment does not have the capacity to serve additional loads. When a home is being remodeled, or additional equipment is being installed, this results in the need for a service upgrade, **Figure 39-1**. The location of the panelboard might depend on the location of the loads. If an addition or remodel is taking place on the opposite end of the dwelling, you may want to consider locating the service closer to those loads. The existing service equipment can be supplied through the installation of a feeder from the new panelboard. This allows most new circuits to be fed from a closer location, which can reduce the cost of installation.

39.1.1 Outdoor Locations

Electrical service upgrades can improve the location of outdoor electrical equipment. Existing equipment is sometimes located where it is inconvenient to work on, unsafe due to improper clearances, or not readily accessible to be maintained, **Figure 39-2**.

One of the first questions you must determine before beginning work on an electrical service upgrade is if the service will be overhead or underground. When service equipment is supplied from the utility through an overhead drop, the overhead drop is subject to damage from ice, wind, and falling trees, **Figure 39-3**. High winds can sometimes cause damage to the conductors. Large accumulations of ice weigh down the conductors and can lead to failure of the conductor support systems. Falling trees and branches can snap lines or pull down conductor supports. Overhead lines can also be less aesthetically pleasing, so homeowners sometimes opt to convert exiting overhead services to underground.

Goodheart-Willcox Publisher

Figure 39-2. Replacement service equipment may not be permitted to be installed in the same location as the existing equipment. This service disconnect and metering equipment is in violation of the working space clearance requirements due to the location of the gas meter and piping.

Steven Belanger/Shutterstock.com

Figure 39-1. Service upgrades are often done to add capacity for additional loads or increased living space. This dwelling is undergoing a significant remodel with the addition of living space that may require a service upgrade.

Picture This Images/Shutterstock.com

Figure 39-3. Falling trees, wind, and ice can all lead to collapse of an overhead service drop. Repair of damaged equipment after the service point is the responsibility of the homeowner.

The resident owns and maintains the equipment after the point of service, which means they are responsible for any damage or repair thereafter. The **point of service** is the point at which the utility wiring ends and the premises wiring begins. The service point is usually at the connection between the conductors leaving the weather head and the overhead drop conductors. If the drop is ripped away from the supporting equipment, the resulting damage can be costly to the homeowner. Homes surrounded by mature trees or other threats may be good candidates for switching from an overhead to an underground service.

The location of outdoor equipment is also based on the installation standards of the local utility. Local utilities provide guidance in locating metering equipment, attachment points, and clearances, **Figure 39-4**. Some utilities may require overhead services to be converted to underground as they plan to transition all customers within a coverage area to an underground distribution system. Connect with your local utility designer or engineer to determine the service location requirements for your service upgrade before beginning work. Authorities having jurisdiction, such as local building inspectors and electrical inspectors, can also be consulted regarding potential issues when determining equipment locations.

Remodeling projects and changes in electrical codes can create compliance issues for existing equipment locations. The equipment may have been properly installed at the time of installation, but the equipment location may no longer be acceptable for the installation of new equipment. Home

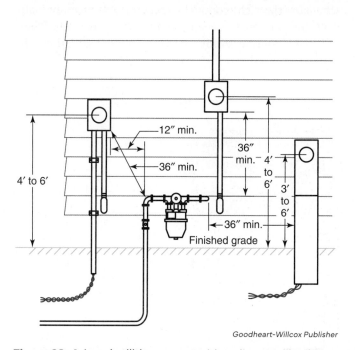

Goodheart-Willcox Publisher

Figure 39-4. Local utilities may provide a diagram like this one that provides information for clearances surrounding electrical service equipment. Mounting heights, clearances from doors or windows, and distances from gas venting regulators may be specified by the utility.

renovations, like the addition of living space, installation of decks, or replacement of roofing or siding, can impact electrical services.

Section 230.24 requires minimum clearances around windows, doors, and over roofs. There are a few implications that can affect updated equipment to these areas. Adding a new door or window opening may create an issue for new equipment in the same location. Adding living space to a home may cause the overhead drop to now travel over the roof of the addition, violating the clearance requirement over roofs. Adding a deck may interfere with working space clearances around equipment.

If a home was re-sided, and the siding contractor sided over the service's riser pipe, then the new riser pipe must be installed on the exterior side of the wall leading to a more complex installation. In addition to the *NEC* working space requirements in *Section 110.26*, the electrical utility may have installation standards that can no longer be met due to the home renovation.

39.1.2 Indoor Locations

When upgrading electrical services, it can be challenging to determine the proper placement of equipment for indoor locations. Working space clearances and dedicated equipment space can be difficult to establish in existing homes. Working space requirements have expanded as the *Code* has changed throughout the years. *Code* changes that have expanded working spaces around electrical equipment have made it safer and more practicable to work on the equipment when needed. A panelboard mounted on the foundation wall of a dwelling's basement may have complied at the time of installation but may no longer meet today's requirements. A common example of a working space clearance violation is the placement of laundry equipment immediately adjacent to or in front of the equipment, **Figure 39-5**.

A new panelboard installed to replace an old service must meet the current working space width, depth, and height requirements in *Section 110.26*. It may be necessary to relocate the equipment to a different space, but there are some limitations to moving the equipment. One limitation outlined in *Section 230.70(A)(1)* requires you to terminate the service entrance conductors in the service disconnect nearest the **point of entrance**, the point at which the conductor, cable, or raceway emerges into the building or structure from the exterior wall or floor. It would be a violation for the service entrance conductors to enter the dwelling at the same point as the old service and run through the dwelling to the new location. Some authorities having jurisdiction provide guidance in determining a physical dimension for the nearest point of entrance.

If it becomes necessary to move the panelboard, a service disconnect could be installed outdoors allowing what is now a set of feeder conductors to enter the building and be routed to the panelboard with no length limitations, **Figure 39-6**.

Goodheart-Willcox Publisher

Figure 39-5. The placement of laundry equipment within the working space of a panelboard is a common code violation that could result in limited access to the equipment for service and repair.

As an alternative option, you can route the service entrance conductors along the exterior surface of the building and then penetrate the building at the new panelboard location.

Another limitation you may encounter when moving the panelboard location is the need to reconnect the existing circuits. Existing circuits can be rerouted to the new location, but inevitably some of the cable and conduit runs may not have enough length to reach the new location. In this case, circuit extensions need to be installed from the old location to the new location. The existing conductors must be spliced to the new extensions, which requires installing boxes to contain the splices.

39.2 Installation Practices

There are many installation practices you are tasked with for a service upgrade, including:

- Selecting the materials best suited for the location that meet the applicable codes and standards
- Determining the best course of action for removing the existing circuits and reattaching them to the new panelboard
- Examining the grounding system for *Code* compliance
- Installing new grounding electrodes and conductors

The following sections will outline these practices and discuss the necessary considerations for each.

39.2.1 Selecting Materials

Choosing the right conductors, cables, and raceway materials for the job is dependent on the application. There are many considerations to cost, ease of installation, and

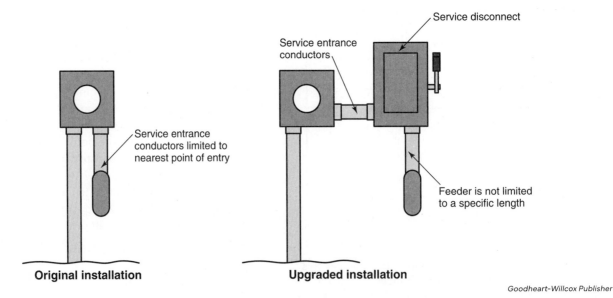

Original installation

Service entrance conductors limited to nearest point of entry

Upgraded installation

Service disconnect

Service entrance conductors

Feeder is not limited to a specific length

Goodheart-Willcox Publisher

Figure 39-6. Installing the service disconnect on the exterior of the building allows for unlimited length feeder conductors to enter the dwelling. This provides greater flexibility when relocating the interior service equipment during a service upgrade.

PROCEDURE **Using the Existing Panelboard as a Junction Box**

If a new panelboard location becomes necessary during a service upgrade, time and material can be saved by using the existing panelboard as a junction box. Follow these steps to complete the task:

1. Convert the old panelboard into a junction box by removing the interior equipment except for the equipment grounding bar.
2. Route the circuit extensions from the old panelboard enclosure to the new panelboard location. This can be done by using several runs of nonmetallic sheathed cable or installing conduit and pulling in conductors.

3. Splice the conductors, ensuring each of the ungrounded conductors is with its corresponding grounded conductor in each cable. You can utilize the existing grounding bar to splice all the equipment grounding conductors.
4. Check that all splices are complete and that the conductors are folded back into the enclosure.
5. Install the panelboard cover. Most panelboard covers have a hinged door to access the breaker handles. This access can be secured by inserting a couple of self-drilling screws to ensure that someone cannot open the cover and expose the splices without removing the screws.

longevity. One of the major material selections is service entrance conductor run. The service entrance conductors can be contained within a cable or a conduit system.

Service entrance cable, Type SE, can be used indoors and outdoors. The cable is not required to be contained with a raceway, but it cannot be subject to physical damage, **Figure 39-7**. Electricians may install a conduit on the exterior side of the dwelling to protect the SE cable and then run the cable without raceway protection once it is indoors. They also usually select cable methods as they consider ease of installation. Cable is more flexible and requires less labor to install. Conduit provides more protection for wiring and reduces the types of materials required for the installation. Overhead service wiring between the service point and the metering equipment is often comprised of conductors in a raceway. If the wiring between the metering equipment and the panelboard is the same type of material less products will be required than if the wiring method changes to a cable type installation.

When conduit is used to connect indoor and outdoor equipment, you must select the raceway type you want to use. PVC conduit is less expensive and easier to install than most metallic raceways. However, it has some limitations because it is more affected by sunlight and thermal contraction and expansion. It also has a lower degree of physical protection. PVC used above grade in outdoor applications should be rated for sunlight resistance. Sunlight can degrade plastics, making PVC brittle and prone to breakage.

When longer PVC runs are used for service upgrades, you must account for thermal expansion and contraction of the raceway. Special expansion fittings may be needed to accommodate the installation, **Figure 39-8**. You will also need to select the wall thickness of the PVC used. Standard PVC is Schedule 40 and can be used in locations where physical protection is not an issue. When PVC is exposed to physical damage, Schedule 80 PVC should be used. Schedule 80 PVC has the same overall outside dimensions as Schedule 40 PVC, but it has a thicker wall that gives the raceway more strength.

Goodheart-Willcox Publisher

Figure 39-7. Service Entrance (SE) cable is not required to be installed within a raceway. This SE cable extends from the metering equipment through the exterior wall and into the dwelling panelboard.

Goodheart-Willcox Publisher

Figure 39-8. An expansion fitting accommodates changes in length of the PVC raceway supplying this electrical service. The fitting allows movement in the conduit for expansion and contraction due to temperature changes and ground movement associated with frost heave.

Metal conduits can also be considered for service upgrades. EMT and RMC are the two types used in residential settings. RMC is often used in overhead services, where a service mast passes through the roof and supports the utility drop to provide the strength needed to support the conductors. RMC is more expensive and harder to work with because it is difficult to cut and thread the conduit in the field.

In the past, it was common for RMC to be installed as a service entrance raceway in services, but most electricians today are selecting other materials. EMT can be used as a

service entrance raceway if any outdoor portion of the raceway is terminated using compression fittings. Meter socket enclosures are often bonded through a connection to the grounded conductor. If metallic raceways are used between a panelboard and meter socket, the metallic raceway creates a parallel pathway with the grounded conductor. Because of this, some neutral current flows in the metallic raceway, **Figure 39-9**.

Section 250.92(B) requires metal pathways on the line side of the service disconnecting means to be bonded through special means. This could include bonding bushings, bonding locknuts, or threaded hubs. The use of PVC raceways simplifies the bonding requirement because no bonding is required when the raceway is nonmetallic. The PVC does

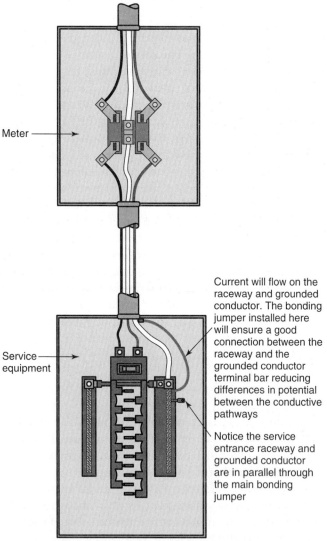

Meter

Service equipment

Current will flow on the raceway and grounded conductor. The bonding jumper installed here will ensure a good connection between the raceway and the grounded conductor terminal bar reducing differences in potential between the conductive pathways

Notice the service entrance raceway and grounded conductor are in parallel through the main bonding jumper

Goodheart-Willcox Publisher

Figure 39-9. The connection of service equipment with a metallic raceway can create a parallel pathway between the grounding system and the grounded neutral conductor. When parallel paths exist, current will flow on all paths, which leads to current flow on the normally non-current carrying grounding conductors and metallic raceways.

not carry current, therefore it does not create a parallel pathway for neutral current to flow on the conduit or service enclosures.

Conductor material also must be selected before starting work. The choice of conductors comes down to selecting copper or aluminum. Copper service conductors are heavier and more expensive than aluminum, but they perform well in areas exposed to moisture like exterior service equipment. Aluminum conductors must be larger than copper conductors to have the same current carrying capacity.

PROCEDURE   **Replacing a Panelboard**

When approaching panelboard replacement, careful attention must be paid to the process. Follow these steps below to properly practice a panelboard replacement:

1. Label each conductor as they are removed from the overcurrent devices.
2. Ensure everything is disconnected and labeled. Remove all the locknuts associated with the cable and conduit fittings.
3. If enough clearance is present, remove the cables from the panel before removing the screws holding the panel in place. Remove as many as possible.
4. Remove the screws securing the panelboard.
5. Slowly pull the enclosure over the remaining conductors being careful not to damage the conductor insulation until all conductors and conduits have been removed.

Putting all the circuits back into the new panelboard can be challenging—most cables provide enough flexibility to facilitate the process, but conduits do not provide as much flexibility. In addition, most residential panelboards are manufactured with premade knockouts, which may not line up with the conduit for entry. Follow the next steps to ensure proper circuitry replacement:

1. Evaluate the panelboard. You might need to remove the first piece of the pipe run to shorten or bend the conduit to line up with a knock out. You might also need to pull the conductors back at the nearest access point to protect them from damage during the reinstallation process. Metallic raceways can be transitioned by adding a short piece of flexible metallic conduit allowing some flexibility for the connection.
2. Complete whatever steps are necessary to reinstall the panelboard properly.
3. Install all locknuts.
4. Ensure the fittings are tight using a wrench. Note: many existing metallic raceways do not contain an equipment grounding conductor, so the grounding connection relies on the bond between the conduit and the enclosure.

PRO TIP   **Two-Piece Conduit Connectors**

Two-piece conduit connectors can help in spaces that do not have a lot of room to work. A two-piece connector allows you to install a conduit into a fixed enclosure without the addition of the threads that are normally contained inside an enclosure. By shortening the length of the conduit without a standard fitting, more space is added to install the conductors without damaging their insulation.

The two-piece connector has a locking retainer ring that screws onto a connector fitting installed from the inside of the enclosure, **Figure 39-10**. Complete the following steps to install the connector:

1. Slide the retainer ring over the conductors and onto the conduit.
2. Route the wires through the enclosure opening and carefully move the conduit into place above the panelboard.
3. Slide the connector fitting over the wires and through the enclosure knockout.
4. Thread the two pieces together using a wrench or pliers.

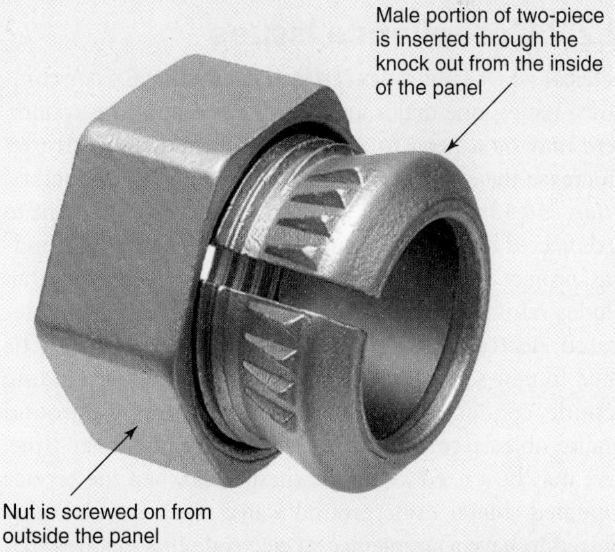

Male portion of two-piece is inserted through the knock out from the inside of the panel

Nut is screwed on from outside the panel

Arlington Industries, Inc.

Figure 39-10. Two-piece connectors can be helpful in reconnecting old piping systems into the new service equipment.

Larger conductors may require larger raceways and termination lugs. Aluminum conductors also expand and contract more than copper conductors, which can lead to termination issues. Aluminum service entrance cables are readily available, while copper service entrance cables are less common. Aluminum conductors are more flexible than copper conductors and are easier to install. Overall, both conductors can do the job, but the choice of which depends on the needs of the installation.

39.2.2 Conduit and Cable Entry

When replacing a panelboard, an electrician must remove the circuits from the old equipment and put them into the new equipment.

Condensation can create moisture problems within an electrical system. *Section 300.7(A)* requires electricians to mitigate the buildup of condensation by sealing conduits that travel between areas of major temperature difference. A service entrance conduit penetrating the exterior wall of building is exposed to large temperature differences. When warm, moist air comes in contact with cool air, the moisture condenses, which leads to a buildup of water within the equipment. Sealing the raceway between the two environments stops the mixture of warm and cold air, which reduces condensation. A seal can be installed where the conduit enters the panelboard. If an LB fitting is used to penetrate the wall, it can be installed in the LB as it enters the dwelling.

Note that not all materials are acceptable for sealing. The sealing material must be identified for use with conductor insulation; thus, always search for products labeled for use with electrical conductors.

39.2.3 Compliance Issues

Electrical service upgrades can lead to additional *Code* compliance issues, one major area being the grounding system. There may be a need to install new grounding electrodes or increase the size of the grounding electrode conductors. *Section 250.50* requires all grounding electrodes present to be connected to the grounding electrode system. If a remodeling project requires installing a concrete footing that includes reinforcing rebar, it may be defined as a concrete-encased electrode. A concrete-encased electrode must be routed to new service equipment by installing a grounding electrode conductor. Ground rods and other in-ground metallic objects corrode and lose effectiveness over time. There may be a need to replace these rods when the service is updated. Metal underground water pipes are also now required to have a supplemental electrode that ensures that a ground connection is in place if the metal water pipe is replaced with a nonmetallic piping system.

When service upgrades are performed, there is usually an increase in current capacity of the service. This increased

capacity requires the installation of larger ungrounded conductors. Grounding electrode conductors are based on the size of the ungrounded service entrance conductors. As the service entrance conductor size increases, so does the size of the grounding electrode conductors, **Figure 39-11**.

Most new circuits within dwelling units require AFCI protection. When service upgrades are performed, the circuit itself is not being modified—rather, it is the service equipment being modified. There is no requirement to install AFCI breakers for existing circuits that require protection in the new equipment. It is a good opportunity to provide arcing fault protection. AFCI circuit breakers can be installed to reduce fire hazards associated with arcing faults at the time of the service equipment replacement.

Pay particular attention to the current rating of the circuits being reinstalled. Even though a circuit was on a 20-amp breaker in the old panelboard, it may not be allowed on a 20-amp breaker in the new equipment. The existing circuits may not have been protected by the proper size

Table 250.66 Grounding Electrode Conductor for Alternating-Current Systems

Size of Largest Ungrounded Conductor or Equivalent Area for Parallel Conductors (AWG/kcmil)		Size of Grounding Electrode Conductor (AWG/kcmil)	
Copper	Aluminum or Copper-Clad Aluminum	Copper	Aluminum or Copper-Clad Aluminum
2 or smaller	1/0 or smaller	8	6
1 or 1/0	2/0 or 3/0	6	4
2/0 or 3/0	4/0 or 250	4	2
Over 3/0 through 350	Over 250 through 500	2	1/0
Over 350 through 600	Over 500 through 900	1/0	3/0
Over 600 through 1100	Over 900 through 1750	2/0	4/0
Over 1100	Over 1750	3/0	250

Notes:
1. If multiple sets of service-entrance conductors connect directly to a service drop, set of overhead service conductors, set of underground service conductors, or service lateral, the equivalent size of the largest service-entrance conductor shall be determined by the largest sum of the areas of the corresponding conductors of each set.
2. If there are no service-entrance conductors, the grounding electrode conductor size shall be determined by the equivalent size of the largest service-entrance conductor required for the load to be served.
3. See installation restrictions in 250.64.

Reproduced with permission of NFPA from NFPA 70, National Electrical Code, 2023 edition. Copyright © 2022, National Fire Protection Association. For a full copy of the NFPA 70, please go to www.nfpa.org

Figure 39-11. *Table 250.66* can be used to determine the minimum size grounding electrode conductor based on the size of the incoming service entrance conductors. Service upgrades requiring larger service entrance conductors may also require the installation of larger grounding electrode conductors.

overcurrent device. This was very common in fused systems where homeowners would overload the circuit, resulting in a blown fuse. They would then replace the fuse with a larger one, violating the ampacity of the conductor and creating a potential hazard. Look closely at the conductor size before terminating on an overcurrent protection device. In general, 14 AWG copper conductors should not be protected by an overcurrent protection device exceeding 15 amps, **Figure 39-12**.

39.3 Final Connections

Service upgrades are not complete until the utility makes its final connections and reenergizes the service. You are responsible for coordinating the final connections of the system. The utility, contractor, and electrical inspector are all involved in the process. Many utilities will not reconnect the electrical service until it has been inspected. In rural areas where inspections are not always readily available, some utilities allow the installer to complete an *affidavit*, a written statement affirming that an installation is compliant with the *NEC* and is safe to energize. Most service upgrades are completed in a single day, but some systems are more complicated and require additional work before final connections can be made.

39.3.1 Temporary Power

Temporary power may be required to complete the tasks associated with an electrical service upgrade. In cases where service equipment is relocated, it is possible to build a new service while the old service is still energized. Outlets from the existing equipment can supply power during the work of installing the new service. When work surrounding the service upgrade is performed on the same day, many contractors supply temporary power using a portable generator. A portable generator can also provide power for homeowners

Section 240.4(D)(3–8) Small Conductor Rules

Conductor Size	Max Overcurrent Protection Device Size for Copper Conductors	Max Overcurrent Protection Device Size for Aluminum Conductors
14 AWG	15 amps	Not Permitted
12 AWG	20 amps	15 amps
10 AWG	30 amps	25 amps

Reproduced with permission of NFPA from NFPA 70, National Electrical Code, 2023 edition. Copyright © 2022, National Fire Protection Association. For a full copy of the NFPA 70, please go to www.nfpa.org

Figure 39-12. Sized 10 AWG and smaller conductors are not permitted to be protected by overcurrent protection devices larger than is listed in the small conductor rules of *Section 240.4(D)*.

that may want to ensure certain equipment like freezers or refrigerators stay running during the outage.

39.3.2 Power Transfer

Power transfer is required on service upgrades that have completely relocated service equipment or are transitioning from overhead to underground. With relocated equipment, a conduit containing feeder conductors or a run of SE cable as a feeder can be run from the new panelboard to the old panelboard. The conduit or cable may not enter the existing panelboard but will be ready to enter as soon as the panelboard is de-energized. If it is not desirable to install a larger feeder, smaller cables can be routed to the old panel location, where they are available to splice onto individual existing circuits, extending them to the new location.

A utility company energizes the new service equipment when it is ready and has been inspected, **Figure 39-13**. The old service is then disconnected by the utility at the same time. Most utilities only allow for a single meter at a residence, so when the new equipment is energized, the meter is transferred from the old equipment, which de-energizes the existing service. Removal of the meter on an energized system should only be done by utility personnel. In some cases, it may be necessary to temporarily feed the old equipment from the new service. This can be done by installing a temporary cable between the panelboards.

One of the last steps of any service upgrade is to properly identify the circuits within the panelboard. *Section 408.4(A)* requires equipment to be labeled in a manner that allows anyone to accurately understand the label. Avoid labels specific to the owner at the time of installation. For instance, instead of labeling it *Christina's bedroom*, label the breaker *2nd floor northeast bedroom*, or some other generic description, which applies regardless of the current owner or resident of the home. Labels should legible and indicate spare breakers if any exist.

When completing service upgrades, finding the existing circuits can be difficult. A circuit tracer is a helpful tool in determining which loads are supplied by each of the circuits, **Figure 39-14**.

Goodheart-Willcox Publisher

Figure 39-13. Service equipment must be inspected before it can be reconnected to the utility supply. The authority having jurisdiction will often leave a written record of the inspection as evidence that the equipment is ready for re-energization.

Goodheart-Willcox Publisher

Figure 39-14. Circuit tracing instruments can make the job of field labeling service equipment more efficient and effective.

Summary

- Electrical service upgrades are often done to increase capacity, increase breaker space, replace legacy equipment, and bring equipment up to date for compliance with industry standards.
- Existing equipment is sometimes located in places where it is inconvenient to work on, unsafe due to improper clearances, or not readily accessible to be maintained.
- When working on a service upgrade, determine if the service will be overhead or underground. When service equipment is supplied from the utility through an overhead drop, the overhead drop is subject to damage from ice, wind, and falling trees. It is also less aesthetically pleasing.
- Converting overhead services to underground reduces the risk of system failure due to weather and environmental aspects.
- Changes in codes that define working clearances may result in the need to move service equipment to a new location during the upgrade process.
- Copper conductors have more current capacity than aluminum conductors, but they are heavier and less flexible.
- If the PVC conduit used to supply service equipment is exposed to physical damage, Schedule 80 PVC should be installed.
- Metallic raceways connected on the supply side of the service equipment are required to have an additional bonding means outside of standard locknuts or fittings.
- Conduits exposed to large temperature differentials should be sealed to reduce condensation in the service equipment.
- Care should be taken to avoid damaging conductors when removing and reinstalling service equipment.
- Increases in current-carrying capacity of a service may require an increase in size of the grounding electrode conductor.
- Replacement of the grounding electrodes may need to be done to bring the service equipment into compliance during a service upgrade.
- Although it will provide extra protection, circuits not already protected by an AFCI are not required to be protected when the circuit breakers supplying the circuits are replaced.
- It may be necessary to provide temporary power through the use of a portable generator throughout the service upgrade installation.
- Coordination between the electrical contractor, the electric utility, and the electrical inspector is required when scheduling a service change.

Know and Understand

1. Which of the following is *not* considered a typical reason for an electrical service upgrade?
   A. To increase energy savings
   B. To provide expanded breaker space
   C. To increase the ampacity of the service equipment
   D. To replace poorly performing legacy equipment

2. Who is responsible for repairing a service mast that is damaged by a tree falling on the overhead service drop?
   A. The owner of the tree
   B. The electric utility
   C. The homeowner
   D. Everyone shares some responsibility

3. Service entrance conductors are permitted to extend _____ into a dwelling and terminate in the service disconnecting means.
   A. no more than 6′
   B. no more than 8′
   C. no more than 10′
   D. no further than nearest the point of entry

4. _____ PVC has a thicker wall and can provide more protection from physical damage than the standard PVC product.
   A. Schedule 30   C. Schedule 80
   B. Schedule 40   D. Schedule 90

5. SE cable is not permitted to be installed _____.
   A. in interior locations   C. in wet locations
   B. below grade   D. above grade

6. Which of the following is an advantage of aluminum conductors over copper conductors?
   A. It is lighter
   B. It is less flexible
   C. It has more current carrying capacity
   D. It allows for smaller raceways

7. Products used to seal raceways must be _____ for use with electrical conductors.
   A. approved   C. identified
   B. listed   D. labeled

8. All _____ grounding electrodes are required to be connected to the grounding electrode system of a dwelling unit.
   A. available   C. existing
   B. present   D. accessible

9. Only _____ should remove the electrical meter from the existing equipment and move it to the new service equipment in another location.
   A. qualified persons   C. licensed electricians
   B. competent persons   D. utility personnel

10. A(n) _____ is a useful tool for the identification of electrical circuits during the process of panelboard labeling.
    A. voltmeter   C. circuit tracer
    B. ohmmeter   D. wattmeter

Apply and Analyze

1. List two reasons why you might recommend that a homeowner change an overhead service to underground.
2. When an electrical service upgrade is performed, why might the new service equipment need to be relocated?
3. What are the advantages of installing service raceways rather than service entrance cables?
4. Other than cost, what factors would you consider when selecting aluminum or copper conductors?
5. Why should you seal service entrance conduits?
6. What is the minimum size grounding electrode for an underground metal water pipe if the electrical service is upgraded to 200 amps and is fed by 4/0 aluminum service entrance conductors?
7. During a service upgrade, what are four requirements for labeling the circuits of a newly installed panelboard?

Critical Thinking

1. A single-story ranch home is supplied by a 100-amp overhead service and is slated for an upgrade to 200 amps. A service mast runs through the roof overhang and terminates in the top of a meter socket mounted on the side wall of the home. The exterior service equipment can be located in the same place, but the panelboard inside the basement has to shift 8′ horizontally from the existing location. Discuss the material selection aspects of this installation, and determine what types of conductors, cables, or conduits you will use in your installation.
2. A remodeling project includes additional living space that will be built slab-on-grade adjacent to one wall of the existing structure. Discuss the factors that need to be considered to bring the grounding system into code compliance when the existing 60-amp service is replaced with a new 200-amp service.

Glossary

A

accent lighting. A type of lighting that illuminates a point of interest. (12)

accessible, as applied to equipment. Capable of being reached for operation, renewal, and inspection. (18)

adjustable wrench. A wrench with a moveable jaw that is adjusted using a thumbscrew. (3)

affidavit. A written statement affirming that an installation is compliant with the *NEC* and is safe to energize. (39)

aggressiveness. The rating of a saw blade based on its teeth per inch. A blade with a higher rating will have fewer teeth per inch. (3)

AIC rating. The level of fault current that a circuit breaker can safely interrupt. In a residential circuit breaker, this rating is typically 10,000 amps. (11)

airtight. A rating given to recessed luminaires that ensures the luminaire will restrict the flow of conditioned air into unconditioned spaces. (37)

alternating current (AC). Electrical current that reverses direction at regular intervals. (1)

ambient temperature. The temperature that surrounds an object on all sides. (8)

American Wire Gauge (AWG). A method for sizing electrical conductors. The larger the AWG number, the smaller the conductor diameter. (8)

amp. The unit of measure for electrical current. (1)

ampacity. The maximum current, in amps, that a conductor can carry continuously under the conditions of use without exceeding its temperature rating. (8, 19)

ampacity adjustment. This process ensures the wire that reaches all the panels and final utilization devices is adequately sized for all the factors that will impact the installation. (19)

anchor. An insert made of plastic or metal that increases the holding power of screws in a variety of materials. (6)

appliance branch circuit. A branch circuit that supplies energy to one or more outlets to which appliances are to be connected and that has no permanently connected luminaires that are not a part of an appliance. (34)

apprentice. Someone who works alongside a skilled master tradesperson to learn a trade firsthand. (1)

approximate estimate. A method of estimating a project based on the physical aspects of the project, such as a price per square foot of floor space. (16)

arc blast. The explosion of the electrical equipment as a result of the arc flash. (2)

arc fault. An unwanted condition in the electrical circuit that causes arcing and sparking between two points. (2)

arc-fault circuit interrupter (AFCI). A safety device that opens the electrical circuit in the event of an arcing fault. (2)

arc-fault circuit interrupter (AFCI) receptacle. A receptacle outlet that protects from the effects of arc faults by recognizing the characteristics unique to arcing and de-energizing the circuit when an arc fault is detected. (10)

arc-fault circuit interrupting (AFCI) breaker. A type of circuit breaker designed to provide arc-fault protection to the entire branch circuit. (11)

arc flash. The light and heat from an arc. (2)

architects' scale. A measuring instrument used on plans for buildings and structures. (16)

architectural detail. An enlarged view of a portion of the plan to provide greater clarity drawn at a larger scale than the floor plan. (16)

architectural drawing. The most basic floor plans and elevations that will be used by all other trades; identified by the letter *A*. (16)

architectural section. A cut-away view into the building. (16)

article. The largest of the chapter subsections. They each cover a general topic. (15)

attached garage. A portion of the house that is designed to allow vehicle parking inside sharing at least one wall and featuring access to the home. (34)

auger bit. A bit used exclusively for drilling into wood. This bit has large, deep flutes to expel the waste chips and a screw tip to pull the bit through the material. (4)

authority having jurisdiction (AHJ). An organization, office, or individual responsible for enforcing the requirements of the *NEC*. They also approve equipment, materials, installations, or procedures. (1, 15)

Note: The number in parentheses following each definition indicates the chapter in which the term can be found.

automatic transfer system. A power system configuration that is powered by a standby power source and use magnetic solenoids to automatically switch the home from the utility to the standby source. (35)

authority having jurisdiction (AHJ). Electrical inspector responsible for enforcing the requirements of the *NEC*. (15)

B

balance of system. In a photovoltaic (PV) system, all components, devices, equipment, and services other than the PV array. (33)

ballast. A device that provides a high voltage start to fluorescent lamps and regulates the current flow while the lamp is in operation. (12)

balloon framing. A method of residential framing in which the studs of the exterior walls run from the top of the foundation to the ceiling of the highest level. (17)

bandwidth. A measure of the rate of data transfer, typically measured in bits per second (bps) modified by a metric prefix such as k, M, or G for kilo, mega, and giga. (32)

bidirectional power. Power that travels in either direction through a system. (33)

bimetallic strip. A strip consisting of two dissimilar metals fused together that exhibit a warping movement when exposed to heat. (11)

bird's mouth. A notch made in the rafter that allows the rafter to sit flush on the top plate of the wall. (17)

bit extender. A drill bit accessory used to increase the reach of a bit or drill deeper holes. (4)

blow-out. The damage that occurs when the bit exits the other side of the workpiece. A term generally applied to concrete and masonry. (4)

blueprint. Building plan used in the construction process and for the planning and installation of the electrical system. (16)

body. The portion of the drill bit between the shank and the point. It has the spirals that remove the chips. (4)

box extender. An accessory attached to the front edge of an outlet box to extend the front edge of the box past wall surfaces. (22)

box fill calculation. Identifying the maximum fill limitations and the properly sized box to provide adequate space for the conductors and devices they contain. (22)

box wrench. A wrench with a fixed head that completely encircles the nut or the bolt's head. They are used to tighten and loosen bolts and nuts. (3)

brad-tip bit. A wood-drilling bit with a center tip that keeps the bit from wandering when starting a hole. (4)

branch circuit. The conductors between the final overcurrent protection device and the outlets being served. (18)

bus bar. An underground copper or aluminum bar located vertically in the center of the panel. (28)

bushing. A conduit fitting used on the ends of the conduit to prevent abrasion of the conductors where they emerge from the conduit. (9)

C

cable connector. An electrical fitting used to secure cables to electrical boxes and equipment. (8)

cable cutter. A tool that cuts cables or large conductors without distorting the cable or conductor's ends. (3)

cable mast. This type of overhead service uses a length of Type SEU cable attached directly to the building as the service mast. (25)

cable ripper. A tool that quickly slices the outer sheathing of NM cable without damaging the inner conductors. (3)

cable system. A wiring method that employs multiple insulated current-carrying conductors and a bare or insulated equipment grounding conductor, all contained in a common outer sheathing. (8)

calculated load. A reduction in the connected load using rules of the *NEC*. (20)

callout. A sequence of numbers that refer to a screw's diameter, threads per inch, and length. (6)

canless luminaire. A very thin luminaire that can be installed just about anywhere. (27)

cat's paw. A type of pry bar used to remove embedded nails from wood. (3)

change order. A revision in the agreed-upon cost of an installation, agreed on by all parties and approved in writing. (16)

chase. A void in the building construction allowing for the installation of pipes, wire, and ductwork vertically through the building. (38)

chuck. The part of a drill that grabs the bit. Some require the use of a tool called a *key* to tighten. Most cordless drills have one that is tightened and loosened by hand. (4)

circuit breaker. A device that protects the conductors in an electrical circuit from the excessive current of a ground fault or short circuit by opening the circuit when the current exceeds the device's rating. These can be reset when tripped. (11)

circuit breaker finder. A two-piece diagnostic tool that uses a plug-in transmitter and a hand-held receiver to determine which circuit breaker protects a given circuit. (5)

circular mil (cmil). A unit of area equal to the area of a circle with a diameter of one-one thousandths of an inch. (8)

circular saw. A power saw that uses a disk-shaped blade and a spinning, rotary action to cut different materials. (4)

clamp meter. A non-contact meter that measures electrical current by sensing the strength of the magnetic field around the conductor. (5)

Class P ballast. A ballast with thermal protection that disconnects it from the power source if it overheats. (12)

clothes closet storage space. The areas in a closet where items may be hung from a rod or stored on a shelf. (27)

clutch. A feature in some drills that sets the amount of torque delivered. Used to prevent screws from driving too deep. (4)

CO/ALR. A rating stamped on receptacles to indicate the device is suitable for copper, aluminum, or copper-clad aluminum conductors. (10)

coarse-thread screw. A type of screw with sharp tips and more space between each thread than a machine-thread screw. (6)

coaxial cable. Cable that consists of concentric layers of conductor and insulation that creates a capacitance which reduces interference from other electrical sources. The most basic construction includes an inner conductor covered by a dielectric (insulator), surrounded by a conductive shield, and covered with an outer insulating jacket. Also called *coax cable*. (32)

Code-Making Panel (CMP). A team of authors from the electrical industry that are responsible for writing portions of the *NEC*. (15)

collar ties. A horizontal framing member that ties opposing rafters together to add strength to the roof frame. (17)

color temperature. A measure, in degrees Kelvin (*K*), of a light's coolness or warmth. (12)

color-rendering index (CRI). A measure, from 0 to 100, of how well a light source reproduces an object's actual color compared to how it appears in natural daylight. Also called *color rendering scale*. (12)

combination device. A device with two devices on the same yoke, such as two switches or a switch and a receptacle. (10)

combination meter socket/disconnect. A pre-assembled, more expensive device that combines a meter socket and service disconnecting means. (24)

combination wrench. A wrench that has one open end and one box end of the same size. (3)

compact fluorescent lamp (CFL). A compact, energy-efficient lamp designed to replace incandescent lamps. This lamp has a built-in ballast at its base. (12)

concrete-encased electrode. A type of electrode that uses the conductivity of concrete and provides the best system grounding with the lowest impedance to earth than other types of electrodes. Also known as the *Ufer ground*. (24)

conduit body. A portion of the conduit system that provides access through a removable cover to its interior. (9)

conduit mast. This type of overhead service uses a length of conduit to protect the overhead service conductors or cables from weather and any physical damage. (25)

conduit-supported. Method for supporting receptacle outlets where metal device boxes are installed on two or more threaded steel conduits that provide support for the box. (30)

connected load. The nameplate rating of an appliance or piece of electrical equipment and represents the actual amount of consumed power. (20)

connector. A conduit fitting used to join a conduit to a junction box, outlet box, or piece of equipment. (9)

constant-power receptacle. A receptacle that is not controlled by a switch but always remains energized. (14)

continuity. An indication of an unbroken connection between two conductive points of circuit. (5, 36)

continuous load. An electrical load where the maximum current is expected to continue for three hours or more. (34)

cordless screwdriver. A small, lightweight power tool used to install and remove screws or to drill small-diameter holes. (4)

corrosive environment. The environment where swimming pool sanitation chemicals are stored and the air contains acid, chlorine, and bromine vapors, which required specific wiring methods. (31)

coupling. A conduit fitting that joins two pieces of conduit together. (9)

crimp connector. A solderless method of terminating electrical conductors. (3)

crimping pliers. A tool used to crimp bare or insulated terminal devices. (3)

cripple. Stud that supports the rough sill. (17)

crosscut. Cutting a length of board across the grain. (17)

crosstalk. Electrical interference in a communication system conductor caused by electromagnetic induction originating from a second communication system conductor. (32)

CU. A rating stamped on receptacles to indicate the device can only be connected to copper conductors. (10)

current. A quantity measurement of the number of electrons passing a point over a period of time. (1)

current rating. The amount of current the device can carry without sustaining damage. (37)

cut-in box. An outlet box with features designed specifically for installation in existing walls. Also called *old work boxes*. (22)

D

damp location. An area subject to moderate degrees of moisture, as in the case of a roofed open porch. (7)

demand factor. The ratio of the maximum demand of a system, or part of a system, to the total connected load of a system or the part of a system under consideration. (20)

demand profile. A graphic representation of energy use over time. (1)

derate. Technique to reduce the amount of current that wires are allows to carry, taking into account ambient temperature. (19)

detached garage. A building or structure which is not part of the main dwelling construction and designed to accommodate parking or storage of one or more vehicles. (34)

detailed estimate. A very accurate estimating method that involves counting and pricing every item to be supplied and installed and knowing with reasonable certainty how long the project will take to complete. (16)

diagonal cutting pliers. Pliers used for cutting cables and conductors in tight spaces. (3)

dimmer switch. A switching device that can also control the lighting output level. (10)

direct current (DC). A type of electricity that does not reverse its direction of current flow. It is typically associated with batteries. (1)

direct lighting. A luminaire that focuses 90% or more of the light produced in the downward direction. (37)

disconnect. A piece of electrical equipment typically used to remove power from a piece of electrical equipment for routine service or replacement. (11)

distribution transformer. A utility transformer, either pole-mounted or pad-mounted, that reduces distribution-level voltage to consumer-level voltages of 240/120 volts, nominal. (1)

dormer. A roofed structure that projects out from a sloped roof. (17)

double-pole branch circuit. Either a 240-volt three-wire circuit to supply an appliance, such as a water heater, electric furnace, or baseboard heater, or a 120/240-volt four-wire circuit, such as a clothes dryer or cooking range. (28)

double-pole circuit breaker. A circuit breaker used for 240-volt and 120/240-volt circuits; rated 15 amps and larger, but typically not over 100 amps; and takes up two spaces in the panelboard. Also called *two-pole breaker*. (11, 28)

driver. An electronic device that converts the incoming ac power into the dc power needed for LED operation. (12)

driver bit. A bit used in drills to install and remove screws and other fasteners. (4)

dry-niche luminaire. A luminaire intended for installation in the floor or wall of a pool, spa, or fountain in a niche that is sealed against the entry of water. (31)

dual-element time-delay fuse. A fuse that uses both a fusible link and a solder link to operate. (11)

dual-purpose circuit breaker. A breaker that provides both AFCI and GFCI protection. (28)

dual-purpose GFCI/AFCI breaker. A circuit breaker that provides both AFCI and GFCI protection to branch circuits. (11)

duplex receptacle. Two receptacle outlets mounted on a common yoke. The most common device installed in a residence. (10)

dwelling unit. A single unit, providing complete and independent living facilities for one or more persons, including permanent provisions for living, sleeping, cooking and sanitation. (18)

E

easement. The right of one party to cross or access another's property for a specific use. (16)

eave. The portion of the roof that extends past and overhangs the outer walls of the house. (17)

Edison base fuse. A plug-type fuse type with interchangeable fuse ratings. (11)

effective ground-fault current path. A low-impedance circuit facilitating the operation of the overcurrent device and capable of safely carrying the maximum ground-fault current likely to be imposed on it from any point on the wiring system where a ground fault may occur back to the electrical supply source. (25)

electric vehicle supply equipment (EVSE). An electrical assembly of electrical apparatus designed to interface with an electric vehicle installed specifically for the purpose of transferring energy between the premises wiring and the electric vehicle. (34)

electrical arc. A condition that occurs in an electrical circuit when electricity bridges a gap in the circuit using the air as a conductor. (2)

electrical box. An enclosure that holds various electrical connections. (7)

electrical conductor. The wires that tie all the other electrical components together to form the residence's electrical system. (8)

electrical conduit. A channel, usually of a circular cross-section, made of metallic or non-metallic material, used to route and protect electric conductors. (9)

electrical construction permit. A document issued by the local municipality to a qualified electrician. It is, in essence, an agreement that the work will be done in a *Code*-compliant manner and will be subject to inspections by the AHJ. (1)

electrical fault. A condition that occurs when electricity follows an unintended path in an electrical circuit. (1)

electrical license. A State-issued document issued to individuals who have proven their electrical systems knowledge through examination and verified work history in the electrical field. (1)

electrical metallic tubing (EMT). An unthreaded thin wall raceway with a circular cross section designed for conductors and cables' physical protection and routing. (9)

electrical plan. A floor plan that uses a set of symbols to convey the locations of switches, receptacles, lighting outlets, and other aspects of the electrical system. (16)

electrical service upgrade. A process of replacing the existing service equipment with new, typically larger, equipment. (39)

electrical troubleshooting. The process of involves systematically examining a nonworking electrical circuit to determine the root problem. (29)

electrician's hammer. A hammer with a long, straight claws and a flat striking face. This tool is used to drive and pull nails, pry electrical boxes from framing members, and strike awls and chisels. (3)

electrician's tool belt. A belt with pouches used to hold an electrician's hand tools. (3)

electrocution. Exposure to a lethal amount of electricity. (2)

elevation. A type of orthographic projection that show the vertical view of the building's exterior sides or an interior feature, such as a built-in bookcase or kitchen cabinet. (16)

end-cutting pliers. A cutting tool used to trim nails and other items close to the surface without damage. Also called *nippers*. (3)

engineered wood products. Fabricated building materials that enhance the natural strength of wood through the manufacturing process. (17)

engineers' scale. A measuring instrument used with larger-scaled land planning projects. (16)

equipment bonding. The connecting of all normally non-current-carrying electrically conductive enclosing electrical equipment or conductors together and to the electrical supply source in a manner that establishes an effective ground-fault current path. (25)

equipment grounding. The connecting of all normally non-current-carrying electrically conductive materials together and to the earth to stabilize the voltage to ground on these materials. (25)

equipment grounding conductor. A conductive path that connects normally non-current-carrying metal parts of equipment together. It is used to help the fuse or circuit breaker clear ground faults. (1)

equipotential bonding. The bonding of all metal parts related to the pool and metal objects in the pool vicinity to reduce voltage gradient in the pool water and the pool area. (31)

ethernet cable. A network cable used for high-speed wired connections between two devices with ethernet ports. (32)

Exception. An alternate *Code*-compliant method or use of materials. (15)

explanatory material. Information included in the *NEC* that may reference additional codes or standards, other sections of the *NEC*, or additional information that helps apply the particular *Code* section. (1)

F

F-clip. An F-shaped device used to secure metallic boxes to the wall's surface. (7)

facia. The front face of the eave. (17)

fastener. Any mechanical component used to join parts together. (6)

feeder cable. The electrical conductors between the service disconnecting means and the branch circuit distribution panel, or from a breaker in the main distribution panel to a subpanel. (8)

feeder conductor. The circuit conductors between the service equipment and the final branch circuit overcurrent device. (24)

female adapter (FA). A PVC fitting that is threaded on the inside. (9)

fiber optic cable. A cable that uses light signals to transmit information through glass or plastic fibers. Also called *optical fiber cable*. (32)

finish nail. A nail with a rounded, dimpled head that is meant to be counter-sunk using a nail setting tool. (6)

fish sticks. Rigid fiberglass rods used to fish electrical wires through walls or ceilings. (3)

fish tape. A length of steel or fiberglass, usually stored in a reel that is used to pull electrical conductors through walls and conduits. (3)

fitting. A part of the electrical system that performs a mechanical rather than an electrical function. (9)

flex bit. A long, flexible bit used to drill holes through several stud-bays or through blocks in walls or ceilings. (4)

floor plan. A type of orthographic projection that shows the layout of the house as viewed from above. (16)

fluorescent lamp. A gas-discharge lamp that produces visible light through a reaction with the lamp's phosphor coating. (12)

flush-mounted luminaire. A luminaire mounted directly to the surface of the ceiling or wall. (27)

foot-candle. A measurement of illumination, or light intensity; equal to one lumen per square foot. (12)

four-way switch. A switch that can control a lighting load from three or more locations. These switches are always installed between a pair of three-way switches. (10)

four-way switched circuit. A switched circuit that controls a lighting load from more than two locations. (13)

four-wire service. When the panelboard is fed from a feeder conductor. The feeder contains a fourth conductor, an equipment grounding conductor. (24)

framing nail. A nail that is pointed on one end and has a smooth, flattened head on the other end. Also called a *common nail*. (6)

fuse. An electrical device that protects the conductors in an electrical circuit from excessive current by opening the circuit when the current exceeds the device's rating. They are generally replaced when blown. (11)

G

ganging. A process of removing the sides of a metallic box to create a larger box. (7)

gem box. A metallic box that has a set of adjustable ears to the hold the box flush with the surface after installation and removable sides which allows multiple boxes to be combined or ganged to provide more openings for multiple devices. (37)

general diffuse light. A luminaire that allows light to spread in all directions. (37)

general illumination. A baseline light level across a space. (37)

general lighting. The overall lighting of a room or space. Also called *ambient lighting*. (12)

general method. Applying several demand factors to the total connected load to arrive at the calculated load. (20)

general-purpose branch circuit. A branch circuit that supplies two or more general purpose receptacles or outlets for lighting and appliances. (18)

general-use drill. A drill used to make holes in various materials and to drive and remove screws. (4)

grain. The direction of the wood cell fibers in a piece of lumber. (17)

grid-tied PV system. A solar photovoltaic system that is interconnected with the utility electrical supply system. (33)

ground fault. An accidental contact between an energized conductor and a grounded conductor or grounded equipment frame. (1)

ground-fault circuit interrupter (GFCI). A shock-prevention device that monitors the current flow in a circuit and acts to open the circuit in the event of a current imbalance. (2)

ground-fault circuit interrupter (GFCI) receptacle. A receptacle that opens the circuit when a ground fault condition is detected. (10)

ground-fault circuit interrupting (GFCI) breaker. A type of circuit breaker designed to provide ground-fault protection to the entire branch circuit. (11)

grounded conductor. The conductor in an electrical circuit that carries the current from the load back to the power source. It may or may not be the "neutral conductor." (1)

grounding electrode. A conductive object that has a direct connection to the earth. (24)

grounding electrode conductor. A conductor that connects the grounded service conductor to the grounding electrode. (24)

H

hacksaw. A hand saw with a fine-toothed, replaceable blade used to cut metal or PVC conduit, strut channel, and larger conductors and cables. (3)

half-split method. A technique for troubleshooting that separates the circuit a point halfway between the source and the load. (36)

hammer drill. A drill that uses a percussive motion to aid in penetrating brick, concrete, or other masonry material. (4)

hand bender. A tool used by electricians to form and bend metal conduit by hand. (3)

handy box. A common type of metallic box with rounded corners that is primarily used for surface mounting. Also called a *utility box*. (7)

hardwood. Wood that comes from slow-growing deciduous trees such as oak or maple. (17)

header. A framing member attached to the sill plate that secures the floor joists. Also the framing member that transfers the weight of the building around openings in walls such as windows and doors. (17)

hex wrench. A tool with a hexagonally shaped cross-section that fits the recessed holes in hex fasteners. Also called an *Allen wrench* or *hex key*. (3)

high-speed steel. A type of steel alloy used in the making of drill bits. (4)

home run. The length of conductor from the panel to the first outlet in the room. (16)

hole saw. A cylindrical saw blade used in power drills to cut holes in thin material such as plywood or siding. (4)

hot tub. A hydromassage pool or tub for recreational use, not located in health care facilities, designed for the immersion of users, and usually having a filter, heater, and motor-driven blower. It may be installed indoors or outdoors, on the ground or supporting structure, or in the ground or supporting structure. Generally, it is not designed or intended to have its contents drained or discharged after each use. These rules also apply to *spas*. (31)

hydromassage bathtub. A permanently installed bathtub equipped with a recirculating piping system, pump, and associated equipment. It is designed so it can accept, circulate, and discharge water upon each use. (31)

I

I-joist. An engineered wood product used for floor joists or rafters that has an *I*-shaped profile. It is constructed of upper and lower chords and webbing of OSB. (17)

impact driver. A tool for installing and removing screws. This tool uses a percussive motion to increase its available torque. (4)

in-line ammeter reading. A method of measuring electrical current using the test leads of the meter in series, or in line, with the load. (5)

incandescent lamp. A lamp that uses a filament suspended in a glass bulb. When current flows through the filament, it glows and produces visible light. (12)

incidental. Small items which would be impractical to charge for individually, such as staples, wire nuts, and cable clamps. (16)

indirect lighting. A luminaire that focuses 90% or more of the light produced in the upward direction. (37)

individual branch circuit. A branch circuit that only supplies a single load. (18)

infinite resistance. A condition of an electrical circuit where there is unlimited opposition to current flow, such as an open switch. (5)

informational note. Explanatory material that aids in understanding or applying the requirements they follow. (15)

inrush current. Momentary excess current that occurs during motor startup and ends when the motor obtains its running speed. (36)

insulated tools. Tools covered in a nonconductive coating that allows an electrician to work on energized circuits more safely. (3)

insulation contact (IC). Rating that indicates if a luminaire can be in contact with thermal insulation. (27)

insulation contact (IC) rated. A rating given to recessed luminaires where the design of the luminaire limits fire hazards associated with heat buildup within the luminaire. (37)

interconnection. The connection of a power source to operate as part of or to parallel with a source of power to a building. (35)

intermediate metal conduit (IMC). A steel threadable raceway with a circular cross section designed for conductors and cables' physical protection and routing. (9)

internal clamp. Device used to secure the wire to a box to prevent the conductor from moving. (22)

intersystem bonding connection. A device for the interconnection of the grounding electrode conductors of other systems such as the telephone, cable or satellite TV, and broadband to the system ground of the dwelling service. (25)

intersystem bonding termination (IBT). A device used to connect communication system bonding conductors to the grounding electrode of the electrical system. (32)

inverse-time breaker. A type of circuit breaker with a time delay that decreases as the magnitude of the current increases. (11)

inverter. An electrical device that converts dc power into ac power. (33)

inverter output circuit. The ac circuit on the output side of an inverter or microinverters. (33)

J

jab saw. A handsaw used to cut holes for device boxes in sheetrock and thinner wood paneling. Also known as a *sheetrock saw* or *keyhole saw*. (3)

jack chain. A metal link chain made of bent loops often shaped in figure eight pattern and used for supporting light objects like suspended luminaires. (38)

jigsaw. A smaller type of reciprocating saw that can make cuts with more precision than the larger reciprocating saw. (4)

joist. A horizontal framing member that supports floors and ceilings. (17)

junction box. An electrical box where two or more wires are joined and there is no device attached. Also called a *J-box*. (7)

K

keystone jack. A female RJ-45 connector that is generally installed in a wall plate. (32)

king stud. The full-length stud on either side of a door or window header. (17)

knob and tube system. A wiring method that consists of single insulated conductors supported by insulating knobs and pass-through framing members through insulated tubes. (37)

knockout. A plug that is partially punched in electrical enclosure designed for easy removal to allow for the installation of wiring and raceways. (38)

L

laminated veneer lumber (LVL). A manufactured wooden beam made of thin veneers with all the grains oriented in the long direction and glued under high pressure. (17)

lamp. A receptacle component that produces light from electric power. (12)

laser plumb. A tool that uses lasers to determine level and plumb. (3)

lath and plaster. A system of wood slats affixed to the wall framing and covered with a layer of plaster. (37)

LED lamp. A lamp that uses light-emitting diodes to produces light. (12)

legend. A table that identifies the symbols used on a building plan. (16)

level. Parallel with the floor or ground. (3)

line termination. Terminal screws for incoming power to the receptacle or other device or appliance. (26)

line/line configuration. Wiring configuration in which only the GFCI device shuts off under a ground-fault condition while other connected devices remain unaffected. (26)

line/load configuration. Wiring configuration that allows a GFCI receptacle to also protect additional outlets wired downstream. (26)

lineman (side cutter) pliers. Pliers used to cut cables, conductors, and small screws, and pull and hold conductors. They may also have additional helpful features like crimping dies, insulation strippers, or a fish tape grip. (3)

liquidtight flexible nonmetallic conduit (LFNC). A raceway of a circular cross section with a seamless inner core and cover bonded together, which may have one or more layers of reinforcement between the core and the covers. This type is generally used in wet locations where flexibility is needed. (9)

load. The portion of the electrical circuit that converts electrical energy into some other form of energy. (1)

load-bearing wall. A wall that supports the weight of the building above it. (17)

load center. A metal enclosed panelboard containing circuit breakers that supply the home's power and lighting circuits. (36)

load diversity. 1. All the connected loads are not energized simultaneously so many loads cycle off and on throughout the day. 2. The total expected power, or load, to be drawn during a peak period. (20, 35)

load termination. Terminal screws that supply power to downstream outlets. (26)

lockout/tagout. A safety practice used extensively in electrical work to prevent the accidental energization of a piece of equipment being serviced. (2)

long-nose (needle-nose) pliers. Pliers with a long, thin nose. They are used to form small conductors, cut, and hold and pull conductors. They are ideal for gripping or retrieving objects in tight spaces. (3)

low-speed, high-torque drill. A type of drill that uses gearing to slow the speed of rotation and provide a considerable turning force or torque. It is used to drill large diameter holes in hardwoods. (4)

low-voltage contact limit. A voltage not exceeding values of 15 volts AC. (31)

lumber. Wood that has been cut to specific dimensions for use in construction. (17)

lumen. The measure of the amount of light a lamp produces. (12)

luminaire. A lighting fixture. (7)

luminaire testing. The process of testing for proper operation of all luminaires and their switching circuits. (29)

luminous efficacy. The ratio of the quantity of light output to the amount of electric power consumed in creating the light. (12)

M

machine-threaded screw. A screw with threads that are designed to fit a matching nut or another threaded component that is generally made of metal. (6)

main bonding jumper. The connection between the grounded circuit conductor and the main equipment grounding conductor at the service; typically a green screw or copper strap in residential wiring. (24)

main circuit breaker. A circuit breaker that feeds and protects an entire panelboard. (11)

major appliance. Appliance that is served by an individual branch circuit supplying no other loads. (26)

mandatory rule. Action in the *Code* that is specifically required or prohibited. (1, 15)

manual transfer system. A power system configuration that requires a person to switch the power source from the utility to the standby system. (35)

masonry bit. A drill bit with a flared cutting tip that is used with a hammer drill to drill into hard surfaces such as concrete, brick, or stone. (4)

masonry chisel. A demolition tool used to break up concrete, brick, or tiles. (3)

messenger wire. A steel cable or a wire that is run along with or integral with a cable or conductor to provide mechanical support for a cable or conductor. (30)

metal file. A tool with closely spaced, hardened-steel grooves that shape, trip, and smooth any material made of metal, wood, or plastic. (3)

metallic box. An electrical box made of metal, usually steel. (7)

meter socket. A device that encloses the service conductors and has a set of terminals for the termination of the service conductors and service entrance conductors. (24)

microgrid. A local, self-sufficient electrical grid separate from the utility grid. (33)

microinverter. A small inverter connected to a single PV module. (33)

multimeter. An all-in-one device that measures volts, values of resistance in ohms, and current in amps. Also called a *VOM (volt-ohm-milliammeter)* or a *DMM (digital multimeter).* (5)

multipurpose tool. A single tool capable of multiple functions, such as cutting, stripping, and crimping of conductors. (3)

multiwire branch circuit. A branch circuit that consists of two or more ungrounded conductors that have a voltage between them, and a neutral conductor that has equal voltage between it and each ungrounded conductor of the circuit and that is connected to the neutral conductor of the system. (34)

N

nail plate. A steel plate that is at least 1/16″ thick that is installed on the face of a framing member to protect a cable from nails or screws. (23)

National Electrical Code (NEC). A standard for the safe installation of electrical wiring and equipment published by the National Fire Protection Association. (1)

National Fire Protection Association (NFPA). A trade association that publishes hundreds of industry codes and standards, including the *National Electric Code.* (15)

NEMA 3R rating. A rating for electrical equipment designed to be installed where exposed to the elements. (11)

neutral conductor. A grounded conductor in an electrical circuit that carries the imbalance between two ungrounded conductors. (1)

new work box. A box that is installed in new construction where there is complete access to the studs. (7)

nominal. In name only. (1)

no-niche luminaire. A luminaire intended for installation above or below the water without a niche. (31)

noncoincident load. Electrical loads that are unlikely to be energized at the same time. (20)

non-contact voltage tester. An electrical device that detects the presence of voltage without having to make direct contact with the conductors. (5)

non-metallic box. An electrical box made of PVC plastic, fiberglass, or other non-metal material. (7)

nut driver. A type of driver used to install or remove nuts, bolts, and hex-head screws. (3)

O

Occupational and Health Administration (OSHA). A division the United States Department of Labor with a mission to ensure safe and healthful working conditions. (2)

off-grid PV system. A solar photovoltaic system that provides all the electrical needs of a residence and is not connected to utility electrical supply. (33)

ohm. The unit of measure of resistance in an electrical circuit. (1)

Ohm's law. An electrical principle that states that current flow in a circuit is directly proportional to the applied voltage and inversely proportional to the resistance of a circuit. (1)

open-end wrench. A wrench similar to the box wrench but with an open head. (3)

old work box. A box that is installed in an existing dwelling where the drywall has already been added. (7)

open web truss. An engineered wood product used for floor joists. It is constructed of top and bottom chords and webbing of vertical and diagonal wood members. (17)

optional method. Applying a single demand factor to the total connected load to determine the calculated load. (20)

oriented strand board (OSB). A sheet lumber material made from various sizes of wood flakes or chips glued together under high pressure. (17)

orthographic projection. A method of rendering a three-dimensional object in only two dimensions, where each view of the building is drawn separately. (16)

oscillating tool. A cutting tool that uses a slight, high-speed side-to-side motion. (4)

outlet. A point in the electrical system where electrical energy is taken to supply utilization equipment. (10)

overcurrent. Any current above the rated current of the equipment or the current-carrying ability of the conductor. Ground faults and short circuits cause these conditions. (1)

overcurrent protective device (OCPD). An electrical device that functions to open the circuit under a fault condition. (11)

overhead conductor span. A method of open wiring installed outdoors using a horizontal conductor or cable strung between supports that insulate the conductors from the support structures. (30)

overhead temporary service. A type of temporary service that uses a utility pole-mounted transformer. (21)

overload. Overcurrents in which low to moderate amounts of current flow in excess of the circuit rating without leaving the normal wiring pathway. (36)

P

P.A.S.S. An acronym to help remember the proper use of a fire extinguisher. First, **p**ull the pin on the extinguisher so you can squeeze the handle, **a**im the nozzle of the fire extinguisher to the base of the fire, and **s**queeze the handle to deploy the extinguisher. Stay six to eight feet from the fire if possible and **s**weep the nozzle back and forth at the base of the fire. (2)

paddle bit. A drill bit with a wide, flat profile and used for rough boring in wood. Also called a *spade* or *butterfly bit*. (4)

panelboard. An electrical enclosure containing fuses or circuit breakers for the protection of branch circuit conductors. Also called *distribution panels*. (11)

parallel arc. An arc across two or more conductors. (2)

part. The first subsection of an *NEC* article. They are designated by Roman numerals. (15)

peak hours. The portions of the day when electrical power consumption is at its highest levels. (1)

pendant chain tool. A tool specifically designed to open and close chains used to hang pendant luminaires. (3)

pendant luminaire. A luminaire hung from the ceiling from several inches to several feet, depending on the installation. (27)

penny system. The method of sizing nails in the United States, abbreviated by the letter "*d*." (6)

permanently installed swimming pool. Constructed in the ground, partially in the ground, or above ground, and they must be capable of holding water in a depth greater than 42″. (31)

permissive rule. Action in the *Code* that is allowed but not required. (1, 15)

personal protective equipment (PPE). Equipment worn to minimize exposure to hazards that cause serious workplace injuries and illnesses. Typically includes eye protection, hearing protection, gloves, proper footwear, and head protection. (2)

photovoltaics (PV). The conversion of light energy into electrical energy. (33)

pigtail. A short length of conductor used for the connection of devices. (10)

pitch. An indication of the steepness of the roof. It is calculated by the number of inches of rise to every 12″ of run. (17)

plaster ring. A device for installing switches and receptacles to a 4″ × 4″ electrical box. They are available in single-gang and two-gang configurations and come in a variety of depths. (7)

platform framing. A framing method where each floor is framed individually and used as the platform for the construction above it. (17)

plumb. Perpendicular, or at a right angle, to the floor or ground. (3)

plumb bob. A weight with a pointed bottom supported at the top by a length of string. Used by electricians to transfer points on the floor to the ceiling or vice-versa. (3)

plywood. A sheet lumber material constructed of several veneers that are cross-stacked according to their grain and glued together under high pressure. (17)

point. The part of the drill bit that does the actual cutting. (4)

point of entrance. The point at which the conductor, cable, or raceway emerges into the building or structure from the exterior wall or floor. (39)

point of service. The point at which the utility wiring ends and the premises wiring begins. (39)

pool. Designed to contain water on a permanent or semipermanent basis and used for swimming, wading, immersion, or therapeutic purposes. (31)

post light. A luminaire with a support pole. (30)

post-and-beam framing. A framing method that uses larger framing members spaced further apart than in other framing methods. (17)

post-supported. Method for supporting receptacle outlets where receptacles use another structure such as a wood post or metal stake to support the boxes. (30)

power. The rate at which work is performed. (1)

power factor. The ratio of power consumed by the resistive loads to the electrical energy required by the generator to operate the circuit. (35)

power loss. Unnecessary heating measured in watts. (19)

precision screwdriver. A screwdriver with tiny tips. Also called a *jeweler's screwdriver*. (3)

pressure treated wood. Lumber that has been processed under pressure with a water/copper solution that acts as a preservative to slow rot. (17)

prybars. A tool used to pull or pry two objects apart. (3)

pull-out disconnect. A disconnecting means that uses a removable block to disconnect an appliance from its source of power. (11)

pump pliers. Pliers with a sliding jaw that locks in place at intervals. They are used to tighten and loosen nuts and bolts, hold and tighten raceway couplings and connectors, and hold and turn conduit and tubing. (3)

punch down tool. A tool used to connect communication wire to a connector by pressing the wire into an opening. (32)

push-in connector. An electrical fitting used to join multiple electrical conductors by pushing the conductor into a port where it is held in place by spring tension. (8)

PV array. An assembly of PV modules connected in series or parallel. (33)

PV module. An assembly of solar cells in a support frame. Also called *solar panel*. (33)

Q

quick-rotating driver. A screwdriver with an offset shaft that is free to rotate in the handle. (3)

R

rafter. A diagonal structural member of a roof that supports the roof decking. (17)

raintight. Constructed or protected so that exposure to a beating rain will not result in the entrance of water under specified test conditions. (30)

raised cover. A type of cover used to install devices in surface-mounted 4″ × 4″ electrical boxes. (7)

rapid shutdown. Disconnection of a PV system in an emergency power off condition. (33)

rasp. A file with coarse, punched teeth made specifically for shaping wood. (3)

ratchet set. A two-piece tool that uses a ratcheting handle and interchangeable socket heads of different sizes to tighten and loosen bolts and nuts. (3)

reactive load. Load that uses magnetic fields to store energy and return some of the stored energy back to the source. (35)

readily accessible. Capable of being reached quickly for operation, renewal, or inspections without requiring those to whom ready access is requisite to take actions such as to use tools (other than keys), to climb over or under, to remove obstacles, or to resort to portable ladders, and so forth. (18)

receptacle. A device installed for the cord and plug connection of a piece of electrical equipment. (10)

receptacle tester. A plug-in device that can diagnose the wiring of a standard receptacle. (5)

receptacle testing. The process used to determine proper wiring and detect circuit conditions, such as open ground, open neutral, open hot, hot/ground reversed, hot/neutral reversed, and correct wiring. (29)

recessed can luminaire. A luminaire composed of two parts: the housing and the trim. (27)

reciprocating saw. A saw in which the blade moves in a back-and-forth motion to make a cut. (4)

request for information (RFI). A written request for clarification when the plans or specifications are incomplete or unclear. (16)

resistance. The opposition to electrical current flow. (1)

resistive load. A load that consumes power by converting electricity to heat and light as a direct result of electrical current. (35)

résumé. A document outlining an individual's education, skills, and work history. Typically used in job searches. (1)

ridge beam. The framing member at the peak of the roof where the top of the rafters attach. (17)

right angle drill. A drill with an angled head used for drilling between framing studs or other tight spaces. (4)

rigid metal conduit (RMC). A steel threadable raceway with a circular cross section designed for conductors and cables' physical protection and routing. (9)

rigid polyvinyl chloride conduit (PVC). A rigid nonmetallic raceway with a circular cross-section for the installation of electrical conductors and cables. (9)

ripping. Cutting a board in the direction of the grain. (17)

rise. The height of the roof; applies to the vertical distance between the top plate of the wall and the ridge beam at the peak of the roof. (17)

rod-type bit. A bit comprised of a shank, body, and point. (4)

roof truss. Engineered roofing frame that is assembled off-site and craned into place on the jobsite. (17)

roofing flange. A type of seal that covers the hole where the service mast penetrates the roof and prevents rainwater intrusion into the roofing structure. Also called *boot*. (25)

rotary hammer drill. A large, powerful drill used in masonry drilling and demolition. (4)

rough sill. The bottom support of the window's rough opening. (17)

run. The horizontal distance from the wall supporting the rafter to the center of the ridge board; one-half the length of the span. (17)

running board. A board attached to the bottom edge of floor joists in an unfinished basement or crawlspace to provide an attachment surface for NM cable. (23)

S

saw bit. A drill bit with the tip of a regular twist bit, but the middle of the shank has a knurled pattern that can cut plywood or paneling. (4)

scale. The relationship of the drawing size to the actual size of the building. (16)

schedule. A table that lists specific materials used in the building process. (16)

sconce. A surface-mounted luminaire installed on a wall. (27)

scratch awl. A pointed tool used to start screw holes, make pilot holes for drilling, and mark metal, wood, and other materials. (3)

screwdriver. A tool used for the installation and removal of screws. (3)

screw-holding driver. A screwdriver designed to hold and release a screw using some form of mechanical means. (3)

section. The text of the *Code*. (15)

self-feed bit. A fluteless bit for drilling large diameter holes in wood. (4)

series arc. An electrical arc is across a break in the same conductor. (2)

service. The conductors and equipment for delivering electric energy from the serving utility to the wiring system of the premises served. (24)

service calculation. A series of calculations that adds all the electrical loads in a house to determine the service size, rated in amps. (20)

service conductor. The electrical conductors from the utility service point to the service disconnecting means. (8)

service disconnecting means. A way to disconnects the entire premise's electrical service from the utility; usually an electrical switch, such as a circuit breaker. (11)

service drop conductor. An overhead conductor that drops from the serving utility transformer to the service point. (21)

service-entrance conductor. The conductors between the terminals of the service equipment and a point usually outside the building where a connection is made to the overhead service drop or to the underground service laterals. (24)

service head. A cap that is installed at the top of the service mast to prevent rain intrusion into the mast and service equipment. Also called *weather head*. (25)

service lateral. The underground conductor between the electric supply system and the service point. (21)

service mast. Serves as a transfer point from the serving utilities overhead power distribution system to a residence. (21)

service point. The point of connection between the serving utility and the premise wiring. (21)

shank. The part of the drill bit that is grabbed by the jaws of the drill chuck. (4)

sheetrock. A sturdy product made of gypsum sandwiched between two layers of paper. (37)

short circuit. An unintentional path where an electrical current flows around, rather than through, the expected current path. This is characterized by very low resistance and very high current. (1)

shrinkage. Items that have been purchased but are no longer useable, including wire scraps and items that are lost, broken, or damaged. (16)

sill plate. The part of the wood framing that is attached to the top of the foundation wall. It is where the foundation ends and the framing begins. (17)

single-pole circuit breaker. A circuit breaker used for 120-volt branch circuits, rated at 15 or 20 amps, and takes up one space in the panelboard. (11, 28)

single-pole switch. A device for controlling a lighting load from a single location; the most common switch used in residential wiring. (10)

single-pole switched circuit. A switched circuit that controls a lighting load from one location. (13)

site plan. A building plan that shows an aerial view of the property and includes the property boundaries and how the building is situated on the lot. (16)

sledgehammer. A hammer with a large, flat, heavy head attached to a long handle. Frequently used in demolition or to drive stakes and the like into the ground. (3)

small appliance branch circuit. This branch circuit requires a minimum of two 20-amp circuits and is intended to supply receptacles in the kitchen, pantry, dining room, or similar areas in a dwelling unit. (18)

soffit. The underside of the eave. (17)

softwood. Wood that comes from fast-growing evergreen trees such as pine or fir. (17)

solar cell. The component of a photovoltaic system that converts light energy into electrical energy. Also called *PV cell*. (33)

sole plate. The portion of the wall frame that attaches to the subfloor. (17)

solenoid voltage tester. A type of voltage tester that uses an electro-magnetic plunger, called a solenoid, and a scale to give an approximate reading of common voltage levels. (5)

span. The distance between supports for joists and beams. (17)

specifications. The written instructions that define the scope of work to be done and the materials used in the construction process. (16)

split-circuit receptacle. A receptacle in which each half of a duplex receptacle is connected to a different circuit. (37)

split-wired. A duplex receptacle that has been wired to provide switch-controlled power to one half of the receptacle while the other receives constant power. (26)

standby power system. A power system intended to provide utility power during equipment failure or natural disaster. (35)

step bit. A stepped, conical-shaped bit that drills a wide range of hole diameters, generally in sheet metal. (4)

stick-built roof. A roof frame constructed by the carpenters on site using individual "sticks" of lumber or engineered wood products as the rafters. (17)

stripped screw extractor. A drill bit used to remove screws when the heads don't respond to the screwdriver's tip. (4)

structural plan. Used as a guide to install the building foundation and any concrete, wooden, or steel structural members, such as the size and spacing of framing members. (16)

stubby screwdriver. A screwdriver with a short shaft for use in tight spaces. (3)

stud. A vertical framing member used to build walls. (17)

subfloor. The lowermost floor layer that serves as the floor during construction. (17)

subdivision. A further explanation of the rules of a section in the *Code*; can contain three levels. (15)

subpanel. A distribution panel in addition to the main panel. A subpanel is fed from a breaker in the main panel. (8)

substation. A portion of the electrical grid that takes the transmission-level voltage and reduces it to the distribution-level voltage. Depending on the population density of an area, these may be placed every five to ten miles. (1)

sunlight resistant. Constructed or protected with ultraviolet absorbing materials which delay the degrading effects of the sun. (30)

supplemental grounding electrode. A supplement the electrode that ensures the system remains grounded if a water pipe is replaced with a non-metallic pipe. (24)

surface mounting. When an electrical installation is exposed and on the wall's surface. (7)

switch. A device used to open or close a part of an electrical circuit. Generally used to control a lighting load. (10)

switch loop. Wiring that creates a path for the ungrounded conductor from the lighting outlet to the switch location, and then back to the lighting outlet. (13)

switch-controlled light. A lighting outlet that is controlled by a switch or a combination of switches. (14)

switched circuit. A circuit that can use a variety of switches to open and close an electrical circuit, usually to control lighting loads. (13)

switched receptacle. A receptacle in which half of a duplex receptacle is connected to an unswitched circuit and the other half is connected to a switch. (37)

system grounding. The connecting of a conductor of the electrical system to ground through one or more grounding electrodes. (25)

system integration. The connection and interaction between two systems, such as a PV system and a home electrical power supply system. (33)

T

take-off. The counting of every item that is to be supplied and installed by the electrical contractor. (16)

tandem circuit breaker. A special type of circuit breaker that contain two separate switches that are designed to provide overcurrent protection for two single pole circuits from a single panelboard opening. (38)

tape measure. A rolled metal ruler that retracts into a case. This tool is used to determine the spacing of outlet boxes along walls, to measure the proper height of outlet box installation, to check measurements on construction prints. (3)

tapping tool. A tool that taps threaded holes for securing equipment to metal, enlarges existing holes, re-taps damaged threads. (3)

task lighting. A type of lighting that illuminates workspace surfaces. (12)

tear-out. The damage that occurs when the bit exits the other side of the workpiece. A term generally applied to wood. (4)

teeth per inch (tpi). On a saw blade, the number of teeth relative to the length of the blade. The higher the number, the finer the cut. (3)

temporary power. The power provided at a jobsite until the project has reached a state in completion. (21)

terminal adapter (TA). The designation given to PVC conduit connectors. (9)

terminal block screwdriver. A screwdriver made for terminating conductors to terminals common in many electrical and electronic components. (3)

terminal loop. A small loop formed at the end of the conductor that wraps around the terminal screw of a device. (26)

thermal generation. A method of energy generation in which fuel is combusted to produce steam that drives utility-scale electric generators. (1)

thread. The raised spirals along a screw's length that provide the holding power. (6)

three-way switch. A type of switch, always installed in pairs, that controls a lighting load from two locations. (10)

three-way switched circuit. A switched circuit that controls a lighting load from two locations. (13)

there-wire service. A distribution panelboard, or service equipment, fed by service-entrance conductors. (24)

through-the-roof mast. An electrical service in which a conduit extends through the roof and provides a point of attachment for the overhead service drop conductors. (25)

tile bit. A drill bit designed for drilling into ceramic tile or glass. (4)

time current curves. Manufacturer's rating that indicates the length of time it will take to open the mechanism on a circuit breaker in relation to the amount of overcurrent that is present. (28)

tin snips. A tool used to cut sheet metal and other thin, rigid material. (3)

title block. A component of the blueprint that contains information such as the project name or owner, the name of the architect or architecture firm, revision number, if any, and required seals or stamps of approval. (16)

top plate. The uppermost portion of the wall frame that supports the ceiling joists. (17)

torpedo level. A small leveling tool used to check electrical conduit, equipment, appliances, and cover plates for switches and receptacles for level and plumb. (3)

torque screwdriver. A tool for tightening screws, lugs, hex head, and bolt-type lugs to manufacturer's recommended torque specifications. (3)

torque wrench. Tool that tightens hex heads, fasteners, and bolt-type lugs to the manufacturer's recommended torque specifications. (3)

track lighting. A type of luminaire comprised of two parts: an electrified track and removable lighting heads that can be positioned anywhere along the track. (27)

transfer panel. A subpanel typically used for essential loads powered by a standby power system. (35)

transfer switch. A device used to disconnect the home power system from the utility wiring. Also called a *microgrid interconnect device*. (35)

transformer. A magnetically operated machine that converts electrical power in an AC system from one voltage or current level to electrical power at another voltage or current level. (1)

trim out. A stage of construction that occurs after the rough-in stage is completed and includes activities such as installing devices and cover plates, connecting major appliances, installing the luminaires, and installing breakers in the distribution panel. Also called *trimming*. (26)

trimmer stud. A stud that supports the window or door header and is nailed to the inside of the king stud. Also called a *jack stud*. (17)

troubleshooting. A series of steps used to identify and repair the cause of a malfunction in an electrical system. (36)

twin circuit breakers. Two single-pole breakers that take one space in the panel. (11)

twist bit. The most common type of drill bit. This bit is used for drilling into soft materials like wood, soft iron, and aluminum. (4)

twist-on connector. An electrical fitting used for joining two to five electrical conductors that are installed by using a screw-on motion. (8)

twisted pair cable. A cable that contains pairs of insulated conductors that are twisted together to reduce the effects of inductive interference (crosstalk) between conductors. (32)

Type NM. Regular, everyday NM cable. Commonly referred to as *Romex*. (23)

Type NM cable. A factory-made assembly of two or more insulated, current-carrying conductors and a bare equipment grounding conductor in an overall non-metallic sheathing. (8)

Type NMC. NM cable with a corrosion-resistant sheathing. (23)

Type NMS. NM cable that includes signaling, data, and communication conductors in addition to the insulated power conductors. (23)

Type S fuse. A plug-type fuse that uses adapter bases that accept only the properly sized fuse. (11)

Type SER cable. A cable assembly consisting of three insulated sunlight-resistant conductors and one uninsulated equipment grounding conductor encased in a gray sunlight-resistant PVC jacket. (8)

Type SEU cable. A cable assembly of two sunlight-resistant, insulated conductors and a bare concentrically wound grounded conductor encased in a gray sunlight-resistant PVC jacket. (8)

Type UF cable. A type of non-metallic cable designed and constructed to be suitable for direct burial in the earth. (8)

Type USE cable. A type of service entrance cable that is constructed of multiple insulated conductors twisted together and has no outer sheathing. (8)

U

underground meter socket. A device that provides a means to route the underground utility wiring from below grade to the meter terminals. (24)

underground temporary service. A type of temporary service that uses a pad-mounted transformer. (21)

underwater luminaire. A type of luminaire designed to be submerged in water. (31)

ungrounded conductor. The conductor in an electrical circuit that carries the current to the load. Also called the "hot" conductor. (1)

ungrounded receptacle. A contact device installed at the outlet for the connection of an attachment plug that does not have a means for attachment of an equipment grounding conductor. (37)

utility knife. A cutting hand tool with a folding or retractable blade. (3)

utility meter. A device that records the amount of electrical usage associated with a dwelling and is used as the basis for the customer's utility bill. (24)

utility service point. The point of a wiring system where the utility's jurisdiction ends and the premises wiring begins. (8)

V

valley. The inside corner where two sloping roofs meet. (17)

vapor barrier. An unbroken layer of material that prevents moisture from transferring from one side of the barrier to the other. (37)

vermiculite insulation. A pebble like mineral product used to insulate wall cavities and attic spaces. (37)

volt. The unit of measure for voltage. (1)

voltage. The pressure that pushes electrons through an electrical circuit. (1)

voltage drop. The loss of voltage in a circuit due to the internal wire resistance combined with the current being drawn by the load. (19)

W

voltage gradient. A difference in electrical potential across a distance or space. (31)

voltage rating. The maximum voltage that should be applied to the receptacle. (37)

voltage tester. An electrical testing device that confirms the presence of voltage without giving a precise readout. (5)

voltmeter. An electrical testing device that reads and displays the precise amount of voltage that is present in a circuit. (2, 5)

W

watt. The unit of measure of power. (1)

wattage. The measure of the lamp's energy consumption. (12)

Watt's law. An electrical principal that states that the power in an electrical circuit is the product of the applied voltage and the amount of current flow in the circuit. (1)

weatherhead. A protective conduit fitting that prevents water from entry into a conduit where a transition is made from open wiring to conduit wiring methods. (30)

weatherproof. Constructed or protected so that exposure to the weather will not interfere with successful operation. (30)

weatherproof box. A sealed enclosure used to install switches or receptacles in wet and damp locations. (7)

wet location. An area that is subject to direct saturation or directly impacted by the weather. (7)

wet-niche luminaire. A luminaire intended for installation in a forming shell mounted in a pool or fountain structure where the luminaire will be surrounded by water. (31)

whip. A short length of flexible conduit between a junction box or disconnecting means and a piece of electrical equipment. (9)

white light. Light where the primary colors are present in proper proportion; used to see an objects true color. (12)

wire stripper. A plier designed to strip insulation from several different wire sizes without damaging the underlying conductor. (3)

work ethics. Personal traits that will make you a more valuable employee. (1)

workplace skills. Professional skills that will make you a more valuable employee. (1)

Y

yoke. The structural framework of a switch, receptacle, or similar device to be installed in an outlet box and generally attached by a screw at the top of the box and a screw at the bottom of the box. (10, 22)

Z

zero resistance. A condition where there is no opposition to current flow, such as a closed switch. (5)

Index

A

abandoned cable, 371
accent lighting, 146
accessible equipment, 213
accessory structures. *See also* detached garages; outbuildings
    solar/EV ready, 398
    supplying power for, 395–396
adjustable wrenches, 46
AFCI circuit breakers, 444–445, 474
affidavit, 475
aggressiveness, saw, 51
AIC rating, 132
air conditioner, room, voltage rating for, 440
airtight, 451
allowable box fill, 252*t*, 298
alternating current (AC)
    defined, 13
    systems, grounding electrode conductor for, 397*t*, 474*t*
ambient lighting, 146
ambient temperature, 104, 220–222
ambient temperature correction factors, 221*t*
American Wire Gauge (AWG), 99–100
amp, 14
Ampacities Table (Table 310.16), 104–105, 105*t*, 208, 220, 220*t*, 221, 222, 223, 224, 274–275
ampacity, 104–105, 219–223
    defined, 104, 220
    Type NM, 315*t*
ampacity adjustments, 220–223
    conductor size selection, 222
    more than three current-carrying conductors, 223*t*
    temperature corrections to ampacity rating, 222
anchor, 83
anchoring hardware, 83–86
anti-island design, 377
appliance branch circuit, 400
appliances
    connecting new, 443
    cooking, demand factors, 237*t*
    major, 299, 325–326
    receptacle spacing, 214
    small, branch circuit requirements, 208
    typical power loads, 411*t*
apprentice, 6
approximate estimates, 185, 188, 190
arc blast, 23–24
arc fault, 23

arc-fault circuit interrupter (AFCI)
    circuit breakers, 134, 316–317
    damaged cable and, 323
    defined, 25
    receptacles, 122
    requirements, 215
    troubleshooting, 434-435
arc-fault circuit protection requirements, 215
arc flash, 23–24
architect's scale, 180
architectural details, 182
architectural drawings, 180
architectural sections, 182
array mounting systems, 378–379
articles, *National Electrical Code (NEC)*, 173–174. *See also*
    National Electrical Code (NEC), articles
asbestos exposure, 445
attached garages, 398–401
    defined, 398
    fire-rated construction, 398–400
    lighting outlets, 400–401
    receptacle outlets, 400–401
augur bit, 66
automatic transfer systems, 412
authority having jurisdiction (AHJ), 10, 170

B

back stabs, 120
baffle trims, 451
balance of system, 386
ballast, 143
balloon framing, 196
band joist, 197–198
bandwidth, data transmission, 363–364
bathroom
    branch circuit requirements, 208
    lighting requirements, 210–211
    luminaire installation in tub and shower, 308–309
    receptacle spacing, 212–213
battery-based inverters, 419
beam, ridge, 200
bell boxes, 95
bell end, 113
bidirectional power, 379
bids. *See* estimates
bimetallic strip, 131

Note: Page numbers followed by *t* indicate tables.

D

W

Y

Z